High-Temperature Electrolysis

From fundamentals to applications

Online at: https://doi.org/10.1088/978-0-7503-3951-3

High-Temperature Electrolysis

From fundamentals to applications

Edited by
Werner Sitte
Chair of Physical Chemistry, Montanuniversitaet Leoben, Leoben, Austria

Rotraut Merkle
Max Planck Institute for Solid State Research, Stuttgart, Germany

IOP Publishing, Bristol, UK

ISBN 978-0-7503-3951-3 (ebook)
ISBN 978-0-7503-3949-0 (print)
ISBN 978-0-7503-3952-0 (myPrint)
ISBN 978-0-7503-3950-6 (mobi)

DOI 10.1088/978-0-7503-3951-3

Version: 20230101

IOP ebooks

British Library Cataloguing-in-Publication Data: A catalogue record for this book is available from the British Library.

Published by IOP Publishing, wholly owned by The Institute of Physics, London

IOP Publishing, No.2 The Distillery, Glassfields, Avon Street, Bristol, BS2 0GR, UK

US Office: IOP Publishing, Inc., 190 North Independence Mall West, Suite 601, Philadelphia, PA 19106, USA

Contents

Preface

This book is intended to give the reader an introduction to the fundamental aspects of high-temperature electrolysis and its applications. It extends to the key technological and economic aspects of employing ceramic electrochemical cells for electrolysis processes, which are decisive for their broad application. The topic of high-temperature electrolysis, i.e. the use of electrolysis cells based on ceramic electrolytes, has already been investigated for many years—often in close relation to solid oxide fuel cells. This topic has recently gained momentum for two reasons: (i) the urgent need to decrease the worldwide CO_2 footprint and the corresponding necessity to convert electrical energy from renewable, fluctuating energy sources into easily storable chemical energy on a large scale; (ii) the recent progress in the field of protonic ceramic fuel cells (with corresponding options for reversible/electrolytic cell operation), which exhibit characteristic differences compared to oxide-ion conductive cells and thus represent an interesting complementary option (although one that is, at present, still less mature).

As introduction to the key aspects of high-temperature electrolysis, this book is intended in particular for researchers, engineers and graduate students entering the field from different disciplines. It covers the current developments including relevant open questions and the methods used to tackle them. This book attempts to bridge the 'language gaps' that may arise between the different scientific communities involved, ranging from materials science to chemical engineering and economic analysis. Emphasis is given to fundamental aspects (which are addressed in more detail than in the usual review articles) that serve as a basis for understanding the continuously improving performance data. Based on knowledge of the key principles, the reader can then follow further progress in the primary literature. Many of the relevant groups active in the field are represented in the author list of this book.

The book comprises 20 chapters, which move from the fundamental and materials aspects towards devices and applications, concluding with systems and economic analysis. **Chapter 1** gives a general overview of high-temperature electrolysis, places it in the context of the global energy supply, considers thermodynamic boundary conditions, and discusses remaining challenges. **Chapters 2–4** present the materials requirements and typical materials compositions of the electrolyte, anode, and cathode of solid oxide electrolysis cells (SOECs). **Chapter 5** is dedicated to interconnects and coatings. Aspects of electrode kinetics, including methods for their characterization as well as the different cell architectures—including metal-supported cells, are treated in **chapters 6–8**. **Chapters 9 and 10** deal with the important issues of electrochemical measurement techniques, microstructural electrode modelling, and advanced data analysis as well as long-term stack tests that include degradation mechanisms. **Chapters 11–14** present specific materials as well as the characteristic differences between protonic ceramic electrolysis cells (PCECs) and solid oxide electrolysis cells (SOECs). These chapters describe the fundamental aspects of protonic electrolytes and mixed-proton conductors; planar proton-conducting electrolysis cells (PCECs), with an emphasis on charge carrier profiles and faradaic efficiency; tubular PCECs, with an emphasis on direct hydrogen compression; and planar PCECs, with a focus on H_2 production and CO_2 conversion. **Chapters 15–18** discuss the application of solid oxide electrolysis and co-electrolysis and their coupling with other processes such as methanation, the high-temperature electrolysis of CO_2, the synthesis of ammonia, and the synthesis of e-fuels via the Fischer–Tropsch route. Finally, **chapters 19 and 20** cover the integration of reversible solid oxide cell (rSOC) systems into the energy system and the economic aspects of the electrolytic and power-to-gas processes.

We hope that the reader will enjoy this book as a reference text, gaining an overview of the current status and the relevant developments of high-temperature electrolysis, and benefiting from it in their future research and device development and application. We thank all the authors for the time and effort taken to share their great expertise and profound insights into high-temperature electrolysis with the reader.

Werner Sitte
Chair of Physical Chemistry, Montanuniversitaet Leoben
Leoben, Austria

Rotraut Merkle
Max Planck Institute for Solid State Research
Stuttgart, Germany

Editor biographies

Werner Sitte

Professor Dr Werner Sitte is full professor of Physical Chemistry at Montanuniversitaet Leoben, Austria. His research interests concern transport and defect chemistry in ionic solids as well as applications towards solid oxide and protonic ceramic fuel and electrolysis cells. He is an associate editor of the journal *Solid State Ionics* and is on the scientific board of *International Conferences on Solid State Ionics*.

Rotraut Merkle

Dr Rotraut Merkle is a scientist at the Max Planck Institute for Solid State Research, Germany. Her research is dedicated to point defect formation and transport in ionic solids, reaction kinetics at oxide surfaces, transport at grain boundaries, and extends to solid oxide and protonic ceramic fuel and electrolysis cells. She is involved in EMRS, SSI and SSPC conference organization, and is an editor for the *Solid State Ionics* journal.

List of contributors

Christian Berger
Max Planck Institute for Solid State Research, Stuttgart, Germany

Hans Böhm
Energieinstitut an der Johannes Kepler Universität, Linz, Austria

Martin Bram
Forschungszentrum Jülich GmbH, Institute of Energy and Climate Research, Jülich, Germany

Ming Chen
DTU Energy, Technical University of Denmark, DK-2800 Kgs. Lyngby, Denmark

Dorela Dhamo
Karlsruhe Institute of Technology, Institute for Micro Process Engineering, Karlsruhe, Germany

Sebastian Dierickx
Institute for Applied Materials—Electrochemical Technologies, Karlsruhe Institute of Technology, Karlsruhe, Germany

Roland Dittmeyer
Karlsruhe Institute of Technology, Institute for Micro Process Engineering, Karlsruhe, Germany

Chuancheng Duan
Department of Chemical Engineering, Kansas State University, Manhattan, USA

Andreas Egger
Chair of Physical Chemistry, Montanuniversitaet Leoben, Leoben, Austria

Qingping Fang
Forschungszentrum Jülich GmbH, Institute of Energy and Climate Research, Jülich, Germany

Christopher Graves
Noon Energy, Palo Alto, CA, USA

Anke Hagen
DTU Energy, Technical University of Denmark, DK-2800 Kgs. Lyngby, Denmark

Per Hjalmarrson
Ceres Power Ltd, Viking House, Foundry Lane, Horsham, West Sussex, UK

John Bøgild Hansen
Haldor Topsoe A/S, Kgs. Lyngby, 2800, Topsoe, Denmark

Dominik Hess
Karlsruhe Institute of Technology, Institute for Micro Process Engineering, Karlsruhe, Germany

Peter Holtappels
Karlsruhe Institute of Technology, Institute for Micro Process Engineering, Karlsruhe, Germany

John T S Irvine
School of Chemistry, University of St Andrews, St Andrews, UK

Soren Hojgaard Jensen
Noon Energy, Palo Alto, CA, USA

Jochen Joos
Institute for Applied Materials—Electrochemical Technologies, Karlsruhe Institute of Technology, Karlsruhe, Germany

Robert J Kee
Mechanical Engineering, Colorado School of Mines, Golden, CO 80401, USA

Thomas Kienberger
Energy Network Technology, Montanuniversitaet Leoben, Leoben, Austria

Dino Klotz
Institute for Applied Materials—Electrochemical Technologies, Karlsruhe Institute of Technology, Karlsruhe, Germany

Andreas Krammer
Chair of Process Technology and Industrial Environmental Protection, Montanuniversitaet Leoben, Leoben, Austria

Yngve Larring
SINTEF Industry, Department of Sustainable Energy Technology, Forskningsveien 1, 0373 Oslo, Norway

Robert Leah
Ceres Power Ltd, Viking House, Foundry Lane, Horsham, West Sussex, UK

Markus Lehner
Chair of Process Technology and Industrial Environmental Protection, Montanuniversitaet Leoben, Leoben, Austria

Fan Liu
Department of Chemical Engineering, Kansas State University, Manhattan, USA

Norbert H Menzler
Forschungszentrum Jülich GmbH, Institute of Energy and Climate Research, Jülich, Germany

Rotraut Merkle
Max Planck Institute for Solid State Research, Stuttgart, Germany

Gurli Mogensen
gummomo consult, DK-3540 Lynge, Denmark

Mogens Bjerg Mogensen
DTU Energy, Technical University of Denmark, DK-2800 Kgs. Lyngby, Denmark

Francesco Mondi
DTU Energy, Technical University of Denmark, DK-2800 Kgs. Lyngby, Denmark

Subhasish Mukerjee
Ceres Power Ltd, Viking House, Foundry Lane, Horsham, West Sussex, UK

Andreas Nenning
TU Wien, Institute of Chemical Technologies and Analytics, Austria

Alexander K Opitz
TU Wien, Institute of Chemical Technologies and Analytics, Austria

David Paczona
Energy Network Technology, Montanuniversitaet Leoben, Leoben, Austria

Sandrine Ricote
Mechanical Engineering, Colorado School of Mines, Golden, CO 80401, USA

Michael Rubin
Karlsruhe Institute of Technology, Institute for Micro Process Engineering, Karlsruhe, Germany

Christoph Sejkora
Energy Network Technology, Montanuniversitaet Leoben, Leoben, Austria

Theis L Skafte
Noon Energy, Palo Alto, CA, USA

Stephen J Skinner
Department of Materials, Imperial College, London, UK

Elena Stefan
SINTEF Industry, Department of Sustainable Energy Technology, Forskningsveien 1, 0373 Oslo, Norway

Belma Talic
SINTEF Industry, Department of Sustainable Energy Technology, Forskningsveien 1, 0373 Oslo, Norway

Robert Tichler
Energieinstitut an der Johannes Kepler Universität, Linz, Austria

Chen-Yu Tsai
Department of Materials, Imperial College, London, UK
Ceres Power Ltd, Viking House, Foundry Lane, Horsham, West Sussex, UK

Einar Vøllestad
SINTEF Industry, Department of Sustainable Energy Technology, Forskningsveien 1, 0373 Oslo, Norway

Shu Wang
DTU Energy, Technical University of Denmark, DK-2800 Kgs. Lyngby, Denmark

André Weber
Institute for Applied Materials—Electrochemical Technologies, Karlsruhe Institute of Technology, Karlsruhe, Germany

Huayang Zhu
Mechanical Engineering, Colorado School of Mines, Golden, CO 80401, USA

Chapter 1

High-temperature electrolysis—general overview

Mogens Bjerg Mogensen, Francesco Mondi and Gurli Mogensen

This overview explains why energy conversion and renewable energy storage are needed in order to make a global change from fossil energy to sustainable energy via the conversion of electricity from photovoltaic solar energy, wind turbines, and hydropower into fuels and chemicals. This is called power-to-X (PtX). Fortunately, plenty of sustainable energy is available. This chapter describes why the electrolysis of H_2O into H_2 is necessary and why that of CO_2 into CO is desirable and possible in order to produce chemicals such as hydrocarbons and ammonia that are suitable for energy storage and the replacement of fossil fuels. Further, strong thermodynamic arguments are provided in favor of the use of high-temperature electrolysis. The technology is briefly described, including some of its weaknesses. Finally, we outline electrochemical tools that will be helpful in further R&D, particularly that of solid oxide electrolysis cells (SOECs), and present recommendations for future work together with a short outlook.

1.1 The need for energy conversion and the storage of sustainable energy

1.1.1 From fossil fuels to sustainable energy

The fear of fast-increasing global warming due to an increase in the concentration of greenhouse gases, such as CO_2 and CH_4, which originate from anthropological activity, has raised an urgent need for a total change of the global energy supply from fossil fuel energy to sustainable energy as soon as possible. This change is a task of gigantic dimensions, but it seems possible, as described in this book. High-temperature electrolysis will most probably become an enabling technology for the transition from fossil fuels to sustainable energy.

doi:10.1088/978-0-7503-3951-3ch1

Arguments in favor of this are presented here, and were previously published in detail in articles elsewhere [1–4].

The present definition of sustainable energy is energy that does not contribute to the increase of greenhouse gases in the atmosphere. Some of the main sustainable power sources are solar photovoltaic (PV), wind, and hydropower. PV, wind, and hydroelectricity will potentially be available in sufficient amounts to supply more than the necessary energy needed for the whole world. The International Energy Agency's (IEA's) estimate of the world's total energy supply was 606.490 EJ, corresponding to 19.3 TWy (terawatt years) in 2019 [5]. 1 TWy = $8.76 \cdot 10^{12}$ kWh = 31.5 EJ. This is a tremendous amount of energy, but the influx of solar energy per year to the outer atmosphere of the Earth is in the order of 170 000 TWy, i.e. about 9000 times higher than our need [6]. Obviously, all the solar influx cannot be harvested. The following represents a quick look at the possibilities.

The average influx of energy from the Sun (corrected for scattering in the atmosphere and absorption by clouds) to the land area between the polar circles corresponds to 24 000 TWy, calculated using data from Tsao *et al* [7], i.e. 1300 times more than the amount needed. Actually, it would require only a small fraction of the world's deserts to supply all the energy needed. Figure 1.1 shows a global horizontal irradiation (GHI) solar resource map, on which a blue square illustrates the limited space it would take to supply the 19.3 TWy that is needed, even when a PV efficiency of only 10% is assumed, i.e. a solar irradiation energy of 193 TWy. In the desert

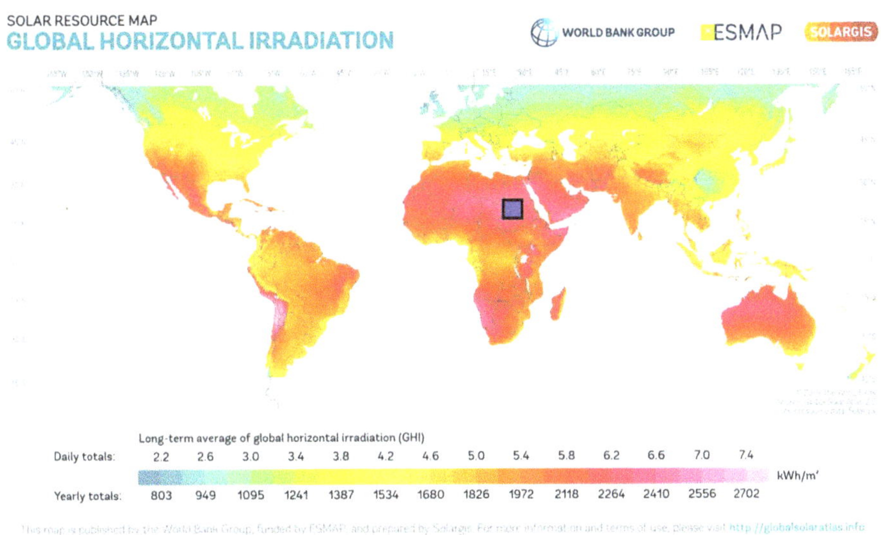

Figure 1.1. Global horizontal irradiation (GHI) solar resource map. The area of PV solar installations required to cover today's world energy demand would be about 700 000 km^2 in total, if established in desert areas with high GHI values year round. This area is represented by the blue square, which is subjectively placed on the map in the Sahara dessert. The efficiency of the solar installations is assumed to be 10%. The map uses solar resource data obtained from the Global Solar Atlas, which is owned by the World Bank Group and provided by Solargis [8].

areas of Libya/Egypt, the GHI yearly total is shown to be 2 500 kWh m^{-2}. It appears that (193 TWy · 8.76 · 10^{12} kWh TWy^{-1})/(2 500 kWh m^{-2} · 10^6 m^2 km^{-2}) ~ only ca. 700 000 km^2 is enough. It should be noted that no use of valuable farmland is necessary for this. Yet, in reality, a much larger area will be needed, as PV has mainly been placed outside deserts to date. Hopefully, other 'unused' areas, such as the roofs of buildings, will be applied for PV.

Parts of the solar energy influx are naturally converted to other forms of power, such as wind power and water streams (hydropower). Even though the potential of these two sources is significantly lower than that of solar radiation, they are sources that may be utilized even north of the Arctic circle. Based on data provided by Tsao *et al* [7], the realistic technical potential of (onshore) wind power is 14 TW (table 2 in [7]) and that of hydropower is 1.6 TW (note 19 in [7]) of electric power on average. Thus, the world will have to rely on solar irradiation as its main sustainable energy source in the future. The intermittent availability of solar irradiation, wind in particular, and to some extent hydropower requires a tremendous capacity for the conversion and storage of electric power in order to maintain a continuous reliable power supply.

Fortunately, electric power obtained from PV, wind, and hydropower may be generally competitive with fossil fuel in the near future, as prices lower than the price of generating electricity from coal have been reported and the cost seems to continue to drop rapidly. The International Renewable Energy Agency (IRENA) [9] states: 'For newly commissioned projects, the global weighted-average levelized cost of electricity (LCOE) of utility-scale solar PV fell by 85% between 2010 and 2020, from USD 0.381/kWh to USD 0.057/kWh, as global cumulative installed capacity of all solar PV (utility scale and rooftop) increased from 40 GW to 707 GW. This represented a precipitous decline, from being more than twice as costly as the most expensive fossil fuel-fired power generation option to being at the bottom of the range for new fossil fuel-fired capacity.' So, if enough conversion capacity can be made available then solar power will generally be competitive with fossil fuels.

1.1.2 Potential conversion and storage technologies

A number of conversion and storage technologies exist, e.g. pumping water into mountain water reservoirs, compressing air, and charging batteries. All of these are already in use, but none of them are suitable for seasonal storage (over periods of several months to years), and they are not suited for fueling heavy transport such as trucks, ships, and airplanes. Therefore, fuels such as hydrogen, CO_2-neutral carbon-based fuels, and ammonia, all of which can be produced from sustainable electricity, are necessary for chemical energy storage, for the transportation of energy, and for the propulsion of vehicles such as airplanes, large ships, and heavy trucks. Conversion technologies that can produce such fuels from sustainable electricity will need electrolysis to convert renewable electrical energy into hydrogen (H_2) from water and to convert carbon dioxide (CO_2) into carbon monoxide (CO). Then, using mixtures of H_2 and CO (syngas), it is possible to produce fuels such as methane, CH_4, also called synthetic natural gas (SNG), methanol (CH_3OH), dimethyl ether

(DME = $(CH_3)_2O$), and many other CO_2-neutral fuels based on electrolysis followed by heterogeneous catalysis. In addition, ammonia (NH_3) and several high-energy-density complex multiatom ammonia compounds, e.g. $CaCl_2(NH_3)_8$ [10] and $LiBH_4NH_3$ [11], are being researched, but the technologies of the complex multiatom compounds have not yet reached a level suitable for large-scale energy storage.

The processes that convert from electricity to chemicals are called PtX, Power-to-Fuel (PtF), etc. A technical evaluation of the various conversion and storage technologies can be performed using the following characteristics: (1) energy density and equivalent power density, (2) manageability, (3) ability to displace CO_2 emissions, and (4) the harmful effects of chemicals and fuels. Naturally, cost will, in practice, be a most important issue, but it will be very much dependent on purpose and circumstances. The general economy of the various conversion and storage technologies is therefore not treated here. However, the economic aspects of Power-to-Gas (PtG) are treated in chapter 20, as PtG is closely related to electrolysis. Investors are in all cases recommended to carefully carry out a life cycle analysis of any big investments before making a decision.

1.1.2.1 Energy density and equivalent power density

Gravimetric energy density values (table 1.1) in particular argue strongly in favor of the conversion into chemical energy storage. In particular, hydrogen stands out, due to its extremely high value of 143 MJ kg^{-1}. However, its very low volumetric energy density and extremely low boiling point make it more difficult to handle than other chemicals, as these properties require heavy containers. This is a major issue for H_2 and also to some extent for CH_4, but liquid natural gas, which consists mainly of CH_4, is handled commercially in huge quantities.

Table 1.1. Data for comparison of the volumetric and gravimetric energy densities (higher heating value (HHV)) and the boiling points of compounds for selected energy storage technologies. The values shown are approximate because of the scatter in the published data. All these technologies are well tested, have existing infrastructure, and seem safe. The values for chemicals are based on data from Perry's tables [12].

Storage type X	MJ L^{-1}	MJ kg^{-1}	Boiling point (°C)
Liquid methane, CH_4	23.6	55.7	−162
Liquid DME, $(CH_3)_2O$	21.2	31.8	− 25
Methanol, CH_3OH	18.0	22.7	+65
Liquid ammonia, NH_3	15.3	22.5	−33
Liquid hydrogen, H_2	10.2	143.0	−253
Li-ion batteries	2.7	0.9	–
Lead acid batteries	0.4	0.15	–
Compressed air—20 MPa	0.1	0.4	–
Water at 100 m elevation	10^{-3}	10^{-3}	–

When fuel is applied for transportation, chemical fuels are strongly favored because of the high equivalent power density. This relates to heavy transport like airplanes, ships, and trucks, and also to some extent for passenger cars. The following short example illustrates this. Assuming a filling rate of 20 L min^{-1} for liquid DME means that it takes 3 min to get 60 L = 1270 MJ on board. This is equivalent to 7.1 MW of power. In comparison, Li batteries require something like 8 h to recharge without unnecessary damage to the battery life. For a 300 kg battery package (1 MJ kg^{-1}) it takes 8 h to get 300 MJ on board. This is equivalent to a power of ca. 10 kW, or more than 700 times slower than pumping DME. Faster charging is possible today, and powers of up to 350 kW have been reported. However, the battery management system, which serves to protect the lifetime of the battery, only allows high voltage 'superchargers' to reach the maximum output power during a short period of charging, resulting in an average power of ~50% of the maximum. Two car models mentioned in [13] reach an 80% state of charge using average powers of 130 kW and 270 kW, respectively, which is 'only' 56–27 times slower than pumping DME.

1.1.2.2 Manageability

It is clear from the boiling points and volumetric energy densities that CH_4, DME, CH_3OH, and NH_3 are the easiest to handle as fuels for transportation. Compressed air and the storage of water at high altitudes do not seem to be good options for storage over long periods, and they are not at all applicable to the transportation sector. Actually, hydropower also needs a seasonal storage technology in some places, because the precipitation of rain may vary seasonally. On the other hand, if extra hydropower storage capacity is available in nearby mountains, then pumping water into the water reservoirs may be very attractive from a commercial perspective.

1.1.2.3 Ability to displace CO_2 emissions

The ability to displace CO_2 emissions is also an important property of a sustainable energy technology. Obviously, H_2, NH_3, batteries, and physical technologies, which do not emit any CO_2, displace CO_2 by 100%, whereas carbon-containing substances such as CH_4 only displace CO_2 by 100% if the CO_2 used in PtX is captured directly from the air or from biogas, which often contains about 40% CO_2. Using CO_2 captured from the exhausts of fossil-fueled power plants and from classical industrial processes—such as cement, iron, and NH_3 production based on fossil fuel sources—would only be a one-time reuse of CO_2. However, as long as the use of fossil fuels is allowed, it makes good sense to reuse CO_2 from such sources, as this is technically and economically possible today. CO_2 capture is still in an early phase of commercialization and is too expensive for commercial use in PtX.

NH_3 is produced from pure H_2 and pure N_2 obtained from natural gas; changing the production method to use pure H_2 and pure N_2 obtained from sustainable sources would also save huge amounts of CO_2 emissions. NH_3 is used for fertilizers and industrial applications and is among the chemicals with the largest production worldwide. The production of over 200 million tons of NH_3 (per annum globally) uses about 2% of the world's consumption of fossil fuel energy and generates over

420 million tons of CO_2 [14]. Several countries have now accepted NH_3 as a promising chemical for the establishment of a CO_2-free energy system, and several major energy agencies (the International Energy Agency, the European Commission, the Japanese Government, and US Scientific Councils) have endorsed this [15]. Thus, NH_3 will become an even more important chemical in the future, sustainable energy system. Power-to-ammonia is treated in chapter 17.

1.1.2.4 Harmful effects of chemicals and fuels

An assessment of harmful effects is difficult, because it is usually not possible to quantify the sum of all harmful aspects in a clear way. The main reason for this is that each technology has its own set of problems, which differ from those of all the other conversion and storage technologies. The storage of hydropower in a huge artificial water reservoir may be harmful to the owners of the land area taken by the reservoir. This is not a problem for CH_4, which may be stored in an underground salt cavern, many of which already exist. Two important harmful aspects of gaseous and liquid energy carriers are toxicity and flammability. H_2, CH_4, and $(CH_3)_2O$ are all non-toxic but very flammable, and some mixtures of these gases with air are explosive. CH_3OH is a toxic and flammable liquid that has a somewhat smaller risk of forming explosive gas mixtures with air. NH_3 has an even lower explosion risk, but it is a more toxic gas at atmospheric pressure. Mestemaker *et al* pointed out that even though the use of NH_3 on work vessels seems to solve the problems of the low volumetric energy density of hydrogen and of the CO_2 emission caused by the use of hydrocarbon fuels, the high toxicity of NH_3 is an extra hurdle to its application in the commercial shipping sector [16].

1.2 Electrolysis cells

1.2.1 Thermodynamics of the electrolysis of H_2O and CO_2

High-temperature electrolysis is, from a thermodynamic point of view, by far the most efficient type of H_2O and CO_2 electrolysis. Figures 1.2(a) and (b) show thermodynamic diagrams for H_2O electrolysis ($H_2O \rightleftharpoons \frac{1}{2} O_2 + H_2$) (figure 1.2(a)) and for CO_2 electrolysis ($CO_2 \rightleftharpoons \frac{1}{2} O_2 + CO$) (figure 1.2(b)), respectively, at atmospheric pressure. The relation between the two *y*-axes simply reflects the fact that the energy measured in units of volts (V) is equal to the energy measured in units of kJ mol^{-1} divided by $2F$ (the factor of 2 arises from the exchange of 2 electrons per molecule in the reactions; F is Faraday's constant). Figure 1.2(a) shows that at standard pressure (1 bar), partial pressure ratio of $pH_2/pH_2O = 1$, and partial pressure of $pO_2 = 1$ bar,

$$E_{eq}(T) = \Delta G_f(T)/2F,$$

where E_{eq} is the equilibrium voltage of the electrolysis cell, and ΔG_f is Gibbs free energy of the formation of the electrolysis products at the given temperature. ΔH_f is the enthalpy of formation of the electrolysis products. This is the total energy that is required for the electrolysis process, and the thermoneutral potential is defined as

$$E_{tn} = \Delta H_f/2F.$$

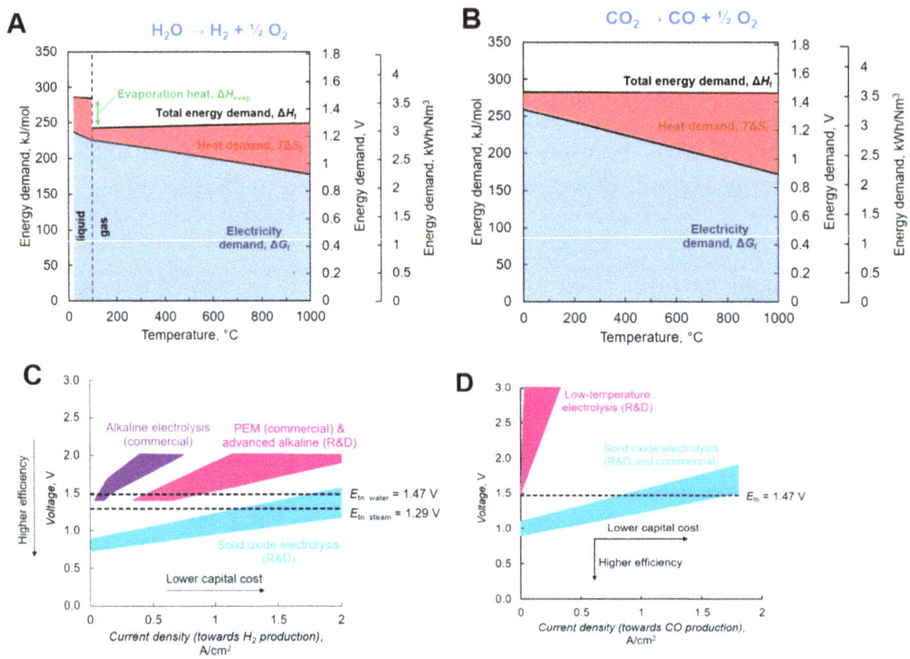

Figure 1.2. Thermodynamics of H_2O and CO_2 electrolysis and a generalized performance illustration of electrolytic technologies. An illustration of the energy demands based on the thermodynamics of H_2O (a) and CO_2 electrolysis (b) at 1 bar, along with typical ranges for the cell voltages of different electrolysis technologies as function of the current density for H_2O (c) and CO_2 (d) electrolysis, respectively [4, 18, 19].

The difference is

$$E_{tn} - E_{eq}(T) = T\Delta S_f/(2F),$$

where T is the temperature in Kelvin and ΔS is the entropic change of the electrolysis reaction.

In other words, when the electrolysis cell is polarized to the cell voltage of E_{tn}, then the Joule heat power, P_J, formed by the loss in the cell equals $T\Delta S_f$ per time unit. Another way to express this is that the area specific Joule heat power, $P_{JA} = i^2 \cdot \text{ASR}$, where i is the current density (in A cm^{-2}) and ASR is the area specific cell resistance (in Ω cm^2), giving P_{JA} the unit W cm^{-2}. Thus, a cell running at E_{tn} produces exactly the power necessary to supply the heat $T\Delta S_f$ for the endothermic electrolysis process at a given current density and temperature. In this case it is not necessary to supply any external heating or cooling to the cell, apart from the heat necessary to compensate the heat loss to the surroundings, which Dönitz *et al* found to be 7% for a system with a stack temperature of 995 °C [17]. The size of heat loss to the surroundings naturally depends on the insulation and the efficiency of the heat exchangers of the electrolysis system. In the interval of $E_{eq} < E_{cell} < E_{tn}$ it is necessary to supply heat in order to maintain the requested cell temperature, and for $E_{cell} > E_{tn}$ the cell needs to be cooled in order to avoid overheating.

Comparing figures 1.2(a) and (b) reveals that the thermodynamics of CO_2/CO and H_2O/H_2 are quite similar in ΔG_f values in the temperature range from about 700 °C to 900 °C. This makes it suitable to perform the co-electrolysis of mixtures of steam and CO_2. If the total polarization of the electrolysis cell is called ΔE_{pol}, then at an applied cell voltage of E_{tn}, $\Delta E_{pol,tn} = E_{tn} - E_{eq}(T)$, the cell polarization is as high as possible if no surplus heat should be produced by the cell. Bearing this in mind, figures 1.2(a) and (b) show that at low temperatures, it is only possible to use a small cell polarization before the electrolysis becomes exothermic and the cell requires cooling. In addition, at low temperatures, the kinetics is slow, and therefore electrolysis at a decent high current density needs a higher polarization and therefore has a lower energy efficiency.

Figures 1.2(c) and (d) are sketches of the typical ranges of cell voltage versus current density from the literature for high-temperature (600 °C–900 °C) SOECs and for two types of low-temperature electrolysis cell, namely alkaline electrolysis cells (AECs) and proton-exchange-membrane electrolysis cells (PEMECs), both of which have operation temperatures that are typically below or around 100 °C. At high temperature, where the kinetics is fast, $\Delta E_{pol,tn}$ is also high, and it is possible to sustain high current densities of more than 1 A cm^{-2} at 800 °C with a cell voltage below E_{tn}. It is clear that the electrical efficiency of high-temperature electrolysis cells, such as SOECs, is significantly better than that of low-temperature AECs and PEMECs, as predicted by the thermodynamics described above. Parts of the data presented in figures 1.2(c) and 1.2(d) are more than 10 years old, but in spite of this, they are—not surprisingly—representative of the situation today. It is not possible for R&D to change the laws of thermodynamics.

1.2.2 Types of electrolysis cell

Many kinds of electrolysis cell have been researched. In particular, cells that can operate at low temperature (in the range up to 200 °C) have been investigated. The AEC is the classical type, which was already commercialized for pure H_2 production around 1920, i.e. more than 100 years ago. Another very popular type is the PEMEC, which is in an early phase of commercialization. However, both types have problems with rather high costs of the H_2 produced, for the reasons discussed in section 1.2.1. In addition to this, the material costs for PEMEC are high, particularly those for the very scarce IrO_2, which is the only known oxygen electrode material that works satisfactorily.

In contrast to the many suggested low-temperature cells, only three types of high-temperature electrolysis cells have been researched: SOECs, molten carbonate electrolysis cells (MCECs), and protonic ceramic electrolysis cells (PCECs), which all operate at temperatures above 500 °C. MCECs have shown promise in the laboratory [20, 21], but seemingly no industrial company has taken an interest in them.

Even though the high-temperature cells are the most energy-efficient cells, the high temperatures have apparently scared many researchers away from them, and it was not until after 1960 that any international significant research was begun. Consequently, the SOEC, and in particular the PCEC, are less mature products compared to the AEC and the PEMEC. The SOEC is, in fact, on the verge of commercialization [3]. However, there is still a need for lower prices, mechanically stronger cells, and improved cell and stack durability in order to achieve the wide commercialization of SOECs. This book treats the SOEC and PCEC types of high-temperature cell, which both contain only solid-state materials, and ceramics play important roles in both of them. Further, both of them are reversible cells, i.e. the very same cell may be operated either as an electrolysis cell or as a fuel cell, depending on the cell voltage/current direction and reactant feed. When the cell is operated in fuel-cell mode, it is referred to as a solid oxide fuel cell (SOFC). The principle of the SOEC/SOFC, abbreviated as solid oxide cell (SOC), is sketched in figure 1.3. Naturally, many cells have to be series connected in stacks or bundles in order to obtain reasonably high voltages. Details of cell and stack designs, cell and stack materials, cell testing, cell degradation, and electrochemical processes are given in chapters 2–10.

The basic difference between the SOEC and the PCEC is that the electrolyte in an SOEC is an oxide ion (O^{2-}) conductor and in a PCEC it is a proton (H^+) conductor. This in turn means that in an SOEC, steam (H_2O) is fed to the negative electrode, where H_2 is evolved and ends up being mixed with some H_2O, because it is not practically possible to obtain a 100% conversion of H_2O to H_2. The electrolyte in a PCEC is an oxide-ceramic proton conductor. Steam is fed to the positive electrode, where O_2 is evolved and protons are conducted through the electrolyte to the cathode, where H_2 is evolved at the negative electrode; i.e. the H_2 produced is pure dry H_2. Even though the thermodynamics are the same as for the SOEC (figure 1.2(a)), the internal resistance of the PCEC is higher, and the maturity of the PCEC technology is much lower than the maturity of the SOEC. PCECs are described in detail in chapters 11–14.

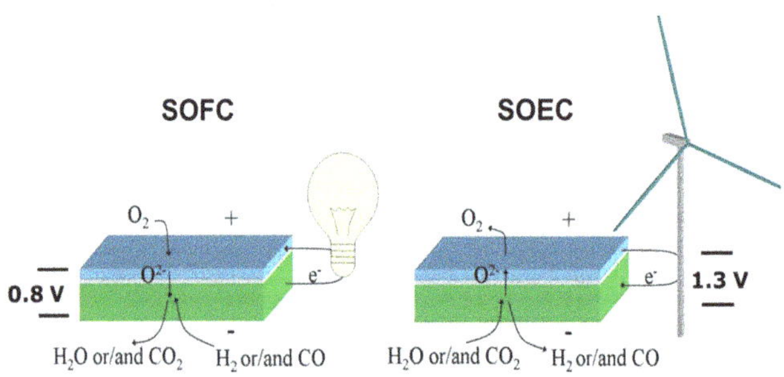

Figure 1.3. Sketch of the SOC principle used in the SOFC and SOEC modes of operation. The blue top layer represents the porous oxygen electrode, the white layer represents the dense oxide-ion-conducting electrolyte, and the green layer represents the porous fuel electrode plus the porous structural support of the cell. The total thickness of a cell may be less than 0.5 mm. The operation temperature is typically around 750 °C–850 °C. The open-circuit voltage (OCV) of an SOC with 50% H_2O + 50% H_2 at the fuel electrode and pure oxygen at the oxygen electrode, at 1 bar and 750 °C, is close to 1.0 V. Details are given in chapters 2–10.

1.3 Useful electrochemical concepts for SOC cells

As the electrochemistry of high-temperature solid oxides, including ceramic proton conductors, is rather different from the low-temperature aqueous electrochemistry, it seems useful to have a brief look at the basic electrochemistry of the non-stoichiometric types of material that play a very important role in SOCs.

1.3.1 Example of SOC structure and materials

Figure 1.4 shows an artificially colored SEM micrograph of a cell of the so-called fuel electrode supported flat plate design. The cell support and the fuel electrode are both made of nickel–yttria stabilized zirconia (Ni–YSZ) cermet. The support is typically made of a composite of 40 vol% Ni and 60 vol% $Zr_{0.94}Y_{0.06}O_{1.96}$ with a relatively coarse structure that has 30%–35% porosity. The active fuel electrode is made of a Ni and $Zr_{0.84}Y_{0.16}O_{1.92}$ composite with the same volumetric ratio (40 vol%/60 vol%), but with a significantly finer structure and a lower porosity than the cell support. $Zr_{0.94}Y_{0.06}O_{1.96}$ is usually denoted by 3YSZ and $Zr_{084}Y_{0.16}O_{1.92}$ by 8YSZ [22]. Both Ni–YSZ cermet layers are prepared and sintered as NiO–YSZ composites and then the NiO is reduced to metallic Ni when exposed to H_2 or polarized to e.g. −1 V versus an O_2 reference electrode (1 bar, 1000 °C) for 1 h. The reason for using 3YSZ in the support is that it is one of the mechanically strongest ceramics due to its martensitic transformation ability (see further details below).

The electrolyte consists of 8YSZ as this composition has the highest oxide ion conductivity of yttria-doped zirconia. Zirconia doped with 10 mol% scandia (Sc_2O_3) plus 1 mol% CeO_2 has even higher ionic conductivity, but is more expensive [23]. The high conductivity is also the reason for the use of 8YSZ in the active fuel electrode. Several other SOC electrolyte materials exist, but all are less mechanically strong and less thermodynamically stable [23, 24].

For the time being, the most popular oxygen electrode materials are the perovskite-structured $(La_{0.75}Sr_{0.25})_{0.95}MnO_3$ (LSM), $(La_{0.6}Sr_{0.4})_{0.98}Co_{0.2}Fe_{0.8}O_{3-\delta}$

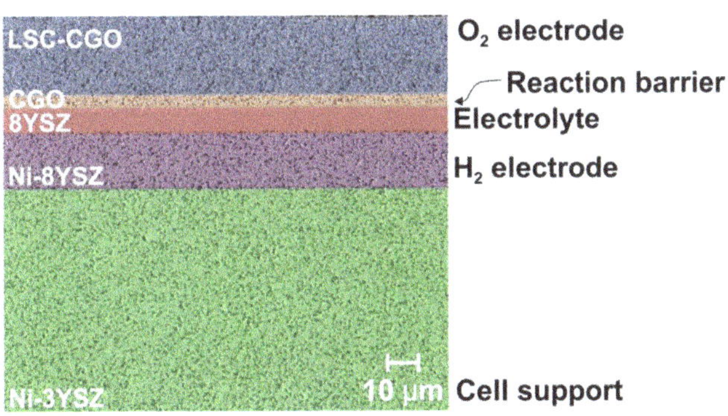

Figure 1.4. Artificially colored scanning electron microscope (SEM) micrograph of the cross-section of an SOC manufactured by the Technical University of Denmark (DTU). Only a sectional view of the active cell parts and a minor part of the cell support are shown with dimensions in accordance with the scale bar in the figure. The full cell support is about 350 μm thick.

(LSCF), and $(La_{0.6}Sr_{0.4})_{0.98}CoO_{3-\delta}$ (LSC) [22, 25]. The *m*ixed oxide *i*on and *e*lectron *c*onductors (MIECs) LSCF and LSC are both excellent oxide ion oxidation and oxygen reduction catalysts [26]. However, LSCF and LSC react with YSZ at the sintering temperature, and therefore, a reaction barrier of $Ce_{0.9}Gd_{0.1}O_{1.95}$ (CGO) is placed between the yttria stabilized zirconia (YSZ) electrolyte and the perovskite. In these cases, the composite electrode does not contain YSZ but CGO mixed with LSCF or LSC (figure 1.4). CGO is an even better oxide ion conductor than 8YSZ. CGO does not react with the perovskites, but CGO is reduced to an MIEC in contact with the usual fuel electrode gases at temperatures above 500 °C. An electrolyte must be close to an electron isolator. Therefore, CGO cannot be used as the main electrolyte in spite of its high oxide ion conductivity.

PCECs may have designs very similar to that of the cell shown in figure 1.4, but with different materials. Their main characteristic is that the electrolyte must be a solid proton conductor, e.g. $BaZr_{0.9}Y_{0.1}H_xO_{2.95+x/2}$ (BZY), which contains some OH^- ions on the O^{2-} sites; protons (H^+) can hop between the O^{2-} ions, facilitating proton conduction. These materials are described in detail in chapters 11–14.

1.3.2 Types of potentials in SOCs

Even though YSZ is generally accepted to be a good oxide-ion-conducting electrolyte, it is also a semiconductor with electron conductivity (n-type) at low oxygen pressures and electron hole conductivity (p-type) in oxidizing atmospheres. Its electronic conductivity is much smaller (1000 times or more) than the oxide ion conductivity [27]. In spite of the small value of the electronic conductivity, σ_e, its existence has important consequences, because it causes a varying redox capability across the YSZ electrolyte in an SOC. This means that the pO_2 in pores in the YSZ varies across the electrolyte layer and that impurities at the YSZ grain boundaries may become oxidized or reduced during operation, depending on the circumstances. Therefore, it is important to know more about at least two types of potentials inside a solid such as YSZ, namely, the Galvani and Fermi potentials.

Figure 1.5 illustrates the types of potentials in a solid conductor relative to the electric potential in vacuum. The Volta potential, or outer potential, ψ, is defined as

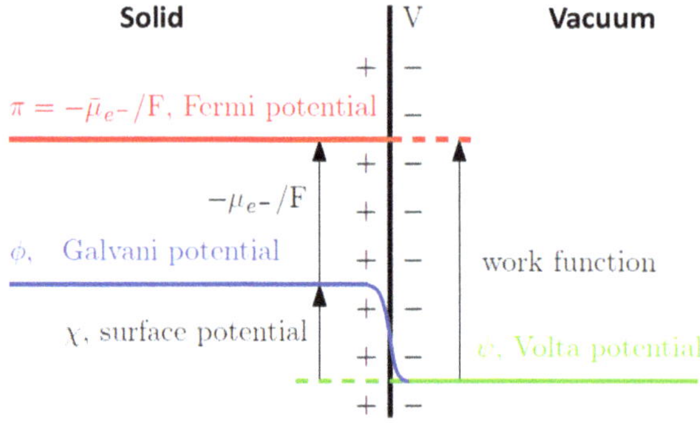

Figure 1.5. Potential types in a solid conductor relative to the outer electric potential in vacuum. Reproduced with permission from [28, 29], copyright The Electrochemical Society.

the work per unit charge needed to transport charge from infinity to the point in question. The electric potential in the interior of the solid, the Galvani potential, ϕ, deviates from the Volta potential by the surface potential, χ. The values of χ and ϕ cannot be determined by experiments, but if the charge distribution is known, they may be calculated using electrostatics. The Fermi potential, π, is defined as $\pi = -\bar{\mu}_e/F$, where $\bar{\mu}_e = \mu_e - F\phi$ is the electrochemical potential of the electrons, μ_e is the chemical potential of the electrons, and F is Faraday's constant [28–31]. The Galvani potential, ϕ, reflects the potential formed by all charged particles in the lattice. In an electrolyte such as YSZ, this means that the influence of the very low free electron concentration on ϕ is small. The lattice ions are the main originators of ϕ, which is the driving potential for the mobile oxide ions. It also affects the electrons, but with the reference point used here, $\pi = (RT/F)\ln(c_e/c_e^0) + \phi$, its effect on the electrons strongly depends on the electron concentration c_e relative to the electron concentration under standard conditions, c_e^0. Thus, π varies strongly across the electrolyte (figure 1.6, the data for calculations are given in table 1.2). Electrons are mainly driven by π. It is worth noting that a voltmeter only measures differences in π, because a voltmeter is only sensitive to electrons.

Figure 1.7 illustrates why it is so important to know the behavior of the potentials π and ϕ across the electrolyte. It shows the calculated values of $\pi - \phi$ (on the left y-axis) as a function of the relative distance x/L, where x is distance from the steam–hydrogen electrode, and L is the electrolyte thickness. The reference potential $(\pi - \phi)^0 = 0$ at 1 bar of O_2 and 1000 °C is chosen as the standard potential for the reaction $O^{2-} \rightleftharpoons \frac{1}{2}O_2 + 2\,e^-$. Figure 1.7 also presents the course of $\log(pO_2)$ (the right y-axis) across the electrolyte, $\pi - \phi = (RT/4)\ln(pO_2/\text{bar})$. From the pO_2 curve in figure 1.7 it can be observed that for -1 A cm^{-2} (corresponding to a 50 mV overpotential), pO_2 inside the YSZ electrolyte close to the oxygen electrode is ca. 7 bar.

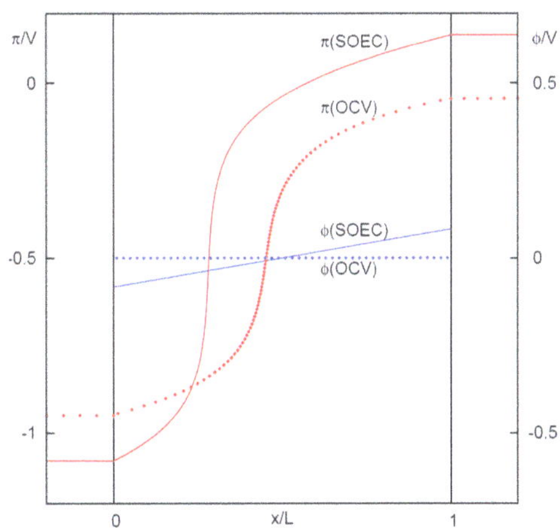

Figure 1.6. Calculated Fermi, π, and Galvani, ϕ, potentials across a YSZ electrolyte in the SOEC mode (-1 A cm^{-2}), compared to open-circuit conditions at 1000 °C. The reference electrode for π = 0 V is the O^{2-}/O_2 potential at 1 bar and 1000 °C. Here, ϕ is arbitrarily taken to be 0 at $\pi = 0.5$ V. The data used to calculate these potentials are given in table 1.2 [28, 29] reproduced with permission, copyright The Electrochemcial Society.

Table 1.2. Data used for the calculations supporting figure 1.6 [29]. The transport and concentration data for 8 mol% YSZ are taken from [27].

Electrode thickness	$L = 200\ \mu m$
Temperature	$T = 1000\ °C$
Oxygen pressure, right	$pO_2 = 0.2$ bar
Oxygen pressure, left	$pO_2 = 1.00 \cdot 10^{-15}$ bar (H_2–H_2O)
SOEC current	$i = -1.00\ A\ cm^{-2}$
Charge-transfer resistances:	
$H_2 + O^{2-} \rightleftharpoons H_2O + 2e^-$	$R_{ct,H} = 0.05$ ohm cm^2
$1/2\ O_2 + 2e^- \rightleftharpoons 2\ O^{2-}$	$R_{ct,O} = 0.1$ ohm cm^2
Electron transfer resistance, hydrogen	$R_{e,H} = 0.01$ ohm cm^2
Electron transfer resistance, oxygen	$R_{e,O} = 0.01$ ohm cm^2

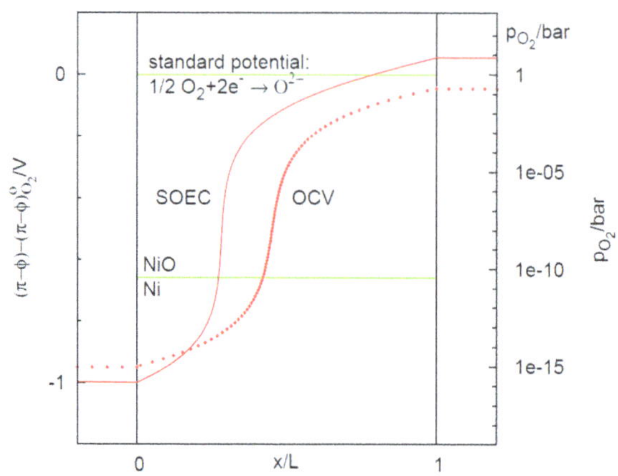

Figure 1.7. Values of $\pi - \phi$ and the local equilibrium oxygen pressure (pO_2) across the YSZ electrolyte in a cell operating in the SOEC mode ($-1\ A\ cm^{-2}$) compared to open-circuit conditions at 1000 °C. The data used for the underlying calculations are given in table 1.2. The reference potential $(\pi - \phi)^{\circ} = 0$ is the O^{2-}/O_2 potential at 1 bar and 1000 °C. Reproduced with permission from [28, 29], copyright The Electrochemcial Society.

This means that O_2 bubbles may form inside the electrolyte at a moderate overpotential, as observed by Knibbe *et al* [32].

Further, the equilibrium potential line of Ni/NiO is marked (the lower green line), and it shows that NiO, which was dissolved in the YSZ during sintering, is reduced to Ni metal at the OCV by exposure to the H_2–H_2O atmosphere. It should be noted that in the electrolysis mode, the Ni is reoxidized in a part of the electrolyte, i.e. through the distance between the crossing of the green line by the OCV curve and that of the SOEC curve. As NiO has a much higher molar volume than Ni, this redox activity inside the electrolyte may be a problem, even though NiO has only a small solubility of about 1%. Its precise solubility is dependent on temperature.

1.3.3 Non-recognized overpotential types in composite electrodes and MIECs

Most workers in the high-temperature electrolysis area assume that charge transfer in both composite electrodes (the fuel electrode and the oxygen electrode) follows the Butler–Volmer equation, and that it contributes significantly to the electrode polarization resistance. However, there seems to be no clear experimental evidence for this assumption. On the contrary, the current density–polarization curve (*i–V* curve) is always very linear in both electrolysis mode and fuel-cell mode if the reactant gas partial pressures (gas concentrations) in both composite electrodes are kept constant [33, 34]. Furthermore, porous composite Ni–YSZ cermet electrodes, other porous composite electrodes, and porous MIEC electrodes are not expected to follow simple Butler–Volmer behavior. They are described by porous electrode theory [35]. In addition, the charge transfer from or to reactant gas molecules happens *on* the electrode surface, i.e. there is no overpotential gradient that drives the charge transfer, which is a premise for Butler–Volmer behavior. Finally, there are no clear experiment-based descriptions in the literature of a non-ohmic behavior of charge transport from the electrode material to the electrolyte and vice versa.

Some SOC researchers argue that in case of low pH_2 in the steam feed in electrolysis mode, or low pH_2O in the hydrogen in fuel-cell mode, the low current density increases more strongly than linearly with increasing cell polarization and take this as an indication in favor of Butler–Volmer electrode kinetics. The observation is correct, but it is not in accordance with Butler–Volmer behavior, which predicts a linear *i–V* curve across the open-circuit potential (OCP) in the region very near OCV.

The reason for the small deviation from *i–V* linearity near the OCV is concentration polarization, and in most cases, the effect known as conversion polarization, which is due to the fact that it is difficult to maintain the partial pressure in the fuel electrode compartment during cell polarization. The conversion polarization is of course not a part of the electrode properties, but it is a result of the test setup and test conditions [36–38].

1.4 Recommendations for future work

The SOC literature presents huge disagreements in the data for almost every possible aspect and every property of cells and electrodes that one can think of. This is particularly the case for model electrodes and cells. For example, the reported three-phase boundary (3pb) line specific resistance (LSR) values of Ni–YSZ are spread over more than four decades, and the reported activation energy for the area specific resistance (ASR) as well as for the LSR may vary by more than a factor of two for nominally equal electrodes [39]. A similar spread in the reported data is seen for oxygen electrodes [40]. This naturally means that the electrodes and cells are in reality not very similar due to a lack of control of the important parameters. Of course, the main recommendation for future work is to control the important parameters or at least try to describe them. Another obvious recommendation, which in fact is a duty, and should be requested by all reviewers, is that a statement of uncertainties should be provided for all measured quantities. Three examples related to the lack of control and the omission of very important parameters from reports follow.

1.4.1 Stoichiometry of materials

An exact knowledge of the stoichiometry of cell materials is necessary, but is most often totally neglected, e.g. A/B ratios in perovskites, ABO_3 (where A are large and B are small metal ions), such as $(La_{1-x}Sr_x)_sCo_yFe_{1-y}O_{3-\delta}$, where $0 < x < 1, 0 < y < 1$, and usually the ratio, s, of A/B ions is in the range of $0.9 < s < 1.1$. For perovskites, $0 \leqslant \delta \leqslant 0.5$. The most popular composition for the oxygen electrode is usually just given as $La_{0.6}Sr_{0.4}Co_{0.2}Fe_{0.8}O_3$, and this totally neglects the fact that this material is non-stoichiometric by nature, not only with respect to its oxygen content but also in terms of its cations. The chance that s is exactly equal to 1 is extremely small, as this would require a method that could measure the amount, e.g. the weight, of all materials without any uncertainty. It is important to know whether $s < 1$ or $s > 1$, because if $s < 1$ then the surface of the perovskite electrode—the electrocatalyst—may be enriched with B-site oxides, here Co and Fe oxides, increasing electrocatalytic activity. If $s > 1$ then the A-site oxide, here mainly SrO, is enriched on the electrode surface and the electrocatalytic activity decreases drastically. In addition, the sinterability and the ability to adhere to the CGO surface are affected. Thus, the general rule is that $s < 1$ is preferred.

Furthermore, $\delta = 0$ means that there are no oxide vacancies in the crystal lattice, but this is not at all the normal situation. At a temperature of 500 °C in air, both the Co and the Fe in LSCF are present mainly in the +3 oxidation state. In the case of $La_{0.6}Sr_{0.4}Co_{0.2}Fe_{0.8}O_3$ in which all Fe and Co are in the +3 state, then $\delta = s{\cdot}x/2 = 0.2$, and if $s = 1$ and $x = 0.4$, this chemical formula should be written as $La_{0.6}Sr_{0.4}Co_{0.2}Fe_{0.8}O_{2.8}$. As one Sr^{2+} can replace one La^{3+}, then for every two Sr^{2+} that replace two La^{3+}, it is necessary to remove one O^{2-} from the basic $LaFeO_3$ lattice (leaving a vacant oxide, O^{2-}, site). However, both Fe and Co may exist in the +2, +3, and +4 oxidation states. Thus, $\delta = 0.2$ only occurs when the average oxidation state of the Fe and Co is +3. δ decreases with increasing pO_2 (an increase in the oxidation states of Fe and Co), and δ increases with increasing temperature and decreasing electrode potential (a decrease in the oxidation states of Fe and Co). It should also be noted that a change in electrode potential is equivalent to a change in pO_2, as given by Nernst's equation. Almost all SOC researchers know this, but often do not write the formulas accordingly; many neither measure δ nor discuss the dependence of δ on temperature, pO_2, and electrode potential. Properties such as ionic and electronic conductivity and molar volume all correlate with the concentration of oxide vacancies.

Thus, both A/B and δ are very important for electrode performance and should be carefully reported in order to be able to compare different electrode and cell tests and to contribute to cell performance optimization through proper adjustment of the composition of the electrode materials.

1.4.2 Impurities and segregations

The effect of impurity concentrations at the ppb level in reactant streams and the ppm level in the materials may be devastating [41], but impurities were, until recently, most often neither mentioned nor analyzed in the SOC literature. However, during the last decade, several published examples have shown how damaging small amounts of gaseous impurities can be, such as impurities in the steam/H_2 and air supplies, such as $Si(OH)_4$, which may originate from the liquid water used for steam production in spite of careful cleaning. A content of only 4–7 μg L^{-1} (4–7 ppb) is enough to damage the SOEC steam/H_2 electrode over time [42]. In addition, $Si(OH)_4$ evaporated from the glass sealant into the incoming steam may have a serious negative impact on the SOC performance [43]. On the oxygen side, $CrO_2(OH)_2$, which may originate from chromium steel pipes or interconnects in the SOC system, impedes the SOC oxygen electrode [44, 45]. Many other compounds of elements such as carbon, sodium, phosphorus, sulfur, chlorine, potassium, and titanium may form oxides which can degrade the SOC electrodes or react directly with the electrode surface, e.g. the chemisorption of S on Ni in the Ni–YSZ fuel electrode.

It is not the effect of a few ppm of e.g. SiO_2 on the properties of Ni that is problematic as such. The problem is the segregation of impurities on the surfaces and interfaces of the electrolytes and electrodes, which blocks the active sites of the electrocatalyst. Similarly, the segregation of a very small amount of an electrode compound such as SrO onto the surface of an $La_{0.6}Sr_{0.4}FeO_3$ (LSF) oxygen electrode may passivate it. In this case there is evidence that this segregation of SrO to the surface is promoted by a reaction at temperatures of 650 °C to 800 °C between SrO on the LSF and CO_2 in air to $SrCO_3$, or of SO_2 and O_2 in air to $SrSO_4$, which blocks the active sites [46]. It is in this case possible to reactivate the LSF electrode by a heat treatment at 1000 °C, during which $SrCO_3$ and $SrSO_4$ decompose.

One method that can be used to remove enough impurities from a gas feed is to pass the gas through a filter made of a porous active electrode material of the electrode to be protected. The filter material should preferably be nanostructured and kept at a temperature slightly lower than that of the operating cell, e.g. 650 °C for electrodes that operate at 700 °C and above. In most cases, such a filter will absorb all types of gas impurity which are toxic to the electrode.

1.4.3 Leaks

In practice, air leaks from the ambient atmosphere into electrode compartments in test setups and from one side of the cell to the other through open porosities or cracks always happen to some extent in electrode tests as well as in cell and stack tests. Furthermore, some leaks may take place through the electrolyte and seals in the form of a current leak due to unintended electron conductivity. Leaks mean loss of energy, which is dissipated in the form of heat and causes unwanted, often local temperature gradients in addition to the loss of electric energy. Therefore, it is important to monitor the leak rate of a test system, at least at the start and end of a test, and, if possible, also regularly through the test, but such data are seldom reported.

The easy way to monitor leaks is to simply compare the actual measured OCV of a cell or stack with the theoretically calculated OCV. There should not be a greater difference between these two numbers than one that can be explained by thermoelectric potential differences.

If there is a leak, the nature and the size of the leak should naturally be examined, e.g. by studying the effect of gas flow rate and small pressure differences between the two electrode compartments. A test for possible electrical leaks may be to simply measure the resistance of the cell at room temperature. At the operation temperature, polarization of the cell may be carried out with a clean inert gas (Ar or N_2) on both electrodes and around the test setup. If any appreciable current density is measured at 0.5 V polarization then it must be electronic or a leak. As there are traces of O_2 and H_2O in even very clean gases, current densities in the order of few $\mu A\ cm^{-2}$ might be expected. More advanced electrochemical leak tests could be performed using the dynamic polarization of cells in electrolysis mode with clean O_2 at the positive electrode and H_2 with, for example, 0.5% H_2O at the negative electrode. If the cell polarization is scanned from 0 to, say, 1.5 V, then for an ideally tight cell this should give a low constant current density. If the current density varies with cell polarization then the i–V curve might indicate which kinds of leak are present. However, this type of investigation has, to the best of our knowledge, not been reported, even though it would assure high-quality measurements in a laboratory test setup.

1.5 Outlook

Strong arguments have been made in favor of high-temperature SOECs as a very efficient and low-cost technology for the future conversion and storage of renewable electricity, but a lot of research is still needed in order to realize the full potential of SOECs [3]. In particular, the durability of both their electrochemical and mechanical performance is important, and the interplay between the mechanical strength over time and the electrochemistry taking place inside the materials is crucial. In this context, it is imperative to remember the old truth of 'the devil is in the details.' Therefore, much more fundamental research, keeping the above recommendations in mind, is still vital, in parallel with the continued technology scale-up into large SOEC electrolysis systems that has already made good progress. PCECs also have great potential, but are still less mature than SOECs. Thus, for PCECs, research is the main objective in the near future. Again, very careful research performed with a deep understanding of both electrochemistry and materials is essential.

It is obvious that the better a researcher understands and controls all the important factors, the more valuable the results of the research will be. If several factors are neglected then the results may be worthless. However, this is apparently often forgotten, but it is really worth remembering. As it may be impossible for one person to know all the relevant scientific disciplines sufficiently well, one of the recommendations is to work in teams, in which each team member masters at least one of the relevant topics in detail.

Acknowledgments

The authors are grateful to colleagues inside and outside the DTU for many helpful, inspiring discussions over the years.

References

[1] Hansen J B 2015 Solid oxide electrolysis–a key enabling technology for sustainable energy scenarios *Faraday Discuss.* **182** 9–48

[2] Mogensen M B *et al* 2019 Reversible solid-oxide cells for clean and sustainable energy *Clean Energy* **3** 175–201

[3] Hauch A, Küngas R, Blennow P, Hansen A B, Hansen J B, Mathiesen B V and Mogensen M B 2020 Recent advances in solid oxide cell technology for electrolysis *Science* **370** 186–94

[4] Graves C, Ebbesen S D, Mogensen M and Lackner K S 2011 Sustainable hydrocarbon fuels by recycling CO_2 and H_2O with renewable or nuclear energy *Renew. Sustain. Energy Rev.* **15** 1–23

[5] IEA 2021 *Key World Energy Statistics 2021* (Paris)

[6] Lewis N S, Crabtree G, Nozik A J, Wasielewski M R, Alivisatos P, Kung H, Tsao J, Chandler E, Walukiewicz W and Spitler M 2005 *Basic research needs for solar energy utilization. report of the basic energy sciences workshop on solar energy utilization, April 18–21, 2005* (DOESC (USDOE Office of Science (SC)))

[7] Tsao J, Lewis N and Crabtree G 2006 *Solar FAQs* (www.sandia.gov/app/uploads/sites/153/2022/02/Solar-FAQs.pdf)

[8] ESMAP 2019 *Global Solar Atlas 2.0 Technical Report* (Washington, DC: World Bank)

[9] IRENA 2021 *Renewable Power Generation Costs in 2020* (Abu Dhabi: International Renewable Energy Agency)

[10] Sørensen R Z, Hummelshøj J S, Klerke A, Reves J B, Vegge T, Nørskov J K and Christensen C H 2008 Indirect, reversible high-density hydrogen storage in compact metal ammine salts *J. Am. Chem. Soc.* **130** 8660–8

[11] Makepeace J W *et al* 2019 Reversible ammonia-based and liquid organic hydrogen carriers for high-density hydrogen storage: recent progress *Int. J. Hydrogen Energy* **44** 7746–67

[12] Speight J 2003 *Perry's Standard Tables and Formulae For Chemical Engineers* (New York: McGraw-Hill Education)

[13] Weiss M, Ruess R, Kasnatscheew J, Levartovsky Y, Levy N R, Minnmann P, Stolz L, Waldmann T, Wohlfahrt-Mehrens M and Aurbach D 2021 Fast charging of lithium-ion batteries: a review of materials aspects *Adv. Energy Mater.* **11** 2101126

[14] Giddey S, Badwal S P S, Munnings C and Dolan M 2017 Ammonia as a renewable energy transportation media *ACS Sustain. Chem. Eng.* **5** 10231–9

[15] Rouwenhorst K H R, Elishav O, Lis B M, Grader G S, Mounaïm-Rousselle C, Roldan A and Valera-Medina A 2020 Future trends *Techno-Economic Challenges of Green Ammonia as an Energy Vector* (New York: Elsevier) pp 303–19

[16] Mestemaker B T W, Castro M B G, van der Blom E C, Cornege H J and Visser K 2019 Zero emission vessels from a shipbuilder's perspective *2nd Int. Conf. on Smart & Green Technology for the Future of Marine Industries (SMATECH 2019)–Conf. Proc.* pp 1–10

[17] Dönitz W, Erdle E and Streicher R 1990 High temperature electrochemical technology for hydrogen production and power generation *Electrochemical Hydrogen Technologies: Electrochemical Production and Combustion of Hydrogen* ed H Wendt (Amsterdam: Elsevier) pp 213–59
[18] Küngas R 2020 Review—electrochemical CO_2 reduction for CO production: comparison of low- and high-temperature electrolysis technologies *J. Electrochem. Soc.* **167** 044508
[19] Mogensen M B 2020 Thermodynamics of high temperature H_2O and CO_2 electrolysis *Figshare* https://figshare.com/articles/online_resource/Thermodynamics_of_high_temperature_H2O_and_CO2_electrolysis/12652322
[20] Hu L, Lindbergh G and Lagergren C 2016 Performance and durability of the molten carbonate electrolysis cell and the reversible molten carbonate fuel cell *J. Phys. Chem.* C **120** 13427–33
[21] Kaplan V, Wachtel E, Gartsman K, Feldman Y and Lubomirsky I 2010 Conversion of CO_2 to CO by electrolysis of molten lithium carbonate *J. Electrochem. Soc.* **157** B552
[22] Hauch A, Brodersen K, Chen M and Mogensen M B 2016 Ni/YSZ electrodes structures optimized for increased electrolysis performance and durability *Solid State Ionics* **293** 27–36
[23] Kilner J, Druce J and Ishihara T 2015 *Electrolytes High-Temperature Solid Oxide Fuel Cells for the 21st Century: Fundamentals, Design and Applications* ed K Kendall and M Kendall (Amsterdam: Elsevier) pp 85–132
[24] Garvie R C, Hannink R H and Pascoe R T 1975 Ceramic steel? *Nature* **258** 703–4
[25] Hauch A, Brodersen K, Chen M, Graves C, Jensen S H, Jørgensen P S, Hendriksen P V, Mogensen M B, Ovtar S and Sun X 2017 A decade of solid oxide electrolysis improvements at DTU energy *ECS Trans.* **75** 3–14
[26] Jiang S P 2019 Development of lanthanum strontium cobalt ferrite perovskite electrodes of solid oxide fuel cells – a review *Int. J. Hydrogen Energy* **44** 7448–93
[27] Park J and Blumenthal R N 1989 Electronic transport in 8 mole percent Y_2O_3–ZrO_2 *J. Electrochem. Soc.* **136** 2867–76
[28] Jacobsen T, Chatzichristodoulou C and Mogensen M 2014 Fermi potential across working solid oxide cells with zirconia or ceria electrolytes *ECS Trans.* **61** 203–14
[29] Jacobsen T and Mogensen M 2008 The course of oxygen partial pressure and electric potentials across an oxide electrolyte cell *ECS Trans.* **13** 259–74
[30] Mogensen M and Jacobsen T 2009 Electromotive potential distribution and electronic leak currents in working YSZ based SOCs *ECS Trans.* **25** 1315–20
[31] Chatzichristodoulou C, Chen M, Hendriksen P V, Jacobsen T and Mogensen M B 2016 Understanding degradation of solid oxide electrolysis cells through modeling of electrochemical potential profiles *Electrochim. Acta* **189** 265–82
[32] Knibbe R, Traulsen M L, Hauch A, Ebbesen S D and Mogensen M 2010 Solid oxide electrolysis cells: degradation at high current densities *J. Electrochem. Soc.* **157** B1209–17
[33] Ebbesen S D, Jensen S H, Hauch A and Mogensen M B 2014 High temperature electrolysis in alkaline cells, solid proton conducting cells, and solid oxide cells *Chem. Rev.* **114** 10697–734
[34] Njodzefon J-C, Graves C R, Mogensen M B, Weber A and Hjelm J 2016 Kinetic studies on state of the art solid oxide cells: a comparison between hydrogen/steam and reformate fuels *J. Electrochem. Soc.* **163** F1451–62
[35] Nielsen J and Hjelm J 2014 Impedance of SOFC electrodes: a review and a comprehensive case study on the impedance of LSM:YSZ cathodes *Electrochim. Acta* **115** 31–45

[36] Primdahl S and Mogensen M 1998 Gas conversion impedance: a test geometry effect in characterization of solid oxide fuel cell anodes *J. Electrochem. Soc.* **145** 2431–8
[37] Primdahl S and Mogensen M 1999 Gas diffusion impedance in characterization of solid oxide fuel cell anodes *J. Electrochem. Soc.* **146** 2827–33
[38] Jensen S H, Hauch A, Hendriksen P V and Mogensen M 2009 Advanced test method of solid oxide cells in a plug-flow setup *J. Electrochem. Soc.* **156** B757–64
[39] Mogensen M and Holtappels P 2013 Ni-based solid oxide cell electrodes *Solid Oxide Fuels Cells: Facts and Figures*ed J T S Irvine and P Connor (London: Springer) pp 25–45
[40] Hansen K V, Norrman K, Traulsen M L and Mogensen M B 2017 Dynamic and impure perovskite structured metal oxide surfaces *ECS Trans.* **80** 91–100
[41] Mogensen M and Hansen K V 2009 Impact of impurities and interface reaction on electrochemical activity *Handbook of Fuel Cells* vol 5 (Wiley Online Library) pp 543–54
[42] Riedel M, Heddrich M P and Friedrich K A 2020 Investigation of the long-term stability of solid oxide electrolysis stacks under pressurized conditions in exothermic steam and co-electrolysis mode *Fuel Cells* **20** 592–607
[43] Hauch A, Jensen S H, Bilde-Sørensen J B and Mogensen M 2007 Silica segregation in the Ni/YSZ electrode *J. Electrochem. Soc.* **154** A619
[44] Jiang S P and Chen X 2014 Chromium deposition and poisoning of cathodes of solid oxide fuel cells—a review *Int. J. Hydrogen Energy* **39** 505–31
[45] Opila E J, Myers D L, Jacobson N S, Nielsen I M B, Johnson D F, Olminsky J K and Allendorf M D 2007 Theoretical and experimental investigation of the thermochemistry of $CrO_2(OH)_2$(g) *J. Phys. Chem.* A **111** 1971–80
[46] Tripković Đ, Wang J, Küngas R, Mogensen M B, Yildiz B and Hendriksen P V 2022 Thermally controlled activation and passivation of surface chemistry and oxygen-exchange kinetics on a perovskite oxide *Chem. Mater.* **34** 1722–36

IOP Publishing

High Temperature Electrolysis
From fundamentals to applications
Werner Sitte and Rotraut Merkle

Chapter 2

Electrolyte materials for solid oxide electrolysis cells

Stephen J Skinner, Chen-Yu Tsai, Per Hjalmarrson, Robert Leah and Subhasish Mukerjee

Electrolyte materials for use in solid oxide electrolysers are critically important to the development of this technology as these components dictate the operating regime for the cells. Several parameters are discussed, notably ionic conductivity, stability and durability, and how these are linked to the overall cell development. The most common electrolytes, those of the fluorite and perovskite structural families, are discussed and the prospects of identifying alternative oxide ion conductors that will meet the cell requirements are reviewed. The contrast in the current density regime in comparison with solid oxide fuel cells is highlighted and the potential impact of this on materials selection is introduced, including the impact on the microstructure of the electrolyte, and ultimately the failure mechanism of these devices. The leading electrolyte materials are identified as the yttria stabilised zirconia (YSZ) phases, with Sc stabilised zirconia, substituted ceria and the perovskite $La_{1-x}Sr_xGa_{1-y}Mg_yO_{3-\delta}$ (LSGM) also under current consideration for device development.

2.1 Introduction

The background theory and contextualisation of solid oxide electrolysis cells (SOECs), including key aspects of design, manufacture and operation are introduced in several subsequent chapters of this book. However, as a brief introduction to the technology, there are a number of critical functional components of SOECs that are essential to enable the required electrochemical processes to occur. These components are the air electrode, fuel electrode and ion conducting electrolyte. Further discussion in several chapters of the book will be devoted to the two main operation modes for high temperature electrolysis: oxide ion and proton conducting based technologies. In this chapter the focus is on the solid oxide electrolytes commonly used in the oxide ion conducting technology, with discussion of protonic electrolytes provided in chapters 11–14.

doi:10.1088/978-0-7503-3951-3ch2

2.1.1 Definition of a solid oxide electrolysis electrolyte

Solid oxide electrolysis cells have considerable overlap in their materials and operation with the solid oxide fuel cell (SOFC) technology. SOFCs have undergone decades of development in which the functional and engineering aspects of the devices have been refined, and the materials requirements in terms of their fundamental ionic and electronic properties resolved. The identification of high temperature steam electrolysis as a viable route for the production of hydrogen was proposed as early as the 1970s [1] with Dornier engaged in electrochemical cell development, with a subsequent report by Doenitz *et al* [2] outlining the thermodynamic advantages of a steam electrolysis cell, that described the necessary requirements for each of the functional components (electrolyte, water vapour-hydrogen electrode and oxygen electrode). From this beginning the authors were able to produce, in 1985, a further report detailing the successful operation of a tubular electrolysis cell operating at 1000 °C [3], achieving current density of 0.37 A cm^{-2} under autothermal operation. Little further attention was devoted to SOECs until in 1992 Eguichi *et al* [4] reported operating a solid oxide cell in both power generation and steam electrolysis modes. This early report of reversible operation directly linked the cell materials for both modes of operation and invigorated research interest in the field. A further option for the use of electrolysis cells is in the reduction of CO_2 as outlined by Isenberg *et al* [5], and in the potential for co-electrolysis of steam and CO_2 [6] which is discussed in detail in chapters 15–16. Clearly the link between the electrochemical operation of both SOECs in all operation modes and SOFCs and the potential for reversible operation is predicated on the materials requirements for both devices being similar, if not identical. Solid oxide electrolysis encompasses a wide range of operational modes, but all are linked through the selection of the electrolyte material. This requirement is discussed below.

2.1.2 Requirements for the electrolyte component

In the case of a generic solid oxide cell, be it operating in hydrogen production (electrolysis) or power generation (fuel cell) mode, the fundamental materials requirements for the electrolyte are effectively identical. As outlined by Doenitz *et al* [2], the electrolyte component of a solid oxide electrolysis cell has to fulfil the following conditions:

- Gas tightness (operation as a separation membrane, preventing gas cross-over);
- Mechanical stability;
- High oxygen ion conductivity;
- Long term stability at the operating conditions of the cell (low degradation rates);
- Low cost raw materials;
- Low cost manufacturing.

These key requirements can be further expanded to include the need for chemical and mechanical compatibility of the selected electrolyte material with the fuel and air electrode materials, meaning that there is limited (no) chemical interaction forming deleterious phases at the electrolyte/electrode interfaces, and that the mechanical compatibility leads to no cell failures through delamination. In order for the components to be compatible, the thermal expansion coefficients of each of the cell components have to be aligned, and it is typically that of the electrolyte component that takes precedence. Finally, of course, it is essential that the electrolyte is stable in both reducing and oxidising atmospheres.

As is common with the SOFC device, the electrolyte component can form the basis of the overall device structure, be that in a planar or tubular architecture, as discussed in chapter 7. In each of these cases the ohmic resistance of the electrolyte is of critical importance, and is directly related to the diffusivity of the ionic species through the membrane. In order to maximise cell performance through maximising current density, the ohmic resistance should be reduced. To achieve this ohmic resistance reduction (ionic conductivity maximum) the solution for developers is to reduce the membrane thickness. This leads to the conclusion that the cell architecture can then be either electrode-supported (air or fuel electrode) or self-supporting. This feature is less of a concern when operating at higher temperatures (>800 °C), given that the oxide ion transport is a thermally activated process, but for lower temperature devices (500–800 °C), reducing ohmic losses through membrane thickness reduction is critically important. This lower operation temperature regime requires the SOEC to be an electrode supported architecture, as with the reduced thickness these membranes will not meet the requirement of mechanical stability previously discussed. To ensure that a device operates effectively in electrolysis mode, careful consideration is needed to select the appropriate electrolyte membrane material. The materials that are most commonly encountered as single-phase electrolytes in SOEC operations are considered in the following section.

2.2 Materials in common use

The materials that are commonly used as solid oxide electrolysis cell electrolytes are required to be able to conduct oxygen ions, as discussed above, and for these materials to be effective the stated value that meets cell developer needs is at least 0.01 Scm^{-1} [7]. This minimum value is achieved at a range of different temperatures depending on the material selected with the most studied electrolytes being based on the fluorite oxides. Fluorite structured oxides, as detailed in figure 2.1, have been the materials of choice for SOFC applications [8], and were the logical initial choice for SOEC devices. In addition to the ionic transport, it is essential that the electronic conductivity of the electrolyte is negligible (ionic transport number, $t_o = 1$) to avoid any short-circuit and the related loss of cell efficiency [9, 10].

Depending on the oxygen ionic conductivity of the electrolyte material selected, a minimum operating temperature is required to ensure that the electrolyte performs effectively. In the case of the fluorite based electrolytes, the yttrium stabilised zirconia (YSZ) family requires a minimum temperature of at least 700 °C to have an acceptable performance [9]. In addition, a thin dense robust electrolyte is usually preferred as it will decrease total cell resistance and prevent gas crossover from the

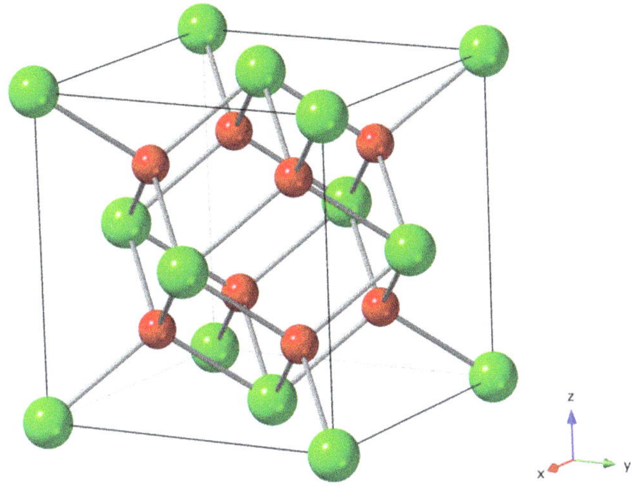

Figure 2.1. Schematic representation of the fluorite structure adopted by the zirconia and ceria based electrolytes used in SOECs. Red spheres—oxygen, green spheres—Zr, Ce.

fuel and air side electrodes [9, 10]. There are some niche developments of thin film devices using nm scale supported electrolytes that envisage operation at ~400 °C, but these are in early development stages[1].

As is common in solid oxide fuel cell electrolyte development maximising the oxide ion conductivity is a central requirement for SOECs. In order to achieve this, host structures such as the fluorite structure depicted in figure 2.1, are typically substituted with aliovalent cations, that require the creation of oxygen vacancies to maintain electroneutrality on the lattice. This creation of point defects within the lattice, leading to higher carrier concentration and thus higher ionic conductivity is described using the Kröger–Vink notation. In Kröger–Vink notation the lattice is considered as consisting of a series of atoms located at lattice points. In ionic crystals, each lattice point is fully occupied by the nominally charged ionic species, and any defect charges are counted relative to the regular ions. Therefore, an oxygen ion located at an oxygen site would have an effective charge of 0, whilst an oxygen vacancy would have an effective charge of +2 (a 2- ion has been removed from the lattice, leaving an effective +2 charge on the lattice). This formalism allows for defect equations to be created, and the variation of the concentration of charged species as a function of, for example pO_2, to be presented as a Brouwer diagram. To illustrate the use of Kröger–Vink notation let's consider the substitution of Y_2O_3 into ZrO_2, creating the YSZ family of electrolytes. In this case we can generate an equation:

$$Y_2O_3 + O_O^x + 2Zr_{Zr}^x \leftrightarrow 2Y'_{Zr} + V_O^{\bullet\bullet} + 2ZrO_2 \tag{2.1}$$

where Y'_{Zr} represents a substitutional Y atom on a Zr lattice site with an effective lattice charge of −1, $V_O^{\bullet\bullet}$ represents an oxygen vacancy with an effective charge of

[1] www.epistore.eu.

+2 and O_O^x represents a charge neutral oxygen ion located on an oxygen lattice site. Further discussion of the defect chemistry of oxides can be found in [11].

As aliovalent substitution increases it may be expected that ionic conductivity would continue to track substituent content, but many authors have reported that there is a peak in the ionic conductivity at between 4–11 mol% substituent [12, 13]. The origin of this conductivity maxima has been explored by Koettgen *et al* [14], who report that this effect can be explained by the formation of defect associates, ordering of the generated oxygen vacancies, or by the modified jump probabilities, and that these could be independent or mutually dependent. Further explanation of this phenomenon identified that trapping, blocking and vacancy–vacancy interactions are the primary factors that influence ionic conductivity, and that it is trapping that limits the conductivity maximum.

All of the electrolyte materials considered in this chapter are developed and used as polycrystalline ceramics, which means that the microstructure of the ceramic electrolyte consists of a combination of grains and grain boundaries. Ionic conductivity in electrolytes can be shown to consist of transport through the grains and across the grain boundaries, but also potentially along the grain boundaries. In this instance the grain boundaries have been proposed to be either blocking or fast transport pathways, with the electrolyte materials most commonly exhibiting blocking grain boundaries [15, 16]. The characterisation of the grain and grain boundary contributions to the total conductivity is typically determined through the use of electrochemical impedance spectroscopy (EIS) [17] and its interpretation using the bricklayer model, in which the ceramics are viewed as analogous to a series of bricks separated by mortar (i.e. the grain boundary), as discussed by Kidner *et al* [18].

2.2.1 Zirconia-based electrolytes

The most common and abundantly researched electrolyte materials for SOEC applications are those based on stabilised zirconia, with yttria substitution providing the most common electrolytes, due to their advantages in terms of low cost, low toxicity, abundance, chemical stability and mechanical strength [9, 10]. Stabilisation of the tetragonal or cubic polymorphs of zirconia (ZrO_2) with aliovalent substituents, introducing ionic defects that facilitate oxide ion transport are well known in the fuel cell community, with yttria at levels of between 3 and 8 mol% typical in devices (3YSZ and 8YSZ, respectively), 3 mol% having lower ionic conductivity but greater mechanical stability [19]. The relatively low ionic conductivity of YSZ, e.g., 0.10 S cm^{-1} and 0.03 S cm^{-1} at 1000 °C and 800 °C, respectively for 8YSZ limits its application. Despite this limitation there have been numerous studies demonstrating the capability of devices based on YSZ electrolytes. One key aspect to consider in applying electrolytes in the electrolysis mode is that the cell voltages that are experienced are typically significantly higher (1.1–1.4 V) than those in the corresponding fuel cell mode (0.6–1.0 V). This fundamental difference has been the cause of concern in terms of long-term durability, with the potential for induced electronic conduction through reduction influencing cell behaviour. Schefold *et al* [20] investigated the influence of electronic conduction in 8YSZ, figure 2.2, under a

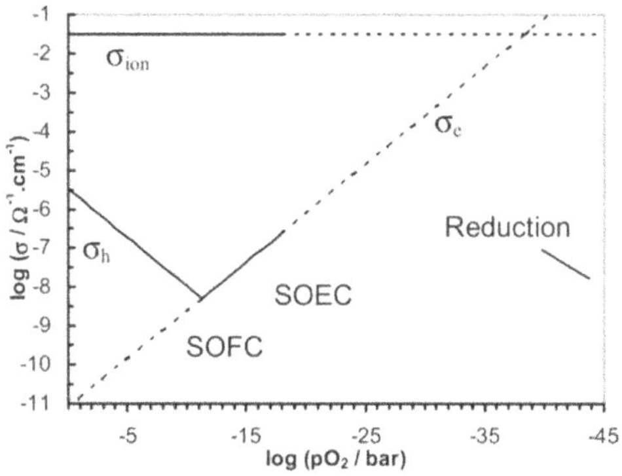

Figure 2.2. pO_2 regime under which electronic conduction becomes significant in 8YSZ. Reprinted from [20], copyright (2009) with permission from IOP.

range of operating conditions, both with and without steam supply. They note that a typical YSZ cell would not reach voltage saturation until 1.9 V, well in excess of the cell operating conditions, but also note that electrolyte electronic conduction could be a significant factor in the case of cell failure/steam supply failure. Indeed, it is proposed that the onset of electronic conduction in these situations of cell failure could in fact be beneficial in protecting the cell through impeding the electrolyte reduction. Sohal *et al* [21] compiled an extensive summary of the degradation issues surround SOECs based on YSZ electrolytes, finding a number of degradation phenomena, but the loss of ionic conductivity of the electrolyte was the only process associated with the membrane, with the remaining issues associated with electrodes, or with the electrode/electrolyte interface. It is therefore clear that for the YSZ based electrolytes the overall cell performance is dictated by the ionic transport within the membrane and the activity of the electrodes. To enhance the electrolyte performance there are few options available—altering membrane thickness, increasing operating temperature or changing the material to one with higher ionic conductivity.

Other studies have suggested that the long term operation of YSZ based devices over extended time periods results in significant degradation in ionic transport, noting that after only 5000 min there was up to a 14.2% decrease in conductivity [22], figure 2.3. This could, however, be mitigated through the manipulation of the concentration of the substituting species. It was suggested by the authors that increasing the yttria content to 9.5 mol% reduced the conductivity degradation to <2% over 5000 min However, it should also be noted that these studies were undertaken only on the electrolyte, and not on functioning electrolysis cells. Further studies on YSZ based electrolysis cells have utilised a range of yttria contents, with 8, 9.5, 10 and 10.5% YSZ all receiving attention [23]. A brief overview of the range of cell architectures and operating windows is provided in [6], all of which suggest a tendency to operate at higher temperatures, with the key variations being in the electrode materials, as discussed in chapters 3–4. It is generally concluded that prospects for future development of zirconia based SOEC electrolytes are more likely to rest with the Sc_2O_3 substituted phases given their higher ionic conductivities

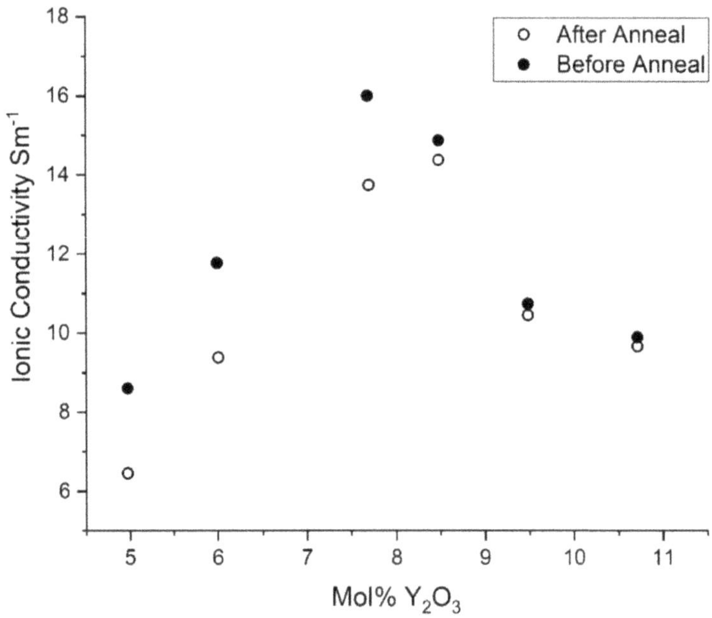

Figure 2.3. Change in ionic conductivity as a function of Y_2O_3 content before and after annealing. Annealing time was 5000 min Reprinted from [22], copyright (1998) with permission from Elsevier.

[10, 23–25], with several reviews reporting successful operation of cells based on these electrolyte compositions [26–28].

Although Sc stabilised zirconia (ScSZ) has higher oxygen ionic conductivity (e.g., 0.12 S cm^{-1} at 800 °C for ScSZ with 10 mol% Sc_2O_3 and 0.24 S cm^{-1} at 1000 °C for ScSZ with 8 mol% Sc_2O_3), conductivity degradation was observed over time at high temperature due to the formation of an ordered phase rich in Sc [9, 10]. For singly substituted ZrO_2, a composition of 9.3mol% Sc_2O_3 was found to provide the lowest conductivity degradation [9], building on earlier reports by Yamamoto *et al* [29]. of the total conductivity of several Yb and Sc zirconia- based electrolytes. Degradation of the conductivity of the ScSZ has been attributed to the presence of secondary phases such as the tetragonal ZrO_2 polymorph and the low conductivity rhombohedral phase of $Sc_2Zr_7O_{17}$. These two impurity phases are found when the Sc_2O_3 content is below 9 mol% and above 10 mol%, respectively [9]. To further stabilise the cubic phase of ScSZ at room temperature and suppress the formation of $Sc_2Zr_7O_{17}$, it was proposed to introduce a small addition of CeO_2 into the system to form 10 mol% Sc_2O_3–1 mol% CeO_2-ZrO_2 (10Sc1CeSZ) [10, 30]. It is believed that Ce, as a larger tetravalent dopant cation, will form eightfold coordination with oxygen and leave oxygen vacancies to interact with the Zr ions [30]. For this CeO_2 substituted composition (10Sc1CeSZ) the oxygen ionic conductivity was found to be 0.057 S cm^{-1} at 700 °C [30]. Whilst the10Sc1CeSZ composition has been suggested as a good alternative to both YSZ and ScSZ as the electrolyte, some studies have also noted the degradation of cells with this composition [31], primarily due to the reduction of the electrolyte at regions close to the Ni-YSZ electrode, but this phenomenon appears to only be observed at high voltages (>1.8 V) and hence far beyond the normal operating parameters for electrolysis cells.

Further adaptations to the ScSZ composition have included the incorporation of Gd as a co-dopant, forming a $Sc_{0.1}Ce_{0.005}Gd_{0.005}Zr_{0.89}O_{2-\delta}$ (SCGZ) phase that is reported to exhibit stable performance over 60 h in comparison to a similarly manufactured cell based on 8YSZ [32], although this phase required sintering at 1723 K, reducing the likelihood of widespread adoption of this electrolyte, despite a reasonable current density of 1 A cm^{-2} being achieved at 1.3 V at 800 °C. It has also been recently proposed that with the 10Sc1CeSZ composition there is the potential to migrate metal cations through the bulk oxide [33], with a consequent reduction in the bulk ionic conductivity. This example was for a symmetrical cell in which the electrolyte was contacted with Ag electrodes and examined using a suite of characterisation techniques, including low energy electron microscopy. However, the polarisation studies were carried out at 12 V and hence could be considered an extreme situation, but this work does build on previous investigations [34] that suggested ScSZ could be used in a low temperature (500 °C) SOEC if Ag was utilised as an electrode. It has been perceived, however, that scandium is a higher cost option than yttria as a substituent and that this could limit the usage of Sc-based electrolytes in large-scale applications [9, 10].

2.2.2 Ceria-based electrolytes

The second family of fluorite oxides that have been considered as SOEC electrolytes are those based on ceria. In general, ceria-based electrolytes such as samarium-doped ceria (SDC) and gadolinium-doped ceria (GDC) have a higher ionic conductivity than the conventional YSZ electrolytes and have a good chemical and mechanical compatibility with common electrodes [28, 35]. From the extensive literature regarding the ionic conductivity of substituted ceria [36–41] it is well established that aliovalent substituted ceria has higher ionic conductivity at lower temperatures than the comparable zirconia compositions. For example at 600 °C, the oxygen ionic conductivity of GDC and SDC are 5×10^{-3} and 3×10^{-2} S cm^{-1}, respectively [9]. Considering the oxygen ionic conductivity and its conduction range, GDC with 10 mol% of gadolinium (GDC10) is generally selected as the best composition [35]. One significant major drawback with the substituted ceria electrolytes is the possible reduction of cerium from +4 to +3 under highly reducing operating conditions, or high potential, which generates n-type electronic conduction and significant chemical expansion of the lattice, leading to microcracks [10, 35]. The reduction of cerium will occur under low local oxygen partial pressure, high temperature and high applied voltages [42]. For this reason Chen and Jiang [43] suggest that doped cerias are unsuitable as SOEC electrolytes. The high applied voltage is an operating feature in the SOEC mode that lowers the oxygen partial pressure at the fuel electrode/electrolyte interface leading Zhu *et al* [42] to report that these chemical processes introduce mechanical instability in Sm-doped ceria. As a result of these issues ceria-based electrolytes have been viewed as inappropriate for SOEC operation [35, 42].

The effect of using other trivalent ions as aliovalent substituents in CeO_2 have been investigated, with Sumi *et al* [44] comparing the GDC10 and YDC10 compositions, finding that YDC10 has lower conductivity, but greater sinterability, but in both cases the open circuit voltage of the cells was lower than theoretical due to electronic leakage. To effectively use these ceria compositions as electrolytes the authors discuss the use of a $BaCe_{0.8}Y_{0.2}O_{3-\delta}$ (BCY) blocking layer at the electrolyte/fuel electrode interface. With this configuration they report a successful reversible SOEC/SOFC operation at a temperature of only 500 °C. In addition, it was reported that an addition of praseodymium to GDC with 20 mol% of gadolinium (GDC20) will increase stability at intermediate temperatures but that the impact is insignificant for GDC10 [9, 10]. The use of Pr substituted CeO_2 (PDC) as an electrolyte has also been discussed in several reports, but these have been mainly confined to their application in CO_2 electrolysis [45], a topic beyond the scope if this chapter.

In further work, Temluxame *et al* [32] compared the performance of ScSZ based electrolysers with those based on GDC, with both devices being prepared as electrolyte supported cells. They found that the GDC-based cell exhibited high performance in comparison to those based on zirconia, but had relatively low stability, and that the bilayered GDC/YSZ electrolyte suffered significant degradation in performance when operating in electrolysis mode. As such, the use of ceria-based electrolytes in electrolysers is viewed as inferior to those utilising the zirconia electrolyte system.

To further enhance the oxygen ionic conductivity of the ceria-based electrolytes, ceria-based composites have been developed with one example composition being $20SDC\text{-}Li_2CO_3\text{-}Na_2CO_3$ [9, 46]. Whilst these composite electrolytes have been widely reported in the SOFC literature, there are few reports of their application in electrolysis, with Zhu and Mat [47] discussing the electrolysis performance of the SDC-20wt% carbonate composition operating at 650 °C, but with ideal Pt electrodes. From the data presented [47, 48] it is argued that these composite electrolytes exhibit both oxide ion and protonic conduction. It is also noted that the maximum current density is only 0.8 A cm^{-2} at 650 °C [48]. There are few further studies reporting the use of carbonate composites as electrolytes in contrast to the many reports of these being used in fuel cell mode [49, 50].

By far the most common use of ceria phases in SOECs is as an electrolyte barrier layer, rather than as the electrolyte itself. Following the fuel cell literature it is relatively common to introduce a ceria barrier layer, most often GDC, in the YSZ based cell architecture [43, 51]. Kim *et al* [51] applied a GDC10 interlayer on YSZ to prevent Sr diffusion from the electrode layer, with the interlayer thicknesses varying from 2–11 μm. In these studies, a range of dense and porous interlayers were evaluated and it was concluded that a pinhole free film of 2 μm thickness provided the lowest voltage degradation (8%) compared to 28% in the 11 μm thick film interlayer case, suggesting that stable voltages in SOECs with zirconia electrolytes could be achieved with only the thinnest pinhole free films. Further advanced characterisation of ceria thin films in solid oxide electrochemical cells using ambient pressure X-ray photoelectron spectroscopy [46], with the ceria (CeO_2) being deposited as a thin film on a YSZ electrolyte, demonstrated that the reaction of

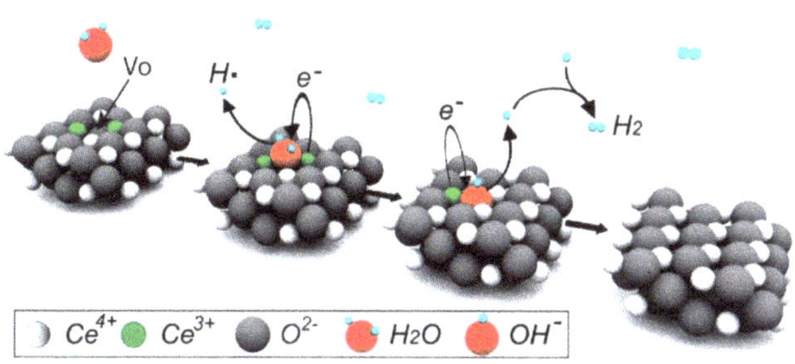

Figure 2.4. Proposed mechanism for electrolysis process on a ceria surface. Reprinted with permission from [38], copyright 2013 American Chemical Society.

H_2O with Ce^{3+} surface species is the rate limiting step for electrolysis, generating an electrical double layer (induced surface dipole layer). These authors propose a surface electrochemical reaction on ceria, figure 2.4, but in this case the ceria is acting as an electrode rather than an electrolyte or interlayer.

Further studies utilising GDC barrier layers on ScSZ electrolytes have reported exceptional performance [52] with high current densities (1.4 A cm^{-2} at 1.3 V) operating at 900 °C, with only a 1 μm thick GDC layer acting to prevent Sr diffusion from the electrode. This represents a significant advance on the report of 0.63 A cm^{-2} at 850 °C on a similar configuration cell with a GDC barrier layer, also operating at 1.3 V [53]. It is therefore most likely that ceria-based compositions will only feature as barrier layers, and not as single-phase electrolytes in electrolysis mode, and hence it is unlikely that reversible cells would be viable based on the substituted ceria compositions.

Further discussion of the materials used as electrolytes will now focus on non-fluorite structure types, beginning with the ABO_3 (A = Lanthanide, B = Transition metal) perovskite structured oxides.

2.2.3 Lanthanum gallate-based perovskite electrolytes

As with the previously discussed fluorite structured oxide electrolytes, the perovskite type electrolyte development has stemmed from the extensive studies of these phases as SOFC materials [54–56]. The A-site (La) and B-site (Ga) of lanthanum gallate ($LaGaO_3$) was substituted with lower valence strontium and magnesium ions, respectively, to increase the oxygen ionic conductivity of the material through the introduction of charge compensating anion vacancies. Typically, there are two compositions that have received the majority of attention: $La_{0.8}Sr_{0.2}Ga_{0.8}Mg_{0.2}O_{3-\delta}$ (LSGM8282) and $La_{0.9}Sr_{0.1}Ga_{0.8}Mg_{0.2}O_{3-\delta}$ (LSGM9182). For the LSGM9182 the oxygen ionic conductivity is 0.10 S cm^{-1} at 800 °C and 0.12 S cm^{-1} for the LSGM8282 [57], which is significantly higher than YSZ and is comparable to the ceria-based electrolytes [10]. Increasing the substitution at A and B sites tends to increase the oxygen ionic conductivity. The electrical conductivity of the LSGM phases is predominantly oxygen ionic conducting under an oxidising atmosphere, but a small electronic conduction will occur under a reducing atmosphere due to the reduction of gallium from +3 to +2 [9, 10]. In general, the electronic conductivity is negligible when compared to its oxygen ionic conductivity and hence has little impact on the cell performance.

The main disadvantages of LSGM electrolytes are their reactivity with NiO at the fuel electrode and their poor mechanical strength [58]. The reaction between LSGM and NiO forms $LaNiO_3$ secondary phase which significantly increases the electronic conductivity of the electrolyte and the electrode overpotential [9]. Additionally, it has been shown that Ga volatility is an issue, and that secondary phases such as $La_4Ga_2O_9$ may form at grain boundaries on sintering. However, these issues have been mitigated in the fuel cell field through adaptation of the synthesis process [58].

Ishihara *et al* [59, 60] were among the earliest proponents of LSGM as an electrolysis electrolyte, developing a cell with either Ni-Fe or $Ba_{0.6}La_{0.4}CoO_3$ electrodes, with Ni-SDC interlayers. From their studies they find excellent performance of the cell with the Ni-SDC interlayers at 600 °C, achieving a current density of 150 mA cm^{-2} at 1.8 V. Further work by these authors [61] identified that the LSGM electrolyte exhibits high chemical stability, but that overall performance of the device was limited by diffusion of the steam to active sites, not the electrolyte conductivity. Further development of the LSGM electrolyte-based cells enabled an increase in current density to 1.07 A cm^{-2} at 1.6 V at 600 °C [62], but once again the advance in performance is attributed to development of electrodes, rather than any improvement with the electrolyte. In this case the high performance is partly achieved through the ability of the authors to manufacture cells with electrolytes of only 30 μm thickness. A further study of LSGM as an electrolyte [63], using perovskite cathodes also achieved high electrolysis performance, with 1.31 A cm^{-2} at 1.3 V and 800 °C. Due to the issues of cation migration identified in the fuel cell development, the use of barrier interlayers such as ScSZ, GDC, and La_2O_3 doped ceria were proposed to avoid La and Ni cation interdiffusion in the electrolysis mode [10]. Indeed, the poor mechanical strength of LSGM electrolytes has limited its adoption in commercial applications. A comparison of the conductivity of LSGM, compared with typical SOEC electrolytes, is shown in figure 2.5, highlighting that each of the electrolytes discussed has ionic conductivity in the range of ~0.01–~0.1 Scm^{-1} at 800 °C, and a range of $\sim 5 \times 10^{-4}$–$\sim 8 \times 10^{-3}$ Scm^{-1} at 500 °C.

2.2.4 New electrolyte compositions

Beyond the traditional electrolytes discussed above there have been few new possible electrolytes proposed. One such composition that is proving interesting is the $LaNb_{1-x}W_xO_{4+\delta}$ fergusonite/scheelite composition first proposed by Laguna-Bercero *et al* [65]. Further studies of this composition have highlighted its excellent compatibility with existing electrode materials, particularly the $La_{1-x}Sr_xMnO_3$ (LSM) material [66]. Following these initial reports Kawaguchi *et al* [67] have undertaken an investigation of the Gd analogue, but found that the ionic conductivity of this potential electrolyte was only 10^{-5} Scm^{-1} at 600 °C, significantly below that required for commercial devices. One further possibility in the niobate family is that of Mo substitution, with Auckett *et al* [68] and Toyoura *et al* [69] confirming higher oxide ion conduction in comparison with the W substituted

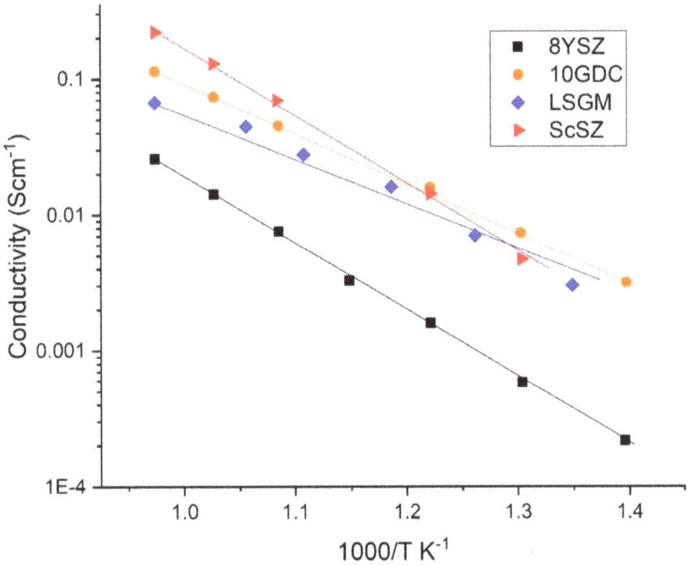

Figure 2.5. Arrhenius plot comparing the ionic conductivity of the leading candidates for use as solid oxide electrolyser electrolytes. Reprinted from [64], copyright (2020), with permission from Elsevier.

$LaNbO_4$ with ionic conductivity reaching 10^{-2} S cm^{-1} at 800 °C. A range of alternative compositions for use as electrolyte materials, particularly on solid oxide fuel cells, have been proposed, including δ-Bi_2O_3 [70–72], $A_2B_2O_7$ pyrochlore phases [73, 74], $La_2Mo_2O_9$ (LAMOX) family [75–77], Melilite [78] ($La_{1.54}Sr_{0.46}Ga_3O_{7.27}$) and the $Bi_4V_2O_{11}$ (BIMEVOX) [79, 80] series of substituted compositions. Whilst in many of these cases the oxide ion conductivity is exceptional (~1 Scm^{-1} for δ-Bi_2O_3 at 750 °C) the stability and compatibility with cell components, and their ability to function effectively under the demanding environment of solid oxide cells, has limited their application. In addition, the overall ion transport properties of many of these competing oxides do not significantly outperform the conventional electrolyte materials.

2.3 Electrolyte degradation mechanisms

The materials discussed in section 2.2 are suitable for application as SOEC electrolytes due to their ionic conductivity and mechanical stability, and have been demonstrated to function well with appropriate electrodes. However, SOEC devices are expected to operate for extended time periods and at high temperature, and thus degradation in performance is expected. Indeed, the most significant degradation processes in SOECs are typically associated with the electrodes and interfaces. Regardless, each of the electrolyte compositions discussed will undergo some form of degradation process during cell operation. The extent and significance of this will vary. Perhaps the most evident failure/degradation mode in SOECs is observed with the ceria-based systems. In this case the reduction of cerium at the fuel electrode side at high temperature and/or applied voltages results in the Ce^{4+}-to- Ce^{3+}

reduction as commonly observed in fuel cell applications [38, 81]. Given the relatively sparse use of ceria as an electrolyte there are no studies dedicated to ceria degradation in electrolysis mode. Any studies of ceria degradation are associated with the air side of the cell in which doped ceria is used as a barrier/interlayer. In the case where thin GDC layers have been utilised as diffusion barrier layers, it is clear that the microstructure of the layer is of importance, but that even in the case of the highest performing barrier layer the GDC layer undergoes electrical degradation (8% voltage degradation at 0.8 A cm^{-2}) attributed to the formation of insulating phases [51].

In terms of degradation of ceria the most significant studies have been associated with the use of the 10Sc1CeSZ electrolytes, and again the exposure of the Ce species to the fuel environment is the primary source of performance reduction [31, 43, 82, 83]. Laguna-Bercero has shown, using spectroscopic techniques, that at high voltage operation there were significant spectral changes through the thickness of the electrolyte, apart from at a point close to the air electrode. Some of these changes in the Raman spectra were attributed to the formation of the rhombohedral β phase, and this phase transformation proceeds from the fuel electrode interface through the electrode thickness. This transition behaviour was also associated with the progression of Ce^{4+} reduction.

For the majority of SOECs the electrolytes that have seen the most intensive investigation of degradation are those based on the YSZ electrolyte. The source of the electrolyte degradation is generally viewed as being associated with the high current densities experienced, leading to significant increases in the electrolyte (series) resistance [84]. Further studies of degradation phenomena have included the development of accelerated testing protocols, combining electrical testing with microstructural investigations. In this case a commercial cell was probed, but with a deliberate Si impurity introduced. For these studies [84] it was observed that significant electronic conduction in the YSZ occurred. This was accompanied by the evolution of microstructural changes, including the formation of pores and cracks, and that cation species were transported from the electrodes to the YSZ electrolyte. The impact of these ion migration mechanisms has not been clarified to date.

Microstructural changes in the YSZ electrolyte have also been probed using microscopy techniques, with the degradation of the electrolyte attributed to oxygen bubble formation [85] and intergranular fracture [86] at the air–electrode interface, as shown in figures 2.6 and 2.7. In these cells operating at high current density Knibbe *et al* [85] observed that gaseous oxygen evolved close to the air–electrode interface, forming oxygen bubbles, leading to porosity at the YSZ grain boundaries, schematically illustrated in figure 2.6(b).

By contrast, but also at the air–electrode–YSZ interface, figure 2.7, Kim *et al* [86] observed electrolyte failure through intergranular fracture when operating a cell at 1.5 A cm^{-2} at 750 °C. This failure was observed after a period of only 120 h, raising concerns over the long-term durability of this combination of materials. Further extensive investigation of degradation phenomena carried out by Chatzichristodoulou *et al* [87] highlights the complexity of degradation in bilayer electrolyte cells, and demonstrates that degradation can be avoided through operation within a window that avoids extremes of low and high oxygen activity. It has also been suggested [88] that the

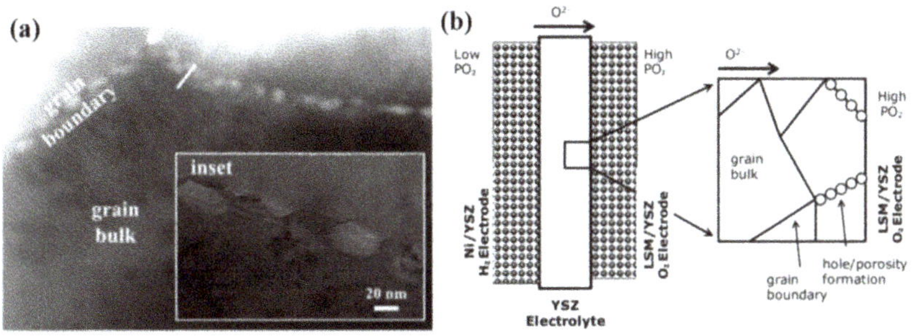

Figure 2.6. (a) TEM micrograph of YSZ at the electrolyte/electrode grain boundary and (b) illustration of the nucleation and oxygen bubbles formation in the YSZ grain boundaries close to the air electrode side after operating under SOEC mode. Reprinted from [85], copyright (2010) with permission from IOP.

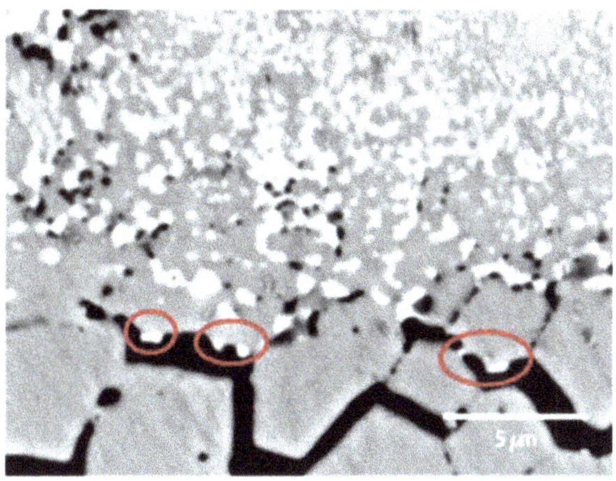

Figure 2.7. Evidence of YSZ intergranular fracture close to LSM-YSZ air–electrode after applying anodic current of 1.5 A cm^{-2} for 120 h at 750 °C. Reprinted from [86], copyright (2013) with permission from Elsevier.

microstructural degradation that arises at the air– electrode/electrolyte interface can be eliminated by cycling the device between fuel cell and electrolysis modes. It is noted that this study has focussed only on microstructural degradation, and the authors acknowledge that their work is in conflict with several previous studies, although they argue that the systematic elimination of competing degradation processes lends greater credibility to this study of microstructural degradation.

2.4 Concluding remarks

In selecting an electrolyte material for use in a solid oxide electrolysis cell, it is clear that there are a relatively small number of options that are currently available. By far the majority of studies are focussed on the stabilised zirconia (Y or Sc) families, with developers adopting a ceria-based diffusion barrier to prevent electrode interactions. Each of these electrolytes are considered for operation with a wide variety of electrode materials, as discussed elsewhere. The use of ceria electrolytes has received some interest but the reducibility of Ce^{4+} has limited the use of this phase, particularly at higher operation temperatures and current densities. Finally, the use of the perovskite LSGM has been only briefly investigated, but this does demonstrate interesting potential performance, but with similar durability issues as identified in the fuel cell mode of operation.

References

[1] Erdle E, Donitz W, Schamm R and Koch A 1992 Reversibility and polarization behaviour of high temperature solid oxide electrochemical cells *Int. J. Hydrogen Energy* **17** 817–9

[2] Doenitz W, Schmidberger R, Steinheil E and Streicher R 1980 Hydrogen production by high temperature electrolysis of water vapour *Int. J. Hydrogen Energy* **5** 55–63

[3] Dönitz W and Erdle E 1985 High-temperature electrolysis of water vapor-status of development and perspectives for application *Int. J. Hydrogen Energy* **10** 291–5

[4] Eguchi K, Hatagishi T and Arai H 1996 Power generation and steam electrolysis characteristics of an electrochemical cell with a zirconia- or ceria-based electrolyte *Solid State Ionics* **86–88** 1245–9

[5] Isenberg A O and Cusick R J 1988 Carbon dioxide electrolysis with solid oxide electrolyte cells for oxygen recovery in life support systems *SAE Tech. Pap.* **97** 360–73

[6] Zheng Y, Wang J, Yu B, Zhang W, Chen J, Qiao J and Zhang J 2017 A review of high temperature co-electrolysis of H_2O and CO_2 to produce sustainable fuels using solid oxide electrolysis cells (SOECs): advanced materials and technology *Chem. Soc. Rev.* **46** 1427–63

[7] Minh N Q 1993 Ceramic fuel cells *J. Am. Ceram. Soc.* **76** 563–88

[8] Nechache A and Hody S 2021 Alternative and innovative solid oxide electrolysis cell materials: A short review *Renew. Sustain. Energy Rev.* **149** 111322

[9] Gómez S Y and Hotza D 2016 Current developments in reversible solid oxide fuel cells *Renew. Sustain. Energy Rev.* **61** 155–74

[10] Zhang L, Hu S, Zhu X and Yang W 2017 Electrochemical reduction of CO_2 in solid oxide electrolysis cells *J. Energy Chem.* **26** 593–601

[11] Smyth D M 2000 *The Defect Chemistry of Metal Oxides* (Oxford: Oxford University Press)

[12] Nakamura A and Wagner J B 1986 Defect structure, ionic conductivity, and diffusion in yttria stabilized zirconia and related oxide electrolytes with fluorite structure *J. Electrochem. Soc.* **133** 1542–8

[13] Wang D, Park D, Griffith J and Nowick A 1981 Oxygen-ion conductivity and defect interactions in yttria-doped ceria *Solid State Ionics* **2** 95–105

[14] Koettgen J, Grieshammer S, Hein P, Grope B O H, Nakayama M and Martin M 2018 Understanding the ionic conductivity maximum in doped ceria: trapping and blocking *Phys. Chem. Chem. Phys.* **20** 14291–321

[15] Guo X, Sigle W and Maier J 2003 Blocking grain boundaries in yttria-doped and undoped ceria ceramics of high purity *J. Am. Ceram. Soc.* **86** 77–87

[16] Zhou X-D, Huebner W, Kosacki I and Anderson H U 2004 Microstructure and grain-boundary effect on electrical properties of gadolinium-doped ceria *J. Am. Ceram. Soc.* **85** 1757–62

[17] Barsoukov E and Macdonald J R 2005 *Impedance Spectroscopy:Theory, Experiment and Applications* 2nd edn (Hoboken, NJ: Wiley)

[18] Kidner N J, Perry N H, Mason T O and Garboczi E J 2008 The brick layer model revisited: introducing the nano-grain composite model *J. Am. Ceram. Soc.* **91** 1733–46

[19] Ghatee M, Shariat M H and Irvine J T S 2009 Investigation of electrical and mechanical properties of 3YSZ/8YSZ composite electrolytes *Solid State Ionics* **180** 57–62
[20] Schefold J, Brisse A and Zahid M 2009 Electronic conduction of yttria-stabilized zirconia electrolyte in solid oxide cells operated in high temperature water electrolysis *J. Electrochem. Soc.* **156** B897–904
[21] Sohal M S, O'Brien J E, Stoots C M, Sharma V I, Yildiz B and Virkar A 2012 Degradation issues in solid oxide cells during high temperature electrolysis *J. Fuel Cell Sci. Technol.* **9** 011017
[22] Gibson I R, Dransfield G P and Irvine J T S 1998 Influence of yttria concentration upon electrical properties and susceptibility to ageing of yttria-stabilised zirconias *J. Eur. Ceram. Soc.* **18** 661–7
[23] Ni M, Leung M and Leung D 2008 Technological development of hydrogen production by solid oxide electrolyzer cell (SOEC) *Int. J. Hydrogen Energy* **33** 2337–54
[24] Kazlauskas S, Kazakevičius E and Kežionis A 2017 Electrical properties of scandia- and ceria-stabilized zirconia ceramics *Solid State Ionics* **310** 143–7
[25] Dasari H P, Ahn J S, Ahn K, Park S-Y, Hong J, Kim H, Yoon K J, Son J-W, Lee H-W and Lee J-H 2014 Synthesis, sintering and conductivity behavior of ceria-doped scandia-stabilized zirconia *Solid State Ionics* **263** 103–9
[26] O'Brien J E, Stoots C M, Herring J S, Lessing P A, Hartvigsen J J and Elangovan S 2005 Performance measurements of solid-oxide electrolysis cells for hydrogen production *J. Fuel Cell Sci. Technol.* **2** 156–63
[27] O'Brien J E, Stoots C M, Herring J S and Hartvigsen J 2006 Hydrogen production performance of a 10-cell planar solid-oxide electrolysis stack *J. Fuel Cell Sci. Technol.* **3** 213–9
[28] Laguna-Bercero M A 2012 Recent advances in high temperature electrolysis using solid oxide fuel cells: a review *J. Power Sourc.* **203** 4–16
[29] Yamamoto O 1995 Electrical conductivity of stabilized zirconia with ytterbia and scandia *Solid State Ionics* **79** 137–42
[30] Laguna-Bercero M A A, Skinner S J J and Kilner J A A 2009 Performance of solid oxide electrolysis cells based on scandia stabilised zirconia *J. Power Sourc.* **192** 126–31
[31] Laguna-Bercero M A and Orera V M 2011 Micro-spectroscopic study of the degradation of scandia and ceria stabilized zirconia electrolytes in solid oxide electrolysis cells *Int. J. Hydrogen Energy* **36** 13051–8
[32] Temluxame P, Puengjinda P, Peng-ont S, Ngampuengpis W, Sirimungkalakul N, Jiwanuruk T, Sornchamni T and Kim-Lohsoontorn P 2021 Comparison of ceria and zirconia based electrolytes for solid oxide electrolysis cells *Int. J. Hydrogen Energy* **46** 24568–80
[33] Mosiałek M *et al* 2020 Changes in properties of scandia-stabilised ceria-doped zirconia ceramics caused by silver migration in the electric field *Electrochim. Acta* **338** 135866
[34] Sasaki K, Muranaka M and Terai T 2010 Compatibility analysis of Ag and electrolyte materials for LT-SOFCs and LT-SOECs *Solid State Ionics* **181** 1303–7
[35] Preux N, Rolle A and Vannier R N 2012 Electrolytes and ion conductors for solid oxide fuel cells (SOFCs) *Functional Materials for Sustainable Energy Applications* (Amsterdam: Elsevier), 370–401
[36] Andersson D A, Simak S I, Skorodumova N V, Abrikosov I A and Johansson B 2006 *Optimization of ionic conductivity in doped ceria Proc. Natl Acad. Sci.* **103** 3518–21
[37] Kilner J A 2008 Defects and conductivity in ceria-based oxides *Chem. Lett.* **37** 1012–5

[38] Inaba H 1996 Ceria-based solid electrolytes *Solid State Ionics* **83** 1–16
[39] Wang S, Kobayashi T, Dokiya M and Hashimoto T 2000 Electrical and ionic conductivity of Gd-doped ceria *J. Electrochem. Soc.* **147** 3606–9
[40] Koettgen J and Martin M 2020 The ionic conductivity of Sm-doped ceria *J. Am. Ceram. Soc.* **103** 3776–87
[41] Schmitt R, Nenning A, Kraynis O, Korobko R, Frenkel A I, Lubomirsky I, Haile S M and Rupp J L M 2020 A review of defect structure and chemistry in ceria and its solid solutions *Chem. Soc. Rev.* **49** 554–92
[42] Zhu S, Wang Y, Rao Y, Zhan Z and Xia C 2014 Chemically-induced mechanical unstability of samaria-doped ceria electrolyte for solid oxide electrolysis cells *Int. J. Hydrogen Energy* **39** 12440–7
[43] Chen K and Jiang S P 2016 Review—materials degradation of solid oxide electrolysis cells *J. Electrochem. Soc.* **163** F3070–83
[44] Sumi H, Suda E and Mori M 2017 Blocking layer for prevention of current leakage for reversible solid oxide fuel cells and electrolysis cells with ceria-based electrolyte *Int. J. Hydrogen Energy* **42** 4449–55
[45] Kumari N, Haider M A, Anjum U and Basu S 2020 Identifying operating mechanism in the electrochemical reduction of CO_2 on thin-film praseodymium-doped ceria electrodes *Ionics* **26** 5673–84
[46] Zhang C *et al* 2013 Mechanistic studies of water electrolysis and hydrogen electro-oxidation on high temperature ceria-based solid oxide electrochemical cells *J. Am. Chem. Soc.* **135** 11572–9
[47] Zhu B and Mat M D 2006 Studies on dual phase ceria-based composites in electrochemistry *Int. J. Electrochem. Sci.* **1** 383–402
[48] Zhu B, Albinsson I, Andersson C, Borsand K, Nilsson M and Mellander B-E 2006 Electrolysis studies based on ceria-based composites *Electrochem. Commun.* **8** 495–8
[49] Fan L, He C and Zhu B 2017 Role of carbonate phase in ceria-carbonate composite for low temperature solid oxide fuel cells: a review *Int. J. Energy Res.* **41** 465–81
[50] Zhu B, Fan L and Lund P 2013 Breakthrough fuel cell technology using ceria-based multi-functional nanocomposites *Appl. Energy* **106** 163–75
[51] Kim S J, Kim S W, Park Y M, Kim K J and Choi G M 2016 Effect of Gd-doped ceria interlayer on the stability of solid oxide electrolysis cell *Solid State Ionics* **295** 25–31
[52] Bernadet L, Moncasi C, Torrell M and Tarancón A 2020 High-performing electrolyte-supported symmetrical solid oxide electrolysis cells operating under steam electrolysis and co-electrolysis modes *Int. J. Hydrogen Energy* **45** 14208–17
[53] Fan H, Keane M, Singh P and Han M 2014 Electrochemical performance and stability of lanthanum strontium cobalt ferrite oxygen electrode with gadolinia doped ceria barrier layer for reversible solid oxide fuel cell *J. Power Sourc.* **268** 634–9
[54] Feng M, Goodenough J B, Huang K and Milliken C 1996 Fuel cells with doped lanthanum gallate electrolyte *J. Power Sourc.* **63** 47–51
[55] Ishihara T, Shibayama T, Honda M, Nishiguchi H and Takita Y 2000 Intermediate temperature solid oxide fuel cells using $LaGaO_3$ electrolyte II. Improvement of oxide ion conductivity and power density by doping Fe for Ga site of $LaGaO_3$ *J. Electrochem. Soc.* **147** 1332–7
[56] Jacobson A J 2010 Materials for solid oxide fuel cells *Chem. Mater.* **22** 660–74

[57] Stevenson J W, Armstrong T R, McCready D E, Pederson L R and Weber W J 1997 Processing and electrical properties of alkaline earth-doped lanthanum gallate *J. Electrochem. Soc.* **144** 3613–20
[58] Singh P and Minh N Q 2005 Solid oxide fuel cells: technology status *Int. J. Appl. Ceram. Technol.* **1** 5–15
[59] Jirathiwathanakul N, Matsumoto H and Ishihara T 2007 Intermediate temperature steam electrolysis using doped lanthanum gallate solid electrolyte (2) effects of CeO_2 interlayer on activity *Eco-Materials Processing and Design VIII—ISEPD-8, Proc. of the 8th Int. Symp. on Eco-Materials Processing and Design* ed H Kim, J Hojo and S W Lee 544–545 1005–8
[60] Ishihara T and Kannou T 2011 Intermediate temperature steam electrolysis using $LaGaO_3$-based electrolyte *Solid State Ionics* **192** 642–4
[61] Ishihara T, Jirathiwathanakul N and Zhong H 2010 Intermediate temperature solid oxide electrolysis cell using $LaGaO_3$ based perovskite electrolyte *Energy Environ. Sci.* **3** 665–72
[62] Tan Z, Song J T, Takagaki A and Ishihara T 2021 Infiltration of cerium into a NiO-YSZ tubular substrate for solid oxide reversible cells using a LSGM electrolyte film *J. Mater. Chem. A* **9** 1530–40
[63] Jun A, Kim J, Shin J and Kim G 2016 Achieving high efficiency and eliminating degradation in solid oxide electrochemical cells using high oxygen-capacity perovskite *Angew. Chemie Int. Ed* **55** 12512–5
[64] Shi H, Su C, Ran R, Cao J and Shao Z 2020 Electrolyte materials for intermediate-temperature solid oxide fuel cells *Prog. Nat. Sci. Mater. Int.* **30** 764–74
[65] Laguna-Bercero M A, Bayliss R D and Skinner S J 2014 $LaNb_{0.84}W_{0.16}O_{4.08}$ as a novel electrolyte for high temperature fuel cell and solid oxide electrolysis applications *Solid State Ionics* **262** 298–302
[66] Canu G, Giannici F, Chiara A, Confalonieri G, Longo A, Buscaglia M T, Dapiaggi M, Buscaglia V and Martorana A 2021 Characterisation of scheelite $LaW_{0.16}Nb_{0.84}O_{4.08}$ ion conductor by combined synchrotron techniques: Structure, W oxidation state and inter-diffusion *J. Alloys Compd.* **857** 157532
[67] Kawaguchi R, Akizawa R, Shan Y J, Tezuka K and Katsumata T 2020 Synthesis and examination of $GdNb_{1-x}W_xO_{4+\delta}$ new scheelite-type oxide-ion conductor *Solid State Ionics* **355** 115415
[68] Auckett J E, Lopez-Odriozola L, Clark S J and Evans I R 2021 Exploring the nature of the fergusonite–scheelite phase transition and ionic conductivity enhancement by Mo^{6+} doping in $LaNbO_4$ *J. Mater. Chem. A* **9** 4091–102
[69] Toyoura K, Sakakibara Y, Yokoi T, Nakamura A and Matsunaga K 2018 Oxide-ion conduction via interstitials in Scheelite-type $LaNbO_4$: a first-principles study *J. Mater. Chem. A* **6** 12004–11
[70] Bayliss R D, Cook S N, Kotsantonis S, Chater R J and Kilner J A 2014 Oxygen ion diffusion and surface exchange properties of the α- and δ-phases of Bi_2O_3 *Adv Energy Mater.* **4** 1301575
[71] Zhong G H, Wang J L and Zeng Z 2006 Ionic transport properties in doped δ-Bi_2O_3 *J. Phys. Conf. Ser.* **29** 106–9
[72] Takahashi T and Iwahara H 1978 Oxide ion conductors based on bismuthsesquioxide *Mater. Res. Bull.* **13** 1447–53
[73] Wilde P and Catlow C R A 1998 Defects and diffusion in pyrochlore structured oxides *Solid State Ionics* **112** 173–83

[74] Feighery A J, Irvine J T S and Zheng C 1997 High oxide ion conductivity in non-stoichiometric pyrochlores and fluorites in the ternary system ZrO_2 - Gd_2O_3 - TiO_2 *Ionics* **3** 30–5

[75] Georges S, Skinner S J, Lacorre P and Steil M C 2004 Oxide ion diffusion in optimised LAMOX materials *Dalton Trans.* 3101–5

[76] Georges S, Goutenoire F, Bohnke O, Steil M C, Skinner S J, Wiemhöfer H D and Lacorre P 2004 The LAMOX family of fast oxide-ion conductors: overview and recent results *J. New Mater. Electrochem. Syst.* **7** 51–7

[77] Jacquens J, Farrusseng D, Georges S, Viricelle J P, Gaudillère C, Corbel G and Lacorre P 2010 Tests for the use of $La_2Mo_2O_9$-based oxides as multipurpose SOFC core materials *Fuel Cells* **10** 433–9

[78] Kuang X, Green M A, Niu H, Zajdel P, Dickinson C, Claridge J B, Jantsky L and Rosseinsky M J 2008 Interstitial oxide ion conductivity in the layered tetrahedral network melilite structure *Nat. Mater.* **7** 498–504

[79] Vannier R N, Chater R J, Skinner S J, Kilner J A and Mairesse G 2003 Characterisation of the oxygen transfer in BIMEVOX membranes under applied current conditions *Solid State Ionics* **160** 327–34

[80] Vannier R N, Skinner S J, Chater R J, Kilner J A and Mairesse G 2003 Oxygen transfer in BIMEVOX materials *Solid State Ionics* **160** 85–92

[81] Kharton V V, Figueiredo F M, Navarro L, Naumovich E N, Kovalevsky A V, Yaremchenko A A, Viskup A P, Carneiro A, Marques F M B and Frade J R 2001 Ceria-based materials for solid oxide fuel cells *J. Mater. Sci.* **36** 1105–17

[82] Omar S, Belda A, Escardino A and Bonanos N 2011 Ionic conductivity ageing investigation of 1Ce10ScSZ in different partial pressures of oxygen *Solid State Ionics* **184** 2–5

[83] Laguna-Bercero M A 2012 Recent advances in high temperature electrolysis using solid oxide fuel cells: a review *J Power Sourc.* **203** 4–16

[84] Nechache A, Boukamp B A, Cassir M and Ringuede A 2019 Accelerated degradation of yttria stabilized zirconia electrolyte during high-temperature water electrolysis *J. Solid State Electrochem.* **23** 871–81

[85] Knibbe R, Traulsen M L, Hauch A, Ebbesen S D and Mogensen M 2010 Solid oxide electrolysis cells: degradation at high current densities *J. Electrochem. Soc.* **157** B1209–17

[86] Kim J, Ji H-I, Dasari H P, Shin D, Song H, Lee J-H, Kim B-K, Je H-J, Lee H-W and Yoon K J 2013 Degradation mechanism of electrolyte and air electrode in solid oxide electrolysis cells operating at high polarization *Int. J. Hydrogen Energy* **38** 1225–35

[87] Chatzichristodoulou C, Chen M, Hendriksen P V, Jacobsen T and Mogensen M B 2016 Understanding degradation of solid oxide electrolysis cells through modeling of electro-chemical potential profiles *Electrochim. Acta* **189** 265–82

[88] Graves C, Ebbesen S D, Jensen S H, Simonsen S B and Mogensen M B 2015 Eliminating degradation in solid oxide electrochemical cells by reversible operation *Nat. Mater.* **14** 239–44

Chapter 3

Anode materials for solid oxide electrolysis cells

Christian Berger and Andreas Egger

Air electrodes in solid oxide cells have to fulfill many requirements to ensure high-performance operation and long-term stability. One of the most important requirements for solid oxide electrolysis cell (SOEC) applications is the efficient release of oxygen migrating through the electrolyte in order to avoid a pressure build-up of oxygen at the interface, which can lead to delamination of the air electrode or crack formation in the adjacent electrolyte layer. In this chapter, two of the most promising material classes are reviewed with respect to their application as air electrodes in SOECs: perovskites and Ruddlesden–Popper-type nickelates. Both types of material provide mixed ionic–electronic conductors that feature high electrocatalytic activity for the oxygen reduction/oxidation processes as well as fast oxygen surface exchange and transport properties, which are of great importance for the transport of oxide ions to and from the electrolyte under current flow.

3.1 Solid oxide electrolysis cell anodes

The anodes of SOECs must have a number of properties which are discussed in this chapter. In high-temperature electrolysis, gaseous oxygen and electrons are formed at the anode (the positive electrode or positrode). The oxygen ions required for oxidation are formed at the cathode (negative electrode or negatrode) by the reduction of water vapor. The electrolyte (densely sintered oxygen ion conductor, see chapter 2) then transports these to the positrode. The first important property for a promising positrode is therefore high electronic ($\sigma_{h^\cdot}$) and ionic conductivities ($\sigma_{V_O^{\cdot\cdot}}$), so that the total electrode area of the porous electrode is active (the so-called 'bulk path'). Using perovskites, $\sigma_{h^\cdot}$ values in the range of 10–1000 S cm^{-1} can be realized, depending on the host material, the degree of substitution, the temperature, and pO_2. The Co-containing perovskite $La_{0.8}Sr_{0.2}Co_{0.8}Fe_{0.2}O_{3-\delta}$ is known to have high electronic conductivity ($\sigma_{h^\cdot}$ = 1000 S cm^{-1} at 800 °C in air) [1]. The electronic conductivities of layered nickelates are in the range of 50–100 S cm^{-1}.

doi:10.1088/978-0-7503-3951-3ch3

The conductivity of oxygen ions is in the range of 10^{-5}–1 S cm^{-1} for both classes of material. In addition to the conductivity, the catalytic activity for the formation of oxygen from oxide ions is also important. The diffusion coefficients of selected materials are given in section 3.3.3. Regarding these materials properties, perovskite and perovskite-related materials (in particular, Ruddlesden–Popper phases (RP phases)) have proven to be particularly suitable. In terms of stability, the nickelates have the advantage that they do not require acceptor doping with alkaline-earth elements, which tend to react with zirconia-based electrolyte materials and impurities in the gas phase. A drawback of nickelates appears to be their reactivity with commonly used electrolyte materials.

3.2 Perovskites: a material scientist's playground

The class of perovskite materials was first established by the discovery of a Ca- and Ti-rich mineral in 1839 in the Russian Ural mountains by Gustav Rose [2]. Victor Goldschmidt provided the first description of their crystal structure in 1926. The first published perovskite structure, based on X-ray diffraction data, was reported by Helen Dick Megaw in 1945 [3, 4]. Perovskite is named after the Russian mineralogist Count Lev Alekseevich Perovski (1792–1856). Since the publication of the crystal structure, all crystal structures similar to $CaTiO_3$ which have an ABX_3 stoichiometry (where A and B are cations and X represents an anion), are known as perovskites.

Most of the elements of the periodic table can form perovskites, as seen in figure 3.1. Keeping this in mind, the number of potential ABX_3 perovskites could reach tens of thousands. If we consider doping or partial substitution at the A or B sites, the number of theoretically accessible perovskites could easily exceed a factor of 10^7 [5, 6]. Approximately 1000 perovskites have been experimentally investigated to date [7].

Many elements are able to form perovskite structures. Figure 3.2 defines classes based on the chemical elements commonly used for solid oxide cells and solar cells.

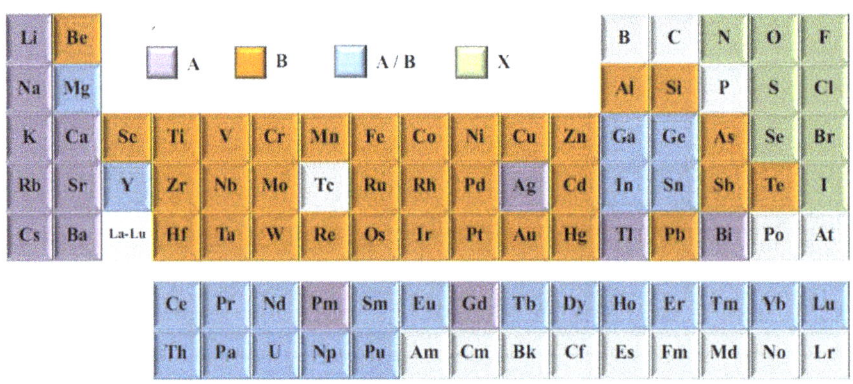

Figure 3.1. Elements of the periodic table that have a high probability of forming perovskite structures; the A cation is significantly larger than the B cation. This figure is reproduced with permission from [8].

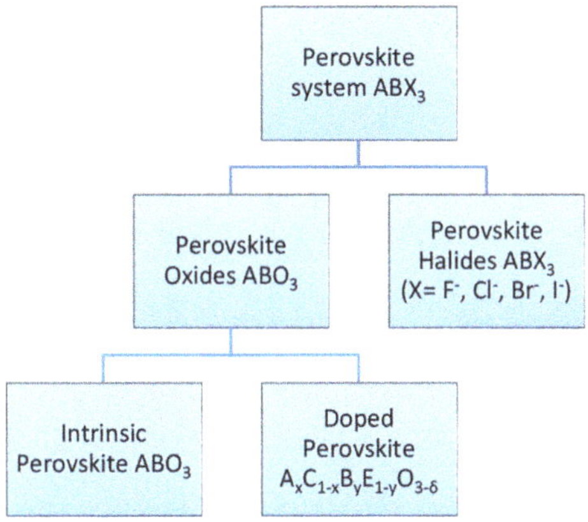

Figure 3.2. Different material classes that have an ABX_3 perovskite structure.

Halide perovskites, which typically have A = alkali or organic cation and B = Pb^{2+} or Sn^{2+}, are of particular interest for the development of modern solar cells because it is possible to fine-tune their bandgaps by varying the cations and the anion. They can easily be processed in solution and exhibit large photocarrier lifetimes and mobilities. Special consideration must be given to achieving broad and strong absorption of light in the visible and near infrared range [9].

Perovskite oxides have a wide range of properties, depending on their elemental composition, and are therefore suitable for a broad range of applications in the field of renewable energy [10–16]. The crystallographic structure of perovskites and the structure–property relationship of mixed conductors are explained in more detail in the following sections to provide a fundamental understanding of high-temperature SOEC anode materials.

3.2.1 Crystal structure of perovskites

One of the strengths of perovskites is their flexibility in the arrangement of the different ions. Depending on the desired application, the ions can be positioned at various lattice sites (see figure 3.1), which has a great impact on the crystal structure. The resulting structural differences depend not only on the composition and arrangement of the ions but also on pressure and temperature.

3.2.1.1 The cubic perovskite

The well-known compound $BaTiO_3$ is used as a prototypic example of a cubic perovskite. The standard crystallographic description places the larger A-site ion (in this case Ba^{2+}) at the origin of the unit cell. Figure 3.3(a) shows the cubic unit cell with Ba^{2+} ions located on the corners and the smaller B-site ion (Ti^{4+}) six-fold coordinated by the surrounding oxygen ions (O^{2-}), forming a TiO_6 octahedron (see figure 3.3(b)). The connection of the structural units has a decisive effect on the material's properties. In the present example, the TiO_6 octahedra are all connected via their corners. In the cubic system, this implies that all the Ti^{4+}–O^{2-} bond lengths

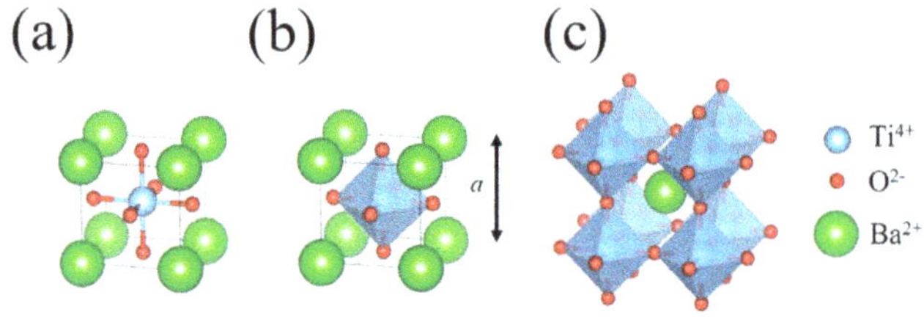

Figure 3.3. Different views of the idealized cubic perovskite structure of $BaTiO_3$: (a) a simple unit cell (black lines) with Ba^{2+}-ions at the cell origin; (b) a TiO_6 coordination octahedron; (c) an extended polyhedral framework centered around a Ba^{2+} ion.

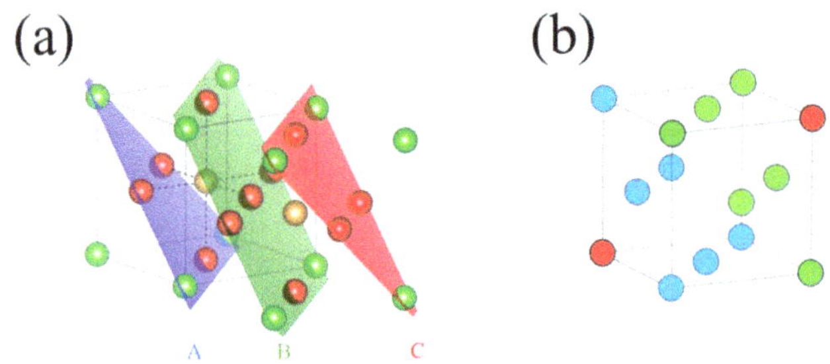

Figure 3.4. Different views of the ABC stacking within a cubic fcc lattice (single unit cell outlined with black lines). The dashed lines in (a) represent the *B–X* bonds. The colors of the ions in (b) correspond to those of the ABC planes on the left.

are equal and the O^{2-}–Ti^{4+}–O^{2-} bonds are linear. The combined arrangement of the large A-cations and O^{2-} anions (which have similar sizes) corresponds to a face-centered cubic (fcc) lattice (figure 3.3 (b)). The fcc structure is built of (111) planes that are located normal to the unit cell diagonal [111] in an ABCABC stacking sequence (see figure 3.4). The smaller B cations occupy 50% of the octahedral voids (those which have only O^{2-} as the coordination partner). This connectivity allows for good hybridization of transition-metal 3*d* and oxygen 2*p* orbitals and thus enables a high mobility of electronic carriers.

3.2.1.2 Deviations from the cubic structure

Since the unit cell of simple cubic perovskites has no free parameters with respect to atomic positions, under- or oversized cations lead to a destabilization of the ideal cubic structure. Goldschmidt analyzed the geometrical conditions of the formation of a cubic perovskite phase [3]. It is assumed that for a stable structure, the cations (X) touch the surrounding anions (A). From figure 3.3 we see that the lattice parameter a is twice the B–X bond length, i.e. 2(B–X) = a, while the bond length A–X is given by 2(A–X) = $\sqrt{2}a$. Thus, for an ideal cubic perovskite structure, the ratio of the bond lengths should be $\sqrt{2}$. This yields the (dimensionless) Goldschmidt tolerance factor t.

$$t = \frac{(A - X)}{\sqrt{2}(B - X)} = 1 \tag{3.1}$$

Applying Goldschmidt's considerations to (3.1) would lead to a change from bond lengths to the ionic radii of anions and cations. This is described by (3.2), where r_A is the ionic radius of the A-site ion, r_B is the ionic radius of the octahedrally coordinated B-site ion and r_X is the ionic radius of the anion. It is important to use the actual valence and coordination numbers to determine the ionic radii and thus obtain correct t-values [17].

$$\frac{(r_A + r_X)}{(r_B + r_X)} = \sqrt{2} \Leftrightarrow t = \frac{(r_A + r_X)}{\sqrt{2}(r_B + r_X)} \tag{3.2}$$

According to the previous equation, a stable perovskite structure is formed for a t value close to unity. Figure 3.5 shows different crystal structures with their ranges of tolerance factors.

In the cubic perovskite, all unit cell lengths (a, b, c) are equal and the angles (α, β, γ) are equal to 90°. For $t > 1$, the cubic unit cell is replaced by a hexagonal one that has an increased length in the *c-direction* and a modified angle of γ = 120°. This represents a completely different structure in which the octahedra are partially face sharing. This different connectivity has a negative influence on the mobility of ionic and electronic carriers. An example of the slow transition to the hexagonal phase is $Ba_xSr_{1-x}Co_{0.8}Fe_{0.2}O_{3-\delta}$ [18]. When t is just slightly less than one, a tetragonal phase

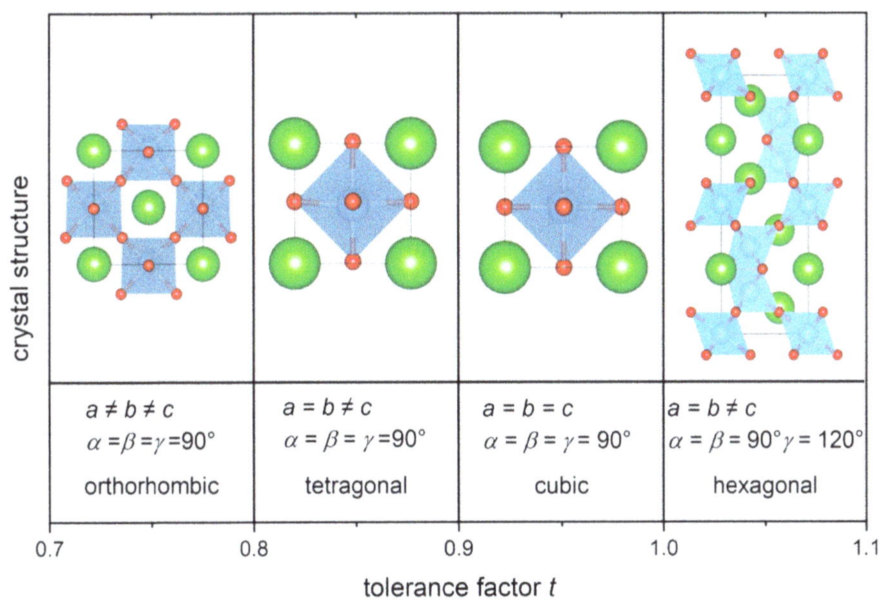

Figure 3.5. Crystal structures of perovskites as a function of the tolerance factor. All structures are oriented along the *crystallographic a*-axis. The unit cells are marked as black lines and the A- and B-site ions are displayed as shown in the legend of figure 3.3 (c).

can form. This phase is very similar to the cubic one and only differs in the *c*-axis length. The phase transition from cubic to tetragonal is often induced by temperature changes. Here, the B cation is slightly shifted and therefore the transition from cubic to tetragonal and vice versa is a fast process (e.g. in $BaTiO_3$). A further reduction of t leads to a further decline in the symmetry. The orthorhombic system is characterized by different axis lengths ($a \neq b \neq c$) and often shows tilting of the octahedra and correspondingly larger unit cells. It is the basis of many mixed ionic–electronic conducting perovskites.

3.2.2 The influence of different A- and B-site ions on selected materials properties

In the previous section, the influence of the different A- and B-site cations on the crystal structure was addressed. Here, the variation and substitution of these ions and their impact on the materials properties relevant to SOEC anodes are discussed.

Doping with aliovalent ions, i.e. ions with a different nominal valence than those of the host material, is a well-established way to adjust the concentrations of electronic and ionic defects. Dopants are called donor dopants if they provide electrons, i.e. if they lead to an increase in *n*-type conductivity. In the range of high pO_2, where SOEC anodes are used, *p*-type conductivity is the dominant electronic transport process. This can be increased by acceptor doping. It is important to note that, in principle, doping always changes electronic as well as ionic defect concentrations; the exact contribution depends on the material, pO_2, and temperature. When a defect is formed, it is charge compensated by an oppositely charged one (electroneutrality).

3.2.2.1 Donor doping

An aliovalent ion acts as a donor dopant if its valence is higher than that of the parent ion; for example, in the incorporation of Sb_2O_5 into TiO_2, Sb^{5+} substitutes for Ti^{4+} by forming an effectively positive defect, $Sb^{\bullet}_{Ti}$. This has a major impact on the stoichiometry of the phase and on the charge balance. In this example, there are three possible ways to compensate for the positive charge. These can be concisely represented in Kröger–Vink notation (see chapter 2) as a defect chemical reaction [19]:

(1) Cation vacancies $V^{''''}_{Ti}$

$$2Sb_2O_5 + 5Ti^{x}_{Ti} \rightleftharpoons 4Sb^{\bullet}_{Ti} + V^{''''}_{Ti} + 5TiO_2 \tag{3.3}$$

(2) Anion interstitials $O^{''}_{i}$

$$Sb_2O_5 + 2Ti^{x}_{Ti} \rightleftharpoons 2Sb^{\bullet}_{Ti} + O^{''}_{i} + 2TiO_2 \tag{3.4}$$

(3) Electrons e'

$$Sb_2O_5 + 2Ti^{x}_{Ti} \rightleftharpoons 2Sb^{\bullet}_{Ti} + 2e' + \frac{1}{2}O_2 + 2TiO_2 \tag{3.5}$$

The occurrence of cation vacancies (if the defect concentration is sufficiently large) can be detected by density measurements. The formation of oxygen interstitials is unfavorable in perovskites, in contrast to RP phases and materials with a fluorite structure. Oxygen interstitials O_i'' lead to oxygen overstoichiometry $\delta_{O_i''}$ and give rise to the tracer diffusion coefficient of oxygen D_O^*. Due to the lower mobility of ionic carriers, the measured total conductivity is determined by the contribution of the electrons. In reality, all three processes contribute, depending on pO_2 and temperature. In any case, the overall electroneutrality condition must be fulfilled.

3.2.2.2 Acceptor doping

An aliovalent ion acts as an acceptor dopant if its valence is lower than that of the parent ion; for example, in the incorporation of Al_2O_3 into TiO_2, Al^{3+} substitutes for Ti^{4+} by forming an effectively negative defect, Al'_{Ti}. There are three possible ways to compensate for the negative charge:

(1) Cation interstitials $Al_i^{\bullet\bullet\bullet}$ (rather unfavorable in perovskites)

$$2Al_2O_3 + 3Ti_{Ti}^x \rightleftharpoons 3Al'_{Ti} + Al_i^{\bullet\bullet\bullet} + 3TiO_2 \tag{3.6}$$

(2) Oxygen vacancies $V_O^{\bullet\bullet}$

$$Al_2O_3 + 2Ti_{Ti}^x + O_O^x \rightleftharpoons 2Al'_{Ti} + V_O^{\bullet\bullet} + 2TiO_2 \tag{3.7}$$

(3) Holes $h^\bullet$

$$Al_2O_3 + \frac{1}{2}O_2 + 2Ti_{Ti}^x \rightleftharpoons 2Al'_{Ti} + 2h^\bullet + 2TiO_2 \tag{3.8}$$

The introduction of aliovalent ions into the cation sublattice of the perovskite oxide not only changes the cation sublattice but also the oxygen sublattice. This often leads to a competition between electronic and ionic charge compensation, which results in mixed ionic–electronic conductivity. This property is essential for a material that is used for the positrodes of SOECs and SOFCs, such that the active zone for oxygen reduction is expanded beyond the gas/electrolyte/positrode triple-phase boundary [20].

In the following sections, the influences of A-site and B-site acceptor doping on state-of-the-art perovskite positrode materials are discussed.

3.2.2.3 Influence of the A cation

$LaCoO_3$ perovskites have been investigated in numerous studies as a particularly promising class of materials for SOEC anodes. These materials are usually synthesized in air, resulting in the nominal composition $La^{3+}Co^{3+}O_3$, in which all oxygen positions are fully occupied ($\delta = 0$). Here, δ represents the deviation from fully occupied oxygen sites. If treated at high temperature and/or low pO_2, or if doped with acceptors, this material acquires an oxygen non-stoichiometry.

The oxygen site is no longer fully occupied, as denoted by (3−δ). Divalent ion substitution (for example, using alkaline-earth ions such as Sr^{2+}, Ca^{2+}, Ba^{2+}) at the A site creates oxygen vacancies $V_O^{\bullet\bullet}$, reducing the oxygen content:

$$2SrO + 2La_{La}^{x} + O_O^{x} \rightleftharpoons 2Sr'_{La} + La_2O_3 + V_O^{\bullet\bullet}. \tag{3.9}$$

The concentration of the doubly charged oxygen vacancies is half of the concentration of the negatively charged Sr dopant (in the notation for defect concentrations, the individual defects are written in square brackets):

$$[Sr'_{La}] = 2\,[V_O^{\bullet\bullet}];\ \text{with}\ [V_O^{\bullet\bullet}] = \delta. \tag{3.10}$$

If the charge compensation is electronic in nature, then the oxygen vacancies are filled with atmospheric oxygen:

$$\frac{1}{2}O_2 + V_O^{\bullet\bullet} \rightleftharpoons 2h^{\bullet} + O_O^{x} \tag{3.11}$$

$$2SrO + \frac{1}{2}O_2 + 2La_{La}^{x} \rightleftharpoons 2Sr'_{La} + 2h^{\bullet} + La_2O_3. \tag{3.12}$$

If one assumes that the holes are largely located at the Co ions, this corresponds to the following reaction equation:

$$2SrO + \frac{1}{2}O_2 + 2La_{La}^{x} + 2Co_{Co}^{x} \rightleftharpoons 2Sr'_{La} + 2Co_{Co}^{\bullet} + La_2O_3. \tag{3.13}$$

The tetravalent Co^{4+} ion ($Co_{Co}^{\bullet}$) is then equimolar to the divalent Sr^{2+} ion. This corresponds to a solid solution between $LaCoO_3$ (Co^{3+}) and $SrCoO_3$ (Co^{4+}), i.e. $La_{1-x}Sr_xCo^{3+}{}_{1-x}Co^{4+}{}_xO_3$.

Figure 3.6 shows (a) the effect of the dopant content on the concentration of oxygen vacancies and (b) the change of the oxygen content (3−δ) as a function of temperature and pO_2.

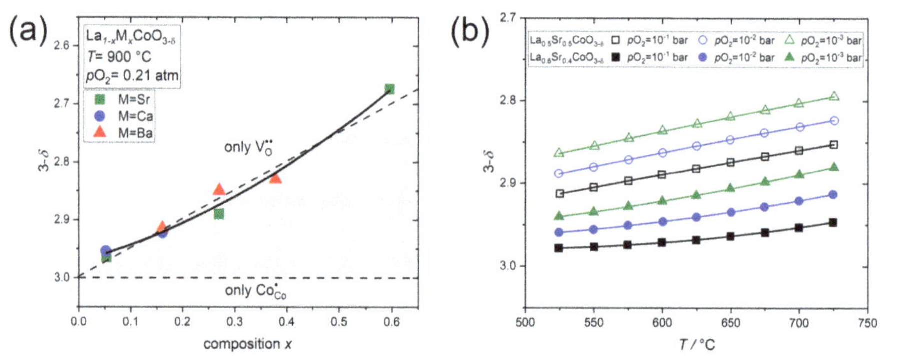

Figure 3.6. (a) Values for the oxygen content as a function of sample composition. The experimental points are taken from [21]. The gray area represents the region of mixed conduction. (b) Oxygen deficiency as a function of temperature and oxygen partial pressure for selected Sr-doped La cobaltites. Data points were taken from [22].

3.2.2.4 Influence of the B cation

One method used to create mixed ionic–electronic conductors is to replace an insulator (e.g. $SrTiO_3$) with a redox-active transition metal (e.g. Fe). This acceptor doping (some Ti^{4+} is replaced by Fe^{3+}) leads to an increase in the ionic conductivity, as mobile oxygen vacancies ($V_O^{\bullet\bullet}$) are introduced. The *p*-type electronic conductivity is increased as well. In particular, when the substitution by the redox-active cation exceeds approx. 30 mol%, a separate 'impurity band' replaces the bandgap [23].

This can be written as:

$$Fe_2O_3 + 2Ti_{Ti}^{x} + O_O^{x} \rightleftharpoons 2Fe'_{Ti} + 2TiO_2 + V_O^{\bullet\bullet}. \tag{3.14}$$

The following defect equations and the Brouwer diagram (figure 3.7) show the dependence of the total conductivity on the oxygen partial pressure [24]. If one formulates the mass action constant K_o (the subscript o in K_o denotes oxidation) for (3.11), the following expression for the electronic conductivity results (3.15):

$$K_o = \frac{[h^\bullet]^2}{[V_O^{\bullet\bullet}]\sqrt{pO_2}} \text{ with } \sigma \propto [h^\bullet] \propto pO_2^{1/4} \text{ (as long as } V_O^{\bullet\bullet} \approx \text{ const.)} \tag{3.15}$$

$$Fe'_{Ti} + h^\bullet \rightleftharpoons Fe_{Ti}^{x}. \tag{3.16}$$

At low pO_2, all Fe is in the 3+ state (Fe'_{Ti}), and further reduction leads to the formation of e′ according to

$$O_O^{x} \rightleftharpoons \frac{1}{2}O_2 + V_O^{\bullet\bullet} + 2e' \tag{3.17}$$

$$Fe'_{Ti} + e' \rightleftharpoons Fe''_{Ti} \quad \text{where } Fe''_{Ti} = Fe^{2+}. \tag{3.18}$$

The mass action constant K_r (the subscript r in K_r denotes reduction → low pO_2) from (3.17) yields the following expression for the conductivity:

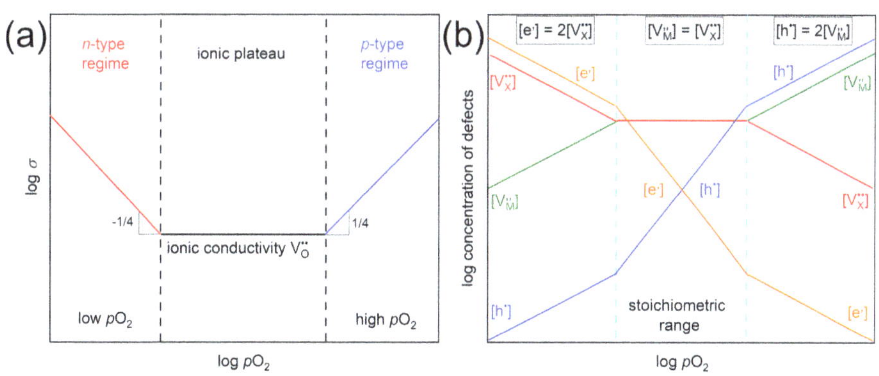

Figure 3.7. Schematic representation of the three regimes of a mixed-conduction perovskite at constant temperature demonstrated by (a) the change in conductivity (b) the change in defect concentration (Brouwer diagram).

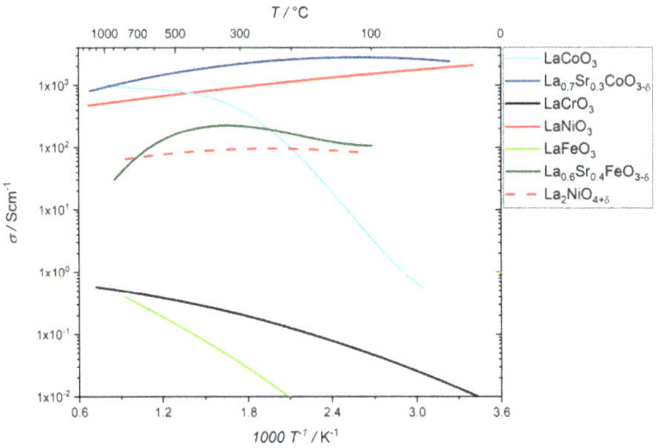

Figure 3.8. Electrical ≈ electronic conductivity for $LaCoO_3$ [26], $LaNiO_3$, $LaCrO_3$ (both from [25]), $La_{0.7}Sr_{0.3}CoO_{3-\delta}$ [27], $La_{0.6}Sr_{0.4}FeO_{3-\delta}$ [28], $LaFeO_3$ [29] and $La_2NiO_{4+\delta}$ [30] as a function of inverse temperature at $0.01 \leqslant pO_2/\text{bar} \leqslant 1$.

$$K_r = [e']^2 \cdot [V_O^{\bullet\bullet}] \cdot \sqrt{pO_2} \text{ with } \sigma \propto [e'] \propto pO_2^{-1/4}. \tag{3.19}$$

An important point regarding figure 3.7 is that the mobility of the electronic defects shown in the Brouwer diagram of figure 3.7(b) is two to three orders of magnitude greater than the ionic mobility.

Until now, we have discussed the influence of partial substitution at the B site on defect concentrations and on conductivities. In the following example, the influence of complete B-site substitution is discussed using $LaFeO_3$, $LaNiO_3$, $LaCrO_3$, and $LaCoO_3$. Although all four materials are perovskite oxides, they differ significantly in their physicochemical properties. $LaCrO_3$ is a highly thermostable semiconducting oxide (*p*-type), while $LaNiO_3$ is thermally unstable and behaves as a metallic conductor [25]. $LaFeO_3$ and $LaCoO_3$ are thermally stable and also *p*-type semiconductors. The differences in their electronic conductivities are shown in figure 3.8.

The major difference here is the conductivity of $LaNiO_3$, which is several orders of magnitude larger and shows metallic characteristics (a decrease in σ with increasing temperature). $LaFeO_3$ and $LaCoO_3$ show high activation energies at lower temperatures due to the combination of thermally activated migration (polaron hopping) and strongly temperature-dependent carrier concentration. The activation energies are lower (and conductivities higher) in Ca- or Sr-doped $LaFeO_3$ and $LaCoO_3$ because of the strongly increased hole concentration (with decreased *T*-dependence) [22, 31–33]. At higher temperatures (above 500 °C), a decrease of conductivity is caused by reduced $[h^{\bullet}]$, which is related to increasing oxygen deficiency. The electronic conductivity of the RP phase $La_2NiO_{4+\delta}$ is approximately one order of magnitude lower than that of $LaNiO_3$, but similar to that of $La_{0.6}Sr_{0.4}FeO_{3-\delta}$. More data for promising SOEC positrode materials can be found in the literature [34].

In the following section, ion transport in the bulk of a solid is discussed. Mechanisms for surface incorporation are covered in chapter 6.

3.3 Diffusion in the solid state

3.3.1 Definitions of diffusion coefficients

This section provides an overview of the fundamental aspects of bulk diffusion in solids, including the different diffusion coefficients. For a detailed consideration, we refer the reader to specific literature [35–38]. The origins of the different diffusion coefficients are discussed and the parameters of promising SOEC anodes described in the literature are listed.

Transport in solids occurs via mobile defects. On an atomistic level, there are several possibilities. The most important for perovskites and related oxides are the vacancy and interstitial mechanisms as illustrated in figure 3.9. Other mechanisms can be found in the literature [36].

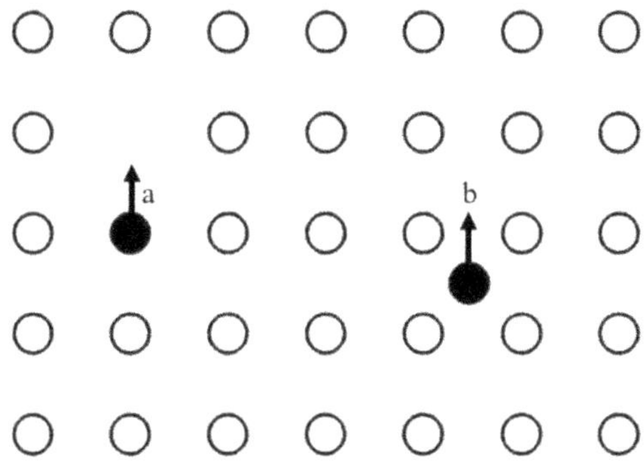

Figure 3.9. Different mechanisms of diffusion in the crystal lattice of perovskites and perovskite-related structures: (a) vacancy, (b) interstitial. Reproduced from [39].

For a fundamental understanding of transport processes, Fick's first law of diffusion has to be considered:

$$\vec{J} = -D\nabla c, \tag{3.20}$$

where the mass flux of a substance $\vec{J}$ is caused by a concentration gradient ∇c and the transport coefficient D. Equation (3.20) represents the definition of the diffusion coefficient D. The actual driving force is a gradient of the chemical potential μ equation (3.21)

$$\vec{J} = -\lambda\nabla\mu, \tag{3.21}$$

where λ is a transport parameter. Depending on the experimental approach, several mutually related diffusion coefficients have to be distinguished, which are discussed in the following. The process of oxygen stoichiometry relaxation after a pO_2 step leads to the diffusion of O as a neutral component via the coupled transport of ionic and electronic defects. This is described by the chemical diffusion coefficient D_{O}^{δ}. In this case, the flux equation becomes

$$\vec{J} = -D_{\mathrm{O}}^{\delta}\nabla c_{\mathrm{O}} = -\frac{\sigma_{\mathrm{O}}}{4F^2}\nabla\mu_{\mathrm{O}} = -\frac{\sigma_{\mathrm{O}}}{4F^2}\frac{\partial\mu_{\mathrm{O}}}{\partial c_{\mathrm{O}}}\nabla c_{\mathrm{O}}, \tag{3.22}$$

in which the transport parameter λ contains the ambipolar conductivity σ_O

$$\sigma_O = \frac{\sigma_{ion}\,\sigma_{eon}}{\sigma_{ion} + \sigma_{eon}} = \sigma_{ion} \quad \text{for} \quad \sigma_{eon} \gg \sigma_{ion} \tag{3.23}$$

and D_O^δ comprises a term $\frac{\partial \mu_O}{\partial c_O}$ which is the reciprocal of the chemical capacitance C_O^δ

$$D_O^\delta \equiv \frac{\sigma_O}{4F^2}\frac{\partial \mu_O}{\partial c_O}. \tag{3.24}$$

A tracer experiment (typically, isotope exchange without a chemical concentration gradient) is described by the tracer diffusion coefficient D^*, which covers the transport of the labeled ion. The driving force is the concentration gradient ∇c^*, and the chemical potential gradient of the tracer is $\nabla\mu^* = RT\nabla c^*/c^*$. This results in

$$\vec{J^*} = -D^*\nabla c^* = -\frac{RT}{4F^2}\frac{\sigma^*}{c^*}\nabla c^*. \tag{3.25}$$

Inserting $\sigma^*/c^* = \sigma_{O^{2-}}/c_{O^{2-}} = \sigma_{V_O^{\bullet\bullet}}/c_{O^{2-}} = \sigma_{V_O^{\bullet\bullet}}/c_O$ and

$$\sigma_{V_O^{\bullet\bullet}} = \frac{4c_{V_O^{\bullet\bullet}} D_{V_O^{\bullet\bullet}} F^2}{RT} \tag{3.26}$$

yields the tracer diffusion coefficient D^* as follows:

$$D^* = \frac{c_V}{c_O} D_{V_O^{\bullet\bullet}}. \tag{3.27}$$

Here, $D_{V_O^{\bullet\bullet}}$ is the defect diffusion coefficient. Defect diffusion coefficients related to oxygen are usually orders of magnitude higher than self-diffusion coefficients. This is caused by the term $\frac{c_V}{c_O}$ in (3.27). As long as pronounced correlation effects are absent, the tracer diffusivity is close to the self-diffusion coefficient, i.e. $D^* \approx D^{self}$ [38]. The relation between $D_{V_O^{\bullet\bullet}}$ and D_O^δ contains the thermodynamic factor γ_O, which is inversely proportional to the chemical capacitance C_O^δ:

$$D_O^\delta = \gamma_O\, D^{self} \approx \gamma_O\, D^* \tag{3.28}$$

$$\gamma_O = \frac{c_O}{RT}\frac{\partial \mu_O}{\partial c_O} = \frac{1}{2}\left(\frac{\partial \ln pO_2}{\partial \ln c_O}\right)_T = \frac{1}{2}\left(\frac{\partial \ln pO_2}{\partial \ln (3-\delta)}\right)_T. \tag{3.29}$$

A collection of γ_O values for various perovskites can be found in table 3.1.

The relation between D_O^δ and the defect diffusion coefficient is given by (3.30), where t_{eon} and t_{ion} are the transference numbers of the electronic and the ionic contributions [47]:

$$D_O^\delta = t_{eon} D_{V_O^{\bullet\bullet}} + t_{ion} D_{h^\bullet}. \tag{3.30}$$

If trapping is relevant, this equation is further modified [48].

Table 3.1. Potential perovskite-structured SOEC anode materials and their ranges of γ_O under oxidizing conditions. $SrTiO_{3-\delta}$ with low Fe doping is an example of a compound that has an exceptionally high γ_O value.

Compound	γ_O	Reference
$La_{1-x}Ca_xFeO_{3-\delta}$ $0.1 \leqslant x \leqslant 0.25$	250–2500 (T = 800 °C–600 °C)	[40]
$Ba_{0.5}Sr_{0.5}Co_{0.8}Fe_{0.2}O_{3-\delta}$	125 at 600 °C	[41]
$La_{0.6}Sr_{0.4}Co_{0.6}Fe_{0.4}O_{3-\delta}$	100–300 (T = 982 °C–732 °C)	[42]
$La_{0.6}Sr_{0.4}Co_{0.2}Fe_{0.8}O_{3-\delta}$	100 at 800 °C	[43]
$La_{0.9}Sr_{0.1}FeO_{3-\delta}$	20 at 800 °C	[44]
$SrTi_{0.65}Fe_{0.35}O_{3-\delta}$	100–1000 at 600 °C (bulk or thin film)	[45]
$SrTiO_{3-\delta}$ (with low Fe doping)	1×10^5 at 600 °C	[46]

3.3.2 Measurement of diffusion coefficients and ionic conductivity

In tracer diffusion the transport takes place along an isotope gradient. In order to obtain a tracer profile, the sample is kept at a constant temperature until it is in equilibrium with the surrounding pO_2 to ensure the absence of chemical driving forces. The sample is then quenched and brought into contact with the tracer gas enriched in $^{36}O_2$. The reintroduction of the sample at a constant temperature then ensures the penetration of the tracer into the bulk and thus a purely isotopic concentration gradient results. The isotope concentration profile is examined using a suitable technique, e.g. ion-exchange depth profiling secondary ion mass spectroscopy (IEDP-SIMS) or line-scan SIMS, and the corresponding transport parameters (i.e. bulk diffusion D^* and surface exchange k^*) can be extracted simultaneously [35, 49].

For dense samples exposed to a pO_2 step, or for oxygen-permeable membranes, a gradient (indicated by the superscript δ) builds up in the oxygen chemical potential. This gradient induces an oxygen diffusion process characterized by the chemical diffusion coefficient D_O^δ. Experimentally, D_O^δ can be obtained from *in situ* conductivity relaxation experiments on densely sintered samples [50]. The electrical conductivity response of the sample to a step-wise change of the oxygen partial pressure is recorded, and D_O^δ can be obtained from nonlinear least-squares fits of the solution of the diffusion equation to the conductivity relaxation data [51–53]. The solution of the diffusion equation assumes that the oxygen surface exchange is very large, and that D_O^δ does not change during the pO_2 step. The latter point, however, is only fulfilled for small changes of the oxygen chemical potential (small pO_2 steps).

The ionic conductivity is another important material parameter for SOEC anodes. The ionic conductivity σ_{ion} is defined as the movement of ions under the influence of an external electric field (3.31):

$$\sigma_{ion} = |z|\, neu, \tag{3.31}$$

where z is the charge number, n the number of mobile ions in the unit cell volume (m^{-3}), e the elementary charge (1.6022×10^{-19} C) and u the mobility of the charge carrier (m^2 V^{-1} s^{-1}). Direct measurement of the ionic conductivity in mixed conductors is often difficult and requires a suitable measurement cell (e.g. ion-selective electrodes according to Wagner [54] and Hebb [55]); otherwise, the *p*-type charge carriers dominate the total conductivity. Alternatively, if D_O^{δ} or D^* is known, σ_{ion} can be calculated from the Nernst–Einstein relation (3.32):

$$\sigma_{ion} = \frac{nz^2D^*e^2}{k_bT}. \tag{3.32}$$

The next chapter summarizes the different diffusion coefficients of promising SOEC anodes with perovskite and Ruddlesden–Popper-type structures.

3.3.3 Diffusion coefficients of relevant positrode materials

As can be seen from table 3.2, the respective diffusion coefficients differ by several orders of magnitude, depending on the chemical composition and to a lesser degree on pO_2 (T = const. in table 3.2). The defect diffusion coefficient (here, for oxygen vacancies $D_{V_O^{\bullet\bullet}}$) is quite similar for all (La,Sr)(Co,Fe)$O_{3-\delta}$ perovskites [56]. The activation energy for the vacancy migration barriers is between 0.8 and 1 eV and therefore lower than the migration barriers calculated using tracer experiments (2–2.5 eV). This difference is caused by the temperature dependence of the defect concentration. Only $Ba_{0.5}Sr_{0.5}Co_{0.8}Fe_{0.2}O_{3-\delta}$ has a lower $D_{V_O^{\bullet\bullet}}$ activation energy of 0.5 eV. $La_{1-x}Sr_xMnO_3$ (LSM), a well-investigated SOFC cathode material, is less suitable for application in SOECs because of its extremely low oxygen vacancy concentration.

Table 3.2. Oxygen tracer diffusion coefficients and oxygen chemical diffusion coefficients of promising perovskite-structured SOEC anode materials estimated at 700 °C.

Compound	D_O^*/cm^2 s^{-1}	D_O^{δ}/cm^2 s^{-1}	Reference
$La_{0.8}Ca_{0.2}FeO_{3-\delta}$	1×10^{-8} at 0.1 bar O_2 (D^* calculated from D_O^{δ})	8×10^{-6} at 0.1 bar O_2	[32]
$La_{0.6}Sr_{0.4}Co_{0.2}Fe_{0.8}O_{3-\delta}$	1×10^{-11} at 1 bar O_2 (interpolated)	1×10^{-8} at 1 bar O_2 (D_O^{δ} calculated from D^*)	[57]
$La_{0.9}Sr_{0.1}CoO_{3-\delta}$	5×10^{-11} at 0.05 bar O_2 (extrapolated)	5×10^{-9} at 0.05 bar O_2 (extrapolated)	[56]
$La_{0.6}Sr_{0.4}CoO_{3-\delta}$	2×10^{-8} at 0.1 bar O_2 (D^* calculated from D_O^{δ})	1×10^{-6} at 0.1 bar O_2	[22]
$Ba_{0.5}Sr_{0.5}Co_{0.8}Fe_{0.2}O_{3-\delta}$	1×10^{-6} at 0.02 bar O_2	1×10^{-5} at 0.02 bar O_2	[41]

Table 3.3. Oxygen tracer diffusion coefficients of promising SOEC anodes that have Ruddlesden–Popper-type structures estimated at 700 °C and pO_2 = 0.2 bar.

Compound	$D_O^*/cm^2\ s^{-1}$	Reference
$La_2NiO_{4+\delta}$	3×10^{-8}	[58]
$La_2Ni_{0.9}Co_{0.1}O_{4+\delta}$	2×10^{-8}	[58]
$Pr_2NiO_{4+\delta}$	1×10^{-7}	[59]
$Pr_{1.9}Ni_{0.75}Cu_{0.25}O_{4+\delta}$	2×10^{-6}	[60]
$Pr_{1.9}(Ni_{0.75}Cu_{0.25})_{0.95}Ga_{0.05}O_{4+\delta}$	2×10^{-6}	[60]

Table 3.3 lists the oxygen tracer diffusion coefficients of simple RP phases that have a K_2NiF_4 structure, which have diffusivities similar to, or even higher than, those of perovskites.

3.4 Compatibility with electrolyte materials

Chapter 2 presents and discusses state-of-the-art electrolyte ceramics for SOEC technology in detail. Two fluorite-structured materials are particularly interesting: yttria-stabilized zirconia (which usually has 8 mol% yttria), $Zr_{0.85}Y_{0.15}O_{2-\delta}$ (8-YSZ), and Gd-doped ceria (which usually has 10 mol% Gd), $Ce_{0.9}Gd_{0.1}O_{2-\delta}$ (GDC10). In contrast to Ruddlesden–Popper phases (see the examples in table 3.3 and the next section), perovskite oxides require alkaline-earth acceptor doping to ensure sufficient mixed ionic and electronic conductivity. As shown in figure 3.8 and table 3.2, Sr^{2+} and Ca^{2+} are the most common substituents in perovskite oxides. Unfortunately, YSZ reacts with La to form $La_2Zr_2O_7$, and with alkaline-earth metals to form e.g. $SrZrO_3$ [61]. Both impurity phases have low conductivity and thus reduce the performance of the cell [62]. There are different ways to suppress this unwanted phase formation, e.g. (i) substitution with 2 mol% Cu at the B site of the perovskite to improve the sintering behavior at reduced temperatures (below 1000 °C to reduce secondary phase formation) [62], or (ii) $La_{0.8}Ca_{0.2}FeO_{3-\delta}$ exhibits reduced substitution of Zr at the B site of the perovskite and can be sintered at 1000°C–1100°C without an impurity phase [63]. The most common approach is to use a gadolinium-doped ceria (GDC) barrier layer between the positrode and the electrolyte [64]. Wilde *et al* [65] showed that the polarization resistance of A-site cation-deficient $La_{0.58}Sr_{0.4}Co_{0.2}Fe_{0.8}O_{3-\delta}$ with a GDC buffer layer decreases by up to three orders of magnitude compared to the polarization resistance without a buffer layer. Since GDC does not form impurity phases with either ferrates or cobaltates, it is particularly suitable for increasing the ionic conductivity in composite electrodes. In addition to the transport properties, composite formation also improves the mechanical stability of the cell by matching the thermal expansion coefficient to that of the electrolyte [66].

3.5 Layered rare-earth nickelates

3.5.1 Introduction

Layered rare-earth nickelates of the Ruddlesden–Popper structural type are another class of materials that show interest and promise for application as positrodes in high-temperature fuel cells and electrolysis cells. They are well-established materials with performance characteristics that approach those of state-of-the-art perovskite materials more closely than any other material class currently available. Nickelates are mixed ionic–electronic conductors with electrical conductivities of up to 100 S cm^{-1} at around 800 °C, which is considered to be sufficient for SOFC/SOEC applications. They feature high electrocatalytic activities for oxygen reduction/oxidation processes as well as fast oxygen surface exchange and transport properties, which is of high importance for the transport of oxide ions to and from the electrolyte under current flow. Their thermal expansion coefficients (TECs) are in the range of 13–15 ppm K^{-1}, which is quite close to those of current electrolyte materials such as stabilized zirconium oxide and doped cerium oxide.

Layered nickelates are similar to perovskites in that they offer a wide range for doping (i.e. substitution) with various elements for both rare-earth elements and nickel, thus allowing for the tailoring of materials properties. However, many rare-earth nickelates are oxygen-excess compounds with oxygen transport via interstitials, which renders acceptor doping with alkaline-earth ions (in particular, strontium) unnecessary. In general, positrodes are susceptible to several modes of degradation processes, such as limited phase stability of the material itself or reactivity with the electrolyte or gas species in the oxidant. Many of these degradation processes in positrode materials are known to originate from—or to be aggravated by—the presence of alkaline-earth cations, for example segregation of Sr to the surface, the formation of poorly conducting Sr zirconates at the electrode–electrolyte interface, the formation of carbonates or hydroxides at the surface, the formation of Sr sulfates from SO_2 in air or other sources, poisoning by chromium from stainless steel interconnects, or Si poisoning from sealant materials. In this respect, the application of unsubstituted rare-earth nickelates such as $La_2NiO_{4+\delta}$ or $Pr_2NiO_{4+\delta}$ as positrodes appears to be very promising.

Nickelate compounds have been considered especially advantageous as positrodes in electrolysis applications due to their oxygen transport mechanism, which differs from that of perovskite compounds. In mixed ionic–electronic conducting perovskites, oxygen transport is based on oxygen vacancies, which have to be induced by acceptor doping, i.e. substitution by lower-valent cations. To a first approximation (i.e. neglecting effects such as defect interactions and vacancy ordering), the ionic conductivity should be directly proportional to the concentration of oxygen vacancies. Thus, in SOEC mode, in which oxygen ions are transported from the negatrode through the electrolyte to the positrode material, the number of available vacancy sites is reduced. In contrast, oxygen ion conduction in nickelates (if not too highly doped with acceptor elements) is based on an interstitial-type mechanism, which means that the ionic current is mainly carried by surplus oxygen ions. Thus, increasing the oxygen content during SOEC operation is expected to benefit the ionic transport by increasing the concentration of ionic charge carriers (i.e. oxygen interstitials). Indeed, it has been often observed in the literature that rare-earth nickelates perform better in electrolysis mode than in fuel-cell mode [67]. Table 3.3 indicates that the tracer diffusion coefficients of layered nickelates are comparable to or even higher than those of mixed-conduction perovskites (table 3.2).

Current issues affecting the application of nickelates appear to be their rather pronounced reactivity with common electrolyte materials during conventional cell manufacturing (i.e. the sintering of screen-printed electrodes)—especially when applied as composites with electrolyte materials—as well as the limited thermodynamic stability of Pr-rich compounds, which currently yield the best cell performances. It should also be mentioned that even for alkaline-earth-free nickelate compounds, degradation is still an issue, and reactivity with Cr, Si, SO_2, or electrolyte materials such as yttria-stabilized zirconia (forming pyrochlore zirconates with rather low ionic conductivity) has been reported.

It should be mentioned that other 3*d* transition metals such as Co, Fe, Cu, and Mn can also form Ruddlesden–Popper-type compounds, some of which have been investigated for use as electrode materials in solid oxide cells. The undoped compounds suffer from various drawbacks such as limited phase stability or low electrical conductivity. Of course, these deficiencies can be remedied by partial substitution on the A and B lattices, but this topic lies outside the scope of this chapter. Conversely, many 3*d* transition metals can be applied as doping elements for the partial substitution of Ni in order to modify the materials properties of the parent nickelate compound, which is addressed in section 3.5.3.

3.5.2 Crystal structure

Rare-earth nickelates crystallize in the so-called K_2NiF_4 structure, which is a structural type first described by Balz and Plieth for the compound K_2NiF_4 [68]. This structural type gained prominence in the 1980s, when high-temperature superconductivity was observed in compounds derived from the cuprate homolog La_2CuO_4 [69]. The K_2NiF_4 structure is often viewed as a perovskite-related structure because it can be thought of as being constructed from single perovskite sheets (or slabs), which are stacked at some distance along their perpendicular direction. In order to reduce coulombic repulsion between equally charged ions, the perovskite sheets are shifted in-plane, so that the cations and anions of two neighboring sheets face each other. The ionic arrangement between two adjacent sheets matches that of a rock-salt structure, which is why nickelates can also be considered as intergrowth compounds between the perovskite and rock-salt structures. While the perovskite structure is rather tightly packed and offers little space for interstitial sites, the rock-salt layers are more open and thus allow for the accommodation of oxygen interstitials (with some ion displacements in their immediate environment).

A more general structural type can be obtained by considering that the stacked sheets can not only consist of single perovskite layers, but of double layers, triple layers, and so on. Hence, the K_2NiF_4-structure may be viewed as the first member of the so-called Ruddlesden–Popper (RP) series, whose members differ in the number (n) of perovskite layers that make up the perovskite sheets of the structure. Although the second member of the higher RP homologs of K_2NiF_4 does exist as a stable compound, the series was discovered in the Sr titanate system by Ruddlesden and Popper, who published the crystal structures of $Sr_3Ti_2O_7$ ($n = 2$) and $Sr_4Ti_3O_{10}$ ($n = 3$) [70, 71]. The generic formula of the RP oxide series is $A_{n+1}B_nO_{3n+1}$ (= $(AO)(ABO_3)_n$) and the most symmetric crystallographic space group is *I4/mmm* (#139) for all members. In general, 'A' stands for a large ion from the lanthanides, alkaline, or alkaline-earth metals, while 'B' denotes a small (3*d*) transition-metal ion. Accordingly, the terms 'A site' and 'B site' are often used when referring to the corresponding atomic positions in $A_{n+1}B_nO_{3n+1}$ RP structures (even though an 'A site' may consist of several crystallographically non-equivalent sites).

Figure 3.10 gives an overview of the first three members of the RP nickelate series, which are those that have been synthesized with a sufficient level of phase purity. Fourth- and higher-order members of the nickelates have not yet been prepared as bulk compounds, but have been identified as 'stacking faults' in lower-order RP structures by transmission electron microscopy (TEM) analysis [72, 73]. An obvious exception is the perovskite structure, which may be considered as a limiting case of the RP series for $n \rightarrow \infty$.

A very important property of nickelates is their ability to tolerate deviations from stoichiometry with respect to the oxygen lattice, as quantified by the 'δ' in $La_2NiO_{4+\delta}$. The sign and magnitude of δ depend on the order of the RP structure and its

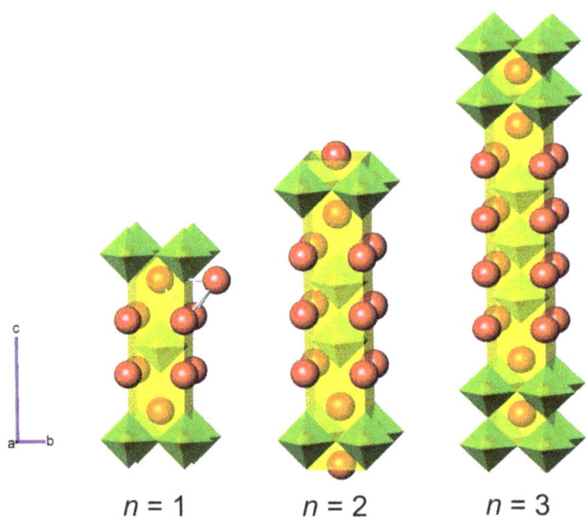

Figure 3.10. Crystal structures of first-order (i.e. K_2NiF_4-type, $n = 1$), second-order ($n = 2$) and third-order ($n = 3$) RP phases, $A_{n+1}B_nO_{3n+1}$. A ions are shown as red spheres, BO_6 coordination units are plotted as green octahedra, unit cells are colored yellow. The image of the K_2NiF_4 structure shows an A_4 tetrahedron, whose center is a potential interstitial site that can accommodate interstitial oxygen ions [74].

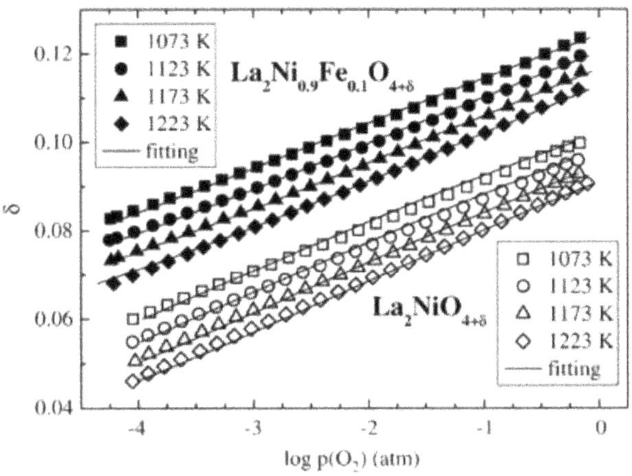

Figure 3.11. Dependence of the oxygen overstoichiometry (δ) of the first-order RP compound $La_2NiO_{4+\delta}$ on temperature, pO_2, and B-site doping (adapted with permission from [76]).

composition as well as parameters such as temperature and oxygen partial pressure (figure 3.11). For example, unsubstituted rare-earth nickelates $Ln_2NiO_{4+\delta}$ (Ln = La, Pr, Nd) are oxygen overstoichiometric (also called hyperstoichiometric) compounds, for which the value of δ falls in the range of $0 \leqslant \delta \leqslant 0.25$. This oxygen excess is realized by accommodating additional oxygen ions on interstitial sites within the rather loosely packed rock-salt layers of the intergrowth structure. Their actual crystallographic location has been elucidated by Jorgensen *et al* [75] using neutron diffraction measurements. As shown in figure 3.10 for the K_2NiF_4-structure, oxygen interstitials are located at the center of a tetrahedron formed by four A-site cations within the rock-salt layer. Due to the exothermic nature of the oxygen incorporation reaction, nickelates release interstitial oxygen at elevated temperatures (provided the exchange kinetics is sufficiently fast); see figure 3.11. For higher-order RP phases, e.g. unsubstituted rare-earth nickelates $Ln_4Ni_3O_{10+\delta}$ (Ln = La, Pr, Nd), δ is often found to be negative—thus usually written as $Ln_4Ni_3O_{10-\delta}$—and oxygen vacancies are located within the perovskite slabs.

The Ni ion is octahedrally coordinated by six oxygen ions, which is similar to the case of perovskites. Due to the layered structure of nickelates, however, the oxygen sites are not all symmetry equivalent (as is the case for the perfect cubic perovskite structure) but are distinguished as being in equatorial or apical positions. Equatorial oxygen ions bridge Ni ions along the perovskite slabs (i.e. parallel to the crystallographic *a*–*b* plane), while apical oxygen ions have their Ni–O bonds oriented parallel to the crystallographic *c*-axis. As in the case of perovskites, the actual crystal symmetry of nickelate compounds is often found to be lower than the highest-symmetry space group *I4*/*mmm* (i.e. the aristotype) due to tilting and/or distortion of the Ni–O octahedra. Such deviations are caused by the imperfect match of the ionic radii of the constituents and also depend crucially on the actual oxygen content—and thus on the temperature and oxygen partial pressure in the surrounding atmosphere.

Due to their layered structure, the materials properties of nickelates are expected to be highly anisotropic, which has indeed been reported for oxygen diffusivities experimentally obtained for single crystals [77]. Molecular dynamics simulations and density functional calculations confirm these findings and have also aided in elucidating the prevalent ionic conduction mechanism for oxygen (see e.g. [78]). According to these studies, ion conduction in oxygen overstoichiometric nickelates is mostly carried by interstitial oxygen through direct (interstitial) or indirect (interstitialcy) jump mechanisms. When the direct interstitial mechanism is used, an oxygen ion jumps directly from an interstitial site into a neighboring empty one, whereas in the case of the interstitialcy mechanism, an interstitial oxygen ion displaces an apical oxygen ion, which in turn moves into an adjacent vacant interstitial site. Since conventionally prepared SOEC electrodes (e.g. those prepared by screen printing or tape casting) are polycrystalline structures, anisotropy in ionic conduction and other properties is of little concern for such applications. In case of electrodes prepared by thin-film techniques such as pulsed laser deposition (PLD), more or less strictly oriented structures are generated, for which the anisotropy of materials properties may have to be considered.

3.5.3 First-order Ruddlesden–Popper phases

Within the first-order RP phases of rare-earth nickelates, $La_2NiO_{4+\delta}$, $Pr_2NiO_{4+\delta}$, and $Nd_2NiO_{4+\delta}$ have been synthesized as bulk materials at high phase purities. Structurally, the rare-earth metal ions are a bit too small as compared to the Ni–O network, which leads to deformation and/or cooperative rotation of the Ni–O octahedra in order to compensate for the size mismatch. Attempts have been made to quantify the internal strain by defining a mismatch factor in analogy to the Goldschmidt tolerance factor used for perovskite compounds [79] (see section 3.2.1). The size mismatch is thought to be the driving force for the incorporation of excess oxygen on interstitial sites within the rock-salt layers, since it can diminish the interlayer strain for two reasons. First, the incorporation of oxygen is accompanied by the partial oxidation of Ni^{2+} to Ni^{3+} (assuming localized charges) and thus leads to a decrease in the average Ni–O bond distance. Second, interstitial oxygen increases the average coordination number of the rare-earth ions, which results in larger average Ln–O bond lengths. Due to the lanthanide contraction, the size mismatch increases for lanthanides with higher atomic numbers, which explains why there are no stable unsubstituted nickelates for Sm and later rare-earth elements. This is also the reason that the compound $Ce_2NiO_{4+\delta}$ has not been successfully synthesized, since under normal ambient (i.e. oxidizing) conditions, the prevalent oxidation number of Ce is 4+ and the ionic radius of Ce^{4+} is much too small to yield a stable nickelate structure. Of course, partial A-site substitution with larger cations relieves the internal strain, and thus compounds such as $(Sm, Sr)_2NiO_{4+\delta}$ and $(Gd, Sr)_2NiO_{4+\delta}$ have been synthesized with high phase purity [80, 81].

The electrical conductivities of K_2NiF_4-type rare-earth nickelates show a broad maximum at elevated temperatures [59]. This is a consequence of the decrease in electron hole concentration due to the loss of interstitial oxygen upon heating, which compensates the increase in hole mobility with rising temperatures (for details see e.g. [82]). Exemplary data for $La_2NiO_{4+\delta}$ are included in figure 3.8.

$La_2NiO_{4+\delta}$ has been extensively investigated as for use as a positrode material in SOFCs and SOECs, whereas less focus was given to $Pr_2NiO_{4+\delta}$ and $Nd_2NiO_{4+\delta}$. $La_2NiO_{4+\delta}$ and $Nd_2NiO_{4+\delta}$ are often reported to exhibit moderate electrode performance, while $Pr_2NiO_{4+\delta}$ is commonly found to show excellent performance characteristics. Unfortunately, $Pr_2NiO_{4+\delta}$ is unstable at SOEC operating temperatures around 800 °C and decomposes into Pr oxides (typically Pr_6O_{11}) and the third-order RP phase $Pr_4Ni_3O_{10\pm\delta}$ [83] (at lower decomposition temperatures, the perovskite compound $PrNiO_{3-\delta}$ has also been identified). It is important to mention, however, that these decomposition products are themselves electrocatalytically active components, which could possibly render the observed instability tolerable or even beneficial by further enhancing the electrode activity.

The fundamental reason for the excellent electrode performance of $Pr_2NiO_{4+\delta}$ is not quite clear. It has been suggested that the redox variability of praseodymium ions, i.e. the Pr^{3+}/Pr^{4+} redox couple (as opposed to the fixed-valent lanthanide ions La^{3+} and Nd^{3+}), may play a role. Results obtained using X-ray spectroscopy measurements, however, show that praseodymium appears to exist primarily in its trivalent state, Pr^{3+}, over a wide temperature range in $Pr_2NiO_{4+\delta}$ [84]. It may also be the case that the abovementioned intrinsic instability of Pr nickelate and the *in situ* generation of catalytically highly active compounds may enhance the electrode performance. In particular, it is well known that Pr oxides are very efficient catalysts that promote the oxygen reduction and oxygen evolution reactions. This has been impressively demonstrated by Nicollet *et al*, who infiltrated a porous Gd-doped ceria backbone with Pr_6O_{11} and obtained excellent electrode performance at 600 °C [85].

3.5.3.1 A-site doping

Due to the very promising electrode performance of $Pr_2NiO_{4+\delta}$, there have been several attempts to stabilize its structure through the use of A-site doping with rare-earth and alkaline-earth elements. There appears to be a wide solid solubility range between La-, Pr-, and Nd-nickelates, i.e. $(La,Pr,Nd)_2NiO_{4+\delta}$, so that partial substitution of Pr by La or Nd can be used to improve the phase stability while attempting to retain the excellent electrode properties of the $Pr_2NiO_{4+\delta}$ parent compound. Vibhu *et al* extensively investigated the $La_{2-x}Pr_xNiO_{4+\delta}$ system, focusing on compositions with $x = 2$, 1.5, 1.0, 0.5, and 0 [86–89]. Their results showed that the electrode performance of the material decreases with increasing La content, as would be expected, and that the proper amount of La needed to stabilize Pr nickelate requires the composition $La_{1.5}Pr_{0.5}NiO_{4+\delta}$ (within the resolution of A-site doping studied). Similar results were reported by Sharma *et al*, who proposed that the composition $LaPrNiO_{4+\delta}$ is the best compromise between electrochemical performance and thermodynamic stability, even though decomposition was observed at 800 °C [90]. Montenegro-Hernández *et al* investigated the effect of Nd doping in $Nd_{2-x}Pr_xNiO_{4+\delta}$ and found the stability limit to be at $x = 1$ [91]. As in the case of

La, increasing the levels of Nd doping in $Pr_2NiO_{4+\delta}$ results in a significant decrease in electrode performance [92].

A-site doping with alkaline-earth metals, in particular with strontium, has been extensively applied to $La_2NiO_{4+\delta}$, $Nd_2NiO_{4+\delta}$, and $Pr_2NiO_{4+\delta}$. Doping with lower-valent cations is a simple way to increase the electronic conductivity of nickelates, but it also reduces their oxygen overstoichiometry. Moreover, in the case of $Pr_2NiO_{4+\delta}$, it seems to be a way to stabilize its structure and suppress the decomposition reaction [93, 94]. Unsubstituted rare-earth nickelates feature *p*-type electronic conductivity, meaning that their electronic conductance is based on electron holes (defect electrons), whose concentration can be increased by A-site substitution with lower-valent elements (i.e. A-site acceptor doping). Moderate acceptor doping has been found to increase the electronic conductivity of nickelates within a factor of ~3 as compared to the undoped parent compound, while oxygen transport properties are negatively affected, mainly because of the reduction in the oxygen excess. In any case, it should be carefully considered whether it is expedient to introduce alkaline-earth elements into the system, since this removes one of the principal advantages of rare-earth nickelates for electrode applications and potentially gives rise to the whole range of Sr-based degradation modes known from perovskite materials.

3.5.3.2 B-site doping

The effect of B-site doping with various 3*d* transition metals such as Co, Cu, Fe, Mn, and others has been investigated for rare-earth nickelates. Drawing on the wealth of research experience obtained from perovskites, substitution with cobalt seems to be particularly promising, considering the well-known effects of increased electrical conductivity, enhanced electrocatalytic activity, and improved sintering activity induced by Co-doping. Unfortunately, Co also raises the TEC. Surprisingly, Co-substitution is found to act somewhat differently in nickelates. Partial replacement of Ni by Co reduces the electronic conductivity, even though it increases the oxygen overstoichiometry. Some authors, however, report higher electronic conductivities for Co-doped nickelates at elevated temperatures [95–97]. It appears that Co decreases the mobility of electron holes, but at the same time increases their concentration due to higher levels of oxygen overstoichiometry. Since both carrier mobility and oxygen content are temperature dependent, this could give rise to distinct changes in the conductivity–temperature relationship. The sintering activity of $La_2NiO_{4+\delta}$ is lowered with increasing Co content [95], while the electrocatalytic activity for oxygen reduction/oxidation seems to be enhanced. For example, Kilner *et al* applied Co-doping to $La_2NiO_{4+\delta}$ and observed an increase of one to two orders of magnitude in the oxygen surface exchange coefficient—an important parameter for mixed-conducting positrodes, as it measures the rate at which the material can exchange oxygen with the surrounding atmosphere [58].

Doping with copper does not seem to offer any striking improvements to the electrochemical properties, but it lowers the amount of oxygen excess, which is expected to have a negative impact on the ionic conductivity of oxygen. On the other hand, Cu enhances the sintering activity of the materials [98], which may allow for lower sintering temperatures (or shorter sintering times) during cell fabrication and thus potentially reduce detrimental electrode–electrolyte interactions.

Fe doping has been reported to increase oxygen excess in $La_2NiO_{4+\delta}$ (see figure 3.11), but oxygen ion conductivity in $La_2Ni_{0.9}Fe_{0.1}O_{4+\delta}$ does not seem to be favorably affected, despite the increase in ionic charge-carrier concentration [76, 99]. Moreover, the electrical conductivity of $La_2Ni_{0.9}Fe_{0.1}O_{4+\delta}$ was found to be lower than that of undoped La-nickelate. This is surprising, considering that higher oxygen overstoichiometry results in a higher concentration of electron holes for charge compensation, and has been explained by the trapping of electron holes on Fe^{3+} ions [100].

In conclusion, studies of B-site-substituted rare-earth nickelates have greatly expanded the detailed understanding of this material class with respect to crystallography, thermodynamics, mass and charge transfer, etc. However, with respect to their practical application as solid oxide cell (SOC) positrodes, significant improvements of these materials over the nickel-only parent compounds—with the possible exception of cobalt doping—have yet to materialize.

3.5.4 Compatibility with electrolyte materials

As mentioned above, rare-earth nickelates do not require acceptor doping in order to exhibit sufficient mixed ionic–electronic conductivity. The absence of alkaline-earth elements, in particular strontium, would lead one to expect good compatibility with common electrolyte materials such as stabilized zirconia, since the formation of electrically insulating reaction phases such as $SrZrO_3$ cannot occur. Unfortunately, however, alkaline-earth-free nickelates also react with zirconia-based electrolytes, forming zirconate pyrochlores $(La,Pr,Nd)_2Zr_2O_7$ with rather low ionic conductivity. Thus, protective barrier layers (also called diffusion barriers) between the nickelate electrode and the zirconia electrolyte appear to be a requirement, as is the case for all mixed-conduction perovskite compounds. For this purpose, ceria-based barrier layers (ceria doped with Gd (GDC), Sm (SDC), Y (YDC), La (LDC) etc.) are commonly applied. The absence of alkaline earths in nickelates may still prove to be an advantage. Although rare-earth elements have a reduced tendency to migrate through the—often porous—ceria barrier layer to the zirconia electrolyte (as is frequently observed for the more volatile Sr), careful adjustment of the sintering profile is required even when using a diffusion barrier [101]. It is therefore important to establish the chemical compatibility between nickelates and ceria-based electrolytes at SOEC operating temperatures as well as at the sintering temperatures used during cell preparation.

Many groups have conducted reactivity studies between nickelates and GDC. The most comprehensive study was published by Montenegro-Hernández *et al*, who investigated the chemical compatibility between $Ln_2NiO_{4+\delta}$ (Ln = La, Pr, Nd) and YSZ as well as GDC [102]. They found that material reactivity—apart from the obvious parameters temperature and contact time—depends significantly on powder particle size and morphology as well as on the degree of powder compaction (thus varying the contact area and accessibility for oxygen), which might explain discrepancies between results reported by different groups. As expected, X-ray diffraction (XRD) analysis of nickelate–YSZ powder mixtures confirmed the formation of rare-earth zirconates at 900 °C in all cases. For compatibility tests with GDC, however, decomposition products of the nickelate phase were detected at 900 °C, but no clear reaction phases could be identified for $La_2NiO_{4+\delta}$ and $Pr_2NiO_{4+\delta}$. This might be an indication that the underlying chemistry between these nickelates and ceria is not a specific chemical reaction but rather a continuous interdiffusion process, in which the rare-earth element is diffusing into the ceria while the nickel remains in the electrode phase (Ni appears to be insoluble in the ceria structure). Whether this is considered to be harmful for electrode performance and stability is probably simply a matter of its extent (as determined by the sintering profile and the operating temperature), since the ionic conductivity in ceria-based electrolytes is expected to remain sufficiently high upon minor rare-earth doping and the nickelate structure can tolerate some A-site deficiency. As a concrete example, consider a $Pr_2NiO_{4+\delta}$ positrode with some diffusion of Pr into the GDC barrier during sintering [103]. Pr-doped ceria is a mixed ionic–electronic conductor, which itself has significant catalytic activity for the oxygen exchange reaction and has been proposed for use as an active component in composite electrode functional layers. On the other hand, Pr nickelate can tolerate quite a substantial understoichiometry of ~10% on the A site [104] or—in order to accommodate the loss in Pr—may be transformed into a higher-order RP nickelate with a lower A/B-site ratio, such as $Pr_4Ni_3O_{10\pm\delta}$, which has itself catalytic properties that sustain the electrode reaction.

Interestingly, a quite different behavior was reported by Montenegro-Hernández *et al* for the compatibility of $Nd_2NiO_{4+\delta}$ with common electrolyte materials [102]. Reaction phases or decomposition products were only observed after the annealing temperature of the powder mixtures was raised to 1000 °C. For the $Nd_2NiO_{4+\delta}$–YSZ couple, the expected zirconate phase $Nd_2Zr_2O_7$ was formed. For GDC, however, the cerate pyrochlore compound $Nd_2Ce_2O_7$ was detected in the powder mixture after 72 h at 1000 °C, together with minor amounts of Nd_2O_3 and NiO, but without higher-order RP phases of the nickelate. Such enhanced compatibility might be enough for the successful preparation of nickelate positrodes, provided that sintering temperatures of 1000 °C are sufficient to establish a good-quality interface (i.e. strong adhesion, low interfacial resistance) with the GDC layer. Sayers *et al* recommended even lower sintering temperatures of <900 °C in order to suppress cation diffusion from rare-earth nickelates into GDC [105].

Experimental compatibility studies of $La_2NiO_{4+\delta}$ with lanthanum gallate-based electrolyte materials ($La_{1-x}Sr_xGa_{1-y}Mg_yO_{3-\delta}$, LSGM) indicate reactivity at temperatures above 1100 °C, while no reactivity was observed at 1000 °C [67, 105]. The results of thermodynamic modeling, however, suggest that $La_2NiO_{4+\delta}$ is not chemically compatible with lanthanum gallate-based electrolytes, even at operating temperatures of 700 °C [106].

Viewed from the standpoint of thermomechanical compatibility, the above-mentioned interdiffusion processes may be detrimental to cell integrity by inducing electrode delamination, which has indeed often been reported for nickelate positrodes. It thus seems to be quite essential to find ways to mitigate detrimental interactions between nickelates and electrolyte materials. One obvious way to tackle delamination issues is to use a different electrode architecture, such as infiltrated electrodes, in which a porous electrolyte backbone is impregnated with a precursor solution. The precursor is then thermally converted into the target material, usually at much lower temperatures than those required for screen-printed electrodes. Another strategy for improving chemical compatibility, which is well known for perovskite materials, is the synthesis of nickelate compounds with A-site deficiency, i.e. $Ln_{2-x}NiO_{4+\delta}$ ($x \leqslant \sim 0.1$), which may reduce or even suppress the interdiffusion or reactivity of rare-earth ions with the electrolyte. Indeed, Mauvy *et al* reported an improvement in the chemical stability of $Nd_{1.95}NiO_{4+\delta}$ with YSZ [107]. In general, the success of this approach may be limited by the capability of K_2NiF_4-type nickelates to transform into higher-order RP phases upon rare-earth loss. In any case, more systematic studies are needed to further explore this topic. Another approach is to develop new barrier-layer materials (or to modify existing barrier-layer materials) to obtain lower reactivity towards the electrode material. For example, Laguna-Bercero *et al* used a $Pr_2NiO_{4+\delta}$-GDC composite as a diffusion barrier between a $Pr_2NiO_{4+\delta}$ positrode and YSZ and observed very good cell performance, which even improved over time [108]. Dogdibegovic *et al* tested cells with Pr/Gd-co-doped ceria buffer layers and $Pr_2NiO_{4+\delta}$-based electrodes and reported enhanced stability as compared to Pr-free GDC interlayers [109–111]. The observed effect may be due to a reduced driving force for Pr diffusion into the ceria barrier if the barrier layer already contains some Pr. The proposed material combination would be especially advantageous, considering the highly active nature of both $Pr_2NiO_{4+\delta}$ and Pr-doped ceria, as mentioned above. However, this strategy is expected to require very careful adjustment of the Pr/Gd-contents, as Pr-doped cerias have very high TECs due to chemical expansion above 400 °C [112].

3.5.5 Higher-order Ruddlesden–Popper phases

Higher-order RP nickelates have been much less investigated as electroactive materials than their K_2NiF_4-type homologs. The characterization of these materials has been hampered by the fact that it is difficult to obtain sufficiently dense samples by sintering—a prerequisite for obtaining reliable materials parameters. For the preparation of porous structures, such as SOC electrodes, this is obviously less of an issue. In some cases, it has also turned out to be difficult to achieve high phase purity, especially for the second-order RP phases. This, again, makes it challenging to obtain reliable materials parameters, but may be of less concern for practical applications. An advantage of higher-order RP phases is their higher electronic conductivity, which increases with n in the rare-earth nickelate series $Ln_{n+1}Ni_nO_{3n+1}$ (Ln = rare-earth element). Another potential advantage may be decreased reactivity or interdiffusion with electrolyte materials due to the lower Ln/Ni ratio. In the case of Pr nickelates, the compound $Pr_4Ni_3O_{10\pm\delta}$ may be of special interest, since it is one of the decomposition products of $Pr_2NiO_{4+\delta}$ and is thus expected to be thermodynamically stable at SOC operating temperatures. Consequently, $Pr_4Ni_3O_{10\pm\delta}$ electrodes have been used in cell tests and performance values close to those of $Pr_2NiO_{4+\delta}$ have been reported [113]. However, if the material is to be kept in its phase-pure state, the maximum sintering temperature in air is limited to temperatures below 1050 °C, above which $Pr_4Ni_3O_{10\pm\delta}$ is converted into $Pr_2NiO_{4+\delta}$ and NiO [113].

There appears to be some inconsistency in the literature in terms of whether higher-order RP nickelates feature oxygen deficiency or oxygen excess. Usually, higher members of RP phases are found to display oxygen deficiency (i.e. $Ln_3Ni_2O_{7-\delta}$ and $Ln_4Ni_3O_{10-\delta}$). In some instances, however, oxygen excess has been reported. This may be due to differences in the preparation method, the temperature profile and atmosphere used for calcination, the quenching rates after annealing, etc. considering the intrinsic property of the materials that allows for both positive and negative deviations from oxygen stoichiometry. The oxygen excess might also be an artifact caused by insufficient phase purity of the analyzed compounds, especially when such impurities are below the detection limit of conventional X-ray diffraction methods.

3.5.6 SOEC positrode performance

The SOEC performance of many K_2NiF_4-type rare-earth nickelates, both doped and undoped, has been reported in the literature. Several reviews of SOECs or reversibly operated cells are also available, which cover RP-type nickelates in some detail [34, 114–117]. The application of nickelates in proton-conducting electrolyzers (and fuel cells) has been thoroughly reviewed by Tarutin *et al* [118]; see also chapter 11. Some instructive examples of high-temperature steam electrolysis in single cells with nickelate positrodes are discussed below.

Ogier *et al* [119] studied $Pr_2NiO_{4+\delta}$ and $La_2NiO_{4+\delta}$ as positrodes in hydrogen electrode-supported cells with a YSZ electrolyte, both with and without ceria-based barrier layers (GDC and YDC). In agreement with the expected trends, they found higher electrolysis performance for cells with positrodes made of $Pr_2NiO_{4+\delta}$ as compared to $La_2NiO_{4+\delta}$ and also observed a significantly lower polarization resistance with the use of GDC/YDC diffusion barriers than for cells without such an interlayer.

$Nd_2NiO_{4+\delta}$ has been investigated for use as the positrode in planar and micro-tubular SOECs. Chauveau *et al* [120] tested planar electrolyte-supported cells with an active electrode area of 3.14 cm^2 in SOEC mode and observed significantly better cell performance for $Nd_2NiO_{4+\delta}$ positrodes as compared to LSM/YSZ between 750 °C and 850 °C. In a subsequent study, Chauveau *et al* [121] tested cells with positrodes made of $Nd_2NiO_{4+\delta}$, $La_2NiO_{4+\delta}$, and LSM/YSZ in electrolysis mode and again found that the lowest area specific resistance (ASR) was that of the cell with a $Nd_2NiO_{4+\delta}$ positrode (figure 3.12).

Vibhu *et al* [122] tested $Pr_2NiO_{4+\delta}$-based positrodes with different Co-contents and found excellent current–voltage characteristics for $Pr_2Ni_{0.8}Co_{0.2}O_{4+\delta}$ at 800 °C

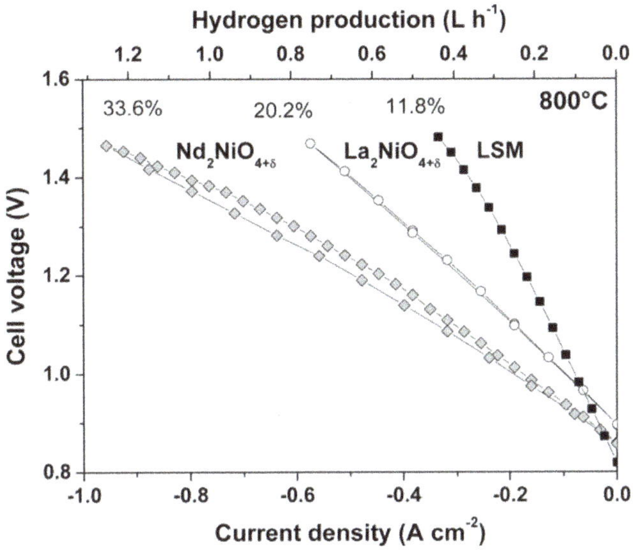

Figure 3.12. SOEC performance of electrolyte-supported cells with $Nd_2NiO_{4+\delta}$, $La_2NiO_{4+\delta}$, and LSM/YSZ positrodes operating at 800 °C (reprinted with permission from [121]).

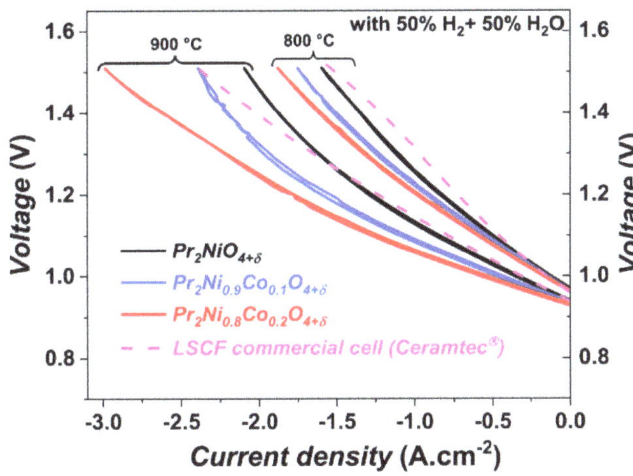

Figure 3.13. SOEC current–voltage analysis of hydrogen electrode-supported cells with $Pr_2(Ni,Co)O_{4+\delta}$ positrodes operating at 800 °C and 900 °C with 50% H_2O (reproduced from [122]).

and 900 °C (figure 3.13). Very encouraging results were also obtained for (La, Pr)$_2$NiO$_{4+\delta}$ when it was used as the SOEC positrode, while strong degradation occurred in SOFC mode [89]. Laguna-Bercero *et al* [108, 123] investigated $Pr_2NiO_{4+\delta}$ electrodes in reversibly operated microtubular cells and obtained quite promising results using a $Pr_2NiO_{4+\delta}$-GDC composite functional/barrier layer.

3.6 Concluding remarks

Perovskites are by far the most applied material class for positrodes in solid oxide cells. The vast extent of their compositional variability makes them very flexible materials that can be tailored to the specific requirements of the application. Most perovskite materials used as positrodes are mixed ionic–electronic conductors with fast oxygen transport and high electrocatalytic activity for the oxygen reduction/oxidation reactions. High ionic conductivity is especially important for SOEC anode materials, as large gas pressures can evolve at the electrode–electrolyte interface if oxygen cannot be liberated fast enough, resulting in delamination issues. Layered rare-earth nickelates are an alternative material class; they are suitable for positrode applications in solid oxide cells and have similar materials properties to those of perovskite compounds. They may be especially promising for SOEC applications, since their ionic conductivity is carried by oxygen interstitials as opposed to the vacancy-based oxygen transport in perovskites. Both classes of materials have high flexibility in terms of doping, which allows one to tailor the required properties for a specific application.

Acknowledgments

Christian Berger extends thanks to R Merkle, G Raimondi, M F Hoedl, and J Maier (MPI Stuttgart) for numerous valuable discussions. Andreas Egger acknowledges the support of the Austrian Research Promotion Agency FFG (project HydroMetha, no. 864578). The financial support of the 'Klima- und Energiefonds' and AVL List GmbH (Austria) within the program 'Energieforschung (e!MISSION)' is also gratefully acknowledged.

References

[1] Jiang S P 2019 Development of lanthanum strontium cobalt ferrite perovskite electrodes of solid oxide fuel cells—a review *Int. J. Hydrogen Energy* **44** 7448–93

[2] Rose G 1839 Beschreibung einiger neuen mineralien des urals *Ann. Phys.* **124** 551–73

[3] Goldschmidt V M 1926 Die Gesetze der Krystallochemie *Naturwissenschaften* **14** 477–85

[4] Megaw H D 1945 Crystal structure of barium titanate *Nature* **155** 484–5

[5] Yin W-J, Weng B, Ge J, Sun Q, Li Z and Yan Y 2019 Oxide perovskites, double perovskites and derivatives for electrocatalysis, photocatalysis, and photovoltaics *Energy Environ. Sci.* **12** 442–62

[6] Li C, Hao H, Xu B, Zhao G, Chen L, Zhang S and Liu H 2020 A progressive learning method for predicting the band gap of ABO_3 perovskites using an instrumental variable *J. Mater. Chem.* C **8** 3127–36

[7] Sun Z, Li Z, Cheng G, Xu Q, Hou Z and Yin W 2019 Research progress and perspective of machine learning in material *Chin. Sci. Bull.* **64** 3270–5

[8] Tao Q, Xu P, Li M and Lu W 2021 Machine learning for perovskite materials design and discovery *NPJ Comput. Mater.* **7** 23

[9] Xiang W, Liu S and Tress W 2021 A review on the stability of inorganic metal halide perovskites: challenges and opportunities for stable solar cells *Energy Environ. Sci.* **14** 2090–113

[10] Gautason B and Muehlenbachs K 1993 Oxygen diffusion in perovskite: implications for electrical conductivity in the lower mantle *Science* **260** 518–21

[11] Ishihara T, Eto H, Zhong H and Matsumoto H 2009 Intermediate temperature solid oxide fuel cells using $LaGaO_3$ based perovskite oxide for electrolyte *Electrochemistry* **77** 115–22

[12] Iwahara H 1992 Oxide-ionic and protonic conductors based on perovskite-type oxides and their possible applications *Solid State Ionics* **52** 99–104

[13] Kreuer K D, Adams S, Münch W, Fuchs A, Klock U and Maier J 2001 Proton conducting alkaline earth zirconates and titanates for high drain electrochemical applications *Solid State Ionics* **145** 295–306

[14] Huang K, Wan J and Goodenough J B 2001 Oxide-ion conducting ceramics for solid oxide fuel cells *J. Mater. Sci.* **36** 1093–8

[15] Geary T C and Adler S B 2013 Oxygen nonstoichiometry and defect chemistry of the mixed conductor $La_{0.9}Ca_{0.1}FeO_{3-\delta}$ at low oxygen partial pressure *Solid State Ionics* **253** 88–93

[16] Horita T, Yamaji K, Sakai N, Yokokawa H, Weber A and Ivers-Tiffée E 2001 Electrode reaction of $La_{1-x}Sr_xCoO_{3-\delta}$ Cathodes on $La_{0.8}Sr_{0.2}Ga_{0.8}Mg_{0.2}O_{3-y}$ electrolyte in solid oxide fuel cells *J. Electrochem. Soc.* **148** A456–62

[17] Shannon R D 1976 Revised effective ionic radii and systematic studies of interatomic distances in halides and chalcogenides *Acta Crystallographica* A **32** 751–67

[18] Švarcová S, Wiik K, Tolchard J, Bouwmeester H J M and Grande T 2008 Structural instability of cubic perovskite $Ba_xSr_{1-x}Co_{1-y}Fe_yO_{3-\delta}$ *Solid State Ionics* **178** 1787–91

[19] Kröger F A and Vink H J 1956 *Solid State Phys.* ed F Seitz and D Turnbull (New York: Academic) pp 307–435

[20] Kreuer K-D 2013 *Fuel Cells – Selected Entries from the Encyclopedia of Sustainability Science and Technology* (New York: Springer)

[21] Petrov A N, Cherepanov V A and Zuev A Y 2006 Thermodynamics, defect structure, and charge transfer in doped lanthanum cobaltites: an overview *J. Solid State Electrochem.* **10** 517–37

[22] Egger A, Bucher E, Yang M and Sitte W 2012 Comparison of oxygen exchange kinetics of the IT-SOFC cathode materials $La_{0.5}Sr_{0.5}CoO_{3-\delta}$ and $La_{0.6}Sr_{0.4}CoO_{3-\delta}$ *Solid State Ionics* **225** 55–60

[23] Rothschild A, Menesklou W, Tuller H L and Ivers-Tiffée E 2006 Electronic structure, defect chemistry, and transport properties of $SrTi_{1-x}Fe_xO_{3-y}$ solid solutions *Chem. Mater.* **18** 3651–9

[24] Brouwer G 1954 A General asymmetic solution of reaction equations common in solid state chemistry *Philips Res. Rept.* **9** 366–76

[25] Höfer H E and Schmidberger R 1994 Electronic conductivity in the $La(Cr,Ni)O_3$ perovskite system *J. Electrochem. Soc.* **141** 782–6

[26] Øygarden V, Lein H L and Grande T 2012 Structure, thermal expansion and electrical conductivity of Nb-substituted $LaCoO_3$ *J. Solid State Chem.* **192** 246–54

[27] Mineshige A, Kobune M, Fujii S, Ogumi Z, Inaba M, Yao T and Kikuchi K 1999 Metal–insulator transition and crystal structure of $La_{1-x}Sr_xCoO_3$ as functions of Sr-content, temperature, and oxygen partial pressure *J. Solid State Chem.* **142** 374–81

[28] Bucher E and Sitte W 2004 Defect chemical analysis of the electronic conductivity of strontium-substituted lanthanum ferrite *Solid State Ionics* **173** 23–8

[29] Hung M-H, Rao M V M and Tsai D-S 2007 Microstructures and electrical properties of calcium substituted $LaFeO_3$ as SOFC cathode *Mater. Chem. Phys.* **101** 297–302

[30] Shen Y, Zhao H, Liu X and Xu N 2010 Preparation and electrical properties of Ca-doped $La_2NiO_{4+\delta}$ cathode materials for IT-SOFC *Phys. Chem. Chem. Phys.* **12** 15124–31

[31] Berger C, Bucher E and Sitte W 2017 Mass and charge transport properties of $La_{0.9}Ca_{0.1}FeO_{3-\delta}$ *Solid State Ionics* **299** 46–54

[32] Berger C, Bucher E, Windischbacher A, Boese A D and Sitte W 2018 Strontium-free rare earth perovskite ferrites with fast oxygen exchange kinetics: experiment and theory *J. Solid State Chem.* **259** 57–66

[33] Kozuka H, Yamagiwa K, Ohbayashi K and Koumoto K 2012 Origin of high electrical conductivity in alkaline-earth doped $LaCoO_3$ *J. Mater. Chem.* **22** 11003–5

[34] Gómez S Y and Hotza D 2016 Current developments in reversible solid oxide fuel cells *Renew. Sustain. Energy Rev.* **61** 155–74

[35] Crank J 2011 *The Mathematics of Diffusion* (Oxford: Oxford University Press)

[36] Tilley R J D 1987 *Defect Crystal Chemistry and Its Applications* (Glasgow: Blackie)

[37] Maier J 2004 *Physical Chemistry of Ionic Materials – Ions and Electrons in Solids* (Chichester: Wiley)

[38] Mehrer H 2007 *Diffusion in Solids* vol 1 (Berlin: Springer)

[39] Cambridge U o 2004-22 Dissemination of IT for the Promotion of Materials Science (DoITPoMS) https://doitpoms.ac.uk/tlplib/diffusion/diffusion_mechanism.php

[40] Berger C, Bucher E, Gspan C and Sitte W 2019 Crystal structure, oxygen nonstoichiometry, and mass and charge transport properties of the Sr-free SOFC/SOEC air electrode material $La_{0.75}Ca_{0.25}FeO_{3-\delta}$ *J. Solid State Chem.* **273** 92–100

[41] Bucher E, Egger A, Ried P, Sitte W and Holtappels P 2008 Oxygen nonstoichiometry and exchange kinetics of $Ba_{0.5}Sr_{0.5}Co_{0.8}Fe_{0.2}O_{3-\delta}$ *Solid State Ionics* **179** 1032–5

[42] ten Elshof J E, Lankhorst M H R and Bouwmeester H J M 1997 Chemical diffusion and oxygen exchange of $La_{0.6}Sr_{0.4}Co_{0.6}Fe_{0.4}O_{3-\delta}$ *Solid State Ionics* **99** 15–22

[43] Wang S, van der Heide P A W, Chavez C, Jacobson A J and Adler S B 2003 An electrical conductivity relaxation study of $La_{0.6}Sr_{0.4}Fe_{0.8}Co_{0.2}O_{3-\delta}$ *Solid State Ionics* **156** 201–8

[44] ten Elshof J E, Lankhorst M H R and Bouwmeester H J M 1997 Oxygen exchange and diffusion coefficients of strontium-doped lanthanum ferrites by electrical conductivity relaxation *J. Electrochem. Soc.* **144** 1060–7

[45] Perry N H, Kim J J and Tuller H L 2018 Oxygen surface exchange kinetics measurement by simultaneous optical transmission relaxation and impedance spectroscopy: $Sr(Ti,Fe)O_{3-x}$ thin film case study *Sci. Technol. Adv. Mater.* **19** 130–41

[46] Claus J, Leonhardt M and Maier J 2000 Tracer diffusion and chemical diffusion of oxygen in acceptor doped $SrTiO_3$ *J. Phys. Chem. Solids* **61** 1199–207

[47] Kingery W D, Pappis J, Doty M E and Hill D C 1992 Oxygen lon mobility in cubic $Zr_{0.85}Ca_{0.15}O_{1.85}$ *J. Am. Ceram. Soc.* **75** 483–8

[48] Maier J 1993 Mass transport in the presence of internal defect reactions—concept of conservative ensembles: II, evaluation of electrochemical transport measurements *J. Am. Ceram. Soc.* **76** 1218–22

[49] De Souza R A and Kilner J A 1998 Oxygen transport in $La_{1-x}Sr_xMn_{1-y}Co_yO_{3\pm\delta}$ perovskites: Part I. oxygen tracer diffusion *Solid State Ionics* **106** 175–87

[50] Preis W, Holzinger M and Sitte W 2001 Application of the van der Pauw method to conductivity relaxation experiments on $YBa_2Cu_3O_{6+\delta}$ *Monatsh. Chem.* **132** 499–508

[51] den Otter M W, Bouwmeester H J M, Boukamp B A and Verweij H 2001 Reactor flush time correction in relaxation experiments *J. Electrochem. Soc.* **148** J1–6

[52] Preis W, Bucher E and Sitte W 2002 Oxygen exchange measurements on perovskites as cathode materials for solid oxide fuel cells *J. Power Sources* **106** 116–21

[53] Preis W, Bucher E and Sitte W 2004 Oxygen exchange kinetics of $La_{0.4}Sr_{0.6}FeO_{3-\delta}$ by simultaneous application of conductivity relaxation and carrier gas coulometry *Solid State Ionics* **175** 393–7

[54] Wagner C 1959 Ionen- und elektronenleitung in Silberbromid und Abweichungen von der idealen stöchiometrischen Zusammensetzung *Zeitschrift für Elektrochemie, Berichte der Bunsengesellschaft für physikalische Chemie* **63** 1027–30

[55] Hebb M H 1952 Electrical conductivity of silver sulfide *J. Chem. Phys.* **20** 185–90

[56] Ishigaki T, Yamauchi S, Kishio K, Mizusaki J and Fueki K 1988 Diffusion of oxide ion vacancies in perovskite-type oxides *J. Solid State Chem.* **73** 179–87

[57] Benson S J, Chater R J and Kilner J A 1998 Oxygen diffusion and surface exchange in the mixed conducting perovskite $La_{0.6}Sr_{0.4}Fe_{0.8}Co_{0.2}O_{3-\delta}$ *Proc. 3rd Int. Symp. on Ionic and Mixed Conducting Ceramics* ed T A Ramanarayanan *et al* (Paris: The Electrochemical Society) pp 596–609

[58] Kilner J A and Shaw C K M 2002 Mass transport in $La_2Ni_{1-x}Co_xO_{4+\delta}$ oxides with the K_2NiF_4 structure *Solid State Ionics* **154** 523–7

[59] Boehm E, Bassat J M, Dordor P, Mauvy F, Grenier J C and Stevens P 2005 Oxygen diffusion and transport properties in non-stoichiometric $Ln_{2-x}NiO_{4+\delta}$ oxides *Solid State Ionics* **176** 2717–25

[60] Hyodo J, Tominaga K, Ju Y-W, Ida S and Ishihara T 2014 Electrical conductivity and oxygen diffusivity in Cu- and Ga-doped Pr_2NiO_4 *Solid State Ionics* **256** 5–10

[61] Yokokawa H 2003 Understanding materials compatibility *Annu. Rev. Mater. Res.* **33** 581–610

[62] Simner S, Anderson M, Bonnett J and Stevenson J 2004 Enhanced low temperature sintering of (Sr, Cu)-doped lanthanum ferrite SOFC cathodes *Solid State Ionics* **175** 79–81

[63] Anderson M D, Stevenson J W and Simner S P 2004 Reactivity of lanthanide ferrite SOFC cathodes with YSZ electrolyte *J. Power Sources* **129** 188–92

[64] Wang D, Wang J, He C, Tao Y, Xu C and Wang W G 2010 Preparation of a $Gd_{0.1}Ce_{0.9}O_{2-\delta}$ interlayer for intermediate-temperature solid oxide fuel cells by spray coating *J. Alloys Compd.* **505** 118–24

[65] Wilde V, Störmer H, Szász J, Wankmüller F, Ivers-Tiffée E and Gerthsen D 2018 $Gd_{0.2}Ce_{0.8}O_2$ diffusion barrier layer between $(La_{0.58}Sr_{0.4})(Co_{0.2}Fe_{0.8})O_{3-\delta}$ cathode and $Y_{0.16}Zr_{0.84}O_2$ electrolyte for solid oxide fuel cells: effect of barrier layer sintering temperature on microstructure *ACS Appl. Energy Mater.* **1** 6790–800

[66] Seeharaj P, Berenov A, Raj E, Rudkin R and Atkinson A 2011 Mixed-conducting LSC/CGO composites for passive oxygen separation membranes *Solid State Ionics* **192** 638–41

[67] Fawcett L, Kilner J A and Skinner S J 2012 Compatibility and electrochemical behavior of $La_2NiO_{4+\delta}$ on $La_{0.8}Sr_{0.2}Ga_{0.8}Mg_{0.2}O_3$ *10th European SOFC Forum* ed F Lefebvre-Joud (Lucerne, Switzerland: European Fuel Cell Forum AG) B0409

[68] Balz D and Plieth K 1955 Die Struktur des Kaliumnickelfluorids, K_2NiF_4 *Zeitschrift für Elektrochemie* **59** 545–51

[69] Bednorz J G and Müller K A 1988 Perovskite-type oxides—the new approach to high-T_c superconductivity *Rev. Mod. Phys.* **60** 585–600

[70] Ruddlesden S N and Popper P 1957 New compounds of the K_2NiF_4 type *Acta Crystallogr.* **10** 538–9

[71] Ruddlesden S N and Popper P 1958 The compound $Sr_3Ti_2O_7$ and its structure *Acta Crystallogr.* **11** 54–5

[72] Greenblatt M 1997 Ruddlesden–Popper $Ln_{n+1}Ni_nO_{3n+1}$ nickelates: structure and properties *Curr. Opin. Solid State Mater. Sci.* **2** 174–83

[73] Ram R A M, Ganapathi L, Ganguly P and Rao C N R 1986 Evolution of three-dimensional character across the $La_{n+1}Ni_nO_{3n+1}$ homologous series with increase in *n J. Solid State Chem.* **63** 139–47

[74] Egger A 2013 Rare earth nickelates as cathodes for solid oxide fuel cells, *PhD Thesis* University of Leoben

[75] Jorgensen J D, Dabrowski B, Pei S, Richards D R and Hinks D G 1989 Structure of the interstitial oxygen defect in $La_2NiO_{4+\delta}$ *Phys. Rev.* **B 40** 2187

[76] Tsipis E V, Naumovich E N, Shaula A L, Patrakeev M V, Waerenborgh J C and Kharton V V 2008 Oxygen nonstoichiometry and ionic transport in $La_2Ni(Fe)O_{4+\delta}$ *Solid State Ionics* **179** 57–60

[77] Bassat J M, Odier P, Villesuzanne A, Marin C and Pouchard M 2004 Anisotropic ionic transport properties in $La_2NiO_{4+\delta}$ single crystals *Solid State Ionics* **167** 341–7

[78] Chroneos A, Parfitt D, Kilner J A and Grimes R W 2010 Anisotropic oxygen diffusion in tetragonal $La_2NiO_{4+\delta}$: molecular dynamics calculations *J. Mater. Chem.* **20** 266–70

[79] Ganguly P and Rao C N R 1984 Crystal chemistry and magnetic properties of layered metal oxides possessing the K_2NiF_4 or related structures *J. Solid State Chem.* **53** 193–216

[80] Li Q, Fan Y, Zhao H, Sun L P and Huo L H 2007 Preparation and electrochemical properties of a $Sm_{2-x}Sr_xNiO_4$ cathode for an IT-SOFC *J. Power Sources* **167** 64–8

[81] Pikalova E, Kolchugin A, Zakharchuk K, Boiba D, Tsvinkinberg V, Filonova E, Khrustov A and Yaremchenko A 2021 Mixed ionic–electronic conductivity, phase stability and electrochemical activity of Gd-substituted $La_2NiO_{4+\delta}$ as oxygen electrode material for solid oxide fuel/electrolysis cells *Int. J. Hydrogen Energy* **46** 16932–46

[82] Egger A, Bucher E and Sitte W 2011 Oxygen exchange kinetics of the IT-SOFC cathode material $Nd_2NiO_{4+\delta}$ and comparison with $La_{0.6}Sr_{0.4}CoO_{3-\delta}$ *J. Electrochem. Soc.* **158**

[83] Sullivan J D, Buttrey D J, Cox D E and Hriljac J 1991 A conventional and high-resolution synchrotron X-ray diffraction study of phase separations in $Pr_2NiO_{4+\delta}$ *J. Solid State Chem.* **94** 337–51

[84] Ogier T, Prestipino C, Figueroa S, Mauvy F, Mougin J, Grenier J C, Demourgues A and Bassat J M 2019 *In-situ* study of cationic oxidation states in $Pr_2NiO_{4+\delta}$ using X-ray absorption near-edge spectroscopy *Chem. Phys. Lett.* **727** 116–20

[85] Nicollet C, Flura A, Vibhu V, Rougier A, Bassat J-M and Grenier J-C 2016 An innovative efficient oxygen electrode for SOFC: Pr_6O_{11} infiltrated into Gd-doped ceria backbone *Int. J. Hydrogen Energy* **41** 15538–44

[86] Vibhu V, Rougier A, Grenier J C and Bassat J M 2013 Mixed nickelates $Pr_{2-x}La_xNiO_{4+\delta}$ used as cathodes in metal supported SOFCs *ECS Trans.* **57** 2093–100

[87] Vibhu V, Rougier A, Nicollet C, Flura A, Grenier J-C and Bassat J-M 2015 $La_{2-x}PrxNiO_{4+\delta}$ as suitable cathodes for metal supported SOFCs *Solid State Ionics* **278** 32–7

[88] Vibhu V, Bassat J-M, Flura A, Nicollet C, Grenier J-C and Rougier A 2015 Influence of La/Pr ratio on the ageing properties of $La_{2-x}Pr_xNiO_{4+\delta}$ as cathodes in IT-SOFCs *ECS Trans.* **68** 825–35

[89] Vibhu V, Flura A, Rougier A, Nicollet C, Fourcade S, Hungria T, Grenier J-C and Bassat J-M 2020 Electrochemical ageing study of mixed lanthanum/praseodymium nickelates

$La_{2-x}Pr_xNiO_{4+\delta}$ as oxygen electrodes for solid oxide fuel or electrolysis cells *J. Energy Chem.* **46** 62–70

[90] Sharma R K, Cheah S-K, Burriel M, Dessemond L, Bassat J-M and Djurado E 2017 Design of $La_{2-x}Pr_xNiO_{4+\delta}$ SOFC cathodes: a compromise between electrochemical performance and thermodynamic stability *J. Mater. Chem.* A **5** 1120–32

[91] Montenegro-Hernández A, Vega-Castillo J, Caneiro A and Mogni L 2019 High temperature orthorhombic/tetragonal transition and oxygen content of $Pr_{2-x}Nd_xNiO_{4+\delta}$ ($x = 0$, 0.3, 1, 1.7 and 2) solid solutions *J. Solid State Chem.* **276** 210–6

[92] Dogdibegovic E, Guan W, Yan J, Cheng M and Zhou X-D 2016 Activity and stability of $(Pr_{1-x}Nd_x)_2NiO_4$ as cathodes for solid oxide fuel cells: II. Electrochemical performance and performance durability *J. Electrochem. Soc.* **163** F1344–F9

[93] Bhoga S S, Khandale A P and Pahune B S 2014 Investigation on $Pr_{2-x}Sr_xNiO_{4+\delta}$ ($x = 0.3$–1.0) cathode materials for intermediate temperature solid oxide fuel cell *Solid State Ionics* **262** 340–4

[94] Pikalova E, Kolchugin A, Bogdanovich N, Medvedev D, Lyagaeva J, Vedmid L, Ananyev M, Plaksin S and Farlenkov A 2020 Suitability of $Pr_{2-x}Ca_xNiO_{4+\delta}$ as cathode materials for electrochemical devices based on oxygen ion and proton conducting solid state electrolytes *Int. J. Hydrogen Energy* **45** 13612–24

[95] Kuterbekov K A, Bekmyrza K Z, Aidarbekov N K, Nikonov A V, Pavzderin N B, Khrustov V R, Semenova I V, Medvedev A I and Gyrdasova O I 2021 Effect of the cobalt content on properties of $La_2Ni_{1-x}Co_xO_{4+\delta}$ *Solid State Ionics* **363** 115594

[96] Vibhu V, Vinke I C, Eichel R-A, Bassat J-M and De Haart L G J 2019 Cobalt substituted lanthanide nickelates ($Ln_2Ni_{1-x}Co_xO_{4+\delta}$, Ln = La, Pr; $x = 0$, 0.1, 0.2) as high performance oxygen electrodes for solid oxide cells *ECS Trans.* **91** 1327–39

[97] Vibhu V, Vinke I C, Eichel R A, Bassat J M and de Haart L G J 2019 $La_2Ni_{1-x}Co_xO_{4+\delta}$ ($x = 0.0$, 0.1 and 0.2) based efficient oxygen electrode materials for solid oxide electrolysis cells *J. Power Sources* **444** 227292

[98] Boehm E, Bassat J M, Steil M C, Dordor P, Mauvy F and Grenier J C 2003 Oxygen transport properties of $La_2Ni_{1-x}Cu_xO_{4+\delta}$ mixed conducting oxides *Solid State Sci.* **5** 973–81

[99] Klande T, Efimov K, Cusenza S, Becker K-D and Feldhoff A 2011 Effect of doping, microstructure, and CO_2 on $La_2NiO_{4+\delta}$-based oxygen-transporting materials *J. Solid State Chem.* **184** 3310–8

[100] Kharton V V, Tsipis E V, Naumovich E N, Thursfield A, Patrakeev M V, Kolotygin V A, Waerenborgh J C and Metcalfe I S 2008 Mixed conductivity, oxygen permeability and redox behavior of K_2NiF_4-type $La_2Ni_{0.9}Fe_{0.1}O_{4+\delta}$ *J. Solid State Chem.* **181** 1425–33

[101] Szász J, Wankmüller F, Wilde V, Störmer H, Gerthsen D, Menzler N H and Ivers-Tiffée E 2018 Nature and functionality of $La_{0.58}Sr_{0.4}Co_{0.2}Fe_{0.8}O_{3-\delta}$ /$Gd_{0.2}Ce_{0.8}O_{2-\delta}$ /$Y_{0.16}Zr_{0.84}O_{2-\delta}$ interfaces in SOFCs *J. Electrochem. Soc.* **165** F898–906

[102] Montenegro-Hernández A, Vega-Castillo J, Mogni L and Caneiro A 2011 Thermal stability of $Ln_2NiO_{4+\delta}$ (Ln: La, Pr, Nd) and their chemical compatibility with YSZ and CGO solid electrolytes *Int. J. Hydrogen Energy* **36** 15704–14

[103] Schrödl N, Egger A, Lammer J, Hofer F and Sitte W 2021 Long-term stability of $Pr_2NiO_{4+\delta}$ air electrodes for solid oxide cells against chromium poisoning *J. Electrochem. Soc.* **168** 014509

[104] Sadykov V, Okhlupin Y, Yeremeev N, Vinokurov Z, Shmakov A, Belyaev V, Uvarov N and Mertens J 2014 *In situ* X-ray diffraction studies of $Pr_{2-x}NiO_{4+\delta}$ crystal structure relaxation caused by oxygen loss *Solid State Ionics* **262** 918–22

[105] Sayers R, Liu J, Rustumji B and Skinner S J 2008 Novel K_2NiF_4-type materials for solid oxide fuel cells: compatibility with electrolytes in the intermediate temperature eange *Fuel Cells* **8** 338–43

[106] Solak N, Zinkevich M and Aldinger F 2006 Compatibility of La_2NiO_4 cathodes with $LaGaO_3$ electrolytes: a computational approach *Solid State Ionics* **177** 2139–42

[107] Mauvy F, Lalanne C, Bassat J M, Grenier J C, Zhao H, Huo L H and Stevens P 2006 Electrode properties of $Ln_2NiO_{4+\delta}$ (Ln = La, Nd, Pr) – AC impedance and DC polarization studies *J. Electrochem. Soc.* **153** A1547–A53

[108] Laguna-Bercero M A, Monzon H, Larrea A and Orera V M 2016 Improved stability of reversible solid oxide cells with a nickelate-based oxygen electrode *J. Mater. Chem.* A **4** 1446–53

[109] Dogdibegovic E, Alabri N S, Wright C J, Hardy J S, Coyle C A, Horlick S A, Guan W, Stevenson J W and Zhou X-D 2017 Activity and stability of $(Pr_{1-x}Nd_x)_2NiO_4$ as cathodes for solid oxide fuel cells: Part V. *in situ* studies of phase evolution *J. Electrochem. Soc.* **164** F1115–F21

[110] Dogdibegovic E, Nawf S A, Wright C J, Hardy J S, Coyle C A, Horlick S A, Guan W, Stevenson J W and Zhou X-D 2017 *In situ* studies of phase evolution in $(Pr_{1-x}Nd_x)_2NiO_4$ electrodes with various interlayer chemistries *ECS Trans.* **78** 623–35

[111] Dogdibegovic E, Nawf S A, Tenny K, Wright C J, Hardy J S, Coyle C A, Horlick S A, Guan W, Stevenson J W and Zhou X-D 2017 The role of interlayer on the catalytic activity and performance stability of $(Pr_{1-x}Nd_x)_2NiO_4$ as cathodes for solid oxide fuel cells *ECS Trans.* **78** 983–92

[112] Chiba R, Taguchi H, Komatsu T, Orui H, Nozawa K and Arai H 2011 High temperature properties of $Ce_{1-x}Pr_xO_{2-\delta}$ as an active layer material for SOFC cathodes *Solid State Ionics* **197** 42–8

[113] Vibhu V, Rougier A, Nicollet C, Flura A, Fourcade S, Penin N, Grenier J-C and Bassat J-M 2016 $Pr_4Ni_3O_{10+\delta}$: a new promising oxygen electrode material for solid oxide fuel cells *J. Power Sources* **317** 184–93

[114] Zheng Y, Wang J, Yu B, Zhang W, Chen J, Qiao J and Zhang J 2017 A review of high temperature co-electrolysis of H_2O and CO_2 to produce sustainable fuels using solid oxide electrolysis cells (SOECs): advanced materials and technology *Chem. Soc. Rev.* **46** 1427–63

[115] Nechache A and Hody S 2021 Alternative and innovative solid oxide electrolysis cell materials: a short review *Renew. Sustain. Energy Rev.* **149** 111322

[116] Pandiyan A, Uthayakumar A, Subrayan R, Cha S-W and Krishna Moorthy S B 2019 Review on solid oxide electrolysis cell: a clean energy strategy for hydrogen generation *Nanomater. Energy* **8** 2–22

[117] Laguna-Bercero M A 2012 Recent advances in high temperature electrolysis using solid oxide fuel cells: a review *J. Power Sources* **203** 4–16

[118] Tarutin A P, Lyagaeva J G, Medvedev D A, Bi L and Yaremchenko A A 2021 Recent advances in layered $Ln_2NiO_{4+\delta}$ nickelates: fundamentals and prospects of their applications in protonic ceramic fuel and electrolysis cells *J. Mater. Chem.* A **9** 154–95

[119] Ogier T, Bassat J M, Mauvy F, Fourcade S, Grenier J C, Couturier K, Petitjean M and Mougin J 2013 Enhanced performances of structured oxygen electrodes for high temperature steam electrolysis *Fuel Cells* **13** 536–41

[120] Chauveau F, Mougin J, Bassat J M, Mauvy F and Grenier J C 2010 A new anode material for solid oxide electrolyser: the neodymium nickelate $Nd_2NiO_{4+\delta}$ *J. Power Sources* **195** 744–9

[121] Chauveau F, Mougin J, Mauvy F, Bassat J-M and Grenier J-C 2011 Development and operation of alternative oxygen electrode materials for hydrogen production by high temperature steam electrolysis *Int. J. Hydrogen Energy* **36** 7785–90

[122] Vibhu V, Vinke I C, Eichel R A and de Haart L G J 2021 Cobalt substituted $Pr_2Ni_{1-x}Co_xO_{4+\delta}$ ($x = 0$, 0.1, 0.2) oxygen electrodes: Impact on electrochemical performance and durability of solid oxide electrolysis cells *J. Power Sources* **482** 228909

[123] Morales-Zapata M A, Larrea A and Laguna-Bercero M A 2020 Reversible operation performance of microtubular solid oxide cells with a nickelate-based oxygen electrode *Int. J. Hydrogen Energy* **45** 5535–42

IOP Publishing

High-Temperature Electrolysis
From fundamentals to applications
Werner Sitte and Rotraut Merkle

Chapter 4

Cathode materials for solid oxide electrolysis cells

Peter Holtappels, John T S Irvine and Shu Wang

The cathode in solid oxide electrolysis cells (SOECs) is the electrode at which water and carbon dioxide can be split into hydrogen, carbon monoxide, and oxygen ions. The state-of-the-art material is a Ni cermet, as developed for solid oxide fuel cells (SOFCs). This electrode shows promising electrolysis rates and compatibility with adjacent cell components, and a basic understanding of its electrode processes has also been gained. However, stability issues have recently triggered a search for alternative materials. Electronically conducting oxides, with or without activation by submicron-scale electrocatalysts, are of interest because of their stability in high water contents, as described in the last part of this chapter.

4.1 Fuel electrode processes and requirements

The cathode (also called the fuel electrode or negatrode) in SOECs is the electrode at which water (and/or carbon dioxide) is reduced and split into hydrogen (and/or carbon monoxide) and oxygen ions, which are incorporated into the solid, oxygen-ion-conducting electrolyte and then transported to the SOEC anode (the air electrode or positrode), where they can form oxygen.

The water splitting reaction at the SOEC cathode can be described by

$$H_2O_g + V_O^{\bullet\bullet} + 2e^- \Leftrightarrow H_{2,g} + O_O^x, \quad (4.1)$$

in which H_2O_g and $H_{2,g}$ denote water vapor and hydrogen gas, respectively. O_O^x represents an oxygen ion (oxidation state II) on an oxygen site in the oxide electrolyte lattice, $V_O^{\bullet\bullet}$ is a vacant oxygen site in the electrolyte lattice (using Kröger–Vink nomenclature, see chapter 2).

The corresponding electrochemical reaction for CO_2 reduction is:

$$CO_{2,g} + V_O^{\bullet\bullet} + 2e^- \Leftrightarrow CO_g + O_O^x. \quad (4.2)$$

In the so-called 'co-electrolysis' of water and carbon dioxide, mixtures of H_2 and CO are typically produced and depending on the cathode material, the so-called

doi:10.1088/978-0-7503-3951-3ch4

water gas shift reaction (WGSR) has to be taken into account, which determines the resulting syngas composition:

$$CO_g + H_2O_g \rightleftharpoons CO_{2,g} + H_{2,g}. \tag{4.3}$$

The development of SOEC fuel electrodes builds upon SOFC development, as the solid oxide cell can be reversibly operated in both fuel cell and electrolyzer modes. Thus, materials developed for solid oxide fuel cells are also used for solid electrolysis cells. Two different fuel electrode architectures have been realized as a result of SOFC development, which depend on the electrical properties of the fuel electrode material.

Figure 4.1 illustrates the electrochemical reactions in the fuel electrode. The conversion of chemical energy into electrical energy takes place where the oxygen ions, the electrons in the electrode, and the gas phase are in contact with each other. In case of a pure electronically conducting electrode material, e.g. Ni-cermet electrodes, the contact zone is the so-called 'triple-phase boundary line' (TPL). Depending on the nature of the electrode material, the TPL is located around the contact line at the electrode/solid electrolyte interface, distributed in the porous electrode, or along the surface of the electrode material. The latter case is assumed for mixed oxygen ion and electronically conducting materials, which makes large parts of the electrode active for the surface reaction.

In order to function properly for the electrochemical water splitting reaction, the electrodes have to possess the proper microstructure and fulfill additional chemical and physical properties such as:

- Electronic conductivity to transport the electrons involved in the reaction to an outer circuit
- A sufficient degree of porosity to allow transport of the gaseous reactants and products to and from the electrochemically active area

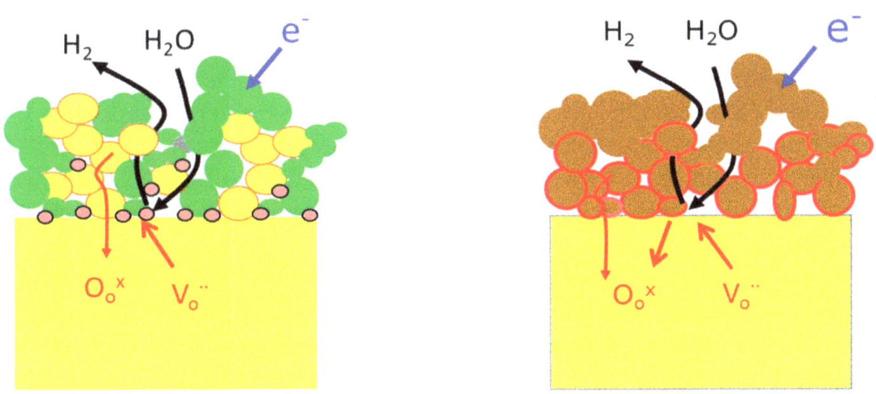

Figure 4.1. Schematic of the electrode processes in the fuel electrode: Electrochemical reaction at the solid-state electrolyte/electrode interface for composite electrodes (e.g. Ni cermet) (left) and mixed electronic–ionic conducting electrodes (right).

- Ideally, oxygen ion transport in the volume of the electrode to create a three-dimensional reaction zone
- The longest possible active area, either the TPL or the surface

The SOEC cathode material has to meet additional requirements, which restricts the choice of SOEC cathode materials. The electrode has to be chemically stable over a wide pO_2 range from reducing conditions in hydrogen and carbon monoxide up to the more oxidizing conditions of high steam/CO_2 feeds but also, for example, during fabrication in an air atmosphere.

The solid oxide cell is a ceramic, all-solid-state multilayer structure, which needs to stay intact during temperature cycles between room temperature and fabrication temperatures of typically more than 1000 °C and during on–off cycles between room temperature and operational temperatures between 600 °C and 900 °C. The coefficient of thermal expansion (CTE) is an important property that has to be matched. Usually, the CTE is adjusted to that of the electrolyte material, to avoid cracking of the solid electrolyte that would result in leakage and destroy the cell.

Furthermore, electrode materials have to be chemically compatible with the electrolyte and other adjacent layers. Since the interface between the electrode and the electrolyte is the central point of energy conversion, interdiffusion between elements and the solid-state reaction must be avoided during both fabrication and operation, as it can inhibit the fuel oxidation and reduction reactions. The following sections present the commonly used materials.

4.2 Ni–YSZ cermet electrodes

Based on the recent development of SOFCs, a widely employed SOEC cathode material is the so-called Ni-cermet electrode. Ni, as a metal, is chemically stable under reducing conditions within the SOEC cathode and highly active in the hydrogen evolution reaction (HER) and the CO_2 reduction reaction (CO_2RR) [1]. At temperatures between 800 °C and 1000 °C, pure nickel metal has a high surface mobility [2], causing it to sinter and agglomerate into a coarse structure with a limited TPL against the electrolyte surface. Furthermore, its coefficient of thermal expansion is far too high. To overcome these drawbacks, the Ni–YSZ cermet was invented by H S Spacil [3]. The Ni cermet is a composite material made of NiO and the electrolyte, typically yttria- or scandia-stabilized zirconia (YSZ and SSZ, respectively), which upon reduction produces Ni and YSZ (or ScYSZ). This results in percolated nickel and ceramic phases as well as pores that allow for electronic conductivity and gas transport within the electrode.

After reduction or the first operation of the NiO–YSZ composite, the NiO is reduced to Ni and the Ni-cermet electrode is formed. The ratio of Ni (in the reduced state) and YSZ has an impact on the electronic conductivity. The electronic conductivity of a simple Ni–YSZ cermet structure with randomly mixed particles of similar size seems to follow the rules of percolation [4, 5]. Gradually varying the Ni content from below 30 vol% of solids to above 50 vol% produces an abrupt increase in the electronic conductivity to around 500–1800 S cm^{-1} in the range of

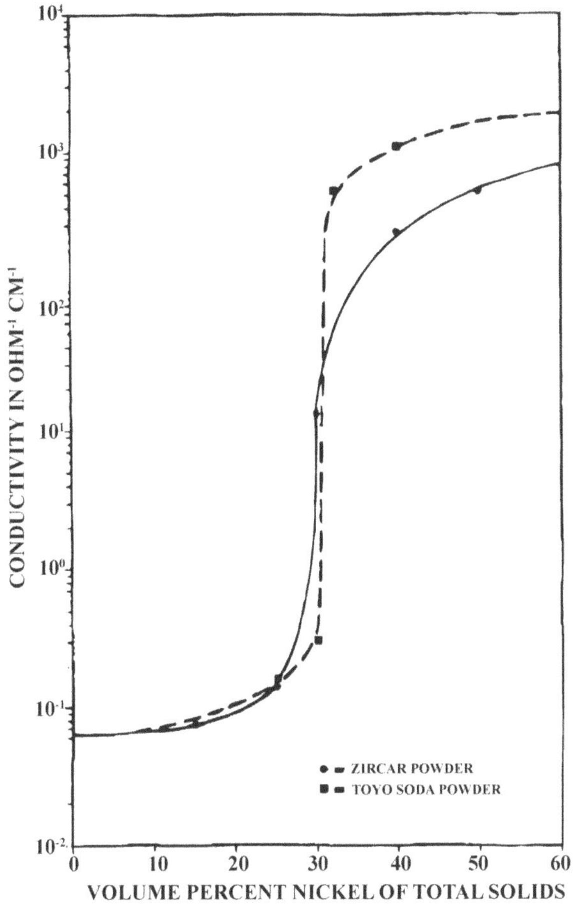

Figure 4.2. Conductivity of two Ni–YSZ cermets as a function of the Ni vol % of total solids at 1000 °C. Reprinted from [5], copyright The Electrochemical Society. Reproduced by permission of IOP Publishing Ltd.

30–40 vol% Ni, as shown in figure 4.2. The exact ratio required to reach percolation also depends on the particle size distribution. For a high fraction of YSZ with much larger particles than the Ni, percolation can be achieved with as little as 15 vol% Ni [6]. Limiting the Ni amount and tailoring the Ni/YSZ ratio is also a possible way to increase the (re-)oxidation tolerance of Ni cermets; however, this is often at the expense of the electrochemically active area due to a reduced TPL.

The composition of the Ni cermet also affects the physical properties, such as the coefficient of thermal expansion. In order to avoid a buildup of destructive stresses and to maintain an intact solid-state cell, it is important to find layers that have suitably matching thermal expansions. While the CTE for YSZ is between $10.6–11.0 \times 10^{-6}\ K^{-1}$ [7], the CTE for NiO is $14 \times 10^{-6}\ K^{-1}$. The development of the CTE as a function of NiO content is shown in figure 4.3. For a Ni–YSZ cermet with 40 vol% NiO, the CTE value was found to be sufficiently close to that of YSZ; this is important during the sintering of the electrode/electrolyte assembly, which is usually done in air. As shown by Müller *et al* [8], the sintering temperature and particle size also affect the CTE match.

For a particle size ratio of NiO/8YSZ = 3:2, the CTE exhibited a minimum of $12.3 \times 10^{-6}\ K^{-1}$, which was attributed to an optimal particle package configuration.

Ni has an even higher CTE of about $17 \times 10^{-6}\ K^{-1}$, but reduced Ni–YSZ cermets show a thermal expansion coefficient close to that of YSZ. It is obvious that a sufficiently strong YSZ network is required to keep the Ni in place and take up stresses arising from the Ni/YSZ CTE mismatch.

The processes of the hydrogen/water reaction in Ni–YSZ cermet electrodes as sketched in figure 4.1 have been investigated for both technical electrodes and model systems. Even though the detailed reaction mechanism is not yet understood, the results clearly indicate that the reaction is indeed limited by the length of the triple-phase boundary line [9, 10]. The fact that the triple-phase boundary is the active reaction site has certain consequences for the electrode design. If we consider Ni spheres closely packed to cover an electrolyte surface, the TPL length is inversely related to the Ni particle radius. The smaller the Ni and electrolyte particles, the longer the TPL.

Based on the above considerations, the Ni-cermet electrodes produced to date have been engineered to have a multilayer structure in order to meet stability and activity requirements (figure 4.4). State-of-the-art Ni–YSZ cermet electrodes consist

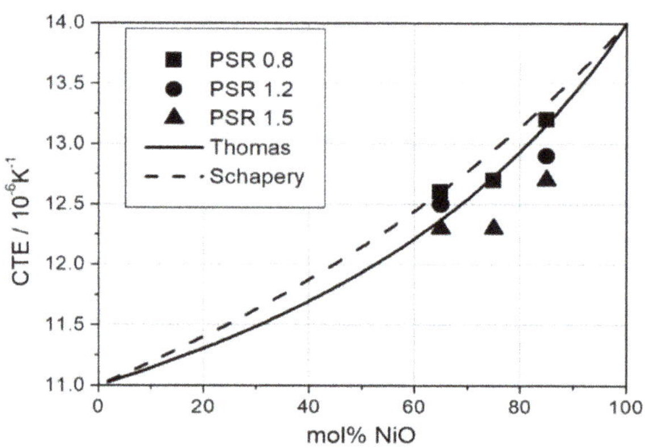

Figure 4.3. Dependence of CTE on NiO content in a mixture with the electrolyte material 8 mol% yttria-stabilized zirconia. Reprinted from [13], copyright (2002), with permission from Elsevier.

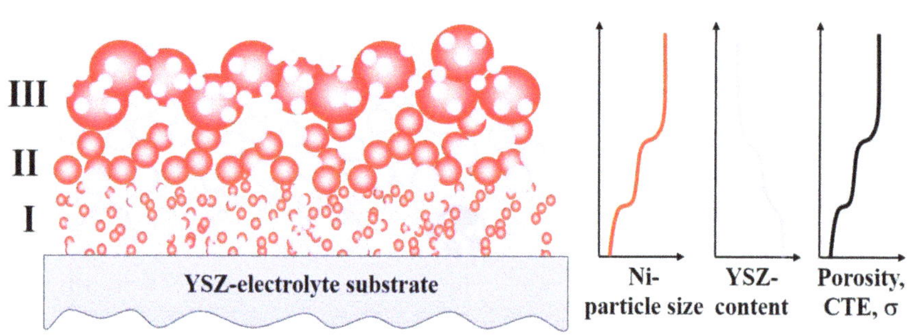

Figure 4.4. Conceptual design of a modern multilayer Ni–YSZ-cermet anode (left); development of particle size, electrolyte (YSZ) content, porosity, CTE, and electronic conductivity, σ, throughout the structure (right). Reprinted from [13] copyright (2002), with permission from Elsevier.

of percolated networks of the three phases (two solids and a gas) to provide triple-phase boundaries (TPBs) throughout the electrode volume, as sketched in figure 4.4(b). Even though, in the original patent, the addition of zirconia particles to the cermet was only intended to prevent the coarsening of the Ni, it turned out that by introducing both electronically and ionically conducting networks, the reaction zone is extended into the volume of the porous electrode. Modeling and experiments have shown that the contribution of the inner electrode volume decreases exponentially with the distance from the solid electrolyte, so that the first 10–20 micrometers can be considered to be the active electrode [11]. A larger TPL increases the current per unit area and is thus inversely proportional to the electrode polarization resistance [12, 13].

The amount and structure of the YSZ affect the Ni agglomeration and thus control the total structure of the composite electrode. Typically, small particles tend to agglomerate more than larger particles and this effect can be utilized to stabilize the cermet structure for reoxidation. Thus, modern Ni-cermet electrodes consist of at least two layers: (1) a fine-grained layer adjacent to the electrolyte and (2) a coarse-grained layer towards the current collector. A gradual transition is often achieved using an additional intermediate layer (see figure 4.4).

A significant fraction of (20%–30%) of fine-grained YSZ (less than 1 μm in diameter) is advantageous to realize a long TPB in the active anode layer in the vicinity of the electrolyte. The even distribution of fine YSZ particles around the Ni particles positively impacts the stability of the Ni-cermet structure, the TPL length, and the conductivity. The porosity also affects the percolation and thus the conductivity. The porosity of the simple cermet with 40 vol% Ni (of the solids) should be lower than about 30%–35% after the reduction of the NiO to Ni [14, 15] in order to retain the connectivity between the Ni particles.

Ni–YSZ cermet has been the state-of-the-art fuel electrode material, which has been developed to a high level of performance for the hydrogen/water reaction. The electrochemical activity of Ni–YSZ electrodes is often expressed as the resistance related to the electrochemical reaction at the open-circuit voltage, the so-called electrode area specific resistance (ASR), which reaches values $\approx 0.15\ \Omega\ cm^2$ at 850 °C and $<0.1\ \Omega\ cm^2$ at 950 °C–1000 °C for the hydrogen/water reaction, including mass transport.

Extensive investigations of the hydrogen reaction in Ni–YSZ cermet electrodes have led to a more advanced understanding of the processes that contribute to polarization resistance in the hydrogen/water reaction. There seems to be general agreement that both the electrochemical reaction at the TPL and also voltage losses due to gas concentration changes induced by diffusion and gas conversion (which is related to the change in the water/hydrogen ratio under load) contribute to the polarization resistance. Curves such as that shown in figure 4.5 are often obtained by electrochemical impedance spectroscopy (EIS) and generally show three well-distinguished contributions. The high-frequency semicircle is usually assigned to the electrochemical reaction at the TPL, while the impedance arc at medium frequencies (peak frequencies at 10–100 Hz) is attributed to mass transport processes such as gas diffusion within the porous electrode. These two contributions are thus influenced by the

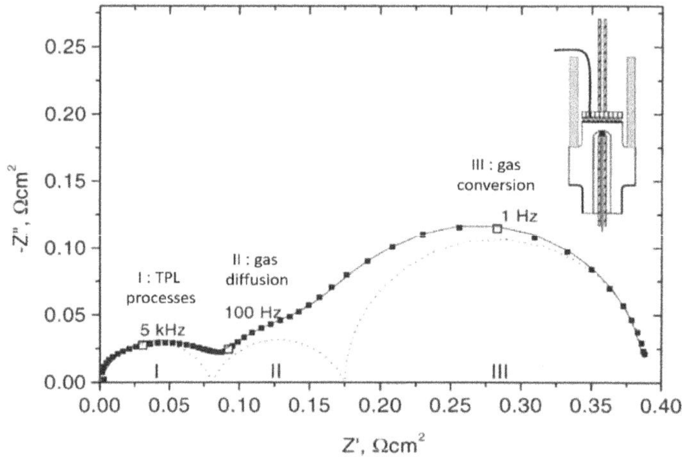

Figure 4.5. Example of an impedance spectrum of technical Ni/8YSZ cermet electrodes in a three-electrode setup at 1000 °C in H_2 + 3% H_2O. Reprinted from [19], copyright The Electrochemical Society. Reproduced by permission of IOP Publishing Ltd.

electrode material and its microstructure. A low-frequency process around (1–10 Hz) is related to the gas composition change outside the porous electrode, and thus depends on the gas flow characteristics in the stack design and test setup.

The lowest reported values of the initial electrode ASR (corrected for mass transport) for the H_2/H_2O/Ni–YSZ electrode are about 20 to 40 mΩ cm^2 at 1000 °C [16]. Typical ranges seen by current SOFC developers are in the range of 40 mΩ cm^2 < R_P < 80 mΩ cm^2 at 850 °C and 60 mΩ cm^2 < R_P < 100 mΩ cm^2 at 750 °C [17, 18].

The porosity of the Ni-cermet electrode is important, as it facilitates gas transport to and from the electrochemically active area. The diffusion resistance of a porous layer 1 mm thick was estimated to be 50 mΩ cm^2, assuming 30% porosity, pores larger than 1 μm, and a tortuosity factor of three [19]. This is around 30% of the acceptable electrode resistance of 0.15 Ω cm^2 [20] and thus might become a significant limitation in high-performance Ni-cermet electrodes.

In order to simplify the Ni–YSZ microstructure for fundamental mechanistic investigations, several attempts have been made to control the length of the TPL using simple, well-defined electrodes, e.g. point electrodes, patterned electrodes, and single-crystal electrolytes. Line specific resistances (LSRs) were obtained from an investigation of the hydrogen/water reaction on model electrodes, as summarized in figure 4.6. The reaction rate, 1/LSR, and the temperature dependence differ significantly. The reason for this might lie in the different levels of impurities and different treatments present at the TPL. An upper limit for a representative Ni–YSZ cermet is also presented in figure 4.6, which was obtained by combining the measured ASR with the length of the TPL next to the bulk electrolyte. This is a rather conservative value, as contributions from TPLs away from the bulk electrolyte further reduce the true LSR. Using 3D imaging methods, the TPL length can now also be derived for real Ni-cermet electrodes [21], revealing both percolated and

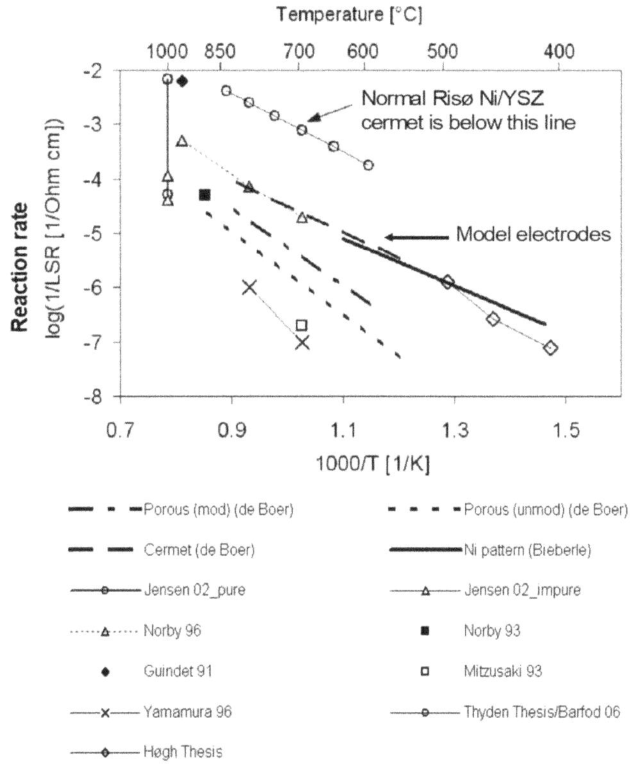

Figure 4.6. Arrhenius plot of the reaction rate at the Ni–YSZ three-phase boundary in the form of the inverse line specific resistance (LSR) reported by various authors summarized in [11]. The 'Normal Risø Ni/YSZ cermet' represents a state-of-the-art Ni–YSZ cermet electrode. Reprinted from [11] with permission, copyright Springer.

non-percolated areas. This is an important tool for analyzing and optimizing cermet production and the choice of raw materials.

There are some indications that the TPL processes in Ni–YSZ are complex around the TPL but determined by the charge-transfer reaction involving the Ni phase [9]. There is also a clear indication that impurities segregate at the TPL and thus that the electrochemical activity is also influenced by the purity of the raw materials [18, 22].

Oxygen ions can also be extracted from other gases/fuels in the SOEC cathode. An important reaction is the CO_2 reduction reaction (CO_2RR), which is relevant for the production of carbon monoxide (CO) and in the co-electrolysis of CO_2 and H_2O to synthesis gas (CO and H_2).

The results of investigating both the hydrogen and carbon monoxide reactions in solid oxide cells are shown for comparison in figure 4.7. From cell tests in various atmospheres, differences in the fuel cell and electrolysis mode are observed for different gas atmospheres [23]. For fuel cell operation (right branch, positive currents), similar cell curves are observed for hydrogen/water as well as co-fed fuels, in contrast to the pure carbon monoxide/carbon dioxide gas mixture, which might indicate that the water gas shift reaction plays an important role in the consumption of the carbon monoxide. However, the situation is different in

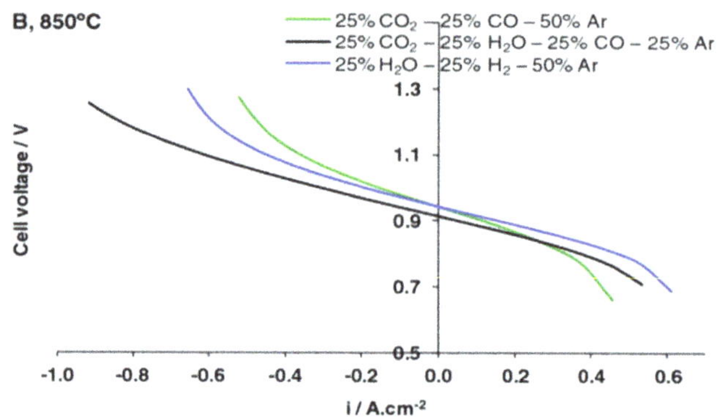

Figure 4.7. Polarization characterization for a planar Ni–YSZ|YSZ|LSM–YSZ SOEC at 850 °C in mixtures of 50% CO_2–25% CO–25% Ar (CO_2 electrolysis), 25% CO_2–25% H_2O–25% CO–25% Ar (co-electrolysis), and 50% H_2O–25% H_2–25% Ar (H_2O electrolysis). Reprinted from [23], copyright The Electrochemical Society. Reproduced by permission of IOP Publishing Ltd.

co-electrolysis mode (left branch, negative currents), indicating that both electrochemical reactions contribute to the cell performance.

Ni–YSZ cermet electrodes exhibit high performance and have been studied extensively for their application in solid oxide fuel cells. The durability of an Ni–YSZ electrode's performance is, to a large extent, linked to its microstructure. While cermets with coarse and fine YSZ fractions sintered together with NiO show negligible degradation over 2500 h at 1000 °C and 200 mA cm^{-2} [24], cermets with fine Ni particles are usually less stable. Poisoning effects due to sulfur and carbon/coke formation which are known to degrade the Ni-cermet performance in fuel cells also affect the SOEC cathode performance in electrolysis mode [25, 26]. Coking is thermodynamically driven by the Boudouard reaction, in particular when operating under high-pressure conditions; however, carbon formation on Ni might occur at lower CO levels than thermodynamically stable levels. Thus, understanding the carbon formation potential as a function of the operational parameters (CO/CO_2 ratio, voltage, and pressure) is important to ensure the safe operation of Ni-cermet cathodes. Currently, *in situ* and in operando characterization techniques (e.g. Raman spectroscopy) are being applied to investigate and better understand coking and other poisoning mechanisms in solid oxide cell (SOC) electrodes [27].

A special degradation phenomenon observed during long-term electrolysis operations is the depletion of Ni from the electrochemically active cermet layer adjacent to the solid electrolyte. This has been observed for various Ni cermets and is illustrated in figure 4.8 (from [28]). The reason seems to be mainly related to a reaction of Ni and the relatively high vapor pressure; however, a more detailed understanding of this degradation process is still missing [29].

Despite the availability of highly engineered Ni-cermet electrodes, the known degradation phenomena and in particular the Ni depletion have triggered the search for alternative SOEC cathode materials.

Benefiting from SOFC development, Ni cermets, due to their high electronic conductivity in the percolated Ni phase, have been widely used as supports in the

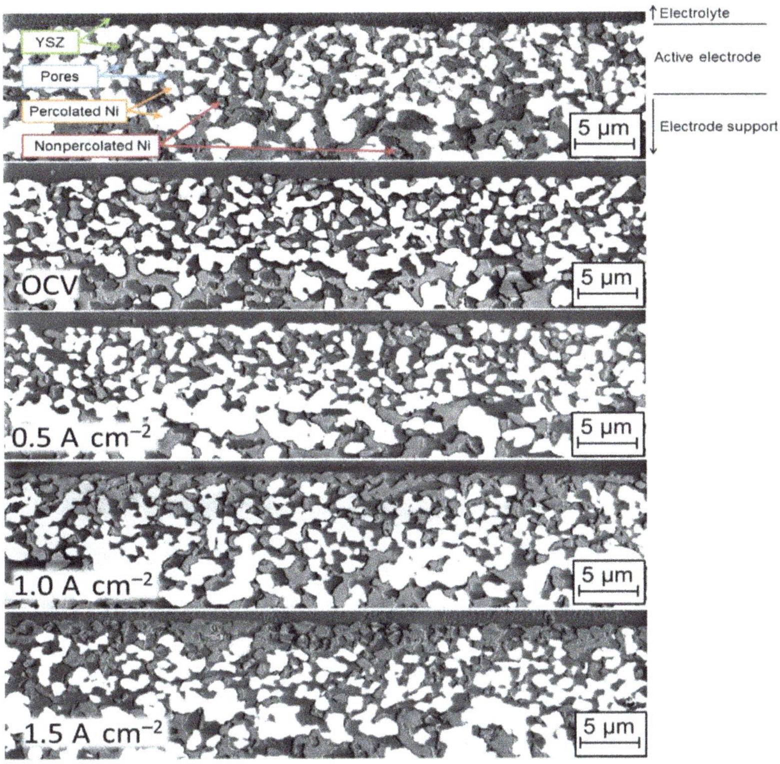

Figure 4.8. Scanning electron microscope (SEM) micrographs of the fuel electrode after operation at 800 °C for 1000 h and with different negative current densities applied (=electrolysis mode). Reprinted from [28], copyright (2018) with permission from Elsevier.

so-called anode-supported cell concept. This concept allows the electrolyte layer to be thin—in the range of a few micrometers to several tenths of a micrometer. The thickness of the anode substrate can vary from 100 micrometers to 2 mm, depending on the design and strength of the components. This concept has also been transferred to SOEC, and the high Ni content in Ni-cermet-supported cells usually defines the product distribution for the co-electrolysis of water and CO_2 due to the water gas shift reaction.

An alternative to Ni–YSZ supports is the use of ferritic steels to produce metal-supported cells [30, 31]. Ferritic steels are inexpensive, highly conductive, and dimensionally stable due to their low volume expansion; however, they suffer from corrosion and other chemical interactions. Furthermore, the manufacture of such cells needs a protective atmosphere with low oxygen partial pressures to prevent corrosion and *in situ* sintering of the cathode layer. Thus, the development of metal-supported cells is still ongoing and has the objective of matching material stabilities and operating conditions, in particular the operating temperature of solid oxide cells (see also chapter 8).

An interesting cell concept for electrolysis is the electrolyte-supported cell concept. Here, the electrolyte is self-supporting and has a thickness of 100–150 microns.

Even though the thicker electrolyte adds to the ohmic resistance and thus to electrical losses in the fuel cell operation, the Joule heat evolved can be utilized in the electrolytic mode of operation to reach the thermoneutral voltage at moderate current densities.

Both the electrolyte-supported cells as well as the metal-supported cells can be flexibly combined with alternative SOEC cathode materials, either as infiltrated coating-type structures in metal supports or as microporous ceramic structures.

4.3 Ceramic electrodes

In the search for alternatives to Ni cermets, ceramic materials have been considered and investigated in detail with respect to their suitability for SOECs. Oxide ceramics with fluorite or perovskite-related structures possess the chemical stability and electronic conductivity, in reducing atmospheres, to act as SOEC cathodes [32]. Depending on their composition and resulting defect chemistry, oxides can be broadly divided into three categories: ionic conductors, mainly electronic conductors, and mixed oxygen ion and electronic conductors. The last two are relevant for SOEC cathodes. A comparison of different oxide ceramic electrodes, including fluorite, perovskites, and Ruddlesden–Popper phases, indicated that in both fuel cell and electrolysis mode, mixed conductivity in the fuel electrode results in higher electrochemical activity for the hydrogen/water reaction compared to mainly electronically conducting oxides [33]. Interestingly, the mainly electronically conducting electrodes show an asymmetric behavior for fuel cell and electrolysis operations, as seen in figure 4.9, favoring the water splitting/HER over the hydrogen

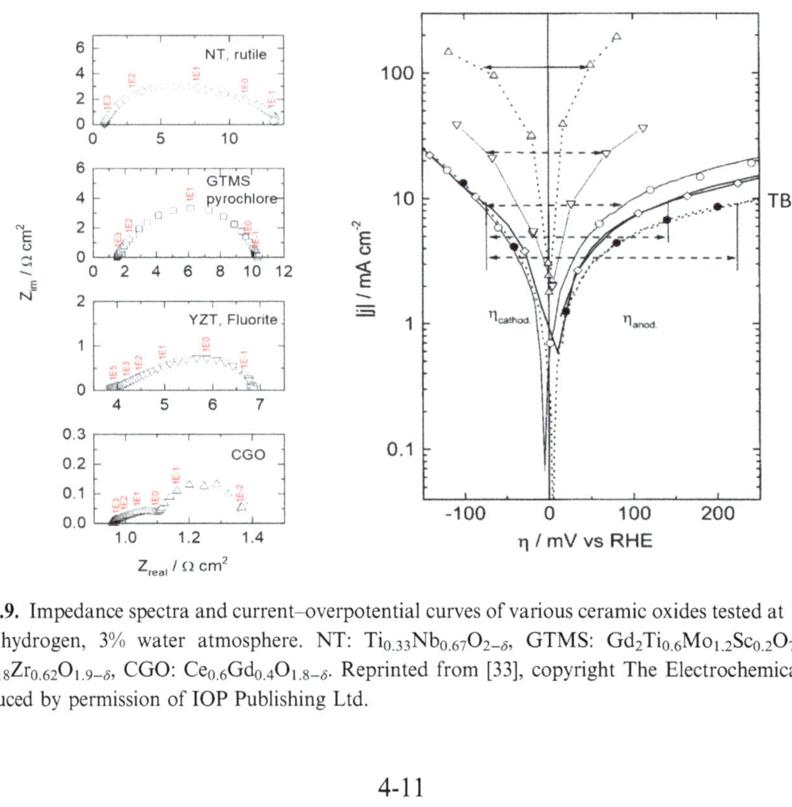

Figure 4.9. Impedance spectra and current–overpotential curves of various ceramic oxides tested at 1000 °C in a 97% hydrogen, 3% water atmosphere. NT: $Ti_{0.33}Nb_{0.67}O_{2-\delta}$, GTMS: $Gd_2Ti_{0.6}Mo_{1.2}Sc_{0.2}O_{7-\delta}$, YZT: $Y_{0.2}Ti_{0.18}Zr_{0.62}O_{1.9-\delta}$, CGO: $Ce_{0.6}Gd_{0.4}O_{1.8-\delta}$. Reprinted from [33], copyright The Electrochemical Society. Reproduced by permission of IOP Publishing Ltd.

oxidation reaction. This might be related to an increase in the electronic n-type conductivity under the more reducing conditions of the SOEC operation.

As outlined for the Ni–YSZ cermet, the SOEC cathode should have a sufficient CTE match and also be chemically compatible with the adjacent electrolyte. An additional important aspect for oxide ceramic materials is dimensional stability under both reducing and oxidizing conditions. Hence, control of the defect chemistry is important to avoid detrimental expansion upon reduction of the oxide. These requirements have limited the number of ceramic materials; the most relevant materials are presented briefly in the following.

4.3.1 Ceria

Ceria modified with lower-valence cations (acceptor doped), such as Gd or Sm, is a widely studied material [34]. Lower-valent doping leads to the formation of oxygen ion vacancies, which facilitate oxygen transport. Under the reduced oxygen partial pressure at the SOEC cathodes and at elevated temperatures, ceria becomes an n-type conductor based on small polaron hopping. Thus, ceria electrodes are mixed ionic electronic conductors at reduced pO_2 and elevated temperatures.

Ceria is a very good oxidation catalyst and has been specifically proposed for hydrocarbon-based SOFCs. It possesses a fluorite structure and has no detrimental chemical reaction with the YSZ electrolyte. However, the CTE of 15.5×10^{-3} K^{-1} is higher than that of YSZ. In addition, ceria expands upon reduction due the fact that a further loss in oxygen is compensated by the reduction of ceria from the 4^{+} oxidation state to the 3^{+} state. The dimensional change is influenced by the 'dopant,' and Gd and Sm show the lowest expansion upon reduction [35]. Modified ceria with mixed electronic–ionic conductivity is, however, not frequently used as an electrode material in SOECs (mainly owing to its limited electronic conductivity), but addition of gadolinia-modified ceria to more dominantly electronically conducting ceramic electrodes by infiltration, for example, is receiving increasing attention.

4.3.2 Lanthanum chromites

Perovskites based on lanthanum chromites form an interesting class of SOEC cathode materials. The addition of Sr increases the stability of these materials at low pO_2 [36] and also increases their catalytic activity [37]. The modification of these metal ions at the B sites using transition-metal ions such as Mn, Fe, Co, Ni, Cu, and Ru increases the concentration of oxygen vacancies, and it has been reported that this positively affects the catalytic activity in perovskites [38].

A comprehensively studied material is based on lanthanum chromite manganites (LSCMs), as proposed by, for example, Sfeir and also Tao [39, 40]. $La_{0.75}Sr_{0.25}Cr_{0.5}Mn_{0.5}O_3$ is a mixed conductor with a total conductivity of 38 S cm^{-1} at 900 °C in air; however, its conductivity decreases to less than 1 S cm^{-1} in hydrogen atmospheres [41].

Its coefficient of thermal expansion matches that of YSZ in air; however, when exposed to reducing conditions, the material expands. This expansion leads to stresses at the sintered electrode/electrolyte interface, which are only acceptable for

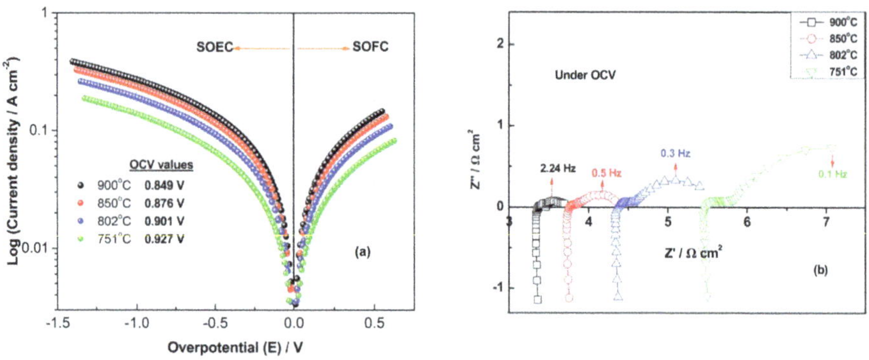

Figure 4.10. Polarization curves (a) and impedance spectra (b) of an SOEC with a GDC-infiltrated LSCM cathode working in a 70% CO_2/30% CO atmosphere at temperatures ranging from 900 °C to 750 °C at intervals of around 50 °C. Reprinted from [42], permission conveyed through Copyright Clearance Center.

relatively small electrode areas and have to be considered in the cell and stack design.

The electrochemical performance of LSCM was tested in hydrogen/water atmospheres and reached ASRs of 0.2 Ω cm^2 at 900 °C. Coking and oxidation-tolerant SOEC cathode materials are especially interesting for carbon dioxide electrolysis and steam electrolysis, in which either carbon monoxide or synthesis gas is the desired product. LSCM infiltrated with ceria has been reported to be a stable SOEC cathode in carbon dioxide electrolysis [42]. Area specific electrode resistances of 0.24 Ω cm^2 have been achieved at 900 °C in a 30% CO/70% CO_2 mixture. (figure 4.10).

4.3.3 Ferrite oxides

As the most abundant transition metal in the Earth's crust, ferrite oxides have been considered to be decent fuel and oxygen electrodes for SOCs due to their low cost and chemical stability. However, unmodified ferrite perovskites ($SrFeO_3$ (SFO) and $LaFeO_3$ (LFO)) are unstable under reducing atmospheres and decompose easily. An important way to stabilize the structure is B-site doping, which maintains the oxygen content in perovskite lattice and reduces the activity of Fe^{3+}/Fe^{2+}; Mo, Cr, and Ti are the most common and popular B-site substitutes.

Mo-modified SFO has been thoroughly studied as a fuel electrode for SOFCs due to its high conductivity in reducing atmospheres, excellent stability, and favorable electrocatalysis. $SrFe_{1-x}Mo_xO_3$ ($0.25 \leqslant x \leqslant 0.6$), which has an ordered arrangement of Fe and Mo cations in double perovskites is a recently popular fuel electrode material. For example, $Sr_2Fe_{1.5}Mo_{0.5}O_{6-\delta}$ shows a remarkable performance for CO_2 electrolysis with pure CO_2 as the feed gas and is thus a promising SOEC cathode. A current density of 1.09 A cm^{-2} was achieved at a cell voltage of 1.5 V at 800 °C [43]. Cr-doped electrodes possess the formula $La_{1-x}Sr_xCr_{1-x}Fe_xO_{3-\delta}$ (x = 0.2–0.67, LSCrF) and their polarization resistances reduce with Cr content. In fuel cell operation, Bi-modified LSCrF, $La_{0.65}Bi_{0.1}Sr_{0.25}Cr_{0.5}Fe_{0.5}O_{3-\delta}$ (LBSCr) shows a significant improvement (360 mW at 800 °C with syngas fuel) compared to the undoped LSCrF [44]. Ti doping reduces the electronic and ionic conductivity of the material in air, but can significantly improve the chemical stability of the material. Ti-modified SFO ($SrFe_{0.7}Ti_{0.3}O_3$) is stable at 650 °C–750 °C under a reducing atmosphere, and possesses a low polarization resistance of 0.17 Ω cm^2 at 800 °C in humidified H_2, which is much lower than the value of 0.39 Ω cm^2 of $SrFe_{0.6}Ti_{0.4}O_3$ and the 3.14 Ω cm^2 of $SrTiO_3$ [45].

4.3.4 Strontium titanates

Another promising perovskite system consists of donor-doped strontium-titanate-based materials. $SrTiO_3$ materials doped at the A site or the B site have attracted attention as fuel electrode materials because of their high electronic conductivity after reduction at high temperatures (e.g. 1400 °C), proven redox stability, and resistance to coking and sulfur poisoning [46]. Among the A-site-modified $SrTiO_3$ materials, La–$SrTiO_3$ and Y–$SrTiO_3$ materials have been widely studied [47, 48, 49, 50, 51]. Nb-doped $SrTiO_3$ materials have been given significant attention for B-site modification [52, 53].

Compared to the two previous examples of LFO and LSCrM, which possess a certain mixed ionic electronic conductivity, lanthanum strontium titanates are mainly electronic conductors [54] (donor doping suppresses the oxygen vacancy concentration). The electronic conductivity reported for these materials varies significantly from one report to another and the variations are attributed to different processing and measurement conditions. An important factor here is the processing atmosphere, which directly influences the defect chemistry and the resulting electronic transport properties of these materials [49, 55]. Composite electrodes consisting of YSZ and doped $SrTiO_3$ materials were also investigated for their suitability as anode materials in a similar manner to the investigations performed for Ni–YSZ cermets [56].

These materials are rather compatible with the electrolyte YSZ in terms of the coefficient of thermal expansion. Since the oxygen exchange kinetics is slow at the operating temperatures of SOCs, expansion upon reduction, as observed in lanthanum chromites, is not usually a severe problem.

However, the main limitation of these materials is their lack of electrocatalytic activity. This requires the incorporation of additional electrocatalysts into these materials to make them active for electrochemical reactions. The following subsections describe two attempts to integrate nanosized electrocatalysts into porous ceramic electrodes.

4.3.5 Integration of nanostructured electrocatalysts by infiltration

A commonly used method is the so-called infiltration method, a direct synthesis of electrocatalyst layers using wet chemistry and related to the wash coat in catalytic filters. Usually, precursors such as nitrate solutions are infiltrated into the porous electrode backbone, and decompose upon heating and reduction to form either ceramic, metallic, or composite metal–ceramic nanoscale surface layers.

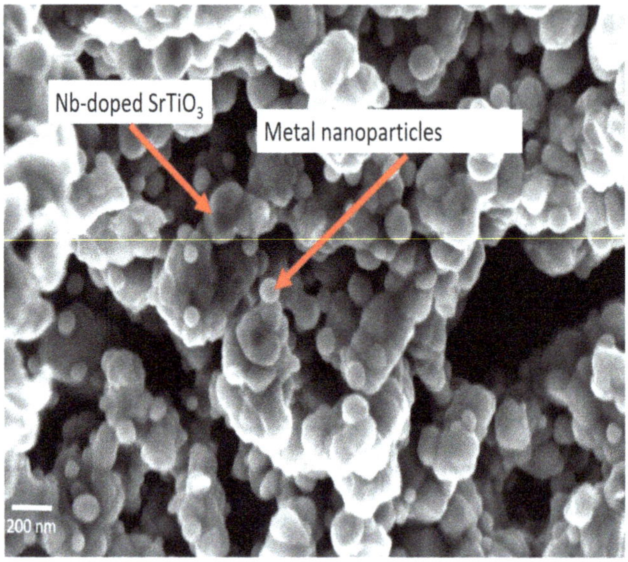

Figure 4.11. SEM image of a Ni catalyst infiltrated into an electrode backbone made of $Sr_{0.94}Ti_{0.9}Nb_{0.1}O_{3-\delta}$.

The backbone can be an electronically conductive ceramic or the electrolyte [57]. In the latter case, however, a continuous coating of an electronically conducting layer is required to provide a sufficient current pathway into the electrode.

Typical infiltrates are Ni-, Cu-, and ceria precursor solutions, which form metallic nanoparticles and ceria on the backbone surface upon decomposition and reduction. Figure 4.11 shows a Ni-impregnated $Sr_{0.94}Ti_{0.9}Nb_{0.1}O_{3-\delta}$ (STN) electrode with isolated, nanosized Ni particles maintained after testing at 900 °C. The separation of important electrode functions such as mechanical stability, compatibility, and electronic conductivity from electrocatalytic activity allows multiple combinations that can tailor the electrode, particularly for temperatures below 600 °C.

The co-infiltration of ceria and metallic nanoparticles into strontium titanates as a redox stable, electronically conducting backbone has been studied in SOCs under various gas atmospheres. Figure 4.12 shows the area specific electrode resistances for pure STN and composite STN–$Ce_{0.6}Gd_{0.4}O_{1.8-\delta}$ (CGO) electrodes with and without Ni–CGO infiltration. While the composite STN-CGO electrode has a slightly lower electrode resistance compared to the pure electrode, the Ni–CGO infiltration increases the electrode activity significantly.

The stability of the nanostructured electrocatalyst and the resulting electrode performance are not fully understood, but metal support interactions are considered crucial. For niobium-modified strontium titanate infiltrated with a Ni–GDC electrocatalyst, a continuous degradation was observed; however, it occurred at different rates. In the case of Ru–CGO-infiltrated STN electrodes, a very low electrode resistance was observed along with stable performance over 400 h [58].

The concept of using nanostructured electrodes has received a lot of attention to date and various backbone/infiltrate combinations are being explored for both fuel

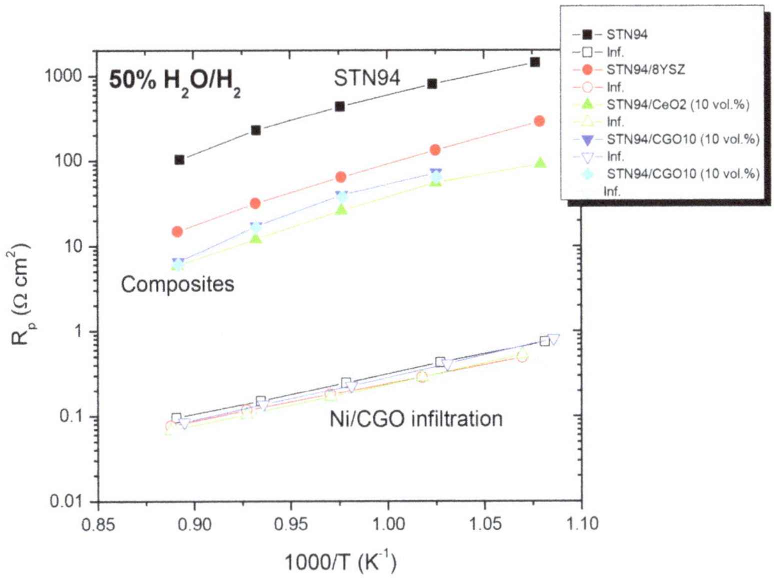

Figure 4.12. Electrode resistance of various STN electrodes: STN_{94} $Sr_{0.94}Ti_{0.9}Nb_{0.1}O_{3-\delta}$, the composites STN94-CGO ($Ce_{0.8}Gd_{0.2}O_{1.9}$) and STN-YSZ ($Zr_{0.84}Y_{0.16}O_{1.92}$), and Ni (1w%)-CGO-infiltrated electrodes.

cell and electrolysis applications. Even though Ru-infiltration has shown stable performances on various backbone materials, more abundant elements such as Co and Fe are currently under investigation. Furthermore, alternative ways to incorporate nanostructured metal nanoparticles are being explored.

4.3.6 Integration of nanostructured electrodes by exsolution

An interesting concept, namely, the decoration of ceramic electrodes with small, nanosized electrocatalysts, uses the exsolution of–mostly—metal particles from host oxides under reducing conditions. The exsolution process is based on the fact that some metal cations incorporated into a host lattice can be reduced under the operating conditions of the SOEC cathode and segregate as metallic particles at the surface of the host ceramic. The generalized exsolution reaction is given in equation 4.4, in which 'Me' is an element that can be reduced to metal under the operating conditions of the SOEC cathode:

$$AB_{1-x}Me_xO_3 + H_2 \rightarrow AB_{1-x}O_{3-\delta} + xMe + H_2O. \quad (4.4)$$

This concept is also interesting because it allows poisoned metal particles to be regenerated by redissolving them in the host oxide upon (re-)oxidation and subsequent re-reduction. In the early research, the exsolution only targeted a limited number of easily reducible, catalytically active cations (Ni^{2+}, Ru^{2+}, Rh^{4+}, Pd^{4+}, and Pt^{4+}), because the exsolution phenomenon was believed to be driven exclusively by their ease of reduction to metals [59]. The exsolved electrocatalysts clearly enhance the electrode performance. However, an overlarge imbalance in the stoichiometry (i.e. cation vacancies) may cause problems that affect long-term stability due to A-site segregation (e.g. SrO) and structural decomposition.

In recent years, harder-to-reduce cations have been exsolved, and in addition, exsolutions preferentially emerge on the surface when highly A-site-deficient perovskites (A/B<1) are employed (for example, $TiO_{2-\delta}$ exsolves from $La_{0.4}Sr_{0.4}TiO_{3-\delta}$, (LST)) [60, 61]. Thus, A-site deficiency can serve as a general driving force to trigger B-site exsolution, producing a wider range of nanoparticle compositions with superior surface distribution and coverage. Moreover, the use of A-site deficiency (A/B<1) can suppress SrO segregation and improve structural stability, as the stoichiometry becomes more balanced. Neagu *et al* [64] designed and developed a series of A-site deficient $(LaSr)_{1-x}Me_xTiO_{3-\delta}$ (Me = Cu, Ni, Fe, etc) perovskites, and found that they not only displayed high catalytic activity but also exhibited a lower propensity for agglomeration, even at high temperatures, which was attributed to the strong interactions between the nanoparticles and the oxide surface. An exsolution of Cu in $La_{0.43}Sr_{0.37}Cu_{0.12}Ti_{0.88}O_{3-\delta}$ (LSCuT) exhibited smaller Cu nanoparticles and a more uniform distribution structure than copper that infiltrated LST after electrochemical reduction in a single cell achieved by applying 2.3 V in a 50% H_2O/N_2 gas atmosphere (see figure 4.13). The exsolved nanoparticles showed a better performance with a lower polarization resistance of 0.04 Ω cm^2 at 900 °C compared to 0.66 Ω cm^2 for infiltrated nanoparticles [62]. La–$SrTiO_3$ electrodes, which also exsolve TiO_x, can also work efficiently and stably for CO_2 electrolysis. $(La_{0.2}Sr_{0.8})_{0.95}Mn_{0.12}Ti_{0.88}O_{3-\delta}$ can last over 100 h at 1.3 V in a 100% CO_2 atmosphere at a stable current of 0.28 A cm^{-2}. Furthermore, it possesses

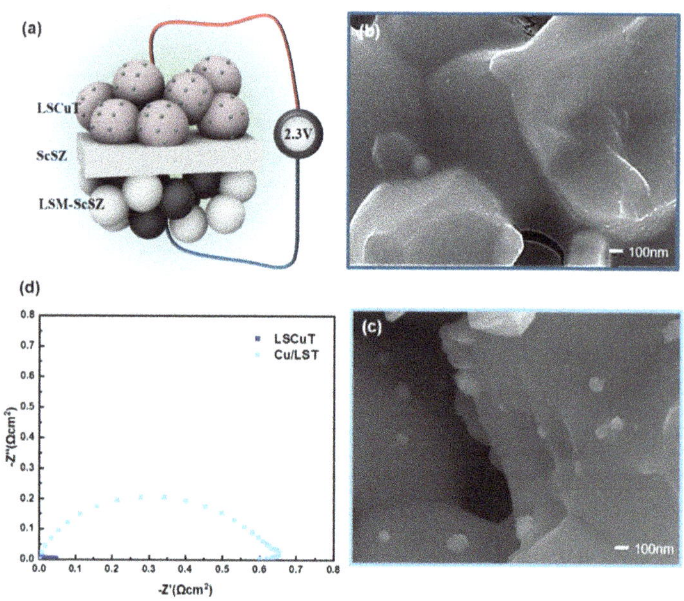

Figure 4.13. (a) Schematic illustration of the electrochemical reduction process in a single cell. SEM images of (b) LSCuT, (c) infiltrated Cu/LST electrodes after electroreduction. (d) Electrode polarization after electro-reduction (900 °C). Reprinted from [62], copyright (2021) with permission from Elsevier.

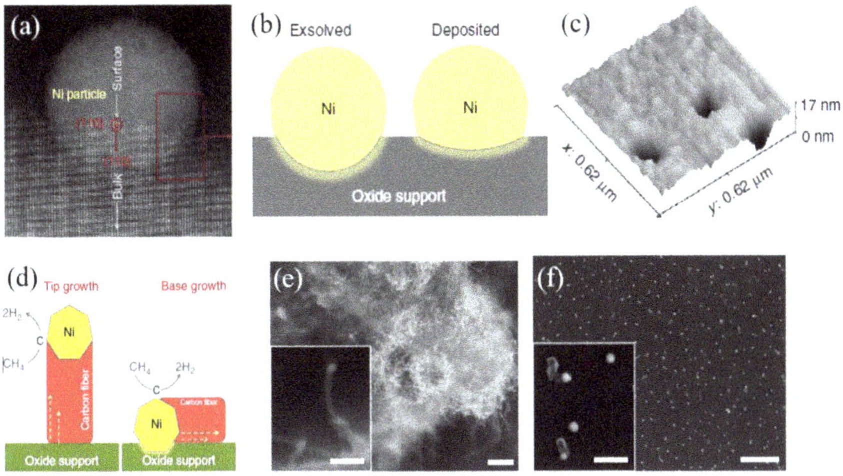

Figure 4.14. (a) TEM micrograph (dark field) of a Ni particle exsolved on a (110) native surface facet. (b) Schematic illustration of the particle–substrate interface for deposited and exsolved nickel particles. (c) Three-dimensional atomic force microscopy (AFM) image of similar sockets. (d) Schematic diagram of two types of carbon fiber growth. (e) Infiltrated Ni particles on $La_{0.4}Sr_{0.4}TiO_3$ after a coking test. (f) Exsolved Ni particles on $La_{0.4}Sr_{0.4}TiO_3$ after a coking test. Scale bars: 1 µm (main image), 100 nm (inset). Reprinted from [64] with permission from Springer Nature.

excellent redox stability, as it can maintain its original performance after 10 cycles of alternate oxidation and reduction [63].

Apart from improving catalyst performance, the exsolved particles can enhance coke resistance. Exsolved nanoparticles appear in pinned structures (figure 4.14(a)–(c)) and are almost 30% submerged into the backbone; this anchor structure inhibits the tip growth of carbon fiber (figure 4.14(d)), thereby significant decreasing the tendency toward coke formation. As shown in figures 4.14(e) and (f), only a few carbon fibers were found on the exsolved $La_{0.52}Sr_{0.28}Ni_{0.06}Ti_{0.94}O_3$ electrode after coke testing, whereas infiltrated Ni was severely coked, as considerable carbon fiber grew [64].

In addition to the lanthanum titanate-based electrodes mentioned above, lanthanum chromite (also used as pure ceramic electrodes as mentioned previously) is also a common A-site deficient exsolution parent perovskite. Compared to electrodes without exsolved electroctalysts, the current density of $(La_{0.75}Sr_{0.25})_{0.9}(Cr_{0.5}Mn_{0.5})_{0.9}Ni_{0.1}O_{3-\delta}$ *in situ* exsolved Ni nanoparticles significantly increased from 0.27 to 0.38 A cm at 2.0 V, and the faradaic efficiency increased from 50% to 80% for CO_2 electrolysis [65]. Very recently, Ni-exsolved $La_{0.65}Sr_{0.3}Cr_{0.85}Ni_{0.15}O_{3-\delta}$ displayed decent electrochemical cell performance with an ASR of 0.676 Ω cm^2, which is comparable to a state-of-the-art Ni–CGO cermet. This cell showed good stability during co-electrolysis ($H_2O/CO_2 = 2$) at 860 °C at a current load of 0.45 A cm^{-2} with a voltage decay of less than 3.5 mV/1000 h, which predicts an operational window of 950 h (see figure 4.15) [66].

Although exsolved Ni shows excellent catalytic activity, such electrocatalysts are prone to failures arising from, for example, impurity poisoning, carbon deposition,

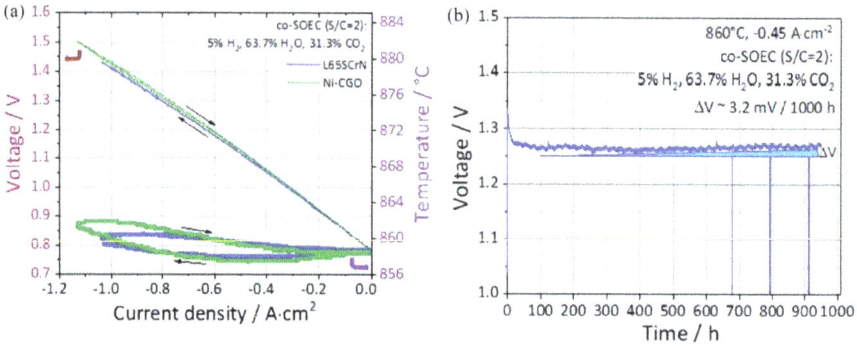

Figure 4.15. IV curves (a) and long-term stability test (b) in co-electrolysis mode (5% H_2, 63.7% H_2O, 31.3% CO_2) of the $La_{0.65}Sr_{0.3}Cr_{0.85}Ni_{0.15}O_{3-\delta}$. Reprinted from [66], with permission from the Royal Society of Chemistry.

reoxidation, and particle agglomeration. As an alternative, Fe metal particle exsolution has recently attracted attention due to its decent stability under reducing conditions and the abundance of iron in the Earth's crust. Comparing $La_{0.4+x}Sr_{0.4-x}Fe_xTi_{1-x}O_3$ (LSFT) with its Ni counterpart, the Fe exsolution requires a higher temperature and a lower oxygen partial pressure because of its higher Fe–O bond energy. The exsolution of a Ni–Fe nanoalloy on titanate perovskite improved CO oxidation at 450 °C and enhanced sulfur tolerance due to the preservation of the initial spacing between particles [67]. Similarly, superior carbon deposition tolerance has been observed for Ni- and Fe-modified La–Sr chromites, e.g. $La_{0.7}Sr_{0.3}Cr_{0.85}Ni_{0.1125}Fe_{0.0375}O_{3-\delta}$ (LSCrFN), compared to chromites modified using Ni only, e.g. $La_{0.7}Sr_{0.3}Cr_{0.85}Ni_{0.15}O_{3-\delta}$ (LSCrN), which was attributed to an increase in the oxide vacancies of chromite introduced by the Fe modification [68].

Iron particles can be exsolved from Fe-modified perovskites such as the above-mentioned LSTF and LSCrFN and can also be obtained from other ferrite-based oxides. Some ferrite-based oxides tend to be unstable in a reducing atmosphere, and a structural phase transition occurs, accompanied by the formation of Fe–Ni nanoparticles, which is therefore called 'structural transformation exsolution.' This phenomenon was initially observed in $(Pr_{0.4}Sr_{0.6})(Co_{0.2}Fe_{0.7}Nb_{0.1})O_{3-\delta}$ (PSCFN) materials, which were found to transform into Ruddlesden–Popper phases and a Co–Fe nanoalloy in a reducing atmosphere. This 'phase-transformed electrode' demonstrated remarkable stability in H_2 with 50 ppm H_2S at a steady constant current load of 0.6 A cm^{-2} over 500 h in fuel cell (SOFC) conditions [69]. Inspired by this observation, researchers have to date developed a series of phase-transition electrode materials, among which the double perovskites $Sr_2Fe_{1.5-x}Mo_{0.5-y}M_{x+y}O_{6-\delta}$ (M= Co, Ni, Cu) have been the focus of recent research. The material $SrFe_{1.5-x}Mo_{0.5-y}Co_{x+y}O_{6-\delta}$ partly transforms to Ruddlesden–Popper phases $Sr_3Fe_{2-x-y}Mo_xCo_yO_{7-\delta}$, while FeCo alloy nanoparticles are simultaneously exsolved from the parent phase (figure 4.16(a)). Compared to the pure parent material electrodes, the FeCo exsolved electrode has better electrochemical performance. The ASR of DC polarization for CO_2 electrolysis at 1.2 V was reduced by 57.8%, from 0.28 Ω cm^2 to 0.118 Ω cm^2, and it

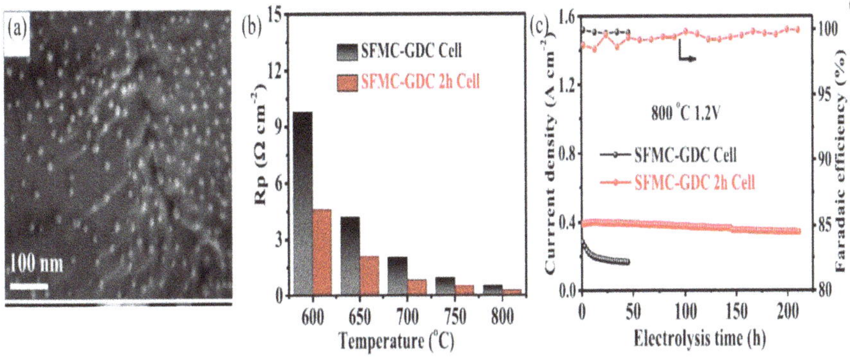

Figure 4.16. (a) Morphology of exsolved nanoparticles. (b) ASR of polarization at 1.2 V. (c) Stability test of exsolved electrode. Reprinted from [70], with permission from John Wiley and Sons.

was able to work steadily for more than 230 h with no distinct degradation (see figures 4.16(b) and (c)) [70].

4.4 Concluding remarks: from the state of the art to advanced materials design

The most important SOEC cathode material for SOCs is the well-developed Ni cermet, which to date is still the material most widely used for SOEC cathodes. However, several limitations, especially the Ni depletion in the active interface layer towards the solid electrolyte and the destructive coke formation potential, especially in the case of carbon dioxide electrolysis, has triggered the development of alternative, notably Ni-free, SOEC cathodes. Since the cell design for SOECs is rather flexible as regards the supporting electrolyte, the decoupling of electrocatalysis from other structural and functional properties (such as structural integrity and current pickup in the electrode by different materials) enables the use of more advanced concepts for supported nanoscale electrocatalysts. Operando characterization of the electrode processes in state-of-the-art and advanced electrode designs is a powerful tool that assists the tailoring of SOEC cathodes for high efficiency and stability.

References

[1] Ebbesen S D, Jensen S H, Hauch A and Mogensen M B 2014 High temperature electrolysis in alkaline cells, solid proton conducting cells, and solid oxide cells *Chem. Rev.* **114** 10697–734

[2] Murphy M M, McEvoy A J and Thampi K R 1994 Electroless deposition of electrodes in solid-oxide fuel cells *J. Electrochem. Soc.* **141** L94–6

[3] Spacil H S 1970 Patent US 3,503,809

[4] Powell M J 1979 Site percolation in randomly packed spheres *Phys. Rev.* B **20** 4194–8

[5] Dees D W, Claar T D, Easler T E, Fee D C and Mrazek F C 1987 Conductivity of porous Ni/ZrO_2–Y_2O_3 cermets *J. Electrochem. Soc.* **134** 2141–6

[6] Itoh H, Yamamoto T, Mori M, Horita T, Saka N, Yokokawa H and Dokiya M 1997 Configurational and electrical behavior of Ni-YSZ cermet with novel microstructure for solid oxide fuel cell anodes *J. Electrochem. Soc.* **144** 641–6

[7] Männer R, Ivers-Tiffée E and Wersing W 1991 Characterization of YSZ electrolyte with various yttria content *Solid Oxide Fuel Cells (SOFCII)* ed F Grosz, P Zegers, S C Singhal and O Yamamoto (Luxemburg: Commission of the European Communities) L. EUR-13564-EN, 715

[8] Müller A C, Herbstritt D and Ivers-Tiffée E 2002 Development of a multilayer anode for solid oxide fuel cells *Solid State Ionics* **152–153** 537–42

[9] Holtappels P, Vinke I C, De Haart L G J and Stimming U 1999 Reaction of hydrogen/water mixtures on nickel-zirconia cermet electrodes. Part 1: D.C. polarization characteristics *J. Electrochem. Soc.* **146** 1620–5

[10] Brown M, Primdahl S and Mogensen M 2000 Structure/performance relations for Ni/yttria-stabilized zirconia anodes for solid oxide fuel cells *J. Electrochem. Soc.* **147** 475–85

[11] Mogensen M and Holtappels P 2013 Ni-based solid oxide cell electrodes *Solid Oxide Fuel Cells: Facts and Figures, Green Energy technology* ed J T S Irvine and P Connor (London: Springer) **2** 25–45

[12] Norby T, Velle O J, Leth-Olsen H and Tunold R 1993 Reaction resistance in relation to three phase boundary length of Ni/YSZ electrodes *Solid Oxide Fuel Cells III* ed S C Singhal and H Iwahara (Pennington NJ: The Electrochemical Society) p 473

[13] Mizusaki J *et al* 1994 Preparation Of nickel pattern electrodes on YSZ and their electrochemical properties in H_2–H_2O atmospheres *J. Electrochem. Soc.* **141** 2129–21344

[14] Mogensen M, Primdahl S, Rheinländer J T, Gormsen S, Linderoth S and Brown M 1995 Relations between performance and structure of Ni-YSZ-cermet SOFC anodes *Solid Oxide Fuel Cells IV* ed M Dokiya, O Yamamoto, H Tagawa and S C Singhal (Pennington NJ: The Electrochemical Society) pp 657–66

[15] Brown M S, Sammes N M and Mogensen M 1997 Relationships between fabrication procedures, structure and conductivity of Ni/YSZ cermet anodes *Solid Oxide Fuel Cells V* ed U Stimming, S C Singhal, H Tagawa and W Lehnert (Pennington NJ: The Electrochemical Society) p 861 Proc. vol 97-40

[16] Singhal S C 1997 Recent progress in tubular solid oxide fuel cell technology *Solid Oxide Fuel Cells V.* ed U Stimming, S C Singhal, H Tagawa and W Lehnert (Pennington NJ: The Electrochemical Society) p 37 Proc. vol. 97-40

[17] Barfod R, Mogensen M, Klemensø T, Hagen A, Liu Y L and Hendriksen P V 2007 Detailed characterization of anode-supported SOFCs by impedance spectroscopy *J. Electrochem. Soc.* **154** B371–8

[18] Hauch A and Mogensen M 2010 Ni/YSZ electrode degradation studied by impedance spectroscopy effects of gas cleaning and current density *Solid State Ionics* **181** 745–53

[19] Primdahl S and Mogensen M 1999 Gas diffusion impedance in characterization of solid oxide fuel cell anodes *J. Electrochem. Soc.* **146** 2827–33

[20] Steele B C H 1996 Survey of materials selection for ceramic fuel cells: II. Cathodes and anodes *Solid State Ionics* **86–88** 1223–34

[21] Wilson J R, Kobsiriphat W, Mendoza R W, Chen H Y, Hiller J M, Miller D J, Thornton K, Voorhees P W, Adler S B and Barnett S A 2006 Three-dimensional reconstruction of a solid-oxide fuel-cell anode *Nat. Mater.* **5** 542–4

[22] Hauch A, Jørgensen P S, Brodersen K and Mogensen M 2011 Ni/YSZ anode – effect of pre-treatments on cell degradation and microstructures *J. Power Sources* **196** 8931–41

[23] Ebbesen S D, Knibbe R and Mogensen M 2012 Co-electrolysis of steam and carbon dioxide in solid oxide cells *J. Electrochem. Soc.* **159** F482–9
[24] Itoh H, Yamamoto T, Mori M, Watanabe T and Abe T 1996 Improved microstructure of Ni-YSZ cermet anode for SOFC with a long term stability *Denki Kagaku* **64** 549–54
[25] Skafte T L, Blennow P, Hjelm J and Graves C 2017 Carbon deposition and sulfur poisoning during CO_2 electrolysis in nickel based solid oxide cell electrodes *J. Power Sources* **373** 54–60
[26] Navasa M, Frandsen H L, Skafte T L, Sundén B and Graves C 2018 Localized carbon deposition in solid oxide electrolysis cells studied by multiphysics modeling *J. Power Sources* **394** 102–13
[27] Traulsen M L, Chatzichristodoulou C, Hansen K V, Kuhn L T, Holtappels P and Mogensen M B 2015 Need for in operando characterization of electrochemical interface features *ECS Trans.* **66** 3
[28] Hoerlein M P, Riegraf M, Costa R, Schiller G and Friedrich K A 2018 A parameter study of solid oxide electrolysis cell degradation: microstructural changes of the fuel electrode *Electrochim. Acta* **276** 162–75
[29] Mogensen M B, Chen M, Frandsen H L, Graves C R, Hauch A, Hendriksen P V and Sun X 2020 Review of Ni migration in SOC electrodes *Proc. 14th European SOFC & SOEC Forum* (Lucerne, Switzerland: European Fuel Cell Forum AG) pp 20–3
[30] Tucker M C, Lau G Y, Jacobson C P, Dejonghe L C and Visco S J 2007 Performance of metal-supported SOFCs with infiltrated electrodes *J. Power Sources* **171** 477–82
[31] Blennow P, Hjelm J, Klemensø T, Ramousse S, Kromp A, Leonide A and Weber A 2011 Manufacturing and characterization of metal-supported solid oxide fuel cells *J. Power Sources* **196** 7117–25
[32] Tao S and Irvine J T S 2004 Discovery and characterization of novel oxide anodes for solid oxide fuel cells *Chem. Rec.* **4** 83–95
[33] Holtappels P, Bradley J L, Irvine J T S, Kaiser A and Mogensen M 2001 Electrochemical characterisation of ceramic SOFC anodes *J. Electrochem. Soc.* **148** A923–9
[34] Mogensen M, Primdahl S, Jørgensen M J and Bagger C 2000 Composite electrodes in solid oxide fuel cells and similar solid state devices *J. Electroceram.* **5** 141–52
[35] Mogensen M, Lindegaard T, Hansen U R and Mogensen G 1994 Physical properties of mixed conductor solid oxide fuel cell anodes of doped CeO_2 *J. Electrochem. Soc.* **141** 2122–8
[36] Sfeir J 2003 $LaCrO_3$-based anodes: stability considerations *J. Power Sources* **118** 276–85
[37] Vernoux P 1997 Lanthanum chromite as an anode material for solid oxide fuel cells *Ionics* **3** 270–6
[38] Doshi R, Alcock C B, Gunasekaran N and Carberry J J 1993 Carbon-monoxide and methane oxidation properties of oxide solid-solution catalysts *J. Catal.* **140** 557–63
[39] Sfeir J, Buffat P A, Mockli P, Xanthopoulos N, Vasquez R, Mathieu H J, Van Herle J and Thampi K R 2001 Lanthanum chromite based catalysts for oxidation of methane directly on SOFC anode *J. Catal.* **202** 229–44
[40] Tao S and Irvine J T S 2003 A redox-stable efficient anode for solid-oxide fuel cells *Nat. Mater.* **2** 320–3
[41] Tao S and Irvine J T S 2004 Synthesis and characterization of $(La_{0.75}Sr_{0.25})Cr_{0.5}Mn_{0.5}O_{3-\delta}$, a redox-stable, efficient perovskite anode for SOFCs *J. Electrochem. Soc.* **151** A252–9
[42] Yue X and Irvine J T S 2017 Modification of LSCM–GDC cathodes to enhance performance for high temperature CO_2 electrolysis using solid oxide electrolysis cells (SOECs) *J. Mater. Chem.* A **5** 7081–90

[43] Li Y H, Hu B B, Xia C R, Xu W Q, Lemmon J P and Chen F L 2017 Novel fuel electrode enabling direct CO_2 electrolysis with excellent and stable cell performance *J. Mater. Chem.* A **5** 20833–42

[44] Wan Y H, Xing Y L, Xu Z Q, Xue S S, Zhang S W and Xia C R 2020 A-site bismuth doping, a new strategy to improve the electrocatalytic performances of lanthanum chromate anodes for solid oxide fuel cells *Appl. Catal.*, B **269** 118809

[45] Cho S, Fowler D E, Miller E C, Cronin J S, Poeppelmeier K R and Barnett S A 2013 Fe-substituted $SrTiO_{3-\delta}$–$Ce_{0.9}Gd_{0.1}O_2$ composite anodes for solid oxide fuel cells *Energy Environ. Sci.* **6** 1850–7

[46] Mukundan R, Brosha E L and Garzon F H 2004 Sulfur tolerant anodes for SOFCs *Electrochem. Solid-State Lett.* **7** A5–7

[47] Slater P R, Fagg D P and Irvine J T S 1997 Synthesis and electrical characterisation of doped perovskite titanates as potential anode materials for solid oxide fuel cells *J. Mater. Chem.* **7** 2495–8

[48] Hui S and Petric A 2002 Electrical properties of yttrium-doped strontium titanate under reducing conditions *J. Electrochem. Soc.* **149** J1–10

[49] Hui S and Petric A 2002 Evaluation of yttrium-doped $SrTiO_3$ as an anode for solid oxide fuel cells *J. Eur. Ceram. Soc.* **22** 1673–81

[50] Pudmich P, Boukamp B A, Gonzalez Cuenca M, Jungen W, Zipprich W and Tietz F 2000 Chromite/titanate based perovskites for application as anodes in solid oxide fuel cells *Solid State Ionics* **135** 433–8

[51] Canales-Vázquez J, Tao S W and Irvine J T S 2003 Electrical properties in $La_2Sr_4Ti_6O_{19-\delta}$: a potential anode for high temperature fuel cells *Solid State Ionics* **159** 159–65

[52] Kolodiazhnyi T T and Petric A 2005 The applicability of Sr-deficient n-type $SrTiO_3$ for SOFC anodes *J. Electroceram.* **15** 5–11

[53] Blennow P, Hansen K K, Wallenberg L R and Mogensen M 2006 Effects of Sr/Ti-ratio in $SrTiO_3$-based SOFC anodes investigated by the use of cone-shaped electrodes *Electrochim. Acta* **2006** 1651–61

[54] Verbraeken M C, Ramos T, Agersted K, Ma Q, Savaniu C D, Sudireddy B R, Irvine J T S, Tietz F and Holtappels P 2015 Modified strontium titanates: from defect chemistry to SOFC anodes *RSC Adv.* **5** 1168–80

[55] Moos R and Härdtl K H 1997 Defect chemistry of donor-doped and undoped strontium titanate ceramics between 1000° and 1400 °C *J. Am. Ceram. Soc.* **80** 2549–62

[56] Gross M D, Carver K M, Deighan M A, Schenkel A, Smith B M and Yee A Z 2009 Redox stability of $SrNb_xTi_{1-x}O_3$–YSZ for use in SOFC anodes *J. Electrochem. Soc.* **156** B540–5

[57] Lee S, Kim G, Vohs J M and Gorte R J 2008 SOFC anodes based on infiltration of $La_{0.3}Sr_{0.7}TiO_3$ *J. Electrochem. Soc.* **155** B1179–83

[58] Ramos T, Veltzé S, Sudireddy B R, Theil Kuhn L and Holtappels P 2014 Effect of Ru/CGO versus Ni/CGO co-infiltration on the performance and stability of STN-based SOFCs *Fuel Cells* **14** 1062–5

[59] Hua B, Li M, Sun Y F, Li J H and Luo J L 2017 Enhancing perovskite electrocatalysis of solid oxide cells through controlled exsolution of nanoparticles *ChemSusChem* **10** 3333–41

[60] Cao T, Kwon O, Gorte R J and Vohs J M 2020 Metal exsolution to enhance the catalytic activity of electrodes in solid oxide fuel cells *Nanomaterials* **10** 2445

[61] Kwon O, Joo S, Choi S, Sengodan S and Kim G 2020 Review on exsolution and its driving forces in perovskites *J. Phys.: Energy* **2** 032001

[62] Jo S, Jeong H G, Kim Y H, Neagu D and Myung J H 2021 Stability and activity controls of Cu nanoparticles for high-performance solid oxide fuel cells *Appl. Catalysis* B **285** 119828
[63] Ye L, Hu X, Wang X, Chen F, Tang D, Dong D and Xie K 2019 Enhanced CO_2 electrolysis with a $SrTiO_3$ cathode through a dual doping strategy *J. Mater. Chem.* A **7** 2764–72
[64] Neagu D, Oh T S, Miller D N, Ménard H, Bukhari S M, Gamble S R and Irvine J T 2015 Nano-socketed nickel particles with enhanced coking resistance grown *in situ* by redox exsolution *Nat. Commun.* **6** 1–8
[65] Ruan C and Xie K 2015 A redox-stable chromate cathode decorated with *in situ* grown nickel nanocatalyst for efficient carbon dioxide electrolysis *Catal. Sci. Technol.* **5** 1929–40
[66] Amaya-Dueñas D M, Chen G, Weidenkaff A, Sata N, Han F, Biswas I, Costa R and Friedrich K A 2021 A-site deficient chromite with *in situ* Ni exsolution as a fuel electrode for solid oxide cells (SOCs) *J. Mater. Chem.* A **9** 5685–701
[67] Papaioannou E I, Neagu D, Ramli W K, Irvine J T S and Metcalfe I S 2019 Sulfur-tolerant, exsolved Fe–Ni alloy nanoparticles for CO oxidation *Top. Catal.* **62** 1149–56
[68] Sun Y F, Li J H, Cui L, Hua B, Cui S H, Li J and Luo J L 2015 A-site-deficiency facilitated *in situ* growth of bimetallic Ni–Fe nano-alloys: a novel coking-tolerant fuel cell anode catalyst *Nanoscale* **7** 11173–81
[69] Yang C, Li J, Lin Y, Liu J, Chen F and Liu M 2015 *In situ* fabrication of CoFe alloy nanoparticles structured $(Pr_{0.4}Sr_{0.6})_3(Fe_{0.85}Nb_{0.15})_2O_7$ ceramic anode for direct hydrocarbon solid oxide fuel cells *Nano Energy* **11** 704–10
[70] Lv H *et al* 2020 *In situ* investigation of reversible exsolution/dissolution of CoFe alloy nanoparticles in a co-doped $Sr_2Fe_{1.5}Mo_{0.5}O_{6-\delta}$ cathode for CO_2 electrolysis *Adv. Mater.* **32** 1906193

IOP Publishing

High-Temperature Electrolysis
From fundamentals to applications
Werner Sitte and Rotraut Merkle

Chapter 5

Interconnects and coatings

Belma Talic, Elena Stefan and Yngve Larring

Interconnects provide the electrical contact between individual cells in an electrolysis stack while physically separating the gases on the anode and cathode sides. This chapter introduces the most important theory for evaluating the performance of interconnect materials in terms of high-temperature oxidation, Cr vaporization kinetics, and electrical conductivity. The discussion focuses on ferritic stainless steel materials and the conditions relevant for operation in high-temperature solid oxide and protonic ceramic electrolysis stacks. Different coatings developed to protect the interconnect steel are introduced and compared in terms of their oxidation resistance, Cr retention capability, and area specific resistance. A brief overview is provided of other degradation mechanisms related to the interconnect material, such as interactions with the contact layers and the sealant.

5.1 Introduction

Interconnects are used to distribute current and provide an electrical contact between individual cells in a planar electrolysis stack, while physically separating the gases on the anode and cathode sides. Interconnects are also important for distributing gases across individual cells via flow fields shaped on the surface. The ideal interconnect material should have high electrical conductivity, low permeability for oxygen and hydrogen, be easy to shape, and have a thermal expansion coefficient (TEC) that matches those of the other stack components. In addition, the material should display sufficient stability at the operating temperature of the stack and in the anode- and cathode-side atmospheres. The relevant operating temperature and atmospheres depend on whether the electrolysis cell is based on solid oxide or protonic ceramic technology and on the mode of operation (steam, CO_2, or co-electrolysis). An overview of the most relevant conditions is given in figure 5.1. One important distinction to notice between steam electrolysis in a solid oxide electrolysis cell (SOEC) and a protonic ceramic electrolysis cell (PCEC) is that oxygen is produced on the air side in the former and on the steam side in the latter. This has

doi:10.1088/978-0-7503-3951-3ch5

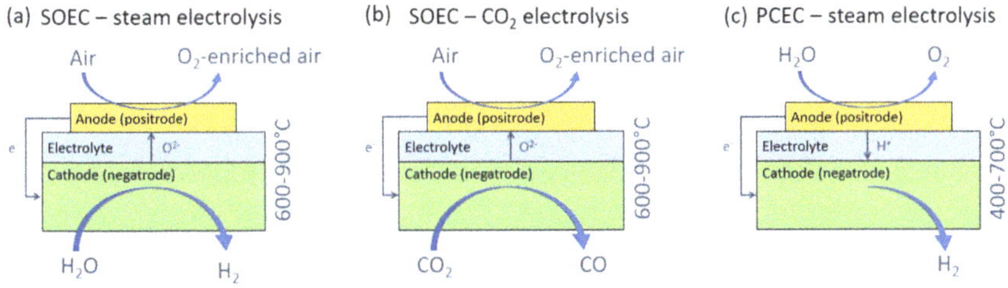

Figure 5.1. Typical operating conditions in different modes of electrolysis operation and for different types of electrolysis cell (SOEC = solid oxide electrolysis cell, PCEC = protonic ceramic electrolysis cell).

important implications for the degradation of the interconnect and the coatings that may be used to protect it, as discussed in section 5.3.

Interconnect materials used for high-temperature electrolysis applications are largely based on materials originally developed for solid oxide fuel cell stacks. For several decades, lanthanum chromite ($LaCrO_3$) doped with Sr, Ca, and other alkaline-earth elements was among the few materials available that displayed appreciable electrical conductivity while being sufficiently stable for use in both the anode and cathode atmospheres at high temperature. However, a great disadvantage of the lanthanum chromite materials is that they are extremely difficult to sinter to full density. Much of the research on lanthanum-chromite-based interconnects therefore focused on improving the sintering capabilities while retaining high electrical conductivity [1].

With the development of solid oxide cells (SOCs) suitable for operation below 900 °C, metallic interconnect materials started to be considered. In addition to being cheaper than lanthanum chromite, metallic interconnects offer the advantages of having high electrical and heat conductivities and are easier to shape. The remainder of this chapter focuses on metallic interconnects, as these are most widely applied by industry today. Ceramic interconnects are still in use, but primarily in niche applications in which the operating temperature remains high.

An early evaluation of candidate metallic interconnect materials in 1992 identified chromium-based alloys and ferritic stainless steels (FSSs) as the most promising alternatives [2]. Conventional chromium-based alloys have high ductile-to-brittle transition temperatures, resulting in low toughness and ductility; however, these can be greatly improved by oxide-dispersion strengthening (ODS). The ODS alloy Cr–5Fe–1Y_2O_3 (CFY, Ducrolloy) was introduced by Plansee specifically for application as an interconnect material [3]. The addition of 5 wt% Fe lowers the TEC to slightly more than that of yttria-stabilized zirconia (YSZ), while the high Cr content provides long-term oxidation resistance. Because ODS alloys rely on powder metallurgical manufacturing routes, they are generally more expensive compared to conventional iron-based alloys such as FSSs.

In the late 1990s/early 2000s, extensive research into FSS for interconnect applications resulted in the launch of several specialized grades, such as Crofer 22 APU, ZMG232L, E-brite, and Sanergy HT [4]. The most important common features of these alloys are a high Cr content (>20 wt%) for long-term stability, the addition of rare-earth elements for improved oxidation resistance and scale adherence, and reduced Si content to prevent the formation of an electrically insulating SiO_2 scale. Among the abovementioned alloys, Crofer 22 APU is probably the one most widely used by the SOC industry today.

A similar grade that was later developed is Crofer 22 H. This alloy has additions of W for solid solution strengthening and Nb to form Laves phase precipitates by reaction with Si, resulting in higher creep and hot tensile strengths than that of Crofer 22 APU [5]. Because the Laves phase precipitates 'bind up' some of the Si in the alloy, Crofer 22 H can contain a larger amount of Si without forming an insulating Si scale. This allows conventional processes to be used in the manufacture of the alloy, making it a less expensive grade compared to Crofer 22 APU, which is manufactured by vacuum induction melting to sufficiently lower the Si content.

Further cost reductions are possible using commodity steels such as AISI 441 and AISI 430. These alloys have lower oxidation resistance than those of the more specialized grades mentioned above, but their use may become increasingly more feasible with the latest development in protective coatings and the further push towards lower operating temperatures [6].

5.2 Theory and characterization methods used to evaluate metallic interconnects

5.2.1 High-temperature oxidation

The following sections summarize some of the thermodynamic and kinetic theory important for understanding the oxidation of interconnect alloys. For a more in-depth treatment of high-temperature oxidation, the reader is referred to specialized textbooks, for example, references [7, 8].

Most metals are unstable in their pure form and oxidize when exposed to oxygen. The minimum partial pressure of oxygen thermodynamically required to oxidize the metal is known as the dissociation pressure and is determined from the Gibbs energy of the reaction between the metal and oxygen. The Ellingham diagram, shown in figure 5.2, provides a convenient way to summarize the Gibbs energy as a function of temperature and oxygen partial pressure (pO_2) for a range of metals. The equilibrium pO_2 values in H_2O–H_2 and CO_2–CO gas mixtures are indicated by straight lines in the diagram. For a SOC operated at more than 700 °C, the relevant atmosphere generally contains sufficient oxygen to oxidize chromium and other elements in the alloy. The only possible exception is the hydrogen side during steam electrolysis in a PCEC, where the atmosphere could theoretically be pure hydrogen (see figure 5.1). However, in practice, the H_2 side most likely contains some steam, because the electrolyte is not purely proton conducting or because steam is intentionally added to improve the proton conductivity of the electrolyte.

While thermodynamics can predict whether a metal will be oxidized or not, kinetic theory is important in order to evaluate how fast the oxide scale grows. The mechanism of high-temperature oxidation can be quite complex and consists of several reaction steps between the metal and oxidant that evolve during the course of oxidation. For

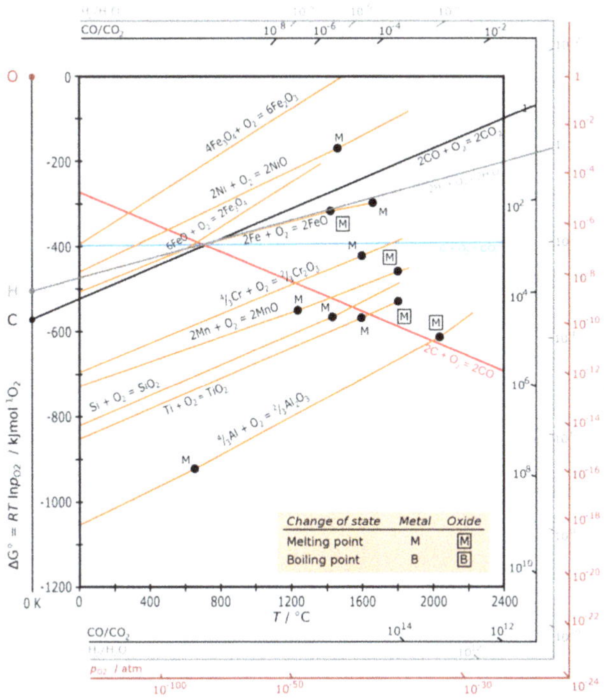

Figure 5.2. Ellingham diagram summarizing the Gibbs energy of reaction as a function of temperature for various metals. Adapted from [9] under the CC BY-SA 4.0 licence.

FSS, the oxide scale growth after an initial, transient period (typically 10–100 h) is, in many cases, found to proceed according to parabolic kinetics, which can be described by

$$x^2 = 2k_p t + C, \tag{5.1}$$

where x is the oxide scale thickness, k_p is the parabolic rate constant, t is the time, and C is a constant reflecting the oxide scale thickness after the initial and transient stages of oxidation. According to the classical oxidation theory developed by Wagner [10], parabolic kinetics reflect the fact that the oxide scale growth is controlled by the diffusion of metal cations or the oxide anion through the oxide scale. The driving force for diffusion is the chemical potential gradient that arises from the increase in pO_2 from the metal/oxide interface to the oxide/gas interface. For an oxide scale that has greater electrical than ionic conductivity (which is usually the case), Wagner showed that the parabolic rate constant is equal to [10]

$$k_p = \frac{1}{2}\int_{p_{O_2}^0}^{p_{O_2}^i}\left(\frac{z_M}{z_A}D_M + D_O\right)\mathrm{d}\ln p_{O_2}, \tag{5.2}$$

where $p_{O_2}^0$ and $p_{O_2}^i$ are the partial pressures of oxygen at the oxide/gas and oxide/metal interfaces, z_M and z_A are the charges of the metal cation and oxide anion, and D_M and D_O are the diffusion coefficients of the metal and oxide ions in the oxide

M_aO_b, respectively. Wagner's model assumes that the oxide scale is continuous, dense, and adherent, which is not always the case in real systems containing voids and microcracks. Nevertheless, the developed theory provides a good basis for evaluating the kinetics of oxide scale growth.

In thermally grown Cr_2O_3, the chromium and oxygen diffusion coefficients at 900 °C are reported to vary from 10^{-17} to 10^{-10} cm^2 s^{-1}, depending on the alloying elements and impurities present in the substrate [11]. Chromium is usually found to migrate faster in the bulk, while oxygen is the fastest migrating species in the grain boundaries. As the diffusivity in grain boundaries is orders of magnitude higher than in the bulk, the grain size can have a significant impact on the growth kinetics and scaling mechanism [12]. The scaling mechanism may also depend on whether the oxidative species is oxygen or steam, and whether the alloy contains or is coated with reactive elements (e.g. Y, Nb, La). It has been suggested that addition of rare-earth elements can change the growth mechanism of the Cr_2O_3 scale from being governed by cation outward diffusion to being governed by oxide ion inward diffusion, the latter of which promotes good adherence between the oxide scale and the metal [13]. In contrast, oxide scale growth dominated by the outward transport of metal cations may result in poor oxide scale adherence due to void formation at the metal/oxide scale interface. However, these theories are frequently based on studies of simple (model) alloys, whereas the oxidation of real alloys is more likely dominated by quick diffusion paths such as grain boundaries, cracks, and pores.

The parabolic oxidation rate constant as defined above can be determined by measuring the thickness of the oxide scale as a function of time. However, it is more practical and less time consuming to evaluate oxidation kinetics by measuring the mass gain due to oxygen uptake. If the composition of the oxide scale is known, the rate constant determined from the mass gain ($k_{p,m}$ [g^2 cm^4 s^{-1}]) can be converted to the parabolic rate constant ($k_{p,t}$ [cm^2 s^{-1}]) using the following equation:

$$k_{p,\,t} = \left(\frac{M_{M_aO_b}}{b \cdot M_O \cdot \rho_{M_aO_b}}\right)^2 \cdot k_{p,m}, \tag{5.3}$$

where $M_{M_aO_b}$ is the molar mass of the oxide, M_O is the molar mass of oxygen, and $\rho_{M_aO_b}$ is the oxide density. Note that this conversion assumes that the oxide scale consists of a single phase, which is seldom the case.

To determine the oxidation rate constant, the mass gain should preferably be continuously recorded using a thermobalance during the oxidation of the sample. However, for practical reasons, the mass gain is commonly measured at room temperature after the sample has been exposed to a higher temperature in a furnace, i.e. after discontinuous oxidation. One should keep in mind that thermal cycling may lead to cracking or spallation of the oxide scale that may have an impact on the mass gain and oxidation kinetics. Furthermore, it is important to take into consideration the potential impact of Cr evaporation (section 5.2.2), which results in a lower measured mass gain relative to the amount of metal that has been oxidized.

To sustain the growth of a chromia scale, the diffusion of Cr to the alloy/oxide scale interface must be fast enough to replace the metal consumed by oxidation.

The minimum content of Cr in the alloy (N_{Cr}) needed to form a Cr_2O_3 scale according to Wagner's diffusional analysis is:

$$N_{Cr} = \frac{V_A}{V_{Cr_2O_3}}\left(\frac{\pi k_p}{2D}\right)^{\frac{1}{2}}, \tag{5.4}$$

where V_A and $V_{Cr_2O_3}$ are the molar volumes of the alloy and Cr_2O_3, respectively; k_p is the parabolic rate constant; and D is the alloy interdiffusion coefficient. The interconnect thickness also has to be considered. Thin interconnects have a relatively smaller reservoir of Cr compared to the surface area and are therefore depleted of Cr faster. Some studies have shown that another reason why thin interconnects may oxidize faster is because of the faster depletion of minor alloying elements such as Mn, Ti, and La and because of a decrease in the creep strength of the component with decreasing thickness [14, 15].

5.2.2 Volatilization of Cr

The oxidation of FSS is influenced by the volatilization of Cr, particularly in wet (>0.1 vol% H_2O), oxidizing conditions, in which the following reaction dominates [16]:

$$1/2Cr_2O_3(s) + H_2O(g) + 3/4O_2(g) \rightleftharpoons CrO_2(OH)_2(g). \tag{5.5}$$

According to the thermodynamics of the above reaction, Cr evaporation is intensified by an increase in the partial pressure of oxygen and water vapor, see figure 5.3. Once a continuous oxide scale is established, the Cr evaporation from FSS typically proceeds according to linear kinetics. The activation energy for Cr evaporation is reported to be ca. 80–90 kJ mol^{-1}, while the activation energy for the oxidation of FSS is 250–280 kJ mol^{-1} [17]. This implies that Cr evaporation has a relatively larger influence on the mass change with decreasing temperature.

The main challenge with Cr evaporation is the reaction between gaseous Cr(VI)-species and the oxygen electrode material, which results in severe degradation of the electrode performance. The degradation of oxygen electrodes due to so-called 'Cr

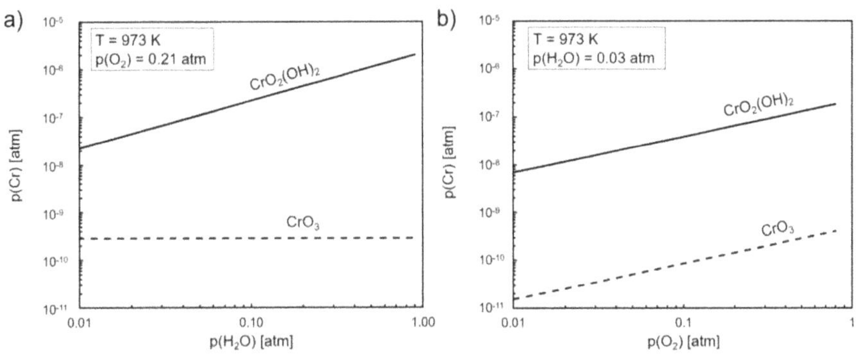

Figure 5.3. Partial pressure of volatile Cr(VI)-species over Cr_2O_3 as a function of steam (a) and oxygen (b) partial pressures. Plotted using thermodynamic data from [16, 18].

poisoning' is discussed in chapter 3. In extreme cases, Cr evaporation may compromise the stability of the interconnect material as well, by hindering the formation of the chromia scale. This may occur if the evaporation kinetics are faster than the oxidation kinetics, or if the Cr content in the steel eventually becomes depleted below the critical level needed to sustain the regrowth of the chromia scale [19]. The consequence may be breakaway oxidation, i.e. the formation of rapidly growing Fe oxides.

Most alloys developed for use as SOC interconnects contain some Mn that leads to the formation of a $(Mn,Cr)_3O_4$ layer on top of the Cr_2O_3 scale. This $(Mn,Cr)_3O_4$ layer reduces the activity of Cr relative to Cr_2O_3 and can thereby reduce the Cr evaporation rate [20]. However, this reduction is not sufficient to avoid 'poisoning' of the oxygen electrode. To achieve a further reduction of Cr evaporation, it is necessary to use protective coatings (see section 5.3.1).

Several methods have been developed to experimentally study Cr vaporization from coated and bare alloys [21]. Among these is a version of the transpiration or entrainment method, in which the hot exhaust gas from a furnace containing the interconnect is passed into a collection tube where the Cr vapors can condense or deposit. For improved collection efficiency, some researchers fill the collection tube (typically fused quartz) with quartz wool [22] or coat it with Na_2CO_3, which readily reacts with the volatile Cr species [23]. The deposited Cr species are removed by washing the tube in concentrated acid or deionized water, and the Cr content in the resulting solution is subsequently analyzed. The Cr evaporation rate from bare FSS in air with 1–3 vol% H_2O and at 800 °C is in the order of 10^{-10}–10^{-9} kg m^{-2} s^{-1} [21]. Under typical operating conditions employing a high flow rate of air, a gaseous boundary layer saturated with Cr(VI)-species is established above the steel surface such that the Cr evaporation rate becomes dependent on the gas flow rate.

Cr evaporation is much lower in the cathode atmosphere (ca. 10^{-14} kg m^{-2} s^{-1} at 800 °C and 20 vol% H_2O–80 vol% H_2) [20]. Nevertheless, observations of Cr depletion from the outer oxide surface of an interconnect after 20 000 h exposure in a stack with a varying H_2O–H_2 ratio indicated that evaporation in the cathode-side atmosphere may be significant during long-term operation [24].

5.2.3 Electrical conductivity

The resistance across a metallic interconnect is typically dominated by the resistivity and thickness of the thermally grown oxide scale. The resistance therefore increases with time due to the growth of the scale. However, because the composition of the oxide scale may also evolve significantly with time, the resistance does not necessarily display a parabolic time dependence, as the oxide scale thickness does. Even for Cr_2O_3 grown on 'pure' Cr metal, the electrical conductivity is found to vary by several orders of magnitude (~10^{-3}–10^{-1} S cm^{-1} at 800 °C) depending on presence and concentration of minor doping elements in the scale [25]. For $(Mn,Cr)_3O_4$, the electrical conductivity varies (4×10^{-3}–0.9 S cm^{-1} at 800 °C) depending on the Mn/Cr ratio [26], which may evolve with time in a thermally grown oxide scale.

The area specific resistance (ASR) of the interconnect should be as low as possible to limit ohmic losses across the stack. In the past, 0.1 Ω cm^2 was considered to be an

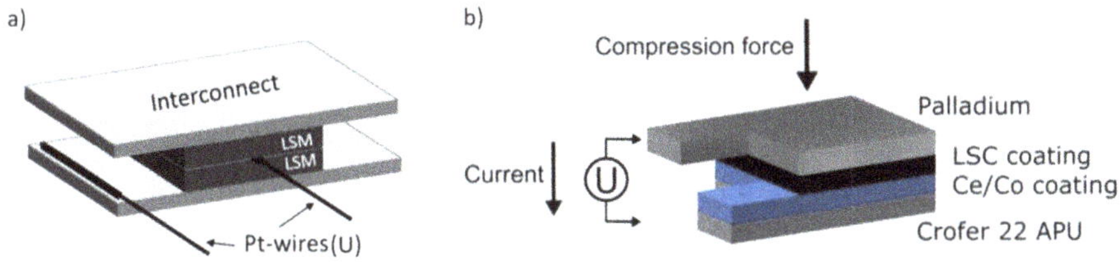

Figure 5.4. Two variations of the setup used to measure the interconnect ASR. The current flows in the vertical direction and the voltage drop is measured by two sensing contacts at the side. Panel (b) is adapted from [28] under a CC BY licence.

acceptable upper limit [27] and is still frequently quoted. However, the value considered acceptable for the interconnect ASR really depends on the operating temperature of the stack and the resistance of the cell. A better rule of thumb is therefore that the interconnect should not dominate the overall stack resistance. The most common way to evaluate the interconnect ASR is to use a four-point setup, as illustrated in figure 5.4. The method by which the interconnect is contacted in this measurement varies. Some researchers use platinum or silver paste to contact the steel surface, while others use ceramic perovskite oxides similar to those employed as contact materials or electrodes in the stack. The contact material can have a significant influence on the measured ASR value, and noble metal contacts generally result in lower ASR values than perovskite oxide contacts.

Noble metals such as Pt are often considered to be inert, but it has been shown that they can lead to anomalous oxidation of the FSS if the ASR is measured over time at a high temperature, thus resulting in ASR values that are not representative of those observed in a real electrolysis stack [29]. Perovskite oxides are, in this respect, a more reliable choice, but the different compositions and densities of the contact materials used by different labs makes it more difficult to compare results with those published in the literature. Perovskite oxides may also react with the oxide scale or coating on the interconnect and thereby influence the measured ASR, but the behavior is, in this case, more representative of what can be expected under real operating conditions in a stack.

It is important to note that even when $La_{1-x}Sr_xMnO_3$ or $La_{1-x}Sr_xCoO_3$ is used as the contact material, the *ex situ* measured ASR may not be equal to the ASR contribution of the interconnect in an actual stack, because of differences in the applied load and fabrication of the contact layer. Some authors have also reported that the current direction may have an influence on the ASR [30]. Nevertheless, *ex situ* ASR measurements are valuable as a tool for comparing different alloys and coatings against each other. For the reasons mentioned above, the comparison is only meaningful if systematic studies are performed using the same setup and contact materials.

5.3 Degradation of interconnects in SOEC atmospheres

The following sections summarize the behavior of metallic interconnects in SOEC-relevant atmospheres and provide an overview of coating materials that have been developed to mitigate degradation.

5.3.1 Oxygen-rich atmospheres

The atmosphere on the anode side of the interconnect can be either oxygen-enriched air, pure oxygen, or a steam–oxygen mixture, depending on whether the cell is based on an oxide ion or a protonic ceramic electrolyte (see figure 5.1). As the much higher steam activity during PCEC operation may have strong implications for the degradation rate of the interconnect, 'dry' (SOEC) and wet (PCEC) conditions are treated separately in the following.

5.3.1.1 SOEC: dry oxygen and oxygen-enriched air

The anode side of the interconnect in a SOEC stack is exposed to oxygen or oxygen-enriched air. The main challenges for the interconnect under these conditions are: (i) oxide scale growth and the resulting increase in ASR, and (ii) Cr evaporation, which results in poisoning of the oxygen electrode. Several coating materials have been developed and evaluated to mitigate these challenges. Initially, perovskite oxides similar to the oxygen electrode material were used for the interconnect coating, e.g. $La_{1-x}Sr_xMnO_3$, $La_{1-x}Sr_xFeO_3$, and $La_{1-x}Sr_xCoO_3$ [31, 32]. Perovskite oxides have many attractive properties that make them suitable for use as protective coatings, such as high electrical conductivity and TECs that closely match those of the interconnect and other cell components. Cation diffusion is typically slow in perovskite oxides, resulting in good protection against Cr evaporation [33]. However, to achieve good oxidation resistance and Cr retention capability, the coating has to be sintered to a high density, which typically requires temperatures in excess of 1000 °C for perovskite oxides. Such high temperatures lead to excessive oxidation of the underlying interconnect steel, while lower sintering temperatures result in more porous and less protective coatings.

Another class of coating materials that has received much attention is the rare-earth metals, such as La, Ce, and Y. These are also used as minor alloying elements in some interconnect alloys, e.g. La is used in Crofer 22 APU. While rare-earth coatings are efficient at reducing the oxidation rate and the ASR of the interconnect [34], they do not provide any significant protection against Cr vaporization [35]. These coatings are therefore typically used in combination with a second coating material.

A third class of coating materials is the spinel oxides. $(Mn,Co)_3O_4$ spinel was first identified as a promising interconnect coating two decades ago and is still considered to be among the best coating materials for oxidizing conditions [32]. Several variations of the coating have been evaluated, for example, those in which some or all of the Co is replaced by Fe or Cu [36, 37]. Like the perovskite coatings, spinel coatings deposited by powder-based methods need to be sintered after deposition, but this can be achieved at lower temperatures than those used for the perovskite oxide coatings by employing a two-step sintering procedure wherein the coating is first reduced to MnO and Co and then re-oxidized to the spinel phase [38]. This results in a high coating density while limiting the oxidation of the alloy.

It has been shown that a dense $(Mn,Co)_3O_4$ coating can reduce the parabolic oxidation rate of FSS by one to two orders of magnitude (from 10^{-14}–10^{-13} $g^2\,cm^{-4}\,s^{-1}$

to 10^{-15} g^2 cm^{-4} s^{-1} at 800 °C) and reduce the Cr vaporization rate by 99% compared to bare FSS [39]. The reduction in oxidation rate obtained using these coatings has been attributed to the formation of a $(Mn,Co,Cr)_3O_4$ reaction layer at the coating/Cr_2O_3 interface. Oxygen isotope exchange measurements indicate that the oxygen diffusivity in a $MnCo_2O_4$ coating is similar to that in a thermally grown Cr_2O_3 scale (~10^{-12} cm^2 s^{-1} at 800 °C), while the oxygen diffusivity in the $(Mn,Co,Cr)_3O_4$ reaction layer is significantly lower [40].

Another coating that has shown promising performance in the anode-side atmosphere is a thin layer of metallic CeCo applied by physical vapor deposition [41]. During operation in the stack, this coating is first oxidized to Co_3O_4 and eventually transformed to $(Mn,Co)_3O_4$ by a reaction with Mn diffusing out of the steel [42].

The various coatings described above have been deposited onto the steel interconnect using a range of different methods, such as electrophoretic deposition, electroplating, spray coating, high-velocity plasma spraying, sputtering, etc [36, 43–45]. The different coating methods result in differences in microstructure and density that may significantly influence the protective capability of the coating. This was nicely illustrated by Bianco *et al* [46] who evaluated $(Mn,Co)_3O_4$ coatings produced by five different deposition methods and reported ASR values ranging from 10 to 80 mΩ cm^2. The best results in terms of ASR were achieved when the coating was fabricated by physical vapor deposition (PVD), presumably because a high coating density was achieved, see figure 5.5. For coatings applied as an oxide powder, the heat treatments used to densify the coating after deposition lead to preoxidation of the steel, which may also have an impact on the subsequent oxidation kinetics due to minor changes in the oxide scale composition and microstructure [47].

Most of the research into interconnects and coatings summarized above has been conducted under conditions simulating the air side of a solid oxide fuel cell stack. These results are, to a large degree, applicable to electrolysis stacks, as current

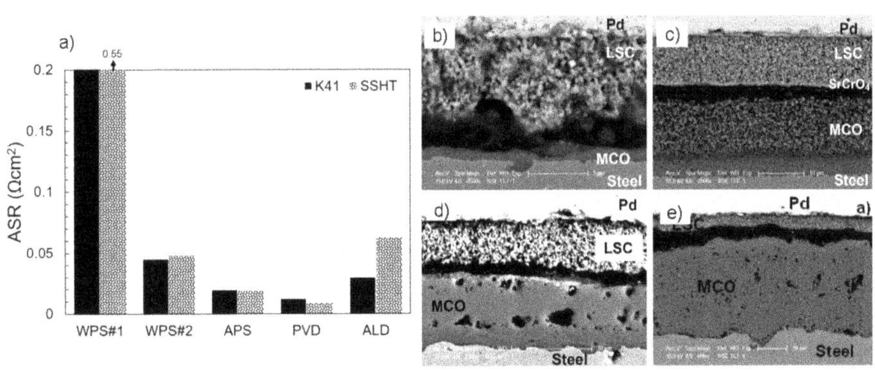

Figure 5.5. (a) ASR and (b)–(e) scanning electron microscope (SEM) cross sectional images of $MnCo_2O_4$ (MCO)-coated AISI 441 (K41) and Sanergy HT (SSHT) prepared using different deposition methods: (b) atomic layer deposition (ALD), (c) wet powder spraying (WPS), (d) physical vapor deposition (PVD), and (e) atmospheric plasma spraying (APS). Figures 5.5(a) and 5.5(b) are reproduced from [46] and [41], respectively, with permission.

practice is to use a high air flow to regulate the temperature of the stack during both modes of operation. However, the oxygen produced as a by-product during electrolysis is potentially valuable and the air sweep used to regulate the temperature can be avoided by operating under thermoneutral conditions. In this case, the interconnect in an electrolysis stack would be exposed to pure oxygen.

Whether or not an increase in the oxygen partial pressure increases the oxidation rate of the interconnect has been debated [48, 49]. Applying the theory of Wagner-type oxidation kinetics introduced above, it can be shown that the oxidation rate should increase with increasing oxygen partial pressure if the chromia scale defect structure is dominated by cation vacancies. If, instead, the defect structure is dominated by interstitial cations, the oxidation rate should be independent of the oxygen partial pressure [12]. The defect structure of chromia depends on pO_2, the presence of impurities, and other factors [25, 50], which may explain the seemingly conflicting results in the literature regarding the pO_2 dependence of the oxidation rate.

5.3.1.2 PCEC: oxygen–steam mixtures

The anode side of the interconnect in a PCEC is exposed to higher steam activities and a lower operating temperature than the interconnect in a SOEC. There have been very few investigations of interconnects and coatings under these conditions, but from more general work on the oxidation of chromia-forming alloys, it is well known that the presence of water vapor can accelerate degradation [8]. Various mechanisms have been proposed to explain this observation; for example, proposals have suggested that H_2O accelerates the oxygen diffusion within the scale or its cavities, modifies the defect structure of the Cr_2O_3 scale, or increases the amount of internal oxidation [51, 52]. In wet, oxygen-rich conditions, the large increase in Cr vaporization with increasing steam content (see figure 5.3) is likely to be an important factor in the accelerated degradation. In extreme cases, the vaporization of Cr may equal the rate of oxide scale formation, leading to a breakdown of the chromia scale and ultimately failure of the interconnect due to breakaway oxidation.

One of the few studies that considered PCEC-relevant atmospheres reported that the interconnect alloy Crofer 22 APU, which normally forms a slow-growing Cr_2O_3 scale in 'dry' (<5 vol% H_2O) air at 650 °C, undergoes breakaway oxidation after less than 100 h of oxidation in air–50 vol% H_2O at the same temperature [53]. Preoxidation of the alloy in dry air at 850 °C to form a continuous Cr_2O_3 scale prior to oxidation at 650 °C was insufficient to protect the steel, while coatings such as $Ce_{0.02}Mn_{1.49}Co_{1.49}O_4$, $CuMn_{1.8}O_4$, and CeCo provided good protection and slow oxidation rates ($1–3 \times 10^{-16}$ g^2 cm^{-4} s^{-1}). Results from this work are shown in figure 5.6.

Thus, coatings originally developed for the protection of the interconnect in relatively dry air (<5 vol% H_2O) may also be applicable at higher steam concentrations. However, more work is needed to evaluate the Cr vaporization rate and to confirm the oxidation resistance of the coated interconnect during long-term exposure to PCEC-relevant atmospheres and temperatures. Another potential challenge caused by applying materials originally developed for SOEC in PCEC stacks is the TEC mismatch between the interconnect/coating ($11–13 \times 10^{-6}$ K^{-1}) and many of the most relevant PCEC materials ($8–11 \times 10^{-6}$ K^{-1}) [54].

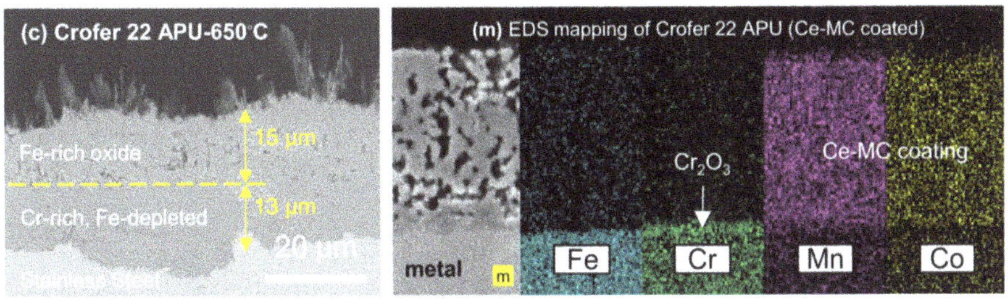

Figure 5.6. SEM cross sectional images of materials oxidized in air–50 vol% H_2O for 500 h at 650 °C. Left: bare Crofer 22 APU. Right: $Ce_{0.02}Mn_{1.49}Co_{1.49}O_4$-coated Crofer 22 APU. Adapted from [53] with permission.

5.3.2 Hydrogen and hydrogen–steam atmospheres

The cathode side of the interconnect in a SOEC stack is exposed to varying steam activity from the inlet (~80–90 vol% H_2O–H_2) to the outlet (~10–20 vol% H_2O–H_2) of the stack. The behavior of the interconnect under these conditions has received considerably less attention than its behavior under anode/air-side conditions. Furthermore, studies that do consider cathode-side conditions have mainly evaluated the behavior under the relatively 'mild' conditions of a low H_2O content, corresponding to SOFC inlet conditions or the SOEC outlet under high steam utilization. Work by Folgner *et al* [55] indicated that there is no significant difference in the oxidation rate of interconnect alloys in 50 vol% H_2O versus 80 vol% H_2O, but it remains unclear whether this also extends to lower steam activities.

The thermodynamic driving force for oxidation is decreased in a H_2O–H_2 atmosphere compared to an air/oxygen atmosphere, and several studies have reported slower oxide scale growth rates in H_2O–H_2 than in air [14, 48, 56]. However, studies have also reported higher oxidation rates in H_2–H_2O atmospheres than in air [57]. Niewolak *et al* [58] found that the oxidation rate of Crofer 22 H at 800 °C was initially slower in air than in Ar–H_2–H_2O, but that with longer exposure (ca. 250 h), the oxidation rate became faster in air. It was suggested that the change in oxidation mechanism could be linked to the formation of a $(Mn,Cr)_3O_4$ spinel layer on top of the Cr_2O_3. Once this layer becomes continuous, the oxidation rate in Ar–H_2–H_2O becomes slower than in air.

Temperature seems to be an important factor that affects whether the oxidation is faster in air or in H_2O–H_2. For example, Stefan *et al* [59] found that at 600 °C and 700 °C, a porous FeCr alloy oxidized faster in 2.5 vol% H_2O–5 vol% H_2–Ar than in 2.5 vol% H_2O–air, while at 800 °C, the oxidation was faster in air. According to Young *et al* [60], the activation energy for parabolic oxidation of high-Cr FSS in air (226 kJ mol^{-1}) is twice the activation energy in 20 vol% H_2O–4 vol% H_2 H_2–Ar (110 kJ mol^{-1}). These values were compared to the activation energy for the interdiffusion of cations in a FeCr alloy (ca. 180 kJ mol^{-1}). The rate of interdiffusion should be faster than the rate of oxidation to avoid void formation at the oxide scale/metal interface and sustain the growth of a protective Cr_2O_3 scale. Comparing the above activation energies and considering equation (5.4) indicates that the critical Cr content required to sustain a Cr_2O_3 scale increases with decreasing temperature in a H_2O–H_2 atmosphere, while the opposite is true in air. This implies that the oxidation mechanism may change from protective to non-protective as the temperature is *decreased* in H_2O–H_2 atmospheres, and as the temperature is *increased* in air.

5.3.2.1 Coatings for the cathode side

Far less work has been done to develop coatings suitable for the cathode side of the interconnect, and in stacks, the cathode side is usually left uncoated [24, 61]. Alternatively, a Ni-layer is applied to improve contact with the Ni-based contact layer or electrode. Many of the coatings commonly applied to the oxygen-rich side of the interconnect, for example, spinel and certain perovskite oxides, decompose on the hydrogen-rich side because of the low pO_2. However, some perovskite oxides are also stable at low pO_2 and appear promising for use as cathode-side coatings. For example, $La(Mn_{0.5}Co_{0.5})_{0,8}O_3$ has been reported to greatly reduce the oxidation rate of porous FeCr substrates in both air and H_2–H_2O atmospheres at 600 °C–800 °C [59].

As Cr evaporation is less critical on the cathode side of the interconnect, reactive rare earth metals may be considered as possible coating materials that can be used on their own. Molin *et al* [62] evaluated the oxidation behavior and electrical resistance of Crofer 22 APU and SUS316 at 650 °C in 90 vol% H_2O–H_2 and reported that good protection was provided by both a Y-coating and a Gd-doped CeO_2 coating. Sachitanand *et al* [63] evaluated the oxidation behavior of Sanergy HT and AISI 441 in 3 vol% H_2O–5 vol% H_2Ar at 850 °C and reported two- to threefold reduction in the oxidation rate when a La or Ce coating was used.

5.3.3 CO_2–CO atmospheres

In atmospheres with high carbon activity, the interconnect steel may be subject to a type of degradation known as internal carburization or metal dusting [64]. Internal carburization is possible if the carbon activity in the atmosphere is higher than the equilibrium carbon activity required to form chromium carbides. The formation of a continuous, dense oxide scale can greatly suppress carbide formation by suppressing the transport of carbonaceous species to the metal surface [8]. However, in an SOEC stack, the formation of a protective scale may be hindered where the interconnect is in contact with Ni [65]. Niewolak *et al* [66] studied the oxidation of interconnect steels in 9.2 vol% CO_2–3.7 vol% H_2–0.2 vol% H_2O H_2O–Ar at 600 °C and found that internal chromium carbides were formed only when the steel surface was coated with Ni (5–10 μm thickness). The amount of carbide formation increased for increasing thicknesses of the Ni coating, increasing temperatures, and decreasing Cr contents of the steel.

Metal dusting involves the disintegration of the alloy into fine metal particles and graphite and can only occur if the carbon activity at the metal surface is >1 [67]. Such high carbon activities are generally avoided in the SOEC stack, as they also present a challenge in terms of C deposition on the Ni-based electrode. Metal dusting is thus unlikely to occur in an electrolysis stack.

5.3.4 Other forms of interconnect degradation

5.3.4.1 Interaction with Ni-based electrodes or contact layers

The cathode side of the interconnect is often left uncoated, so that the alloy is in direct contact with the Ni-based electrode(-support) or a Ni-mesh/foam contact layer. The metal-to-metal contact provides very low contact resistance; however, it also allows for interdiffusion between Ni and the interconnect alloy, as seen in figure 5.7. Ni diffusion into FSS leads to partial transformation of the steel structure from ferritic to austenitic, with a corresponding increase in the TEC from ca. 12×10^{-6} to 18×10^{-6} K^{-1} (room temperature–800 °C) [68]. The large difference in TEC between the bulk of the interconnect and the austenite region increases the risk of contact loss during thermal cycling. Furthermore, Ni diffusion into the steel has been found to accelerate the nucleation and growth of the brittle intermetallic σ phase in the alloy [69]. Cr and Mn from the FSS diffuse in the opposite direction, i.e. into the Ni contact layer or electrode, leading to the formation of oxides that may increase contact resistance.

Preoxidation of the interconnect was initially suggested as a way of impeding interdiffusion between the interconnect and Ni [70]. However, during long-term operation (1000 h), it was observed that Ni was able to dissociate the oxide scale [71]. The same study found that applying a thin layer of CeO_2+Cu to the interconnect provides a more efficient diffusion barrier, although this comes at the expense of a slight increase in the ASR.

5.3.4.2 Interactions with the oxygen-side contact layer

On the anode/oxygen side of the interconnect, perovskite or spinel oxides are commonly applied as the contact/current collection layer. The similarity between the contact material and the interconnect coating provides a basis for good compatibility, but long-term stability may nevertheless be compromised if the elemental concentration gradient between the coating and the contact layer is large. A large chemical gradient (arising from a difference in elemental concentrations) may lead

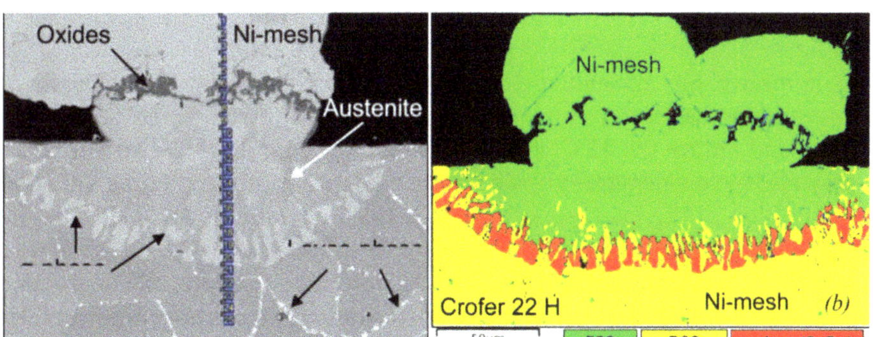

Figure 5.7. SEM backscatter detector images (left) and electron backscatter diffraction (EBSD) phase maps (right) of Crofer 22 APU in contact with a Ni mesh after exposure to 4 vol% H_2–2 vol% H_2OAr for 3000 h at 700 °C. Reproduced from [69] with permission.

to cation diffusion across the different layers. If the coating is relatively thin compared to the contact layer, this cation diffusion may eventually 'consume' the coating, so that it loses its protective effect.

5.3.4.3 Dual atmosphere

Most oxidation studies of metallic interconnects are performed in a single atmosphere, studying the behavior either in the anode or the cathode atmosphere. In a real stack, the interconnect is exposed to the two atmospheres simultaneously. Several studies have found that so-called dual-atmosphere exposure can lead to different oxidation behavior than exposing the same steel separately to the two atmospheres [72, 73]. Alloys that form stable Cr_2O_3 oxide scales in air were found to form Fe-rich oxides in dual atmospheres on the side exposed to air, particularly at lower temperatures (<700 °C). The reasons for this anomalous oxidation behavior are not fully understood, although several possible mechanisms involving hydrogen transport to the air/oxygen side of the interconnect have been proposed [74].

Investigations of interconnects after long-term operation in a stack have so far not reported any anomalous oxidation on the air side that can be attributed to the dual-atmosphere effect. This could be due to the use of coatings on the air side, preoxidation of the steel during coating fabrication and stack assembly, or the stack having been operated at a relatively high temperature (>750 °C). All of these have been found to reduce the extent of the dual-atmosphere effect [75].

5.3.4.4 Reactions with the sealant

Glasses and glass ceramics are commonly used as the sealant materials in planar SOEC stacks, and a wide range of different glass compositions have been evaluated in recent years. Many of these are based on alkaline-earth aluminosilicates with BaO added to increase the TEC to match that of the interconnect and cell materials. Unfortunately, Ba and other alkalis are prone to react with volatile Cr species from the interconnect to form chromates that may negatively impact the mechanical integrity of the joint [61]. Furthermore, glass(-ceramic) sealants have, under certain conditions, been observed to induce breakaway oxidation of the interconnect, as seen in figure 5.8.

An alternative way of sealing electrolysis stacks is by metal brazing. The metal braze is usually based on Ag with some added Cu(O) to improve its wetting properties and reduce the TEC compared to that of pure Ag. However, this sealant type may also have a negative impact on the oxidation resistance of the interconnect steel. Inspection of the interconnects from a stack operated for 34 000 h at 700 °C revealed Fe oxide formation after interaction with Cu(O) in a Ag-based sealant [61].

A promising solution that can mitigate the abovementioned challenges with both glass(-ceramic) sealants and metal brazes is to aluminize the interconnect to form an alumina layer on the surface [77]. The alumina layer provides excellent oxidation resistance and prevents reactions with the sealant material. Because alumina has low electrical conductivity, aluminization must be restricted to the sealing area of the interconnect.

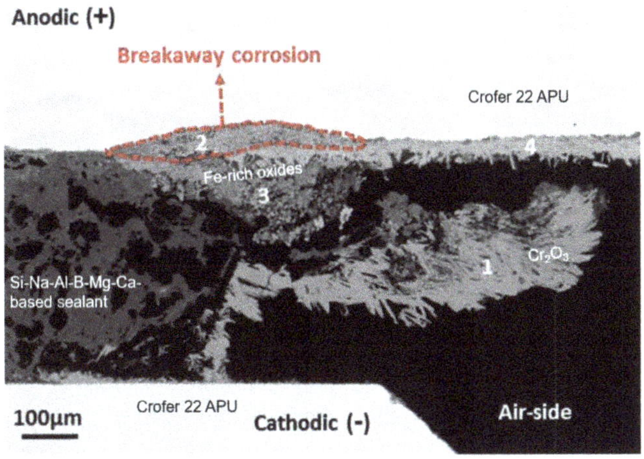

Figure 5.8. Breakaway oxidation of Crofer 22 APU in contact with a Si–Na–Al–B–Mg–Ca-based glass ceramic sealant after a dual-atmosphere test at 800 °C and an applied voltage of 1.6 V. Adapted from [76] with permission.

5.4 Concluding remarks

Several decades of research into interconnects and protective coatings has led to the development of materials that display good durability under typical SOEC operating conditions. However, the shift to lower operating temperatures and protonic ceramic electrolytes may require the development of new materials.

For operation at lower temperatures (<650 °C), more work will be needed to better understand the oxidation mechanism in the cathode-side (H_2O–H_2) atmosphere, how it differs from oxidation in air, and how oxidation in H_2O–H_2 is impacted by reactive rare earth coatings or alloying elements. For operation in PCEC, the oxygen–steam side of the interconnect will likely be the most challenging to handle. Although preliminary results show that coatings developed for SOEC may also be suitable for PCEC, more work is needed to confirm this. This application would also benefit from a better fundamental understanding of the impact of steam on the oxidation kinetics.

Acknowledgment

This chapter was written with financial support provided by WINNER, a project that has received funding from the Fuel Cell and Hydrogen 2 Joint Undertaking under Grant Agreement No (101007165). This Joint Undertaking receives support from the European Union's Horizon 2020 Research and Innovation program, Hydrogen Europe, and Hydrogen Europe Research.

References

[1] Fergus J W 2004 Lanthanum chromite-based materials for solid oxide fuel cell interconnects *Solid State Ionics* **171** 1–15

[2] Kofstad P and Bredesen R 1992 High temperature corrosion in SOFC environments *Solid State Ionics* **52** 69–75
[3] Köck W, Martinz H-P, Greiner H and Janousek M 1995 Development and processing of metallic Cr based materials for SOFC parts *ECS Proc. Vol.* **1995-1** 841
[4] Hojda R, Heimann W and Quadakkers W J 2003 Production-capable materials concept for high-temperature fuel cells *ThyssenKrupp techforum* 20–3
[5] Froitzheim J, Meier G H, Niewolak L, Ennis P J, Hattendorf H, Singheiser L and Quadakkers W J 2008 Development of high strength ferritic steel for interconnect application in SOFCs *J. Power Sources* **178** 163–73
[6] Talic B, Venkatachalam V, Hendriksen P V and Kiebach R 2019 Comparison of $MnCo_2O_4$ coated crofer 22 H, 441, 430 as interconnects for intermediate-temperature solid oxide fuel cell stacks *J. Alloys Compd.* **821** 153229
[7] Kofstad P 1988 *High Temperature Corrosion* (London: Elsevier Applied Science)
[8] Young D J (ed) 2008 *High Temperature Oxidation and Corrosion of Metals* 1 (Amsterdam: Elsevier)
[9] DerSilberspiegel 2022 https://en.wikipedia.org/w/index.php?title=Ellingham_diagram&oldid=1065408429
[10] Wagner C 1952 Theoretical analysis of the diffusion processes determining the oxidation rate of alloys *J. Electrochem. Soc.* **99** 369
[11] Chevalier S 2010 Mechanisms and kinetics of oxidation *Shreir's Corrosion* 1 (Amsterdam: Elsevier) 1.07 132–52
[12] Hindam H and Whittle D P 1982 Microstructure, adhesion and growth kinetics of protective scales on metals and alloys *Oxid. Met.* **18** 245–84
[13] Chevalier S 2014 What did we learn on the reactive element effect in chromia scale since Pfeil's patent? *Mater. Corros.* **65** 109–15
[14] Huczkowski P, Ertl S, Piron-Abellan J, Christiansen N, Höfler T, Shemet V, Singheiser L and Quadakkers W J 2005 Effect of component thickness on lifetime and oxidation rate of chromia forming ferritic steels in low and high pO_2 environments *Mater. High Temp.* **22** 253–62
[15] Asensio-Jimenez C, Niewolak L, Hattendorf H, Kuhn B, Huczkowski P, Singheiser L and Quadakkers W J 2013 Effect of specimen thickness on the oxidation rate of high chromium ferritic steels: the significance of intrinsic alloy creep strength *Oxid. Met.* **79** 15–28
[16] Opila E J, Myers D L, Jacobson N S, Nielsen I M B, Johnson D F, Olminsky J K and Allendorf M D 2007 Theoretical and experimental investigation of the thermochemistry of $CrO_2(OH)_2(g)$ *J. Phys. Chem.* A **111** 1971–80
[17] Falk-Windisch H, Svensson J E and Froitzheim J 2015 The effect of temperature on chromium vaporization and oxide scale growth on interconnect steels for solid oxide fuel cells *J. Power Sources* **287** 25–35
[18] Ebbinghaus B B 1993 Thermodynamics of gas phase chromium species: the chromium oxides, the chromium oxyhydroxides, and volatility calculations in waste incineration processes *Combust. Flame* **93** 119–37
[19] Sachitanand R, Svensson J-E and Froitzheim J 2015 The influence of Cr evaporation on long term Cr depletion rates in ferritic stainless steels *Oxid. Met.* **84** 241–57
[20] Stanislowski M, Wessel E, Hilpert K, Markus T and Singheiser L 2007 Chromium vaporization from high-temperature alloys I. chromia-forming steels and the influence of outer oxide layers *J. Electrochem. Soc.* **154** A295–306

[21] Key C, Eziashi J, Froitzheim J, Amendola R, Smith R and Gannon P 2014 Methods to quantify reactive chromium vaporization from solid oxide fuel cell interconnects *J. Electrochem. Soc.* **161** C373–81
[22] Tatar G, Gannon P, Swain N, Mason R, Remington E and Dansereau S 2019 Investigation of surface interactions between volatile chromium species and ceramics *Surf. Interface Anal.* **51** 506–15
[23] Froitzheim J, Ravash H, Larsson E, Johansson L G and Svensson J E 2010 Investigation of chromium volatilization from FeCr interconnects by a denuder technique *J. Electrochem. Soc.* **157** B1295–300
[24] Ghiara G, Piccardo P, Bongiorno V, Repetto L, Geipel C and Spotorno R 2021 Characterization of metallic interconnects extracted from solid oxide fuel cell stacks operated up to 20,000 h in real life conditions: the fuel side *Int. J. Hydrog. Energy* **46** 23815–27
[25] Holt A and Kofstad P 1994 Electrical conductivity and defect structure of Cr_2O_3. II. Reduced temperatures (<1000 °C) *Solid State Ionics* **69** 137–43
[26] Lu Z, Zhu J, Andrew Payzant E and Paranthaman M P 2005 Electrical conductivity of the manganese chromite spinel solid solution *J. Am. Ceram. Soc.* **88** 1050–3
[27] Zhu W Z and Deevi S C 2003 Opportunity of metallic interconnects for solid oxide fuel cells: a status on contact resistance *Mater. Res. Bull.* **38** 957–72
[28] Tallgren J, Himanen O, Bianco M, Mikkola J, Thomann O, Rautanen M, Kiviaho J and Van herle J 2019 Method to measure area specific resistance and chromium migration simultaneously from solid oxide Fuel cell interconnect materials *Fuel Cells* **19** 570–7
[29] Grolig J G, Froitzheim J and Svensson J-E 2015 Coated stainless steel 441 as interconnect material for solid oxide fuel cells: evolution of electrical properties *J. Power Sources* **284** 321–7
[30] Kawamura K, Nitobe T, Kurokawa H, Ueda M and Maruyama T 2012 Effect of electric current on growth of oxide scale on Fe–25Cr alloy for SOFC interconnect at 1073 K *J. Electrochem. Soc.* **159** B259–64
[31] Montero X, Jordán N, Pirón-Abellán J, Tietz F, Stöver D, Cassir M and Villarreal I 2009 Spinel and perovskite protection layers between crofer22APU and $La_{0.8}Sr_{0.2}FeO_3$ cathode materials for SOFC interconnects *J. Electrochem. Soc.* **156** B188–96
[32] Larring Y and Norby T 2000 Spinel and perovskite functional layers between plansee metallic interconnect (Cr-5 wt% Fe–1 wt% Y_2O_3) and ceramic $(La_{0.85}Sr_{0.15})_{0.91}MnO_3$ cathode materials for solid oxide fuel cells *J. Electrochem. Soc.* **147** 3251–6
[33] Kurokawa H, Jacobson C P, DeJonghe L C and Visco S J 2007 Chromium vaporization of bare and of coated iron–chromium alloys at 1073 K *Solid State Ionics* **178** 287–96
[34] Fontana S, Chevalier S and Caboche G 2012 Metallic Interconnects for solid oxide fuel cell: performance of reactive element oxide coating during 10, 20 and 30 months exposure *Oxid. Met.* **78** 307–28
[35] Grolig J G, Froitzheim J and Svensson J-E 2014 Coated stainless steel 441 as interconnect material for solid oxide fuel cells: oxidation performance and chromium evaporation *J. Power Sources* **248** 1007–13
[36] Wei P, Bateni M R and Petric A 2012 Conversion of copper and manganese metallic films to spinel coating *J. Mater. Sci.* **47** 5205–15
[37] Talic B, Molin S, Wiik K, Hendriksen P V and Lein H L 2017 Comparison of iron and copper doped manganese cobalt spinel oxides as protective coatings for solid oxide fuel cell interconnects *J. Power Sources* **372** 145–56

[38] Yang Z, Xia G, Simner S P and Stevenson J W 2005 Thermal growth and performance of manganese cobaltite spinel protection layers on ferritic stainless steel SOFC interconnects *J. Electrochem. Soc.* **152** A1896–901

[39] Talic B, Falk-Windisch H, Venkatachalam V, Hendriksen P V, Wiik K and Lein H L 2017 Effect of coating density on oxidation resistance and Cr vaporization from solid oxide fuel cell interconnects *J. Power Sources* **354** 57–67

[40] Horita T *et al* 2008 Diffusion of oxygen in the scales of Fe–Cr alloy interconnects and oxide coating layer for solid oxide fuel cells *Solid State Ionics* **179** 2216–21

[41] Tallgren J, Bianco M, Himanen O, Thomann O, Kiviaho J and van Herle J 2015 Evaluation of protective coatings for SOFC interconnects *ECS Trans.* **68** 1597–608

[42] Canovic S, Froitzheim J, Sachitanand R, Nikumaa M, Halvarsson M, Johansson L-G and Svensson J-E 2013 Oxidation of Co- and Ce-nanocoated FeCr steels: a microstructural investigation *Surf. Coat. Technol.* **215** 62–74

[43] Trebbels R, Markus T and Singheiser L 2009 Investigation of chromium vaporization from interconnector steels with spinel coatings *J. Fuel Cell Sci. Technol.* **7** 011013

[44] Puranen J, Pihlatie M, Lagerbom J, Salminen T, Laakso J, Hyvärinen L, Kylmälahti M, Himanen O, Kiviaho J and Vuoristo P 2014 Influence of powder composition and manufacturing method on electrical and chromium barrier properties of atmospheric plasma sprayed spinel coatings prepared from $MnCo_2O_4$ and Mn_2CoO_4 + Co powders on Crofer 22 APU interconnectors *Int. J. Hydrog. Energy* **39** 17246–57

[45] Zanchi E, Molin S, Sabato A G, Talic B, Cempura G, Boccaccini A R and Smeacetto F 2020 Iron doped manganese cobaltite spinel coatings produced by electrophoretic co-deposition on interconnects for solid oxide cells: microstructural and electrical characterization *J. Power Sources* **455** 227910

[46] Bianco M, Tallgren J, Hong J-E, Yang S, Himanen O, Mikkola J, Van herle J and Steinberger-Wilckens R 2019 *Ex-situ* experimental benchmarking of solid oxide fuel cell metal interconnects *J. Power Sources* **437** 226900

[47] Talic B, Molin S, Hendriksen P V and Lein H L 2018 Effect of pre-oxidation on the oxidation resistance of crofer 22 APU *Corros. Sci.* **138** 189–99

[48] Palcut M, Mikkelsen L, Neufeld K, Chen M, Knibbe R and Hendriksen P V 2010 Corrosion stability of ferritic stainless steels for solid oxide electrolyser cell interconnects *Corros. Sci.* **52** 3309–20

[49] Alnegren P, Sattari M, Froitzheim J and Svensson J-E 2016 Degradation of ferritic stainless steels under conditions used for solid oxide fuel cells and electrolyzers at varying oxygen pressures *Corros. Sci.* **110** 200–12

[50] Latu-Romain L, Parsa Y, Mathieu S, Vilasi M, Galerie A and Wouters Y 2017 Towards the growth of stoichiometric chromia on pure chromium by the control of temperature and oxygen partial pressure *Corros. Sci.* **126** 238–46

[51] Essuman E, Meier G H, Żurek J, Hänsel M and Quadakkers W J 2008 The effect of water vapor on selective oxidation of Fe–Cr alloys *Oxid. Met.* **69** 143–62

[52] Hultquist G, Tveten B and Hörnlund E 2000 Hydrogen in chromium: influence on the high-temperature oxidation kinetics in H_2O, oxide-growth mechanisms, and scale adherence *Oxid. Met.* **54** 1–10

[53] Wang R, Sun Z, Choi J-P, Basu S N, Stevenson J W and Tucker M C 2019 Ferritic stainless steel interconnects for protonic ceramic electrochemical cell stacks: oxidation behavior and protective coatings *Int. J. Hydrog. Energy* **44** 25297–309

[54] Løken A, Ricote S and Wachowski S 2018 Thermal and chemical expansion in proton ceramic electrolytes and compatible electrodes *Crystals* **8** 365

[55] Folgner C, Sauchuk V, Megel S, Kusnezoff M and Michaelis A 2017 Interconnect corrosion in steam containing fuel gas *ECS Trans.* **78** 1543–58

[56] Guillou S, Desgranges C and Chevalier S 2013 Study of conductivity of K41X chromia forming alloy in high temperature electrolysis environment *Oxid. Met.* **79** 507–16

[57] Linder M, Hocker T, Holzer L, Friedrich K A, Iwanschitz B, Mai A and Schuler J A 2013 Cr_2O_3 scale growth rates on metallic interconnectors derived from 40,000 h solid oxide fuel cell stack operation *J. Power Sources* **243** 508–18

[58] Niewolak L, Young D J, Hattendorf H, Singheiser L and Quadakkers W J 2014 Mechanisms of oxide scale formation on ferritic interconnect steel in simulated low and high pO_2 service environments of solid oxide fuel cells *Oxid. Met.* **82** 123–43

[59] Stefan E, Denonville C, Larring Y, Stange M and Haugsrud R 2020 Oxidation study of porous metal substrates for metal supported proton ceramic electrolyzer cells *Corros. Sci.* **164** 108335

[60] Young D J, Zurek J, Singheiser L and Quadakkers W J 2011 Temperature dependence of oxide scale formation on high-Cr ferritic steels in Ar–H_2–H_2O *Corros. Sci.* **53** 2131–41

[61] Groß-Barsnick S M, Fang Q, Batfalsky P, Niewolak L, Blum L and Quadakkers W J 2019 Post-test characterization of metallic materials and adjacent components in an SOFC stack after 34,000 h operation at 700 °C *Fuel Cells* **19** 84–95

[62] Molin S, Persson Å H, Skafte T L, Smitshuysen A L, Jensen S H, Andersen K B, Xu H, Chen M and Hendriksen P V 2019 Effective yttrium based coating for steel interconnects of solid oxide cells: corrosion evaluation in steam-hydrogen atmosphere *J. Power Sources* **440** 226814

[63] Sachitanand R, Sattari M, Svensson J-E and Froitzheim J 2016 The oxidation of coated SOFC interconnects in fuel side environments *Fuel Cells* **16** 32–8

[64] Zeng Z and Natesan K 2004 Corrosion of metallic interconnects for SOFC in fuel gases *Solid State Ionics* **167** 9–16

[65] Zurek J, Margaritis N, Naumenko D, Menzler N H and Quadakkers W J 2019 Behaviour of metallic materials in simulated service environments of CO_2/H_2O co-electrolysis systems for Power-to-X application *Oxid. Met.* **92** 353–77

[66] Niewolak L, Wessel E, Hüttel T, Asensio-Jimenez C, Singheiser L and Quadakkers W J 2012 Behavior of interconnect steels in carbon containing simulated anode gas of solid oxide fuel cells *J. Electrochem. Soc.* **159** F725–32

[67] Grabke H J 2003 Metal dusting *Mater. Corros.* **54** 736–46

[68] Malzbender J, Batfalsky P, Vaßen R, Shemet V and Tietz F 2012 Component interactions after long-term operation of an SOFC stack with LSM cathode *J. Power Sources* **201** 196–203

[69] Garcia-Fresnillo L, Shemet V, Chyrkin A, de Haart L G J and Quadakkers W J 2014 Long-term behaviour of solid oxide fuel cell interconnect materials in contact with Ni-mesh during exposure in simulated anode gas at 700 and 800 °C *J. Power Sources* **271** 213–22

[70] Meulenberg W A, Uhlenbruck S, Wessel E, Buchkremer H P and Stöver D 2003 Oxidation behaviour of ferrous alloys used as interconnecting material in solid oxide fuel cells *J. Mater. Sci.* **38** 507–13

[71] Froitzheim J, Niewolak L, Brandner M, Singheiser L and Quadakkers W J 2010 Anode side diffusion barrier coating for solid oxide fuel cells interconnects *J. Fuel Cell Sci. Technol.* **7** 031020

[72] Gannon P and Amendola R 2012 High-temperature, dual-atmosphere corrosion of solid-oxide fuel cell interconnects *JOM* **64** 1470–6

[73] Alnegren P, Sattari M, Svensson J-E and Froitzheim J 2016 Severe dual atmosphere effect at 600 °C for stainless steel 441 *J. Power Sources* **301** 170–8

[74] Gunduz K O, Chyrkin A, Goebel C, Hansen L, Hjorth O, Svensson J-E and Froitzheim J 2021 The effect of hydrogen on the breakdown of the protective oxide scale in solid oxide fuel cell interconnects *Corros. Sci.* **179** 109112

[75] Goebel C, Alnegren P, Faust R, Svensson J-E and Froitzheim J 2018 The effect of pre-oxidation parameters on the corrosion behavior of AISI 441 in dual atmosphere *Int. J. Hydrog. Energy* **43** 14665–74

[76] Sabato A G, Rost A, Schilm J, Kusnezoff M, Salvo M, Chrysanthou A and Smeacetto F 2019 Effect of electric load and dual atmosphere on the properties of an alkali containing diopside-based glass sealant for solid oxide cells *J. Power Sources* **415** 15–24

[77] Chou Y-S, Stevenson J W and Singh P 2008 Effect of aluminizing of Cr-containing ferritic alloys on the seal strength of a novel high-temperature solid oxide fuel cell sealing glass *J. Power Sources* **185** 1001–8

Chapter 6

Electrode kinetics

Alexander K Opitz and Andreas Nenning

Fast electrode kinetics is indispensable for the achievement of highest performing solid oxide cells and therefore an in-depth understanding of the possible electrode processes and their peculiarities is of great importance when aiming at an efficient optimization of these energy conversion devices. This chapter deals with the specific characteristics of electrodes in solid oxide cells, such as different available reaction pathways (triple phase boundary versus bulk path), the role of the defect chemistry of mixed conducting electrode materials, the possible ways in which the electrochemical overpotential acts in the case of mixed conducting electrodes, and the importance of the electrode surface chemistry. Moreover, the approach of using model-type thin film electrodes is discussed, which are a powerful tool to separate and explicitly study the effects relevant to porous SOC electrodes. Finally, a brief overview on electrochemical characterization techniques is given.

6.1 Introduction

The majority of the internal resistance of modern solid oxide cells arises from the limited kinetics of electrode processes. On the one hand, the contribution of the electrolyte to the performance limitation of the cells has been successfully reduced during the last decades by continually decreasing the electrolyte thickness down to only a few micrometers. On the other hand, the operational temperature reduction made possible by this development has a greater effect on electrode processes, because these are often more thermally activated than the ion transport through typical electrolyte materials. As a result, the overall performance of modern solid oxide cells is mainly governed by the reactions at their electrodes. A summary of the most relevant electrode reactions for oxide-ion-conducting and proton-conducting solid oxide cells is sketched in figures 6.1(a) and (b), respectively.

In contrast to conventional heterogeneous catalysis, the possibility of applying an electrical overpotential allows the introduction of an additional driving force, which leads to specific effects on the electrode kinetics. Depending on whether the electrolyte is liquid or solid, these effects on the electrode kinetics can differ noticeably. For example, the effective thermodynamic driving force caused by the overpotential may

doi:10.1088/978-0-7503-3951-3ch6

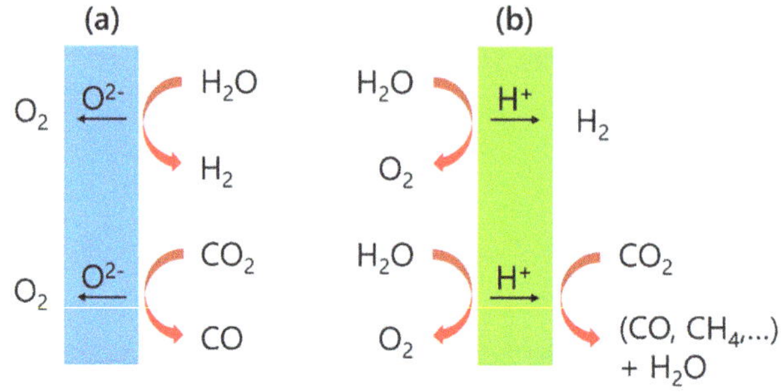

Figure 6.1. Sketch of the electrode reactions that take place in electrolysis cells based on oxide-ion-conducting SOECs (a) and proton-conducting PCECs (b). Since the reactions differ between solid oxide electrolysis cells (SOECs) and protonic ceramic electrolysis cells (PCECs) (e.g. the oxidation of O^{2-} to O_2 versus the oxidation of H_2O to H^+ and O_2), their mechanisms and kinetics may also differ.

be electrostatic in one case and purely chemical in the other. Entirely different reaction pathways may exist for different electrode/electrolyte combinations. These aspects will be discussed in this chapter. In addition, experimental approaches are discussed that show how these specific features can be investigated using model electrodes.

6.1.1 Reaction pathways

Apart from their significantly different operating temperatures, the most obvious difference between electrodes in solid-state electrochemistry and those in liquid electrochemistry is that in the case of a solid electrolyte, the electrodes cannot simply be immersed in it, as is the case with a liquid electrolyte. This difference seems trivial at first glance, but the consequence is that the reaction pathways and thus also the reaction mechanisms on electrodes in solid-state electrochemistry are fundamentally different from those on an electrode in contact with a liquid electrolyte.

In the case of liquid electrolytes, the reactive species of the electrochemical redox reaction are supplied via the electrolyte, and the charge-transfer reaction occurs at the interface between electrode and electrolyte. As a consequence, the electrochemical overpotential, as an external driving force for a net current, affects the electrode reaction by distorting the activation barrier, i.e. it accelerates the forward reaction while slowing down the backward reaction (or vice versa). This role of the overpotential as a parameter that directly affects the energy landscape of the electrode reaction leads to the well-known Butler–Volmer equation (equation (6.1)) describing the I–V relationship of solely charge-transfer-limited kinetics. For further details of charge-transfer-limited electrode kinetics in liquid electrochemistry, please refer to electrochemistry textbooks such as references [1, 2].

$$i = i_0\left(\mathrm{e}^{\frac{\alpha zF}{RT}\eta} - \mathrm{e}^{-\frac{(1-\alpha)zF}{RT}\eta}\right) \tag{6.1}$$

In equation (6.1), i and i_0 are the current density and the exchange current density; α is the so-called symmetry factor, which considers the symmetry of the energy barrier of the reaction at the electrode/electrolyte interface; z is the number of charges transferred; F is Faraday's constant; R the gas constant; T the absolute temperature; and η the overpotential at the respective electrode.

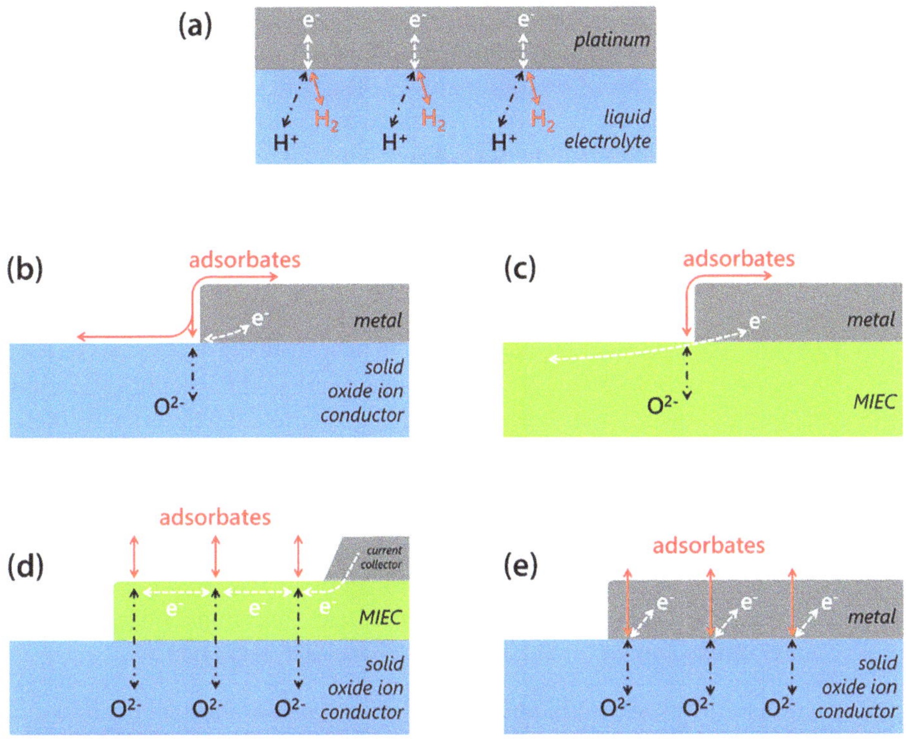

Figure 6.2. Possible pathways of electrode reactions. (a) Charge transfer via the interface between the electrode and the liquid electrolyte sketched for a standard hydrogen electrode. (b) Metal electrode on a solid oxide electrolyte without electronic conductivity, resulting in a surface path with 3PB activity. (c) Metal electrode on a solid mixed ionic and electronic conductor (MIEC). If the MIEC carries no electrochemically active adsorbate species and the spillover length from the metal is very short, 3PB activity results. (d) MIEC covered with reactive adsorbates, resulting in a bulk path with two-phase boundary (2PB) (surface) activity. (e) Neutral adsorbate species penetrate the metal electrode on a solid electrolyte, resulting in an electrochemical reaction at the interface.

In addition to this physically clearly interpretable effect of the overvoltage, there is another important consequence for the electrode kinetics of the electrodes in liquid electrochemistry. From a geometrical point of view, there is virtually only one reaction path possible, which is across the electrode/electrolyte interface—see figure 6.2(a). (Please note that the term 'three-phase boundary' (3PB) can also be found in liquid electrochemistry, but mechanistically this situation is still not comparable to solid-state electrochemistry, as detailed below.)

In contrast, electrodes in solid-state electrochemistry may not obtain all the species involved in the reaction via the electrolyte. Consequently, the electrochemical reaction proceeds at an interface which can be accessed by the gas. Since electronic and ionic charge carriers also have to be provided at this interface for an electrochemical reaction to take place, at least two different possibilities for this electrochemically active interface exist in solid-state electrochemistry (see figure 6.2):

(i) A 3PB as a narrow zone where the electron conductor, ion conductor, and gas phase meet. The restriction to a 3PB can occur for at least two different reasons: a limited electron conductivity of the ionically conducting phase (figure 6.2(b)), or a limited spillover length of an adsorbate (figure 6.2(c)).

Typical examples of predominantly 3PB-active systems with a very short lateral electron transport length, as sketched in figure 6.2(b), are Pt/yttria-stabilized zirconia (YSZ) in an O_2 atmosphere [3–7] and Ni/YSZ in H_2/H_2O [8–11]. However, as elaborated below in more detail, the electrochemical reaction is not exclusively limited to the related surface path, as other reaction pathways may also come into play under certain conditions.

The situation sketched in figure 6.2(c) is much less common on typical electrodes of SOECs. It may be relevant for mixed-conducting fuel electrodes carrying exsolved metal nanoparticles, which provide a spillover species to the surface of an MIEC [12, 13]. However, there is still too little quantitative data on the geometric dependence of the electrocatalytic activity of these surface-decorating particles on MIECs to make a clear statement about the geometric dependence of the corresponding reaction pathway.

(ii) A 2PB, which can be either the surface of an MIEC, as sketched in figure 6.2(d), or the interface between a metal that can be penetrated by a reactive adsorbate (figure 6.2(e)). In both cases, the 2PB activity originates from the ability of the electrode material to conduct two different species—electrons and ions in the first case, electrons and an (atomic) adsorbate species in the latter.

The reaction pathway shown in figure 6.2(d), which passes through the bulk of the MIEC electrode, is the classical situation for current state-of-the-art air electrodes in solid oxide cells. Materials such as $SrTi_{1-x}Fe_xO_{3-\delta}$ (STF), $La_{1-x}Sr_xFeO_{3-\delta}$ (LSF), $La_{1-x}Sr_xCo_{1-y}Fe_yO_{3-\delta}$ (LSCF), and $La_{1-x}Sr_xCoO_{3-\delta}$ (LSC) have been extensively studied during the last two decades and the existence of a bulk path is fully accepted [14–21]. This pathway and its kinetic implications are discussed in detail in section 6.3. Materials such as $LaNiO_{4+\delta}$ and Fe/Co-based Ruddlesden–Popper phases are also examples of this case [22–25].

The second situation leading to a 2PB-active electrode, as depicted in figure 6.2(e), is less common in solid oxide cells, but simple systems such as Ag/YSZ indeed exhibit such a bulk path in which oxygen penetrates through the silver electrode [26]. This type of pathway can also lead to deviations from the ideal 3PB behavior of Pt/YSZ and Ni/YSZ—see the details in section 6.2.

6.1.2 Model-type thin-film electrodes as a tool to identify reaction pathways

The electrochemically active site and the nature of the rate-limiting elementary steps are key issues that must be considered in order to optimize the electrode performance in a real solid oxide cell. To answer these questions, one needs to know the exact geometry of the electrode and be able to modify it in a controlled manner. Hence, although real 3D porous electrodes are the ideal choice for maximizing cell performance, they are only of limited usefulness when it comes to identifying reaction pathways and separating the contributions to their polarization resistance.

For this purpose, geometrically well-defined model electrodes were introduced, since they perfectly fulfill these requirements. To prepare these electrodes, thin films are typically grown, which are, in many cases, also micropatterned using photolithography [3, 6, 16, 21, 27–49]. By this process, electrodes of virtually any shape can be obtained, and it also allows large variations of their geometric parameters such as the 3PB length and the 2PB area. Moreover, the approach of using micropatterned thin-film electrodes allows composite electrodes to be prepared, consisting, for example, of oxide thin films with well-defined metallic current collectors. Figure 6.3 depicts a couple of examples of different model electrodes.

Figure 6.4 shows an example of how the control and variation of the electrode geometry can be used to localize a resistive process that contributes to the electrode polarization resistance R_{pol}. Due to the different scalings of the circumference and the area with the diameter of a circular electrode, the corresponding behavior of R_{pol} (or one of its partial resistances) can be used to infer whether the respective elementary process proceeds at the 3PB (figure 6.4(a)) or at a 2PB such as the surface or the interface (figure 6.4(b)) of the electrode. In case of an area-related resistance, a transport process through the electrode may also be limiting, but in case of thin-film electrodes this situation only appears for very poor ionic conductors.

The situation sketched in figure 6.4 is a very simple example of controlled geometric variation, but especially when multiple mobile charge carriers come into play, it can make sense to use more complex geometries or model composite electrodes and systematically vary their geometric parameters in order to extract information about the relevant elementary electrochemical parameters. A few examples using model electrodes with more complex geometries can be found in references [21, 38, 50, 51].

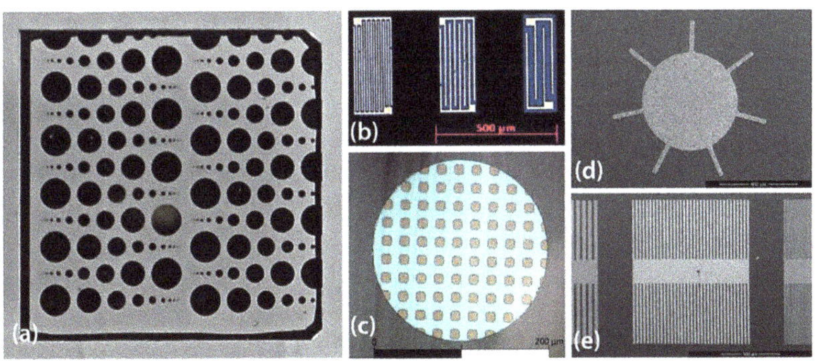

Figure 6.3. Examples of model electrodes. (a) Circular Pt electrodes of various diameters on a single-crystal 10 mm × 10 mm YSZ electrolyte. (b) Rectangular mixed-conducting model electrodes (blue) with interdigitating finger current collectors (bright) at different spacings (the larger bright area is the contact pad). (c) Circular model composite electrode consisting of a mixed-conducting thin film and a buried Pt current-collecting grid. The latter appears light blue in this optical micrograph due to interference effects. (d-e) Scanning electron microscopy images of Ni thin-film electrodes on YSZ single crystals, depicting possible electrode geometries.

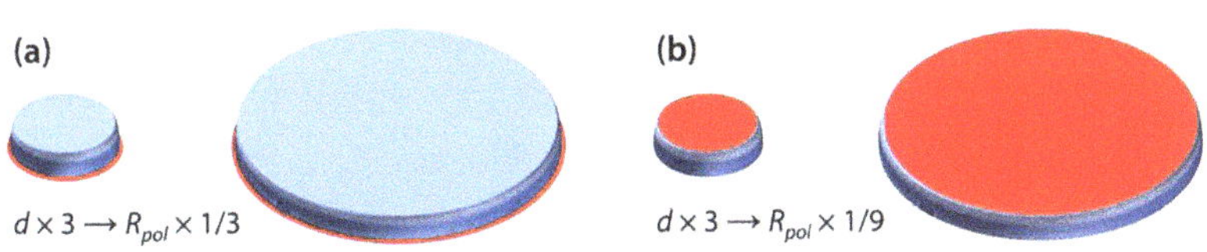

Figure 6.4. Different geometric scalings of a resistive elementary process for the simple case of circular microelectrodes. The location of the respective process is highlighted in red at the 3PB (a) and the 2PB (b).

6.2 Three-phase boundary active electrodes

In this section, the most prominent examples of predominantly 3PB-active electrodes in the field of solid oxide cells are considered. However, 3PB activity is only the sole reaction pathway in rare cases. Separating the pathways and obtaining evidence that the 3PB path is indeed the most relevant are therefore also important topics in this section.

6.2.1 Ni/YSZ as the fuel electrode

Nickel/YSZ ceramic metallic composites (cermets) have a long tradition of being used as fuel electrodes in SOFCs, where they still represent the state of the art in commercial systems. The reasons for this lie in the high catalytic activity of Ni for hydrogen oxidation, but also for a variety of other reactions which become relevant in the case of carbon-containing fuels [52–57]. Due to the extensive experience of the community in the preparation of these cermets, it was thus obvious to also apply this well-known system as the fuel electrodes in SOECs. In the case of electrolyzing H_2O to H_2, the fuel electrode reaction in Ni/YSZ reads

$$H_2O + V_O^{\bullet\bullet} + 2e^-_{(Ni)} \rightleftharpoons O_O^\times + H_2. \tag{6.2}$$

In equation (6.2), $V_O^{\bullet\bullet}$ and $O_O^\times$ denote an oxygen vacancy and a regular lattice oxygen in YSZ, respectively (according to Kröger–Vink notation). The electrons e^- are intentionally not written in this defect chemical notation, as they are assumed to originate from the metallic Ni phase (see figure 6.2(b)).

From a mechanistic point of view, it is very likely that hydrogen adsorption/desorption proceeds on the Ni surface:

$$2H_{ad,\,Ni} \rightleftharpoons H_2. \tag{6.3}$$

For the further reaction steps, however, several different possibilities have been discussed in the literature. Depending on whether the water formation/water splitting part of the reaction proceeds via the YSZ or Ni surface, the mechanism is called hydrogen spillover (equations (6.4) and (6.5)) or oxygen spillover (equations (6.6) and (6.7)), respectively [58, 59]. Beyond these two possibilities, several other mechanisms have been discussed. From the available experimental evidence, hydrogen spillover (equation (6.4) and (6.5)) appears to be the more likely case [59, 60], but unambiguous clarification is still lacking.

$$(OH)_O^\bullet + e^-_{(Ni)} \rightleftharpoons O_O^\times + H_{ad,Ni} \tag{6.4}$$

$$H_2O + V_O^{\bullet\bullet} + O_O^\times \rightleftharpoons 2(OH)_O^\bullet \tag{6.5}$$

$$O_{ad} + V_O^{\bullet\bullet} + 2e^- \rightleftharpoons O_O^\times \tag{6.6}$$

$$H_2O \rightleftharpoons 2H_{ad,Ni} + O_{ad,Ni} \tag{6.7}$$

Since the very beginning of the use of Ni/YSZ fuel electrodes, their 3PB activity was taken for granted. Nevertheless, several studies have been performed since the mid-1990s to prove the 3PB activity of Ni/YSZ. The method of choice in most of these studies was electrochemical impedance measurements on micropatterned Ni thin-film electrodes grown on YSZ single crystals [8–11, 27–29, 40, 48, 61–65], since they allow a broad variation of the 3PB length. As discussed in more detail in section 6.4.2.2, impedance spectroscopy probes the exchange current of the reaction in equation (6.2), yielding the polarization resistance of the electrode reaction. With this approach and the assumption of 3PB-dominated kinetics, data analysis according to equations (6.8a) and (6.8b) becomes possible.

$$R_{\text{pol}} = LSR \cdot l_{3\text{PB}}^{-1} \tag{6.8a}$$

$$\log\left(R_{\text{pol}}\right) = \log\left(LSR\right) - \log\left(l_{3\text{PB}}\right) \tag{6.8b}$$

Therein, R_{pol}, LSR, and $l_{3\text{PB}}$ are the measurable polarization resistance of a pattern electrode, the geometry-independent length specific resistance of Ni/YSZ, and the 3PB length, respectively. In the plot of $\log(R_{\text{pol}})$ versus $\log(l_{3\text{PB}})$, a linear relationship with a slope of −1 is expected, and indeed values between about −0.7 and −1.2 have been measured by several groups. However, the accumulation of impurities at the 3PB and the tendency of Ni thin-film electrodes on YSZ to slightly change their morphology during high-temperature measurements are two well-known deteriorating effects. Hence, the deviations of the experimentally achieved slopes from −1 could not unambiguously attributed either to these irreversible changes or to a true effect due to an additional pathway.

The question of an additional pathway was finally answered by Doppler *et al* [40], again by employing micropatterned Ni thin-film electrodes deposited on YSZ single crystals. Electrochemical measurements were performed on two sets of electrodes: one type was designed such that every electrode had an identical area but a different 3PB length. For the other set of electrodes, the 3PB length was identical but the area was different (see figures 6.5(a) and (d)). For the first electrode type (figure 6.5(a)), a linear relationship between inverse polarization resistance and 3PB length was expected, thus confirming 3PB activity, but the observed *y*-axis intercept suggested an additional source of electrochemical activity (figures 6.5(b) and (c)). The measurements of the second electrode series (figure 6.5(d)) clearly showed this additional activity to be area related, thus indicating that a parallel pathway through the Ni electrode is responsible for this behavior. The two most likely explanations for the existence of this path are based on the transport of either hydrogen or oxygen through the Ni electrode (e.g. along grain boundaries), but further mechanistic details have not yet been clarified.

Due to their different geometric dependences, the relevance of both pathways in real Ni/YSZ cermets depends on the microstructure of these 3D porous fuel electrodes. In case of thin-film electrodes as used in micro SOFCs [5, 66–68], the pathway along Ni grain boundaries may provide significant electrochemical activity. But for typical state-of-the-art Ni/YSZ cermets, the 3PB path can be expected to carry the majority of the current. Thus, classical Ni/YSZ cermets can still be regarded as predominantly 3PB active. This dominance of the 3PB path can also be seen in the susceptibility of Ni/YSZ cermets to morphological changes [69–74], especially under SOEC conditions. Moreover, this morphological degradation is one

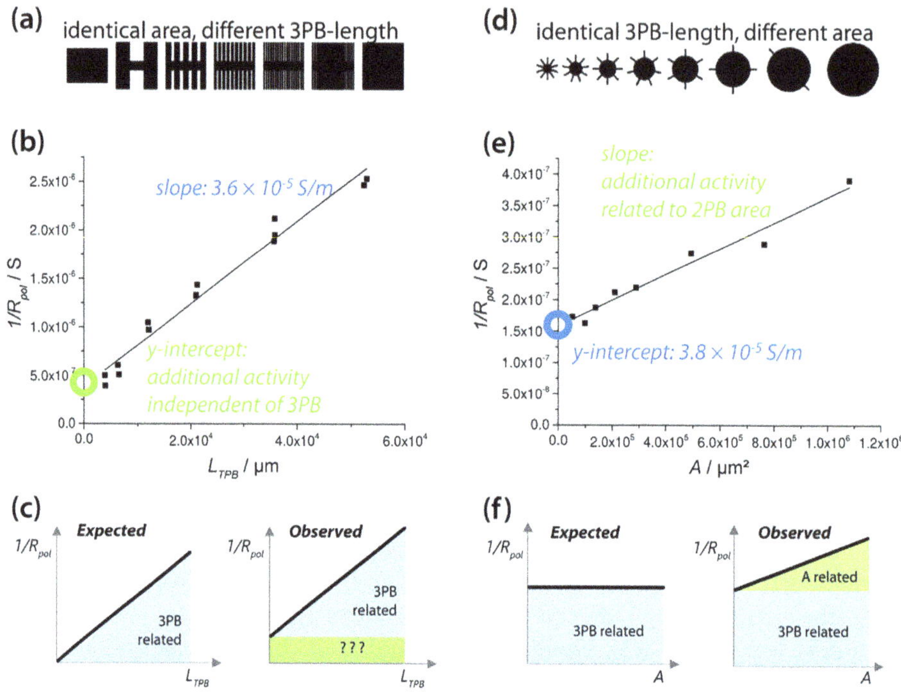

Figure 6.5. Separating reaction pathways on Ni/YSZ by means of geometrically well-defined patterned electrodes. (a and d) Sketches of the electrodes used. (b and e) Plots of inverse polarization resistance versus the 3PB length or 2PB area measured at 800 °C in an atmosphere containing 2.5% H_2 and 0.15% H_2O in Ar. (c and f) Comparison between the expected and observed behaviors, indicating a second active pathway. Data taken from [40].

reason for the currently intensive search for alternative fuel electrode materials, and Ni/gadolinia-doped ceria is a very promising alternative candidate [75–80].

6.2.2 Pt/YSZ in an oxygen-containing atmosphere

The electrode system $Pt(O_2)$/YSZ was among the very first electrodes used in SOECs. Because of this, platinum electrodes on oxide-ion-conducting electrolytes also have a long history of investigation. Application-wise, Pt/YSZ is the state-of-the-art electrode material in the lambda probes used as exhaust gas sensors for automotive emission control [81–84]. Moreover, $Pt(O_2)$/YSZ remains one of the prototypes of a 3PB-active high-temperature electrode, which still presents a couple of unanswered questions for fundamental science.

The first studies that aimed at a mechanistic understanding of the oxygen exchange reactions on Pt typically used porous electrodes prepared by brushing or screen printing platinum paste on oxide-ion-conducting electrolytes (mostly zirconia or ceria based) and subsequent firing. The mechanistic conclusions drawn from electrochemical studies on these porous Pt electrodes, however, differed significantly from each other, lacking conclusive proof for any one of the proposed reaction mechanisms [7, 85–92].

Virtually all of these studies agree that oxygen adsorption occurs via a dissociative process on the Pt surface (see equation (6.9)), which is consistent with the well-studied adsorption behavior of O_2 on Pt [93–97]. Subsequently, the electron transfer takes place, in which the electrolyte is also involved in order to accept the resulting oxide ion (equation (6.10)). As both reactions are slower than their counterparts in an H_2/H_2O atmosphere (equations (6.3)–(6.7)), the polarization resistances of Pt (O_2)/YSZ are commonly also larger than those of Ni(H_2-H_2O)/YSZ.

$$O_2 \rightleftharpoons 2O_{ad} \tag{6.9}$$

$$O_{ad} + V_O^{\bullet\bullet} + 2e^- \rightleftharpoons O_O^{\times} \tag{6.10}$$

As in the case of Ni/YSZ in an H_2/H_2O atmosphere, uncertainties mainly prevail for the detailed steps of equation (6.10), concerning both the reaction pathway and the rate-determining step.

These discrepancies can largely be attributed to differences in the electrodes studied, since the 3D microstructure and surface chemical composition of the electrocatalytically active zones obtained by a paste-based fabrication method can largely differ. For example, different sintering conditions or variations in paste composition, such as different Pt particle size distributions or the use of additives (e.g. sintering aids or fluxes), can have huge effects on the electrochemical properties.

Some modeling approaches have been carried out in an attempt to put the inconsistent data of the real electrodes on a solid theoretical foundation [98–100]. An experimental approach to the avoidance of ill-defined porous paste Pt electrodes used point-contact electrodes, which were produced by pressing a platinum wire, tip, or cone onto the polished electrolyte surface [66, 101–104]. These had the advantage of not needing a reference electrode and thus allowing virtually artifact-free impedance measurements, but the desired control over the geometry was still not entirely achieved, since, for example, controlled variation of the 3PB length could hardly be realized.

Targeted geometry variation became available with the use of micropatterned thin-film electrodes [3, 89, 105–109]. Provided the thin film stays dense and the geometry is well defined, this type of electrode preparation allows exact control of the geometric parameters. Employing this approach, three different faradaic pathways could be identified (figure 6.6): (i) an area-related path through the electrode, (ii) a surface path with a 3PB-length-related rate-determining step, and (iii) an electrolyte surface path with oxygen incorporation at the free YSZ surface close to the Pt electrode [6, 37, 110, 111]. The area-related pathway had already been mechanistically clarified to proceed via oxygen diffusion along Pt grain boundaries and to be relevant only at relatively low temperatures (below ca. 500 °C) [5, 112]. In the case of the 3PB-related pathway, several theoretically possible elementary steps were already excluded as being rate determining, but despite decades of research, a mechanistically unambiguous clarification of the level of elementary processes has not been achieved so far. The remaining candidates for the rate-limiting step are a diffusion process at the 3PB or a so-called co-limitation with a short decay length [7, 15, 37, 86–88, 111, 113, 114].

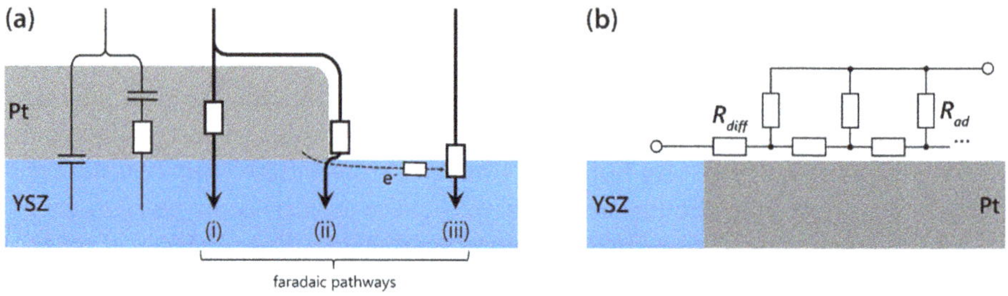

Figure 6.6. (a) Possible current pathways on $Pt(O_2)/YSZ$. The faradaic reaction pathways of O_2 reduction are: a bulk path (i), a surface path via the 3PB (ii), and an electrolyte surface path (iii). (b) Equivalent transmission-line circuit for the DC case of co-limiting oxygen adsorption and lateral diffusion on Pt (i.e. capacitances are neglected).

A co-limitation could, for example, originate from oxygen adsorption and surface diffusion on Pt. The observed 3PB scaling of the polarization resistance may hence be explainable by the very short electrochemical decay length of the corresponding transmission-line circuit. The DC version of this circuit is sketched in figure 6.6(b), and its electrochemical decay length b is given by

$$b = \sqrt{\frac{R_{\mathrm{ad}}}{R_{\mathrm{diff}}}}, \tag{6.11}$$

where R_{ad} and R_{diff} denote the geometry-related adsorption resistance ($\Omega\ \mathrm{cm}^2$) and the lateral diffusion resistance of oxygen on the Pt surface (which is a sheet resistance with unit Ω), respectively. In the case of fast adsorption and slow diffusion—i.e. small R_{ad} and large R_{diff} – b becomes small, which for electrode sizes much larger then b would effectively lead to a polarization resistance that scaled with the 3PB length. However, this particular case is only one of several possible cases—an extension of the transmission line in figure 6.6(b) to the free YSZ surface, for example, is equally conceivable.

6.2.3 $LaMnO_3$-based electrodes for oxygen reduction

A further example of a potentially 3PB-active O_2 electrode material is $La_{1-x}Sr_xMnO_3$ (LSM) [15, 31, 115–118]. However, the reaction mechanism on LSM can be expected to differ significantly from that of its metallic counterpart Pt, since oxygen does not often adsorb as a dissociated neutral species on oxides. Hence, an electron transfer to adsorbed molecular oxygen seems plausible. A pathway via the oxide bulk or grain boundaries may also play a role [15, 113, 116]. This possibility originates from the redox activity of LSM, which can lead to a change of the concentration of point defects in this material upon electrochemical polarization, as indicated in equation (6.12) (Kröger–Vink notation, in which *yte* and *cc* denote the electrolyte and the current collector, respectively).

$$\begin{aligned} &O^{\times}_{O(LSM)} + 2Mn^{\bullet}_{Mn(LSM)} + V^{\bullet\bullet}_{O(yte)} + 2e^{-}_{(cc)} \\ &\underset{\text{anodic}}{\overset{\text{cathodic}}{\rightleftharpoons}} V^{\bullet\bullet}_{O(LSM)} + 2Mn^{\times}_{Mn(LSM)} + O^{\times}_{O(yte)} \end{aligned} \tag{6.12}$$

When LSM is cathodically polarized, it is reduced and thus contains a higher concentration of oxygen vacancies than under equilibrium conditions. This means that the reaction in equation (6.12) is shifted to the right. The oxygen ion moved to the electrolyte is subsequently released as O_2 at the counter electrode of the cell (or in case of an SOFC, as water at its anode), delivering two electrons to the current collector. This reduction can open a pathway for oxygen reduction via the LSM bulk similar to that of surface active electrodes (discussed in more detail in section 6.3 [17]). Since the resulting ionic conductivity is rather low, this additional pathway may still be restricted to the near-3PB region [15, 31]. For polycrystalline LSM, which is typical for 'real' 3D porous electrodes, this effect may be even more pronounced, as grain boundaries in LSM are easier to reduce and also provide faster oxygen diffusion pathways [119, 120].

When operated as an SOEC anode, the effect on the LSM defect chemistry is reversed: the reaction in equation (6.12) is shifted to the left, and the already low concentration of oxygen vacancies becomes further reduced. Hence, a possible pathway via the near-3PB bulk of the LSM becomes suppressed. In accordance with the reduction in electrochemical activity under strongly oxidizing conditions, LSM is known to be an SOEC anode with rather modest oxygen evolution kinetics, which is typically outperformed by truly mixed-conducting materials.

6.3 Surface active electrodes

6.3.1 The role of electrode defect chemistry in electrode reactions

The defect chemistry of an electrode material with mixed conductivity plays an important role in the possible reaction pathways. While a sufficient concentration of electronic charge carriers is self-evident for an electrode material, the presence of oxygen vacancies is the prerequisite for the existence of a bulk path with a surface reaction (see figure 6.2(d)). In this surface reaction, the atmosphere equilibrates with the oxide [121, 122]:

$$V_O^{\bullet\bullet} + \frac{1}{2}O_2 \rightleftharpoons O_O^{\times} + 2h^{\bullet}. \tag{6.13}$$

If the material is a semiconductor, which applies to many MIEC electrode materials, the electron/hole formation reaction also needs to be considered:

$$nil \rightleftharpoons e' + h^{\bullet}. \tag{6.14}$$

If the thermodynamic parameters $\Delta_r H^\circ$ and $\Delta_r S^\circ$ of equations (6.13) and (6.14) are known, the concentrations of oxygen vacancies, electrons, and electron holes can be calculated from the respective mass action laws as a function of the oxygen partial pressure and T. Plotting the logarithm of these concentrations versus $\log(pO_2)$, one obtains the so-called Brouwer diagram of the material. Figure 6.7 depicts the Brouwer diagram of mixed-conducting $La_{0.6}Sr_{0.4}FeO_{3-\delta}$ (LSF64) calculated using the respective thermodynamic data [121, 122], demonstrating that a change in pO_2 directly affects the concentrations of all charge carriers in the material (section 6.2.3).

Since point defects are involved in the oxygen exchange reaction between oxide and atmosphere (equations (6.13) and (6.14)), their concentration consequently also plays an important role in its kinetics. The oxygen partial pressure influences the kinetics of the

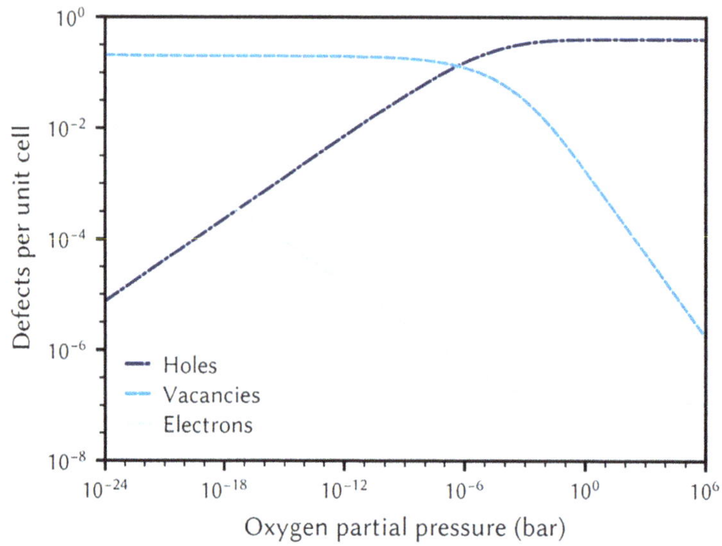

Figure 6.7. Brouwer diagram, i.e. a double logarithmic plot of point defect concentrations versus the oxygen partial pressure of $La_{0.6}Sr_{0.4}FeO_{3-\delta}$ at 600 °C. The thermodynamic data were taken from [121]. For the sake of simplicity, cation vacancies are omitted. (courtesy: Alexander Schmid)

surface exchange reaction in two ways. First, by affecting the concentration of adsorbed oxygen surface species, and second, by affecting the concentration of relevant point defect species according to the Brouwer diagram. The combination of the two can lead to, at first glance, unusual oxygen partial pressure dependencies of the reaction rate of equation (6.13), such as slower kinetics with increasing pO_2 in the atmosphere [123, 124]. To rationalize this, one must understand the role of the point defects of the oxide in the mechanism of the oxygen exchange reaction.

6.3.2 The meaning of the electrochemical overpotential in the case of mixed-conducting electrodes

One of the particularly attractive possibilities of electrochemistry is to influence the rate of an electrode surface reaction not only by changing the concentration (or partial pressures) of the reactants but also by applying a voltage. Such an applied voltage drops by corresponding proportions at the respective resistances of the electrodes and the electrolyte. This is sketched in figure 6.8 for an electrochemical cell with a kinetically slow working electrode (WE) and a comparably fast—i.e. reversible—counter electrode (CE), achieved, for example, by applying a dense thin-film WE and a 3D porous CE. In such a case, the CE overpotential can be virtually neglected and the overpotential drop at the WE is, to a very good approximation, in agreement with that obtained by subtracting the ohmic voltage drop in the electrolyte η_Ω from the applied voltage U_{set}:

$$\eta_{WE} \approx U_{set} - R_\Omega I_{DC}. \tag{6.15}$$

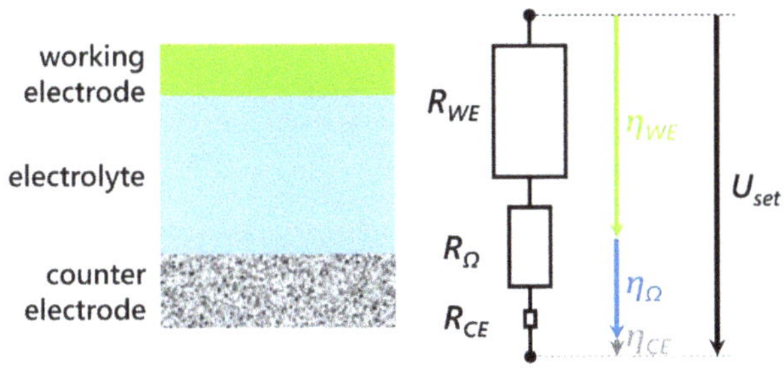

Figure 6.8. Individual contributions to the cell resistance and the corresponding overpotential drops at the partial resistances of the working electrode (η_{WE}), the electrolyte (η_Ω), and the counter electrode (η_{CE}).

I_{DC} is the DC current that flows through the cell upon applying the voltage U_{set} (i.e. $\eta_\Omega = R_\Omega I_{DC}$). In an electrochemical cell with a liquid electrolyte, the electrochemical overpotential η_{WE} drops at the WE/electrolyte interface. There, it affects the energy landscape via a change of the electrostatic (Galvani) potential step and thus the activation barrier of the electrode reaction, bringing it out of equilibrium by creating a difference between the forward and backward reactions. The resulting net current in the case of rate-limiting charge transfer can be described by the well-known Butler–Volmer equation (equation (6.1)). In this equation, due to the effect of η_{WE} on the activation barriers of the forward and backward reactions, the forward and backward currents depend exponentially on the overpotential.

In the case of a solid electrolyte with a mixed-conducting WE, the situation turns out to be very different, since in the solid-state electrochemical system, the reaction takes place at another interface, which is not the WE/electrolyte interface but the WE/gas interface. Consequently, any current–voltage relationship on an MIEC electrode cannot be explained by a comparably simple electrostatic model to that of the Butler–Volmer equation. (It is worth mentioning that for experimental data recorded at moderate overpotentials, the Butler–Volmer equation can, however, give an acceptable fit to the results, since for solid-state electrodes the current also depends exponentially on η_{WE}; nevertheless, one has to be very careful with any mechanistic interpretation of the obtained parameters.) To understand how an applied η_{WE} can provide a driving force for a net reaction on a MIEC electrode, the effect of the overpotential on the defect chemistry of the material needs to be considered. In this respect, an overpotential at an MIEC electrode can act in two ways: (i) as a change in the chemical potential of oxygen in the MIEC, and (ii) as an electrostatic potential step.

6.3.2.1 Interpreting the overpotential as a step change in the chemical potential of oxygen

Let us assume that the kinetics of the WE reaction is limited exclusively by a reaction step at its surface. This assumption is very helpful in simplifying the following considerations, and it represents a very good approximation for many model-type MIEC thin-film electrode systems in solid-state electrochemistry [14]. The charge transport processes of oxide ions and electrons within the MIEC and across the electrode/electrolyte interface are sufficiently fast that they do not contribute to the WE

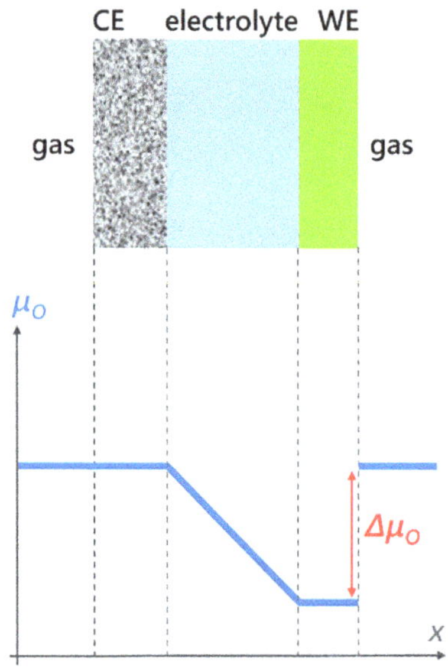

Figure 6.9. Effective chemical potential of oxygen μ_O across an electrochemically polarized solid oxide cell with an O^{2-}-conducting electrolyte, a porous counter electrode (CE) with negligible polarization resistance, and a mixed-conducting working electrode (WE). The rate-limiting electrode reaction step is located at the WE surface. For the sake of simplicity, the sketch shows a single chamber setup, which means that the gas composition at the CE and the WE is identical and hence no μ_O gradients due to different gas compositions need to be considered.

polarization resistance. Hence, as soon as an overpotential drops at the WE (see figure 6.8), it drives the rate-limiting surface reaction step out of its equilibrium, while all other elementary steps remain in equilibrium. This also holds for the charge transport—i.e. the electrochemical potentials of oxide ions $\tilde{\mu}_{O^{2-}}$ and electrons $\tilde{\mu}_{e^-}$ remain constant throughout the MIEC, even upon polarization. From these two potentials, an effective chemical potential of oxygen μ_O within the MIEC can be calculated according to

$$\tilde{\mu}_{O^{2-}} - 2\tilde{\mu}_{e^-} = \mu_O. \tag{6.16}$$

Upon applying an overpotential to the WE (i.e. changing $\tilde{\mu}_{e^-}$), μ_O is homogeneously shifted within the MIEC, thus generating a step change $\Delta\mu_O$ at the surface of the electrode, which is the effective driving force for the surface reaction. The step change is related to the WE overpotential [38, 124, 125]:

$$\Delta\mu_O = 2F\eta_{\mathrm{WE}}. \tag{6.17}$$

The chemical potential profile across a cell with negligible CE resistance is sketched in figure 6.9.

6.3.2.2 Interpreting the overpotential as an electrostatic potential step

To further emphasize that the overpotential at a mixed-conducting WE becomes effective in the form of a chemical driving force at the WE surface rather than as an electrostatic driving force, the location of the electrostatic potential step is elaborated here. Depending on the defect chemical situation, this potential step can be located at different interfaces.

(a) *MIECs with high electronic charge carrier concentration*

In the case of semiconducting MIECs, such as perovskite-type $La_{1-x}Sr_xFeO_{3-\delta}$ (LSF), the predominant point defects compensating the acceptor dopant at high oxygen partial pressures are electron holes ($h^\bullet$)—see the far right part of the Brouwer diagram in figure 6.7. Under these conditions, the MIEC is a good electronic conductor, and applying a small DC voltage cannot cause a significant change in the $h^\bullet$ concentration. Thus, the change in the Fermi level of the MIEC (i.e. the MIEC's electrochemical potential of electrons) imposed by the applied bias voltage cannot be explained by a chemical component but rather by an electrostatic component. Consequently, the electrostatic potential drop is located at the MIEC/electrolyte interface, as sketched in figure 6.10(a) [125]. Moreover, the driving force for the surface reaction is realized mainly by a change in the chemical potential of oxygen vacancies: $2F\eta_{WE} = \Delta\mu_O \approx -\Delta\mu_{V_O^{\bullet\bullet}}$—cf. equation (6.17) and figure 6.7. Even though the electrochemical polarization causes an additional electrostatic potential drop at the WE/electrolyte interface, this interface is still in local equilibrium (see Figure 6.8 together with the consideration that the entire R_{WE} originates from the surface process).

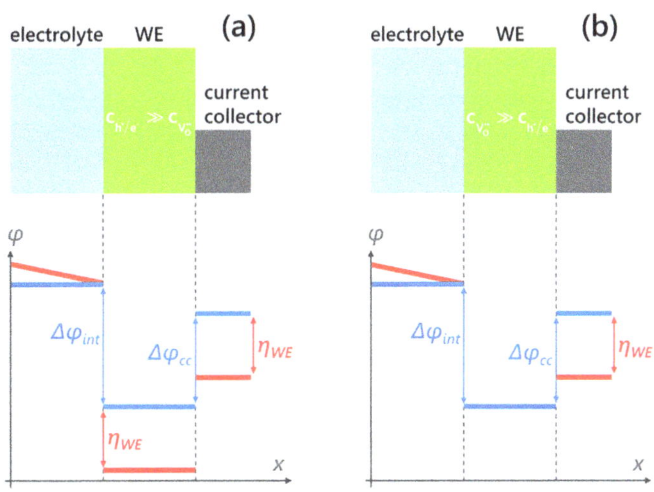

Figure 6.10. Electrostatic potential steps in MIEC electrodes under equilibrium (blue lines) and upon cathodic electrochemical polarization (red lines). The potential steps $\Delta\varphi_{int}$ and $\Delta\varphi_{cc}$ at the WE/electrolyte and WE/current collector interfaces, which are also present under equilibrium, originate from differences in the $\mu_{V_O^{\bullet\bullet}}$ and $\mu_{h^\bullet/e^-}$ of the MIEC and the electrolyte as well as the MIEC and the current collector, respectively. (a) If the majority point defect species in the MIEC is electronic (electrons or holes), the WE overpotential drops at the electrode/electrolyte interface. (b) If ions are the majority defect species in the MIEC, the applied voltage changes the electrostatic potential step at the electrode/current collector interface.

(b) *MIECs with high oxygen vacancy concentration*

As soon as the acceptor dopant of the MIEC electrode material is mainly compensated by oxygen vacancies—which is the case for low pO_2 (see the left part of figure 6.7)—the applied overpotential cannot cause a significant change in the oxygen vacancy concentration. Rather, it translates into a change in the concentration of electronic point defects: $2F\eta_{WE} = \Delta\mu_O \approx 2\Delta\mu_{h^\bullet}$ (i.e. electron holes for the example of LSF used here). Consequently, the change of the MIEC's Fermi level (i.e. its electrochemical potential of electrons) is realized entirely by this change in the chemical potential of electronic charge carriers and virtually no electrostatic component exists. Hence, the electrostatic potential drop at the WE/electrolyte interface is not affected, despite the electrochemical polarization of the electrode. Rather, the electrostatic potential drop under an applied bias can be found at the WE/current collector interface (figure 6.10(b)) [125].

6.3.3 Effect of the electrochemical overpotential on a possible surface potential step

Depending on the adsorbates present on the surface of an MIEC, an additional electrostatic potential step $\chi = \varphi_{MIEC} - \varphi_{adsorbate}$ can exist at this electrode interface. Upon electrochemically polarizing the electrode, the overpotential η_{WE} also modifies the surface potential step by $\Delta\chi$, but the relationship between η_{WE} and $\Delta\chi$ is highly complex and depends on the exact nature of χ (charge of the adsorbates, surface coverage). Possible surface potential step changes can be between almost zero and $4\eta_{WE}$ [124–126]. On ceria-based electrodes, the polarization-induced change of the surface potential step has already been proven by *in situ* near-ambient-pressure X-ray photoelectron spectroscopy (NAP-XPS). On this material, which is a mixed conductor in an H_2/H_2O atmosphere, a large coverage of surface hydroxyls creates a surface potential step χ. Upon electrochemical polarization, a change of χ can be observed via a shift of the O1s binding energy of these surface hydroxyls, which differs from the binding energy shift of bulk oxide ions [127–129].

For perovskite-type MIECs in an oxygen atmosphere, a similar effect has not yet been observed, which to a large extent is due to a surface coverage of electrochemically active oxygen adsorbates that is too low to be detected by NAP-XPS. Hence, it is not possible to unambiguously state whether or not a surface potential step χ exists on perovskite-type oxides in an O_2 atmosphere, and to what extent it is affected by a working electrode overpotential [124]. This uncertainty in terms of detailed knowledge of the surface potential step of MIEC electrodes and its dependence on η_{WE} makes the interpretation of measured *I–V*-curves extremely challenging, as the separation from polarization-dependent concentration changes of the relevant defect species is experimentally complicated. A couple of results point in a direction which suggests that $\Delta\chi$ effects play only a minor role in oxygen exchange on perovskite electrodes such as LSF, electron holes being the limiting defect species for the oxygen evolution, and oxygen vacancies being the limiting point defect for oxygen incorporation [123, 124, 130].

6.3.4 Mechanistic picture of oxygen exchange on MIEC oxide electrodes

Since oxygen exchange on mixed-conducting oxide electrodes depends on the oxygen partial pressure in the atmosphere, the defect chemistry of the oxide, and the possible presence of a surface potential step, a general rate equation for this reaction needs to consider all these parameters. Such an expression for the overall reaction rate r of oxygen incorporation or release is given in equation (6.18) [124, 125]:

$$r = r_0 \theta_{\text{oxygen}}^{\nu_s} \prod_i [\text{i}]^{\nu_i} e^{\frac{\beta F(\chi + \Delta\chi)}{RT}}, \tag{6.18}$$

in which r_0 denotes a pre-factor that considers the relevant activation barrier and mass action constants of the fast elementary steps of the reaction. The exponents ν_s and ν_i are the respective reaction orders of the surface coverage of the relevant oxygen adsorbate θ_{oxygen}, and of the point defects i in the electrode with concentration [i], respectively. The exponential term considers the effect of the possible presence of an electrostatic surface potential step χ as well as its variation $\Delta\chi$ under electrochemical polarization. Therein, β denotes a factor depending on the reaction mechanism; F, R, and T have their usual meanings.

It is worth explicitly mentioning here that the partial pressure of oxygen in the gas atmosphere affects the surface reaction rate in two ways: first, by changing the surface coverage of the electrochemically active oxygen species on the MIEC electrode, and second, by changing the defect concentrations of the material. The same applies to the electrochemical overpotential, which on the one hand affects the concentrations of point defects in the mixed-conducting electrode material according to the change caused in the chemical potential of oxygen (see equation (6.17) and figure 6.7); on the other hand, an overpotential can also affect the concentration of adsorbates—e.g. by influencing the position of an upstream equilibrium or by affecting the number of available adsorption sites.

To separate the defect-independent from the defect-dependent effects, equation (6.18) can be slightly modified to

$$r = r_0 p\text{O}_2^{\nu_p} \prod_k [\text{k}]^{\nu_k} e^{\frac{\beta F(\chi + \Delta\chi)}{RT}}, \tag{6.19}$$

where $p\text{O}_2$ summarizes all the defect-independent effects of the oxygen partial pressure in the atmosphere on the adsorbates and ν_p is the corresponding reaction order (e.g. ½ or 1 for simple atomic or molecular adsorption, respectively). For charged adsorbate species in particular, this reaction order can itself also depend in a rather complicated way on the oxygen partial pressure due to electrostatic interactions between them [131]. The species k in the Π-product term in equation (6.19) are the point defects that depend on both $p\text{O}_2$ and overpotential, and ν_k is the respective reaction order, which in contrast to equation (6.18) now considers the contributions from the point defects of reactants and possibly those from adsorbates, depending on the defect concentrations. For both approaches, it can be concluded that the pO_2 dependence of the current of an electrochemically polarized mixed-conducting oxygen electrode is highly non-trivial.

The situation is even more complicated for the temperature dependence of r. First, a temperature change affects r_0 in at least two ways, as it contains the reaction's relevant activation barrier and mass action constants. Second, the defect concentrations change according to the defect chemistry of the material. Third, the exponential term that considers the effects of a possible electrostatic surface potential step is explicitly temperature dependent. Any measured effective activation energy of the surface reaction on an MIEC electrode is thus a convolution of all these effects, which are experimentally rather difficult to disentangle.

These relatively complex T- and pO_2 dependencies of the oxygen reaction rate on MIEC surfaces are one reason why the rate-determining step of this reaction is still under debate. The long-term evolution of the surface chemistry is the second reason —see section 6.3.5. What has been discussed since the late 1990s is that oxygen vacancies play a crucial role in the rate-limiting surface step of oxygen incorporation, especially at higher pO_2. This was suggested due to an obvious correlation between the surface exchange coefficient k (which is proportional to the rate r in equation (6.18)) and the oxygen diffusion coefficient or the $V_O^{\bullet\bullet}$ concentration (which are both also proportional) [20, 132]. Two examples of such correlations are depicted in figures 6.11(a) and (b). In addition, the crucial role of oxygen vacancies for the rate-determining elementary step has also been supported by theoretical studies [133].

Apart from oxygen vacancies, there is a wide range of experimental evidence that an electron transfer occurs in or before the rate-determining elementary step [20, 33, 124, 134–136]. Together with the above observation that the oxygen vacancies in the rate-limiting step also play a role, the elementary steps in question have already been reduced to a handful. The most recent study on this subject (in which *in situ* impedance measurements were directly conducted in the pulsed-laser deposition chamber to keep the MIEC surface free from any contamination) identified two different rate-determining elementary steps, depending on the pO_2 in the gas atmosphere [130]. Below about 1 mbar of O_2, the elementary reaction step in equation (6.20) is rate determining, while at pO_2 above this value, the elementary step in equation (6.21) is the slowest. The entire reaction mechanism of oxygen exchange on MIEC electrodes, including these two possible rate-determining steps, is sketched in figure 6.11(c). What is also worth mentioning at this point is that these rate-determining steps are the result of fitting the pO_2 dependence of the measured surface exchange coefficients based on a kinetic model and that other mechanisms might also account for the observed behavior.

$$(O_2)_O^{\bullet} \rightleftharpoons O_O^{\times} + O_{ads} + h^{\bullet} \tag{6.20}$$

$$O_{ads} + V_O^{\bullet\bullet} \rightleftharpoons O_O^{\bullet} + h^{\bullet} \tag{6.21}$$

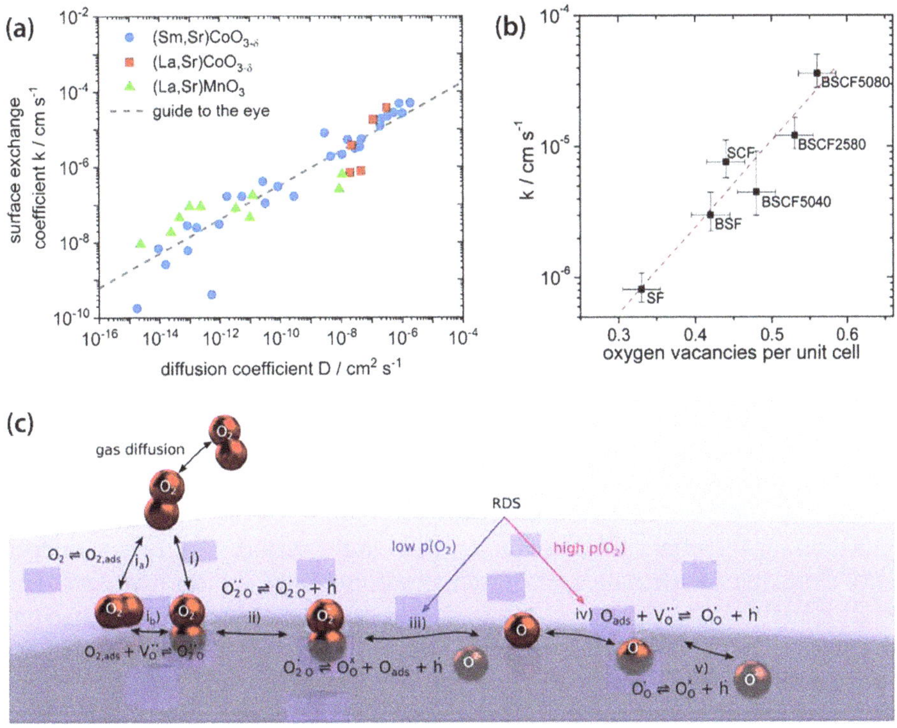

Figure 6.11. Relevance of oxygen vacancies for the oxygen exchange reaction on MIEC electrodes. (a) Correlation between the oxygen exchange coefficient k and the oxygen diffusion coefficient D of different perovskite-type electrode materials (diagram reproduced with permission from Elsevier using data from [132]). (b) Correlation between the surface exchange coefficient k and oxygen vacancies per unit cell of different perovskite-type oxides of the $Ba_xSr_{1-x}Co_yFe_{1-y}O_{3-\delta}$ family. SF = $SrFeO_{3-\delta}$; BSF = $Ba_{0.5}Sr_{0.5}FeO_{3-\delta}$; SCF = $SrCo_{0.8}Fe_{0.2}O_{3-\delta}$; BSCF5040 = $Ba_{0.5}Sr_{0.5}Co_{0.4}Fe_{0.6}O_{3-\delta}$; BSCF2580 = $Ba_{0.25}Sr_{0.75}Co_{0.8}Fe_{0.2}O_{3-\delta}$; BSCF5080 = $Ba_{0.5}Sr_{0.5}Co_{0.8}Fe_{0.2}O_{3-\delta}$ (image reprinted from [20] with permission from IOP Publishing). (c) Visualization of the oxygen exchange mechanism on MIEC oxides suggested in [130]; the dark boxes indicate oxygen vacancies at the surface of the electrode (original image published in [130] under a CC-BY license).

6.3.5 The effect of chemical evolution of the electrode surface

As already mentioned, time-dependent compositional changes of an electrode surface make mechanistic studies of the electrochemical reaction kinetics on MIEC electrodes quite complicated. Such changes can be caused by the segregation of acceptor dopants from the electrode bulk to its surface [18, 39, 137–144], the adsorption of reactive trace components from the atmosphere (such as SO_2) [145–148], poisoning of the MIEC surface by external impurities (such as Si- or Cr-oxide species) [30, 149–151], or a combination of these effects. Due to the evolution of the electrode surface chemistry over time, its electrochemical surface resistance dramatically increases, thus making reversible measurements of I–V-curves or impedance spectra very challenging. Reducing or ideally avoiding these degradation effects is thus of great interest, both for the application of MIEC materials in solid oxide cells and fundamental studies using model systems. While the long-term stability of the cell is of interest in the first case, the latter focuses on an unperturbed investigation of the reaction mechanism. So far, a couple of strategies for addressing these degradation effects have been studied. Among them are selective surface doping to suppress strontium segregation [140, 152], *in situ* impedance measurements within the electrode deposition setup [39, 130], and the application of reactive substances that function as a getter for detrimental foreign species [153].

Another process that modifies the electrocatalytic properties of MIEC electrodes is the formation of metallic nanoparticles on the surface of perovskite-type materials upon their partial decomposition under sufficiently reducing conditions, which is commonly called exsolution [12, 13, 154–159]. Unlike the abovementioned detrimental Sr segregation, this change of the electrode surface chemistry typically improves the fuel electrode kinetics (i.e. hydrogen oxidation/water splitting). Hence, an appreciable number of studies have been conducted to clarify the electrochemical mechanisms behind the observed reaction rate increase due to the exsolution-based surface decoration of MIEC fuel electrode materials [159]. Currently, the most plausible explanation for this effect is that metal particles on the surface of a mixed-conducting electrode provide an alternative adsorption/desorption pathway for hydrogen [12, 13, 160]. This additional pathway is capable of bypassing the rate-determining step on the bare oxide electrode surface, as sketched in figure 6.12. Since this pathway requires a diffusive spillover of H_{ad} species from the metal to the oxide surface, the metal particles are expected to promote the kinetics only in their close vicinity—i.e. up to the decay length of the lateral spillover process. Hence, in terms of reaction pathways, an MIEC electrode decorated with metallic exsolution particles can be regarded as a kind of hybrid electrode in which one reaction mechanism is restricted to the triple-phase boundary (TPB) of the surface particles (figure 6.2(c)) and the other proceeds at the oxide surface further away from the particles (figure 6.2(d)).

A similar behavior was also observed for MIEC electrodes decorated with noble metal particles in an oxygen atmosphere [161, 162]. In addition, a performance enhancement was observed that can very likely be explained by a similar mechanism of the dissociative adsorption of oxygen on the noble metal particles [93–95] and spillover of the resulting catalytically activated adsorbate to the oxide surface.

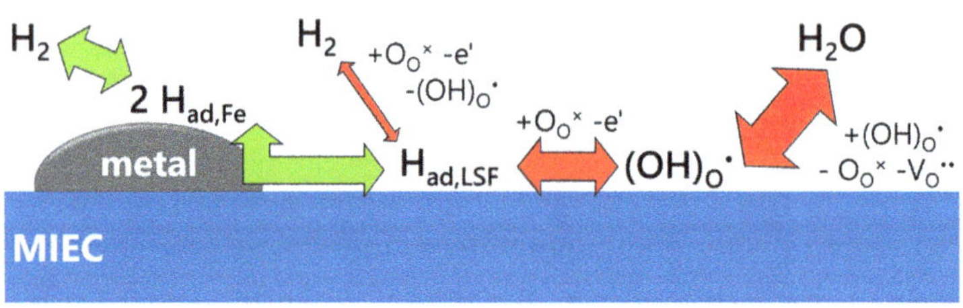

Figure 6.12. Suggested mechanism by which exsolved metal particles establish an additive pathway that bypasses the rate-determining reaction step on the MIEC surface. The green arrows indicate the pathway specific to the metal particles decorating the MIEC surface, while the red arrows show the reaction pathway on the bare oxide surface. The thicknesses of the arrows qualitatively denote the reaction rates (image adapted from [13]).

6.4 Methods used for the characterization of electrode kinetics

6.4.1 Current–voltage curves

The observation of current–voltage (*I*–*V*) characteristics is a classical electrochemical investigation technique with a long tradition. The strength of this method arises from the fact that the recorded current is directly proportional to the net rate r_{net} of the observed electrode reaction (e.g. H_2 oxidation/H_2O splitting, equation (6.2); or O_2 evolution/reduction, equation (6.13)) according to Faraday's law:

$$I_{DC} = zFr_{net}. \tag{6.22}$$

Moreover, the applied overpotential (equation (6.15)) is directly proportional to the thermodynamic driving force which drives this reaction. This holds for electrodes immersed in a liquid electrolyte (e.g. with Butler–Volmer kinetics—equation (6.1)) as well as for electrodes in solid-state electrochemistry. In the case of MIEC electrodes with a rate-limiting surface process, this driving force can be quantified by equation (6.17).

Hence, an I_{DC} versus η_{WE} plot is, in fact, a diagram that connects the net rate of the reaction of interest to the applied thermodynamic driving force. As is the case for *I*–*V* curves in liquid electrochemistry, in solid-state electrochemistry these characteristics also consist of typical features such as exponential current increase, limiting currents, or linear regimes. However, the interpretation of these features can be completely different. While in liquid electrochemistry an exponential *I*–*V* curve is a strong indication of charge-transfer-limited kinetics, in the case of solid MIEC electrodes, changes of the point defect concentrations can also lead to this behavior (in equation (6.19), this is included in the term $\prod_k [k]^{\nu_k}$). A limiting current can, in addition to classical concentration polarization, also result from so-called Wagner–Hebb polarization [163]. One could mention even more examples at this point to illustrate the complexity of the *I*–*V* characteristics of solid-state electrochemical systems. Generally, it is important to note that one needs to be rather careful when physically interpreting such data, and additional experimental measures (such as partial pressure variations) are usually needed to get further insights.

6.4.2 Impedance spectroscopy

Since the interpretation of *I*–*V* curves can be quite tricky, impedance spectroscopy has been used as a very powerful characterization method in solid-state electrochemistry. In contrast to *I*–*V* measurements, which are a DC technique, impedance spectroscopy is an AC method that makes use of the different frequency dependencies of e.g. electrode and electrolyte impedances (impedance = complex AC resistance). As chapter 9 of this book is dedicated to this characterization technique, here, we only briefly elaborate the relevance of impedance spectroscopy for experiments on model systems.

A great strength of impedance spectroscopy is that it is able to distinguish the resistive and capacitive components of the impedance of a sample. For a homogeneous material such as an ion-conducting oxide, this is the case because charge transport through this material can occur via two fundamentally different mechanisms: (i) by the long-range transport of ions, which is commonly referred to as faradaic charge transport and (ii) by displacement polarization. This situation can be represented by the parallel connection of a resistance R and a capacitance C, which is hence often known as an R||C element. In the so-called Nyquist diagram (a plot of the imaginary part versus the real part), the impedance function of an R||C element yields a semicircle—see figure 6.13(a) and chapter 9 for further details. This semicircle has one characteristic point at the apex of the curve, which is called the characteristic angular frequency ω^*:

$$\omega^* = \frac{1}{RC} \tag{6.23}$$

Using a similar approach to that used for the electrolyte, an R||C element can also be obtained for the electrode in the very simplest case, as current can simultaneously flow as a faradaic current via an electrochemical reaction and as a transient current via the electrode capacitance. However, the physical meanings of the parameters R_{ode} and C_{ode} are completely different. R_{ode}, which is called the polarization resistance of the electrode, is not related to a conductivity but is obtained from the fact that impedance spectroscopy measures the slope of the electrode's I–V characteristics (see figure 6.13). It is thus a differential resistance and equals the inverse slope of the I–V curve at the respective DC overpotential. The exact proportionality between R_{ode} and r_{net} depends on the respective reaction mechanism, but in any case, the electrode polarization resistance is a measure of how much rate increase can be gained per additional applied driving force.

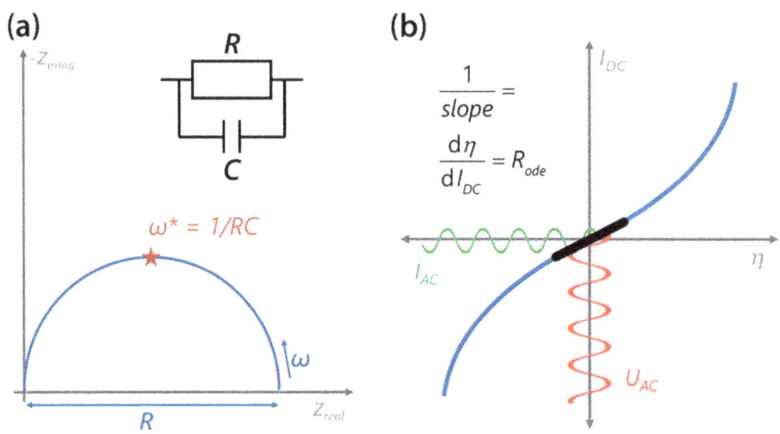

Figure 6.13. (a) Impedance spectrum (Nyquist plot) of an ideal R||C element. (b) Measuring the slope of the I–V curve of an electrode by means of impedance spectroscopy. The perturbation signal is a sine-shaped AC voltage (depicted in red), which leads to a sine-shaped AC current response (shown in green) with an amplitude that depends on the I–V curve of the respective electrode. The AC polarization resistance of the electrode is obtained from the inverse slope of the curve at the respective overpotential.

The connection between R_{ode} and the electrode's I–V curve is also the reason why the impedance characterization of electrode kinetics requires the use of quite low AC voltage amplitudes. Only for small amplitudes can a pseudolinear response be assumed from the commonly nonlinear I–V curves of electrode reactions. Larger amplitudes also trigger higher harmonic oscillations in the current response, which can be the source of artifacts in the case of a standard impedance analysis.

The capacitive behavior of electrodes commonly originates from two main effects. One is the capacitive response of the electrochemical double layer at the electrode/electrolyte interface. For electrolytes with high conductivity—which is the case for common SOEC/SOFC electrolyte materials—the assumption that a Helmholtz double layer is present at the interface can be regarded as a good approximation [1, 2]. Such a double layer behaves like a plate capacitor, and the expected interfacial capacitances can be estimated to be some ten $\mu F\ cm^{-2}$ [64, 164] (assuming $\varepsilon_r \approx 10$ and about half a unit cell of the electrolyte). Indeed, such capacitances can be found in model experiments under certain conditions [14], but in many cases even metal electrodes on solid electrolytes exhibit significantly larger capacitance values [6, 37, 64]. These usually cannot be explained by simple electrostatic models, but are interpreted as chemical capacitances, which are discussed below. For highly doped systems such as YSZ or GDC, more complex space charge models may be needed to correctly model the capacitive behavior of a charged interface [165].

In contrast to electrostatic capacitances, chemical or pseudocapacitances store charge by a faradaic process and accumulation of electrically neutral reactants. MIEC electrodes typically exhibit a chemical capacitance [41, 122, 166, 167], as they change their point defect chemistry (and thus stoichiometry) when a bias voltage is applied (figure 6.7 and equation (6.17)). That means that the charge is effectively stored in the form of electrically neutral oxygen—e.g. by the uptake/release of oxide ions and electron holes. The chemical capacitance is then a measure of how much charge is stored per voltage change. Hence, C_{chem} is a differential capacitance, as given in equation (6.24), and it reflects the ability of the MIEC to change its stoichiometry upon a change in the chemical potential of oxygen caused by the applied overpotential.

$$C_{chem} = \frac{dQ}{dU} \tag{6.24}$$

If a single type of electronic defect is relevant (for SOEC air electrodes, this is typically electron holes) and if the point defects of the material are ideally dilute, the chemical capacitance for a MIEC electrode such as LSF is given by [122, 166, 167]

$$C_{chem} = \frac{F^2}{RT} n_{uc} \left(\frac{1}{4c_{V_O^{\bullet\bullet}}} + \frac{1}{c_{h^\bullet}} \right). \tag{6.25}$$

In equation (6.25), n_{uc} denotes the concentration of unit cells, $c_{V_O^{\bullet\bullet}}$ and $c_{h^\bullet}$ are the concentrations of oxygen vacancies and electron holes in defects per unit cell, respectively, and F, R, and T have their usual meanings. This relationship also very strikingly shows another significance of the chemical capacity of MIEC electrode materials: it allows rather simple and direct access to the minority carrier concentration of the material, which is commonly difficult to access by other methods (such as conductivity measurements).

6.5 Concluding remarks

Compared to electrochemical cells with liquid electrolytes, the reaction kinetics of electrodes in solid oxide electrochemical cells (such as SOFCs and SOECs) have several peculiarities:

First, several different possible reaction pathways exist on solid-state electrodes, and geometrically well-defined patterned electrodes are a powerful tool for separating these pathways. Such thin-film-based model electrodes were used to show that these reaction paths can often coexist—e.g. in the case of $Pt(O_2)$/YSZ and $Ni(H_2–H_2O)$/YSZ, a surface path with 3PB activity parallels a volume path through the electrode.

Second, the point defects in the electrolyte (in the case of 3PB-active electrode systems) or in the electrode itself (in case of 2PB-active electrodes) actively participate in the electrode reaction and are thus of great importance for the electrode kinetics. As these point defect concentrations change upon pO_2 changes, the electrocatalytic properties of the respective materials can be largely affected by compositional changes of the atmosphere. For an in-depth understanding of the reaction kinetics of solid-state electrodes—e.g. for an interpretation of the reaction order with respect to the pO_2—a detailed knowledge of the material's (surface) defect chemistry is thus important.

Third, on MIEC electrodes, the electrostatic voltage drop caused by an applied voltage does not occur at the interface at which the electrochemical reaction proceeds (i.e. the MIEC–gas interface). The driving force for the surface reaction is a step change of the oxygen chemical potential between the electrode and the gas atmosphere, which is proportional to the electrochemical overpotential η_{WE}. The necessary electrostatic voltage step η_{WE} occurs either at the electrode/electrolyte interface or at the electrode/current collector interface, depending on whether electronic or ionic point defects are the majority charge carriers, respectively.

Fourth, if charged adsorbates cause a dipole at the reacting surface, the related electrostatic potential step at this surface also affects the respective electrode reaction rate. This potential step is also affected by an applied overpotential, but the relationship between both is highly complex and yet analytically unavailable for most electrode material and reaction combinations.

Fifth, the importance of the electrode surface is further emphasized by the fact that changes in its chemical composition can have tremendous effects on electrode kinetics. From a performance point of view, these changes can be either highly beneficial or extremely detrimental. For example, the formation of exsolved metal nanoparticles on perovskite-type electrodes can greatly enhance their water splitting/hydrogen oxidation kinetics, while the surface segregation of Sr oxide has been reported to greatly impede oxygen surface exchange.

References

[1] Hamann C H, Hamnett A and Vielstich W 2007 *Electrochemistry* (Weinheim: Wiley-VCH))

[2] Bockris J O M, Reddy A K N and Gamboa-Aldeco M 2002 *Modern Electrochemistry* vol 2A (New York: Kluwer Academic Publishers)

[3] Hertz J, Rothschild A and Tuller H 2009 Highly enhanced electrochemical performance of silicon-free platinum/yttria stabilized zirconia interfaces *J. Electroceram.* **22** 428–35

[4] Mutoro E, Baumann N and Janek J 2010 Janus-faced SiO_2: activation and passivation in the electrode system platinum/yttria-stabilized zirconia *J. Phys. Chem. Lett.* **1** 2322–6

[5] Ryll T, Galinski H, Schlagenhauf L, Elser P, Rupp J L M, Bieberle-Hutter A and Gauckler L J 2011 Microscopic and nanoscopic three-phase-boundaries of platinum thin-film electrodes on YSZ electrolyte *Adv. Funct. Mater.* **21** 565–72

[6] Opitz A K and Fleig J 2010 Investigation of O_2 reduction on Pt/YSZ by means of thin film microelectrodes: the geometry dependence of the electrode impedance *Solid State Ionics* **181** 684–93

[7] Mizusaki J, Amano K, Yamauchi S and Fueki K 1987 Electrode-reaction at Pt,O_2/stabilized zirconia interfaces. Part II: Electrochemical measurements and analysis *Solid State Ionics* **22** 323–30

[8] Mizusaki J *et al* 1994 Kinetic studies of the reaction at the nickel pattern electrode on YSZ in H_2/H_2O atmospheres *Solid State Ionics* **70–71** 52–8

[9] Bieberle A, Meier L P and Gauckler L J 2001 The electrochemistry of Ni pattern anodes used as solid oxide fuel cell model electrodes *J. Electrochem. Soc.* **148** A646–A56

[10] Utz A, Störmer H, Leonide A, Weber A and Ivers-Tiffee E 2009 Degradation effects of Ni patterned anodes in H_2/H_2O atmosphere *ECS Trans.* **25** 2013–21

[11] Yao W and Croiset E 2013 Ni/YSZ pattern anodes fabrication and their microstructure and electrochemical behavior changes in H_2–H_2O environments *J. Power Sources* **226** 162–72

[12] Zhu T, Troiani H E, Mogni L V, Han M and Barnett S A 2018 Ni-substituted Sr(Ti,Fe)O_3 SOFC anodes: achieving high performance via metal alloy nanoparticle exsolution *Joule* **2** 478–96

[13] Opitz A K *et al* 2020 Understanding electrochemical switchability of perovskite-type exsolution catalysts *Nat. Comm.* **11** 4801

[14] Baumann F S, Fleig J, Habermeier H U and Maier J 2006 Impedance spectroscopic study on well-defined (La,Sr)(Co,Fe)$O_{3-\delta}$ model electrodes *Solid State Ionics* **177** 1071–81

[15] Adler S B 2004 Factors governing oxygen reduction in solid oxide fuel cell cathodes *Chem. Rev.* **104** 4791–844

[16] Jung W and Tuller H L 2008 Investigation of cathode behavior of model thin-film $SrTi_{1-x}Fe_xO_{3-\delta}$ (x = 0.35 and 0.5) mixed ionic-electronic conducting electrodes *J. Electrochem. Soc.* **155** B1194–201

[17] Opitz A K, Kubicek M, Huber S, Huber T, Holzlechner G, Hutter H and Fleig J 2013 Thin film cathodes in SOFC research: how to identify oxygen reduction pathways? *J. Mater. Res.* **28** 2085–105

[18] Bucher E, Sitte W, Klauser F and Bertel E 2011 Oxygen exchange kinetics of $La_{0.58}Sr_{0.4}Co_{0.2}Fe_{0.8}O_3$ at 600 °C in dry and humid atmospheres *Solid State Ionics* **191** 61–7

[19] Ringuedé A and Fouletier J 2001 Oxygen reaction on strontium-doped lanthanum cobaltite dense electrodes at intermediate temperatures *Solid State Ionics* **139** 167–77

[20] Wang L, Merkle R and Maier J 2010 Surface kinetics and mechanism of oxygen incorporation into $Ba_{1-x}Sr_xCo_yFe_{1-y}O_{3-\delta}$ SOFC microelectrodes *J. Electrochem. Soc.* **157** B1802–8

[21] Simrick N J, Bieberle-Hütter A, Ryll T M, Kilner J A, Atkinson A and Rupp J L M 2012 An investigation of the oxygen reduction reaction mechanism of $La_{0.6}Sr_{0.4}Co_{0.2}Fe_{0.8}O_3$ using patterned thin films *Solid State Ionics* **206** 7–16

[22] Schrödl N, Egger A, Lammer J, Hofer F and Sitte W 2021 Long-term stability of $Pr_2NiO_{4+\delta}$ air electrodes for solid oxide cells against chromium poisoning *J. Electrochem. Soc.* **168** 014509

[23] Chauveau F, Mougin J, Bassat J M, Mauvy F and Grenier J C 2010 A new anode material for solid oxide electrolyser: the neodymium nickelate $Nd_2NiO_{4+\delta}$ *J. Power Sources* **195** 744–9

[24] Chroneos A, Parfitt D, Kilner J A and Grimes R W 2010 Anisotropic oxygen diffusion in tetragonal $La_2NiO_{4+\delta}$: molecular dynamics calculations *J. Mater. Chem.* **20** 266–70

[25] Kovalevsky A, Kharton V, Yaremchenko A, Pivak Y, Tsipis E, Yakovlev S, Markov A, Naumovich E and Frade J 2007 Oxygen permeability, stability and electrochemical behavior of Pr_2NiO_{4+d}-based materials *J. Electroceram.* **18** 205–18

[26] Van Herle J and McEvoy A J 1994 Oxygen diffusion through silver cathodes for solid oxide fuel cells *J. Phys. Chem. Solids* **55** 339–47

[27] Mizusaki J *et al* 1994 Preparation of nickel pattern electrodes on YSZ and their electrochemical properties in H_2–H_2O atmospheres *J. Electrochem. Soc.* **141** 2129–34

[28] Bieberle A and Gauckler L J 2000 Reaction mechanism of Ni pattern anodes for solid oxide fuel cells *Solid State Ionics* **135** 337–45

[29] Sukeshini A M, Habibzadeh B, Becker B P, Stoltz C A, Eichhorn B W and Jackson G S 2006 Electrochemical oxidation of H_2, CO, and CO / H_2 mixtures on patterned Ni anodes on YSZ electrolytes *J. Electrochem. Soc.* **153** A705–A15

[30] Amezawa K, Shindo Y, Fujimaki Y, Kimura Y, Nakamura T, Iguchi F, Yashiro K, Yugami H and Kawada T 2017 Mechanism of chromium poisoning in SOFC cathode investigated by using pattern thin film model electrode *ECS Trans.* **78** 965–70

[31] Usiskin R E, Maruyama S, Kucharczyk C J, Takeuchi I and Haile S M 2015 Probing the reaction pathway in $(La_{0.8}Sr_{0.2})_{0.95}MnO_{3+\delta}$ using libraries of thin film microelectrodes *J. Mater. Chem.* A **3** 19330–45

[32] Ioroi T, Hara T, Uchimoto Y, Ogumi Z and Takehara Z I 1997 Preparation of perovskite-type $La_{1-x}Sr_xMnO_3$ films by vapor-phase processes and their electrochemical properties *J. Electrochem. Soc.* **144** 1362–70

[33] Jung W and Tuller H L 2011 A new model describing solid oxide fuel cell cathode kinetics: model thin film $SrTi_{1-x}Fe_xO_{3-\delta}$ mixed conducting oxides – a case study *Adv. Energy Mater.* **1** 1184–91

[34] Lu Q, Vardar G, Jansen M, Bishop S R, Waluyo I, Tuller H L and Yildiz B 2018 Surface defect chemistry and electronic structure of $Pr_{0.1}Ce_{0.9}O_{2-\delta}$ revealed in operando *Chem. Mater.* **30** 2600–6

[35] Bliem R, Kim D, Wang J, Crumlin E J and Yildiz B 2021 Hf deposition stabilizes the surface chemistry of perovskite manganite oxide *J. Phys. Chem.* C **125** 3346–54

[36] Brichzin V, Fleig J, Habermeier H U and Maier J 2000 Geometry dependence of cathode polarization in solid oxide fuel cells investigated by defined Sr-doped $LaMnO_3$ microelectrodes *Electrochem. Solid-State Lett.* **3** 403–6

[37] Opitz A K, Lutz A, Kubicek M, Kubel F, Hutter H and Fleig J 2011 Investigation of the oxygen exchange mechanism on Pt/YSZ at intermediate temperatures: surface path versus bulk path *Electrochim. Acta* **56** 9727–40
[38] Nenning A, Opitz A K, Huber T and Fleig J 2014 A novel approach for analyzing electrochemical properties of mixed conducting solid oxide fuel cell anode materials by impedance spectroscopy *Phys. Chem. Chem. Phys.* **16** 22321–36
[39] Rupp G M, Opitz A K, Nenning A, Limbeck A and Fleig J 2017 Real-time impedance monitoring of oxygen reduction during surface modification of thin film cathodes *Nat. Mater.* **16** 640–5
[40] Doppler M C, Fleig J, Bram M and Opitz A K 2018 Hydrogen oxidation mechanisms on Ni/yttria stabilized zirconia anodes: separation of reaction pathways by geometry variation of pattern electrodes *J. Power Sources* **380** 46–54
[41] Chueh W C and Haile S M 2009 Electrochemical studies of capacitance in cerium oxide thin films and its relationship to anionic and electronic defect densities *Phys. Chem. Chem. Phys.* **11** 8144–8
[42] la O' G J, Savinell R F and Shao-Horn Y 2009 Activity enhancement of dense strontium-doped lanthanum manganite thin films under cathodic polarization: a combined AES and XPS study *J. Electrochem. Soc.* **156** B771–B81
[43] Goebel M C, Gregori G and Maier J 2011 Mixed conductivity in nanocrystalline highly acceptor doped cerium oxide thin films under oxidizing conditions *Phys. Chem. Chem. Phys.* **13** 10940–5
[44] Joo J H, Merkle R, Kim J-H and Maier J 2012 Measuring electrical properties of thin film fuel cell electrodes by *In Situ* infrared spectroscopy *Adv. Mater.* **24** 6507–12
[45] Kreller C R *et al* 2013 Origin of enhanced chemical capacitance in $La_{0.8}Sr_{0.2}CoO_{3-\delta}$ thin film electrodes *J. Electrochem. Soc.* **160** F931–F42
[46] Chueh W C, Lai W and Haile S M 2008 Electrochemical behavior of ceria with selected metal electrodes *Solid State Ionics* **179** 1036–41
[47] Mueller D N, Machala M L, Bluhm H and Chueh W C 2015 Redox activity of surface oxygen anions in oxygen-deficient perovskite oxides during electrochemical reactions *Nat. Comm.* **6** 6097
[48] Ehn A, Høgh J, Graczyk M, Norrman K, Montelius L, Linne M and Mogensen M 2010 Electrochemical investigation of nickel pattern electrodes in H_2/H_2O and CO/CO_2 atmospheres *J. Electrochem. Soc.* **157** B1588–96
[49] Gerstl M, Nenning A, Iskandar R, Rojek-Wöckner V, Bram M, Hutter H and Opitz A K 2016 The sulphur poisoning behaviour of gadolinia doped ceria model systems in reducing atmospheres *Materials* **9** 649
[50] Chueh W C, Hao Y, Jung W and Haile S M 2012 High electrochemical activity of the oxide phase in model ceria–Pt and ceria–Ni composite anodes *Nat. Mater.* **11** 155–61
[51] Gao R *et al* 2021 Correlating surface crystal orientation and gas kinetics in perovskite oxide electrodes *Adv. Mater.* **33** 2100977
[52] Weber A, Sauer B, Müller A C, Herbstritt D and Ivers-Tiffée E 2002 Oxidation of H_2, CO and methane in SOFCs with Ni/YSZ-cermet anodes *Solid State Ionics* **152-53** 543–50
[53] Utz A, Leonide A, Weber A and Ivers-Tiffée E 2011 Studying the CO–CO_2 characteristics of SOFC anodes by means of patterned Ni anodes *J. Power Sources* **196** 7217–24
[54] Park S, Vohs J M and Gorte R J 2000 Direct oxidation of hydrocarbons in a solid-oxide fuel cell *Nature* **404** 265–7

[55] Eguchi K, Kojo H, Takeguchi T, Kikuchi R and Sasaki K 2002 Fuel flexibility in power generation by solid oxide fuel cells *Solid State Ionics* **152–153** 411–6
[56] Brandon N P, Skinner S and Steele B C H 2003 Recent advances in materials for fuel cells *Annu. Rev. Mater. Res.* **33** 183–213
[57] Lu J, Zhu C, Pan C, Lin W, Lemmon J P, Chen F, Li C and Xie K 2018 Highly efficient electrochemical reforming of CH_4/CO_2 in a solid oxide electrolyser *Sci. Adv.* **4** eaar5100
[58] Bessler W G, Vogler M, Stormer H, Gerthsen D, Utz A, Weber A and Ivers-Tiffee E 2010 Model anodes and anode models for understanding the mechanism of hydrogen oxidation in solid oxide fuel cells *Phys. Chem. Chem. Phys.* **12** 13888–903
[59] Rossmeisl J and Bessler W G 2008 Trends in catalytic activity for SOFC anode materials *Solid State Ionics* **178** 1694–700
[60] Bieberle A and Gauckler L J 2002 State-space modeling of the anodic SOFC system Ni, H_2–H_2O|YSZ *Solid State Ionics* **146** 23–41
[61] Yao W and Croiset E 2015 Stability and electrochemical performance of Ni/YSZ pattern anodes in H_2/H_2O atmosphere *Canad. J. Chem. Eng.* **93** 2157–67
[62] Utz A, Störmer H, Leonide A, Weber A and Ivers-Tiffée E 2010 Degradation and relaxation effects of Ni patterned anodes in H_2–H_2O atmosphere *J. Electrochem. Soc.* **157** B920–B30
[63] Rao M V, Fleig J, Zinkevich M and Aldinger F 2010 The influence of the solid electrolyte on the impedance of hydrogen oxidation at patterned Ni electrodes *Solid State Ionics* **181** 1170–7
[64] Doppler M C, Fleig J, Bram M and Opitz A K 2016 The capacitance of nickel pattern electrodes on zirconia electrolyte *J. Electrochem. Soc.* **163** H1019–25
[65] Doppler M C, Fleig J, Bram M and Opitz A K 2019 Comparison of electrochemical hydrogen oxidation on different metal/ceramic model anodes and mechanistic implications *J. Phys.: Energy* **1** 035001
[66] Lee B K, Yu Y H, So B S, Kim S M, Kim J, Lee H W, Lee J H and Hwang J H 2006 Electrical characterization of the platinum/YSZ interfaces in SOFCs via micro-contact impedance spectroscopy *J. Electroceram.* **17** 735–9
[67] Hertz J L and Tuller H L 2007 Nanocomposite platinum-yttria stabilized zirconia electrode and implications for micro-SOFC operation *J. Electrochem. Soc.* **154** B413–8
[68] Huang H, Nakamura M, Su P, Fasching R, Saito Y and Prinz F B 2007 High-performance ultrathin solid oxide fuel cells for low-temperature operation *J. Electrochem. Soc.* **154** B20–4
[69] Irvine J T S, Neagu D, Verbraeken M C, Chatzichristodoulou C, Graves C and Mogensen M B 2016 Evolution of the electrochemical interface in high-temperature fuel cells and electrolysers *Nat. Energy* **1** 15014
[70] Hoerlein M P, Riegraf M, Costa R, Schiller G and Friedrich K A 2018 A parameter study of solid oxide electrolysis cell degradation: microstructural changes of the fuel electrode *Electrochim. Acta* **276** 162–75
[71] Laguna-Bercero M A, Campana R, Larrea A, Kilner J A and Orera V M 2011 Electrolyte degradation in anode supported microtubular yttria stabilized zirconia-based solid oxide steam electrolysis cells at high voltages of operation *J. Power Sources* **196** 8942–7
[72] Liu Q, Wang H, Kennouche D, Riscoe C, Butts D and Barnett S A 2018 Effect of reversing-current operation on the structure and electrochemical performance evolution of Ni-YSZ fuel electrodes *J. Electrochem. Soc.* **165** F870–5

[73] Song B, Bertei A, Wang X, Cooper S J, Ruiz-Trejo E, Chowdhury R, Podor R and Brandon N P 2019 Unveiling the mechanisms of solid-state dewetting in solid oxide cells with novel 2D electrodes *J. Power Sources* **420** 124–33

[74] Mogensen M B, Hauch A, Sun X, Chen M, Tao Y, Ebbesen S D, Hansen K V and Hendriksen P V 2017 Relation between Ni particle shape change and Ni migration in Ni–YSZ electrodes – a hypothesis *Fuel Cells* **17** 434–41

[75] Chen J, Ouyang M, Boldrin P, Atkinson A and Brandon N P 2020 Understanding the coarsening and degradation in a nanoscale nickel gadolinia-doped-ceria electrode for high-temperature applications *ACS Appl. Mater. Interfaces* **12** 47564–73

[76] Matsui T, Eguchi K, Shirai K, Okanishi T, Muroyama H and Eguchi K 2017 New Design concept of Ni-doped-ceria cermet anode for solid oxide fuel cells: redox-induced self-modification *ECS Trans.* **78** 1115–23

[77] Kim Y-D, Yang J, Lee J-I, Saqib M, Shin J-S, Park K, Jo M, Song S-J and Park J-Y 2020 Degradation studies of ceria-based solid oxide fuel cells at intermediate temperature under various load conditions *J. Power Sources* **452** 227758

[78] Wang F, Kishimoto H, Ishiyama T, Develos-Bagarinao K, Yamaji K, Horita T and Yokokawa H 2019 Electrochemically induced nickel/ceria interaction in solid oxide electrochemical cells with oxide ion conducting electrolyte *J. Electrochem. Soc.* **166** F587–F93

[79] Green R D, Liu C-C and Adler S B 2008 Carbon dioxide reduction on gadolinia-doped ceria cathodes *Solid State Ionics* **179** 647–60

[80] Rojek-Wöckner V A, Opitz A K, Brandner M, Mathé J and Bram M 2016 A novel Ni/ceria-based anode for metal-supported solid oxide fuel cells *J. Power Sources* **328** 65–74

[81] Zechnal R, Baumann G and Eisele H 1973 Closed-loop exhaust emission control system with electronic fuel injection *SAE Technical Papers* 730566

[82] Riegel J, Neumann H and Wiedenmann H M 2002 Exhaust gas sensors for automotive emission control *Solid State Ionics* **152-53** 783–800

[83] Ramamoorthy R, Dutta P K and Akbar S A 2003 Oxygen sensors: materials, methods, designs and applications *J. Mater. Sci.* **38** 4271–82

[84] Docquier N and Candel S 2002 Combustion control and sensors: a review *Prog. Energy Combust. Sci.* **28** 107–50

[85] Wang D Y and Nowick A S 1979 Cathodic and anodic polarization phenomena at platinum electrodes with doped CeO_2 as electrolyte: I. Steady-state overpotential *J. Electrochem. Soc.* **126** 1155–65

[86] Verkerk M J and Burggraaf A J 1983 Oxygen transfer on substituted ZrO_2, Bi_2O_3, and CeO_2 electrolytes with platinum electrodes: II. AC impedance study *J. Electrochem. Soc.* **130** 78–84

[87] Robertson N L and Michaels J N 1990 Oxygen-exchange on platinum-electrodes in zirconia cells – location of electrochemical reaction sites *J. Electrochem. Soc.* **137** 129–35

[88] van Herle J and McEvoy A J 1993 Impedance characteristics of platinum electrodes on yttria-stabilized zirconia *Ber. Bunsenges. Phys. Chem.* **97** 470–4

[89] Mutoro E, Luerßen B, Günther S and Janek J 2009 The electrode model system $Pt(O_2)YSZ$: influence of impurities and electrode morphology on cyclic voltammograms *Solid State Ionics* **180** 1019–33

[90] Sasaki K, Tamura J and Dokiya M 2001 Pt-cermet cathode for reduced temperature SOFCs *Solid State Ionics* **144** 223–32

[91] Jaccoud A, Falgairette C, Fóti G and Comninellis C 2007 Charge storage in the O_2(g),Pt/YSZ system *Electrochim. Acta* **52** 7927–35
[92] Zheng Y, Sauter U and Moos R 2018 Oxygen transport paths in screen-printed Pt–Al_2O_3 composite model electrodes on YSZ *Solid State Ionics* **316** 53–8
[93] Bonzel H P and Ku R 1973 On the kinetics of oxygen adsorption on a Pt(111) surface *Surf. Sci.* **40** 85–101
[94] Gland J L and Korchak V N 1978 The adsorption of oxygen on a stepped platinum single crystal surface *Surf. Sci.* **75** 733–50
[95] Campbell C T, Ertl G, Kuipers H and Segner J 1981 A molecular beam study of the adsorption and desorption of oxygen from a Pt(111) surface *Surf. Sci.* **107** 220–36
[96] Steininger H, Lehwald S and Ibach H 1982 Adsorption of oxygen on Pt(111) *Surf. Sci.* **123** 1–17
[97] Nolan P D, Lutz B R, Tanaka P L, Davis J E and Mullins C B 1999 Molecularly chemisorbed intermediates to oxygen adsorption on Pt(111): A molecular beam and electron energy-loss spectroscopy study *J. Chem. Phys.* **111** 3696–704
[98] Mitterdorfer A and Gauckler L J 1999 Reaction kinetics of the Pt,O_2(g)c–ZrO_2 system: precursor-mediated adsorption *Solid State Ionics* **120** 211–25
[99] Mitterdorfer A and Gauckler L J 1999 Identification of the reaction mechanism of the Pt, O_2(g)|yttria-stabilized zirconia system. Part I: general framework, modelling, and structural investigation *Solid State Ionics* **117** 187–202
[100] Mitterdorfer A and Gauckler L J 1999 Identification of the reaction mechanism of the Pt, O_2(g)|yttria-stabilized zirconia system. Part II: model implementation, parameter estimation, and validation *Solid State Ionics* **117** 203–17
[101] Jacobsen T and Bay L 2002 Thermal memory effects at the Pt|YSZ interface *Electrochim. Acta* **47** 2177–81
[102] Baker R, Guindet J and Kleitz M 1997 Classification criteria for solid oxide fuel cell electrode materials *J. Electrochem. Soc.* **144** 2427–32
[103] Breiter M W, Leeb K and Fafilek G 1997 Voltammetric studies of electrochemical processes at the interface Pt-YSZ between 300 and 600 °C *J. Electroanal. Chem.* **434** 129–37
[104] Yoon S P, Nam S W, Kim S-G, Hong S-A and Hyun S-H 2003 Characteristics of cathodic polarization at Pt/YSZ interface without the effect of electrode microstructure *J. Power Sources* **115** 27–34
[105] Luerßen B, Mutoro E, Fischer H, Günther S, Imbihl R and Janek J 2006 *In situ* imaging of electrochemically induced oxygen spillover on Pt/YSZ Catalysts *Angew. Chem. Int. Ed.* **45** 1473–6
[106] Radhakrishnan R, Virkar A V and Singhal S C 2005 Estimation of charge-transfer resistivity of Pt cathode on YSZ electrolyte using patterned electrodes *J. Electrochem. Soc.* **152** A927–36
[107] Kishimoto H, Sakai N, Yamaji K, Horita T, Brito M E, Yokokawa H, Amezawa K and Uchimoto Y 2008 Visualization of oxygen transport behavior at metal electrode/oxide electrolyte interface using secondary ion mass spectrometry *Solid State Ionics* **179** 347–54
[108] Hertz J L and Tuller H L 2007 Measurement and finite element modeling of triple phase boundary-related current constriction in YSZ *Solid State Ionics* **178** 915–23
[109] Neumeier J J, Elm M T, Luerßen B and Janek J 2018 Platinum microelectrodes on gadolinia doped ceria single crystals – bulk properties and electrode kinetics *Phys. Chem. Chem. Phys.* **20** 8294–301

[110] Opitz A K, Schintlmeister A, Hutter H and Fleig J 2010 Visualization of oxygen reduction sites at Pt electrodes on YSZ by means of 18O tracer incorporation: the width of the electrochemically active zone *Phys. Chem. Chem. Phys.* **12** 12734–45

[111] Opitz A K, Horlein M P, Huber T and Fleig J 2012 Current–voltage characteristics of platinum model electrodes on yttria-stabilized zirconia *J. Electrochem. Soc.* **159** B502–13

[112] Huber T M, Opitz A K and Fleig J 2015 Oxygen reduction via grain boundary transport in thin film platinum electrodes on yttria stabilized zirconia *Solid State Ionics* **273** 8–12

[113] Van Herle J, McEvoy A J and Ravindranathan Thampi K 1994 Oxygen reduction at porous and dense cathodes for solid oxide fuel cells *Electrochim. Acta* **39** 1675–80

[114] Opitz A K 2011 Oxygen exchange pathways of platinum model electrodes on yttria-stabilized zirconia *Institute of Chemical Technologies and Analytics* (Vienna: TU Wien))

[115] Fukunaga H, Ihara M, Sakaki K and Yamada K 1996 The relationship between overpotential and the three phase boundary length *Solid State Ionics* **86–88** 1179–85

[116] la O' G J, Yildiz B, McEuen S and Shao-Horn Y 2007 Probing oxygen reduction reaction kinetics of Sr-doped $LaMnO_3$ supported on Y_2O_3-stabilized ZrO_2: EIS of dense, thin-film microelectrodes *J. Electrochem. Soc.* **154** B427–B38

[117] Fleig J, Kim H R, Jamnik J and Maier J 2008 Oxygen reduction kinetics of lanthanum manganite (LSM) model cathodes: partial pressure dependence and rate-limiting steps *Fuel Cells* **8** 330–7

[118] Backhaus-Ricoult M, Adib K St, Clair T, Luerßen B, Gregoratti L and Barinov A 2008 In-situ study of operating SOFC LSM/YSZ cathodes under polarization by photoelectron microscopy *Solid State Ionics* **179** 891–5

[119] Navickas E *et al* 2015 Fast oxygen exchange and diffusion kinetics of grain boundaries in Sr-doped $LaMnO_3$ thin films *Phys. Chem. Chem. Phys.* **17** 7659–69

[120] Saranya A M, Pla D, Morata A, Cavallaro A, Canales-Vázquez J, Kilner J A, Burriel M and Tarancón A 2015 Engineering mixed ionic electronic conduction in $La_{0.8}Sr_{0.2}MnO_{3+\delta}$ nanostructures through fast grain boundary oxygen diffusivity *Adv. Energy Mater.* **5** 1500377

[121] Kuhn M, Hashimoto S, Sato K, Yashiro K and Mizusaki J 2011 Oxygen nonstoichiometry, thermo-chemical stability and lattice expansion of $La_{0.6}Sr_{0.4}FeO_{3-\delta}$ *Solid State Ionics* **195** 7–15

[122] Schmid A, Rupp G M and Fleig J 2018 Voltage and partial pressure dependent defect chemistry in $(La,Sr)FeO_{3-\delta}$ thin films investigated by chemical capacitance measurements *Phys. Chem. Chem. Phys.* **20** 12016–26

[123] Schmid A, Rupp G M and Fleig J 2018 How to get mechanistic information from partial pressure-dependent current–voltage measurements of oxygen exchange on mixed conducting electrodes *Chem. Mat.* **30** 4242–52

[124] Schmid A and Fleig J 2019 The current–voltage characteristics and partial pressure dependence of defect controlled electrochemical reactions on mixed conducting oxides *J. Electrochem. Soc.* **166** F831–46

[125] Fleig J, Rupp G M, Nenning A and Schmid A 2017 (Invited) towards an improved understanding of electrochemical oxygen exchange reactions on mixed conducting oxides *ECS Trans.* **77** 93–108

[126] Fleig J 2005 On the current–voltage characteristics of charge transfer reactions at mixed conducting electrodes on solid electrolytes *Phys. Chem. Chem. Phys.* **7** 2027–37

[127] Zhang C *et al* 2012 Multielement activity mapping and potential mapping in solid oxide electrochemical cells through the use of operando XPS *ACS Catal.* **2** 2297–304

[128] Zhang C *et al* 2013 Mechanistic studies of water electrolysis and hydrogen electro-oxidation on high temperature ceria-based solid oxide electrochemical cells *J. Am. Chem. Soc.* **135** 11572–9

[129] Feng Z A, Balaji Gopal C, Ye X, Guan Z, Jeong B, Crumlin E and Chueh W C 2016 Origin of overpotential-dependent surface dipole at CeO_{2-x}/gas interface during electrochemical oxygen insertion reactions *Chem. Mat.* **28** 6233–42

[130] Siebenhofer M, Riedl C, Schmid A, Limbeck A, Opitz A K, Fleig J and Kubicek M 2022 Investigating oxygen reduction pathways on pristine SOFC cathode surfaces by *in situ* PLD impedance spectroscopy *J. Mater. Chem.* A **10** 2305–19

[131] Fleig J, Merkle R and Maier J 2007 The $p(O_2)$ dependence of oxygen surface coverage and exchange current density of mixed conducting oxide electrodes: model considerations *Phys. Chem. Chem. Phys.* **9** 2713–23

[132] Kilner J A, De Souza R A and Fullarton I C 1996 Surface exchange of oxygen in mixed conducting perovskite oxides *Solid State Ionics* **86–88** 703–9

[133] Kuklja M M, Kotomin E A, Merkle R, Mastrikov Y A and Maier J 2013 Combined theoretical and experimental analysis of processes determining cathode performance in solid oxide fuel cells *Phys. Chem. Chem. Phys.* **15** 5443–71

[134] Merkle R and Maier J 2002 Oxygen incorporation into Fe-doped $SrTiO_3$: mechanistic interpretation of the surface reaction *Phys. Chem. Chem. Phys.* **4** 4140–8

[135] De Souza R A 2006 A universal empirical expression for the isotope surface exchange coefficients (k^*) of acceptor-doped perovskite and fluorite oxides *Phys. Chem. Chem. Phys.* **8** 890–7

[136] Nicollet C, Toparli C, Harrington G F, Defferriere T, Yildiz B and Tuller H L 2020 Acidity of surface-infiltrated binary oxides as a sensitive descriptor of oxygen exchange kinetics in mixed conducting oxides *Nat. Catal.* **3** 913–20

[137] Niania M, Podor R, Ben Britton T, Li C, Cooper S J, Svetkov N, Skinner S and Kilner J 2018 *In situ* study of strontium segregation in $La_{0.6}Sr_{0.4}Co_{0.2}Fe_{0.8}O_{3-\delta}$ in ambient atmospheres using high-temperature environmental scanning electron microscopy *J. Mater. Chem.* A **6** 14120–35

[138] Orikasa Y, Crumlin E J, Sako S, Amezawa K, Uruga T, Biegalski M D, Christen H M, Uchimoto Y and Shao-Horn Y 2014 Surface strontium segregation of solid oxide fuel cell cathodes proved by *in situ* depth-resolved X-ray absorption spectroscopy *ECS Electrochem. Lett.* **3** F23–6

[139] Chen Y, Téllez H, Burriel M, Yang F, Tsvetkov N, Cai Z, McComb D W, Kilner J A and Yildiz B 2015 Segregated chemistry and structure on (001) and (100) surfaces of $(La_{1-x}Sr_x)_2CoO_4$ override the crystal anisotropy in oxygen exchange kinetics *Chem. Mat.* **27** 5436–50

[140] Tsvetkov N, Lu Q, Sun L, Crumlin E J and Yildiz B 2016 Improved chemical and electrochemical stability of perovskite oxides with less reducible cations at the surface *Nat. Mater.* **15** 1010–6

[141] Chen Y, Jung W, Cai Z, Kim J J, Tuller H L and Yildiz B 2012 Impact of Sr segregation on the electronic structure and oxygen reduction activity of $SrTi_{1-x}Fe_xO_3$ surfaces *Energ. Environ. Sci.* **5** 7979–88

[142] Opitz A K *et al* 2018 The chemical evolution of the $La_{0.6}Sr_{0.4}CoO_{3-\delta}$ surface under SOFC operating conditions and its implications for electrochemical oxygen exchange activity *Top. Catal.* **61** 2129–41

[143] Rupp G M, Tellez H, Druce J, Limbeck A, Ishihara T, Kilner J and Fleig J 2015 Surface chemistry of $La_{0.6}Sr_{0.4}CoO_{3-\delta}$ thin films and its impact on the oxygen surface exchange resistance *J. Mater. Chem.* A **3** 22759–69
[144] Oh D, Gostovic D and Wachsman E D 2012 Mechanism of $La_{0.6}Sr_{0.4}Co_{0.2}Fe_{0.8}O_3$ cathode degradation *J. Mater. Res.* **27** 1992–9
[145] Bucher E, Gspan C and Sitte W 2015 Degradation and regeneration of the SOFC cathode material $La_{0.6}Sr_{0.4}CoO_{3-\delta}$ in SO_2-containing atmospheres *Solid State Ionics* **272** 112–20
[146] Bucher E, Gspan C, Hofer F and Sitte W 2013 Sulphur poisoning of the SOFC cathode material $La_{0.6}Sr_{0.4}CoO_{3-}$ δ *Solid State Ionics* **238** 15–23
[147] Schmid A, Nenning A, Opitz A, Kubicek M and Fleig J 2020 High oxygen exchange activity of pristine $La_{0.6}Sr_{0.4}FeO_{3-\delta}$ films and its degradation *J. Electrochem. Soc.* **167** 124509
[148] Berger C, Bucher E, Gspan C, Menzel A and Sitte W 2017 Impact of SO_2 on the oxygen exchange kinetics of the promising SOFC/SOEC air electrode material $La_{0.8}Ca_{0.2}FeO_{3-\delta}$ *J. Electrochem. Soc.* **164** F3008
[149] Beez A, Schiemann K, Menzler N H and Bram M 2018 Accelerated testing of chromium poisoning of Sr-containing mixed conducting solid oxide cell air electrodes *Frontiers in Energy Res.* **6** 70
[150] Perz M, Bucher E, Gspan C, Waldhäusl J, Hofer F and Sitte W 2016 Long-term degradation of $La_{0.6}Sr_{0.4}Co_{0.2}Fe_{0.8}O_{3-\delta}$ IT-SOFC cathodes due to silicon poisoning *Solid State Ionics* **288** 22–7
[151] Horita T, Kishimoto H, Yamaji K, Brito M E, Xiong Y, Yokokawa H, Hori Y and Miyachi I 2009 Effects of impurities on the degradation and long-term stability for solid oxide fuel cells *J. Power Sources* **193** 194–8
[152] Tsvetkov N, Lu Q and Yildiz B 2015 Improved electrochemical stability at the surface of $La_{0.8}Sr_{0.2}CoO_3$ achieved by surface chemical modification *Faraday Discuss.* **182** 257–69
[153] Heo S J, Hong J, Aphale A, Hu B and Singh P 2019 Chromium poisoning of $La_{1-x}Sr_xMnO_{3\pm\delta}$ cathodes and electrochemical validation of chromium getters in intermediate temperature-solid oxide fuel cells *J. Electrochem. Soc.* **166** F990–5
[154] Tsekouras G, Neagu D and Irvine J T S 2013 Step-change in high temperature steam electrolysis performance of perovskite oxide cathodes with exsolution of B-site dopants *Energ. Environ. Sci.* **6** 256–66
[155] Sun Y, Li J, Zeng Y, Amirkhiz B S, Wang M, Behnamian Y and Luo J 2015 A-site deficient perovskite: the parent for *in situ* exsolution of highly active, regenerable nano-particles as SOFC anodes *J. Mater. Chem.* A **3** 11048–56
[156] Dimitrakopoulos G, Ghoniem A F and Yildiz B 2019 *In situ* catalyst exsolution on perovskite oxides for the production of CO and synthesis gas in ceramic membrane reactors *Sustain. Energy Fuels* **3** 2347–55
[157] Neagu D *et al* 2019 *In situ* observation of nanoparticle exsolution from perovskite oxides: from atomic scale mechanistic insight to nanostructure tailoring *ACS Nano* **13** 12996–3005
[158] Neagu D, Tsekouras G, Miller D N, Ménard H and Irvine J T S 2013 *In situ* growth of nanoparticles through control of non-stoichiometry *Nat. Chem.* **5** 916–23
[159] Cao T, Kwon O, Gorte R J and Vohs J M 2020 Metal exsolution to enhance the catalytic activity of electrodes in solid oxide fuel cells *Nanomaterials* **10** 2445
[160] Myung J-h, Neagu D, Miller D N and Irvine J T S 2016 Switching on electrocatalytic activity in solid oxide cells *Nature* **537** 528–31

[161] Riedl C, Schmid A, Nenning A, Summerer H, Smetaczek S, Schwarz S, Bernardi J, Opitz A, Limbeck A and Fleig J 2020 Outstanding oxygen reduction kinetics of $La_{0.6}Sr_{0.4}FeO_{3-\delta}$ surfaces decorated with platinum nanoparticles *J. Electrochem. Soc.* **167** 104514

[162] Sahibzada M, Benson S J, Rudkin R A and Kilner J A 1998 Pd-promoted $La_{0.6}Sr_{0.4}Co_{0.2}Fe_{0.8}O_3$ cathodes *Solid State Ionics* **113-15** 285–90

[163] Maier J 2004 *Physical Chemistry of Ionic Materials – Ions and Electrons in Solids* (Chichester: Wiley) pp 454–7

[164] Hendriks M, ten Elshof J E, Bouwmeester H J M and Verweij H 2002 The electrochemical double-layer capacitance of yttria-stabilised zirconia *Solid State Ionics* **146** 211–7

[165] Mebane D S and De Souza R A 2015 A generalised space-charge theory for extended defects in oxygen-ion conducting electrolytes: from dilute to concentrated solid solutions *Energ. Environ. Sci.* **8** 2935–40

[166] Jamnik J and Maier J 2001 Generalised equivalent circuits for mass and charge transport: chemical capacitance and its implications *Phys. Chem. Chem. Phys.* **3** 1668–78

[167] Lai W and Haile S M 2005 Impedance spectroscopy as a tool for chemical and electrochemical analysis of mixed conductors: a case study of ceria *J. Am. Ceram. Soc.* **88** 2979–97

IOP Publishing

High-Temperature Electrolysis
From fundamentals to applications
Werner Sitte and Rotraut Merkle

Chapter 7

Cell architectures

Anke Hagen and Ming Chen

Solid oxide electrolysis cells (SOECs) have two major geometries: planar and tubular. For both geometries, the solid electrolyte can be oxygen ion conducting or proton conducting. The thickest layer in the solid oxide cell determines the cell's configuration, its overall properties (such as the area specific resistance) and how the cell processes contribute to them, its optimum operating temperatures, and mechanical properties. The main configurations are electrolyte supported, fuel electrode supported, and metal supported.

7.1 Cell geometries

7.1.1 Introduction

In an SOEC, electricity is used to split steam (and/or CO_2) into oxygen and hydrogen (and/or CO). As the name suggests, SOECs have a solid oxide as the electrolyte. This electrolyte can be oxygen ion conducting or proton conducting, regardless of the cell architecture. The main requirements for the electrolyte are gastightness, high ionic conductivity, and low to zero electronic conductivity.

The active layers comprise two electrodes—a cathode and an anode—sandwiching the electrolyte. Both electrodes need to be porous in order to allow for gas transport to and from the reaction sites, activity for the desired electrochemical reactions, and conductivity for both ions and electrons. These properties are achieved by either single-phase or mixed-phase electrode compositions.

The following electrochemical reactions occur in SOEC layers for oxygen-ion-conducting electrolytes (see figures 7.1(a) and (b)):

Steam is reduced in the porous cathode, releasing hydrogen and oxygen ions, which travel through the gastight oxide-ion-conducting electrolyte to the porous anode, where they are oxidized to oxygen, thereby releasing electrons to the outer circuit. Please note that both H_2O and CO_2 can be split in an SOEC, either separately or in a mixture. The electrolysis product hydrogen (or CO) emerges at the cathode mixed with steam (or CO_2).

doi:10.1088/978-0-7503-3951-3ch7

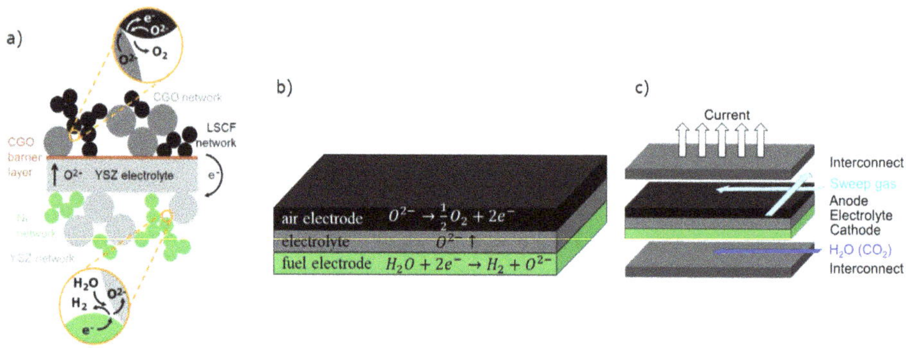

Figure 7.1. Electrolysis processes in an SOEC with an oxide-ion-conducting electrolyte: (a) structural illustration, (b) electrochemical reactions, and (c) the stacking principle of planar cells.

The terms cathode and anode denote the type of redox reaction: reduction occurs at the cathode during the uptake of electrons, while oxidation occurs at the anode during the release of electrons. As these redox reactions are the reverse of those of a solid oxide fuel cell (SOFC), the two electrodes are reversed as well, i.e. the cathode in the SOFC is the anode in the SOEC and vice versa. In order not to confuse the electrode denotations, the term 'fuel electrode' (or negatrode) is often used for the steam/hydrogen electrode, while 'air electrode' (or positrode) is used for the oxygen ion/oxygen electrode, instead of the terms 'anode' and 'cathode.' The fuel electrode is the same electrode in the SOFC and the SOEC, and the air electrode is the same in both modes as well.

The electrode reactions in SOECs with oxygen-ion-conducting electrolytes are as follows:

$$\text{Cathode: } H_2O + 2e^- \rightarrow H_2 + O^{2-} \tag{7.1}$$

$$CO_2 + 2e^- \rightarrow CO + O^{2-} \tag{7.2}$$

$$\text{Anode: } O^{2-} \rightarrow \frac{1}{2}O_2 + 2e^-. \tag{7.3}$$

Figure 7.1(a) illustrates how the electrolysis steps occur in the layered structure of the cell. The example features a Ni/YSZ fuel electrode, a YSZ electrolyte, and a LSCF/CGO–air electrode, separated from the electrolyte by a CGO barrier layer (YSZ = yttria-stabilized zirconia, $Zr_{1-x}Y_xO_{2-x/2}$; CGO = cerium gadolinium oxide, $Ce_{1-x}Gd_xO_{2-x/2}$; LSCF = lanthanum strontium cobaltite ferrite, $La_{1-x}Sr_xCo_{1-y}Fe_yO_{3-\delta}$; Ni = nickel).

The electrochemical reaction sites are located at the three-phase boundary of the fuel electrode, where the pores (gas transport), the electron-conducting phase (Ni), and the oxygen-ion-conducting phase (YSZ) meet. The air electrode, on the other hand, contains a mixed ion–electron-conducting material (LSCF) and the oxygen-ion-conducting (and slightly electron-conducting) phase (CGO), together with pores for gas transport. In this case, the reaction site expands towards the LSCF/CGO–air two-phase boundary. The CGO barrier layer prevents a chemical reaction from taking place between the YSZ electrolyte and the LSCF-containing air electrode. The given materials are typical examples; they can vary for the respective solid oxide electrolysis (SOE) parties.

In the case of a proton-conducting ceramic electrolyte, the following processes occur:

Steam is oxidized in the porous anode, releasing oxygen, electrons, and protons. The protons thus released travel through the gastight proton-conducting electrolyte to the cathode, where they are reduced to molecular hydrogen, thus releasing dry hydrogen.

The electrode reactions in SOECs with proton-conducting ceramic electrolytes (PCECs) are as follows:

$$\text{Anode: } H_2O \rightarrow \frac{1}{2}O_2 + 2H^+ + 2e^- \tag{7.4}$$

$$\text{Cathode: } 2H^+ + 2e^- \rightarrow H_2. \tag{7.5}$$

For both types of electrolyte, the same overall cell reaction occurs:

$$H_2O \rightarrow \frac{1}{2}O_2 + H_2. \tag{7.6}$$

Solid oxide cells have the unique feature of reversibility. The same cell (and stack) can be used as fuel cell and also as an electrolyzer cell (SOFC–SOEC). The overall SOFC reaction is:

$$H_2 + \frac{1}{2}O_2 \rightarrow H_2O(+\text{ heat } + \text{ electricity}). \tag{7.7}$$

Minor modifications can enhance the performance and lifetime in single-mode operation when the cell is used in either mode. Much research and many industrial players started out with solid oxide fuel cells and used the same or very similar cells (and stacks) in electrolysis mode.

7.1.2 Planar cells

The majority of mature SOECs have a planar geometry, with an oxygen-ion-conducting electrolyte. This is a simple design and allows for high current/volume density and short current paths, which are perpendicular to the cellular plane, when combined into a stack (see figure 7.1(c)).

Gas management can be realized using a parallel or cross-flow geometry, as illustrated in figure 7.1(c). Oxygen forms and evolves at the air electrode (the anode in electrolysis). The sweep gases used (e.g. air or oxygen) can differ in cases where oxygen is one of the desired electrolysis products. It is also possible to omit the sweep gas. This is in contrast to using the cell in fuel-cell mode, in which air, which is the gas at the air electrode, has the function of cooling the exothermal fuel-cell reaction via a large flow rate.

The manufacture of planar cells and stacks is straightforward and possible using large-scale compatible technologies, such as tape casting, screen printing, and spraying. For some materials combinations, a thin barrier layer is deposited between the electrolyte and the air electrode in order to prevent interphase reactions. Apart from the technologies mentioned above, physical vapor deposition or other thin-film processes are used. Sealing at the cell edges to avoid the intermixing of gases from anode to cathode might be a challenge in the planar stacking concept, as it is accomplished in the hot zone, potentially leading to the leaching of unwanted impurities into the electrode materials. The typical dimensions of commercial cells are in the range of ca. 10×10 cm^2.

7.1.3 Tubular cells

In contrast to SOFCs, for which several industrial players apply the tubular or flat tubular cell geometries, there are only a few examples of the use of tubular cells for SOECs. The tubular geometry has longer current paths (see figure 7.2). On the other hand, tubes that have one closed end ease the sealing of stacks; the sealant can be moved out of the hot zone, allowing for the use of cheaper materials and enhancing robustness. The most common fabrication technique used for tubular cells is extrusion.

At present, tubular SOECs are mainly used with proton-conducting electrolytes. The tubes can have lengths ranging from few cm (microtubes) to the meter scale.

7.2 SOEC configurations

7.2.1 Introduction

In both geometries, i.e. planar and tubular, the active layers have thicknesses of ca. 10–30 μm, which is too thin for a mechanically self-sustaining unit. Therefore, a thicker supporting layer is needed. The nature of this supporting layer determines the cell configuration.

The main configurations are: (i) electrolyte-supported cells (ESCs), (ii) fuel-electrode-supported cells (or cathode-supported cells (CSCs) in SOE mode), and (iii) metal-supported cells (MSCs) (see figure 7.3). In the first two configurations, one of the active layers is made thicker, sometimes with variations in specific materials compositions or structures, while in the metal-supported case, an inert, metallic layer provides the mechanical strength.

7.2.1.1 Electrolyte-supported cells

The electrolyte thicknesses in ESCs are in the range of ca. 90–250 μm. The electrodes are ca. 10–20 μm thick. The electrolyte is responsible for oxygen ion transport,

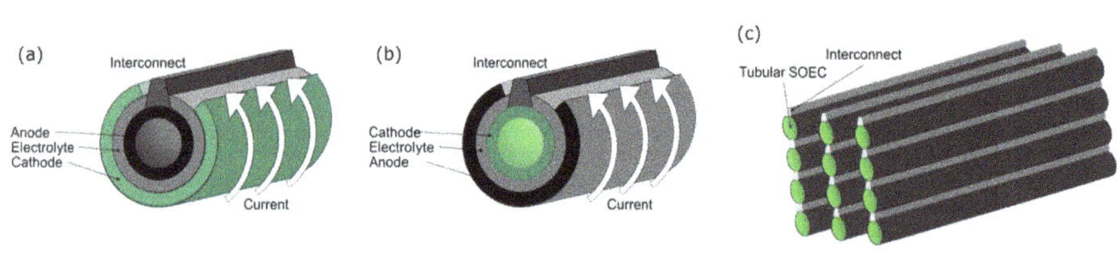

Figure 7.2. Illustration of the tubular cell geometry (a) with the cathode on the outside and (b) inside of the tube, where steam is supplied; (c) stacking principle.

Figure 7.3. Sketches of the three major SOEC configurations: (a) electrolyte supported, (b) fuel electrode supported, (c) and metal supported.

i.e. the ionic conductivity is an important parameter. In fact, in the ESC configuration, the ionic resistivity of the electrolyte is the major contributor to the overall cell resistance. The first versions of ESCs were SOFCs, which were operated at temperatures of around 1000 °C. However, such high temperatures require costly stacks and balance-of-plant components, i.e. the components in the hot part of the unit. Due to lifetime and cost issues, a reduction of the operating temperature was achieved by lowering the electrolyte thickness. Typical operating temperatures are now around 850 °C with the use of today's most mature SOECs. A general advantage of the ESC configuration is its high mechanical integrity (see section 7.4).

Examples of companies using the planar electrolyte-supported configuration are Sunfire, Hexis, and Bloom Energy.

Hexis, which originally developed combined heat and power (μCHP) units with SOFC technology for domestic applications, has now entered the SOEC field for power-to-X applications. The cells have a round shape with a hole in the center. The stack is seal-less, i.e. the gases mix at the outer rim of the cells.

Sunfire works to commercialize industrial solid oxide electrolyzers in the power-to-X and other sectors (such as hydrogen for the steel industry). The cells have a planar, rectangular shape. An example of an ESC cross section is shown in figure 7.4.

Bloom Energy offers green, stable electricity to larger buildings such as supermarkets, hospitals, and office buildings via SOFCs and natural gas fuel. Stacks are also developed for use in SOECs. Their cells have a planar, quadratic shape.

7.2.1.2 Fuel-electrode-supported cells

In the fuel-electrode- (or cathode-) supported configuration, the fuel electrode is the thickest layer; its thickness is in the range between a few hundred μm and 1 mm, most typically from 250–400 μm. Electrolyte layers are a few to ca. 10 μm thick (but can decrease down to 2 μm) and the air electrode is ca. 10–20 μm thick. Due to the thin electrolyte, the ohmic resistance is greatly reduced in this configuration and the major contributions to the total cell resistance originate from the electrodes.

The electrode support and active electrode can either be one merged layer or two distinct layers with larger porosity in the support and sometimes slightly different materials compositions. The latter version provides optimum properties for the electrochemical reaction at the active (thin) fuel electrode, while enabling fast gas transport and mechanical strength in the support.

Among the industrial players with planar fuel-electrode-supported cell configurations in SOEs are Solid Power (recently renamed: SolydEra), Elcogen, FuelCell

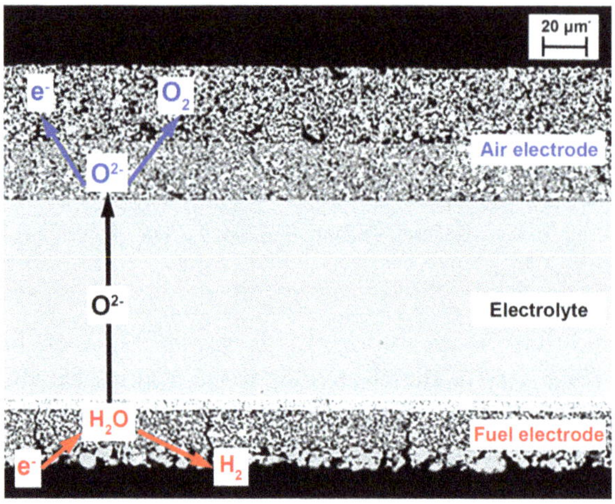

Figure 7.4. Cross section of an electrolyte-supported cell, obtained using a scanning electron microscope (SEM). Reproduced from [1] with permission, © Wiley.

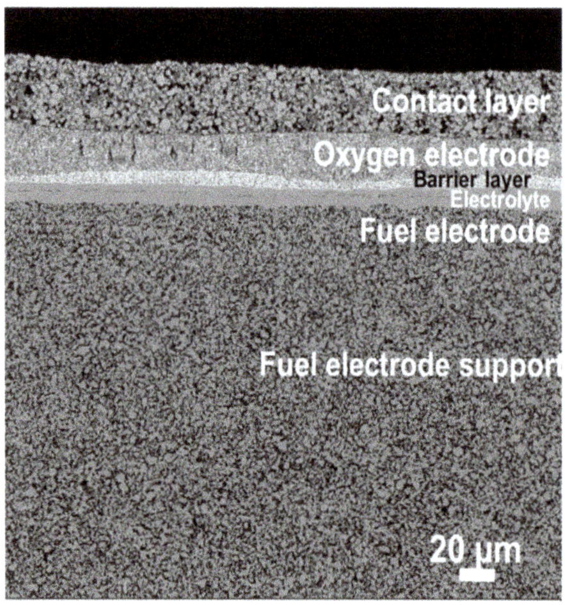

Figure 7.5. Cross section of a fuel-electrode-supported cell, showing the support layer and an active fuel electrode, electrolyte, barrier layer, and air electrode; an electron-conducting contact layer is deposited on top in this example, obtained by a SEM.

Energy, and Topsoe. All these companies started their activities as SOFC technology developers and providers and later extended their activities to dedicated efforts in the electrolysis area. Their cells have a planar, rectangular geometry.

A cross section of a fuel-electrode-supported SOEC is shown in figure 7.5. The support is typically manufactured by tape casting, followed by the deposition of the remaining layers of the half-cell: the active fuel electrode, the electrolyte, and the barrier layer (towards the air electrode) by tape casting and/or thin-film techniques such as physical vapor deposition. The composition and particle size distributions of the support and the active fuel electrode can be the same or tailored in order to achieve both good mechanical properties and high electrochemical performance at the same time. The half-cell is typically sintered before the air electrode is applied by screen printing or spraying, often followed by another sintering step.

7.2.1.3 Metal-supported cells

The metal-supported cell configuration is similar to the fuel-electrode configuration; the support material is, however, different from the electrochemically active layers. The motivations for using a metal support are the typically lower costs than those of ceramics and the enhanced mechanical robustness. This allows, for example, the manufacture of the large-footprint cells required to reach large-scale electrolysis capacities. Furthermore, the metal-supported configuration leads to a higher tolerance towards (thermal) gradients, fast cycling of operating conditions, and potentially also towards higher pressure.

Ceres Power has developed planar metal-supported cells and stacks for application in SOFC mode and recently in SOEC mode as well. In this case, the support is a metal sheet with laser-drilled holes for gas transport. Other metal supports that are still in development use powder metallurgical processing, such as pressing or tape casting, while the active layers are deposited by plasma deposition technologies, pulsed laser deposition, physical vapor deposition, or screen printing [2]. A cross section of a metal-supported cell with a porous ferritic alloy support is shown in figure 7.6.

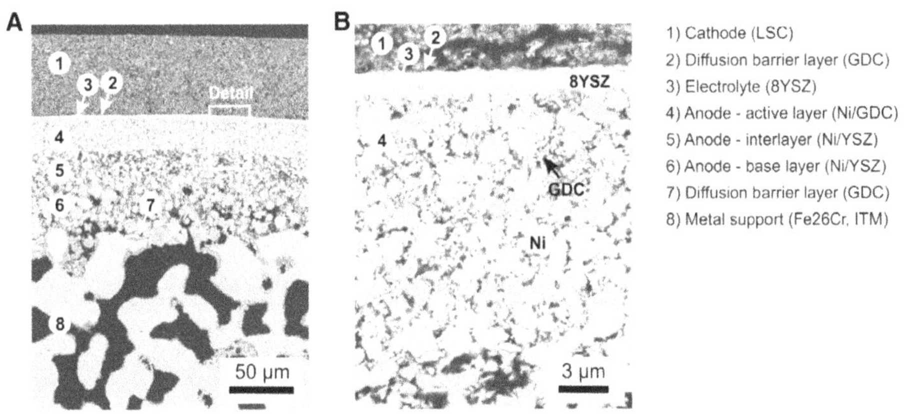

Figure 7.6. Cross section of a metal-supported cell with a porous support layer on the bottom (8), followed by a diffusion barrier layer (7), a three-layered fuel electrode (6–4), a dense electrolyte (3), a diffusion barrier layer (2) and an air electrode (1), obtained using an SEM. Note that the annotation of the electrodes refers to fuel-cell operation, i.e. the fuel electrode is the anode, and the oxygen electrode is the cathode. Reproduced from [3] with permission, © Elsevier.

7.3 Range of operating conditions

Even though materials compositions and structural features of the cell layers allow the cell properties to be tailored for performance and durability, the chosen configuration determines the range of applicable operating conditions.

Depending on the thicknesses of the layers, the resistances arising from them (R_i), both electrochemical reactions, and the gas and the ionic transports contribute differently to the overall area specific cell resistance ASR_{cell} (see equation (7.8)).

$$ASR_{cell} = \sum R_i = R_s + R_p = R_s + R_{FE} + R_{AE} + R_{GT} \tag{7.8}$$

R_i: individual resistance contributions (Ω cm^2)
R_s: serial resistance (Ω cm^2)
R_p: polarization resistance (Ω cm^2)
R_{FE}: electrode resistance caused by the fuel electrode (Ω cm^2)
R_{AE}: electrode resistance caused by the air electrode (Ω cm^2)
R_{GT}: resistance caused by gas transport and conversion (Ω cm^2)

The serial resistance of a cell is mainly due to the resistance of the electrolyte (R_{ohm}). It is inversely proportional to the ionic conductivity σ, which is the oxide ion conductivity in the case of an oxide-ion-conducting electrolyte, as shown in equation (7.9). The conductivity is a strong function of the temperature, according to the Arrhenius relation (equation (7.10)). The higher the temperature, the higher the ionic conductivity and consequently the smaller the ohmic resistance. In the ESC configuration, the purpose of thickness reduction is to reduce this resistance contribution as much as possible without jeopardizing mechanical integrity.

$$R_{ohm} = \frac{1}{\sigma_{electrolyte}} \frac{l}{A} \tag{7.9}$$

R_{ohm}: ohmic resistance of the electrolyte (Ω)
σ: conductivity (Ω^{-1} cm^{-1})
l: electrolyte thickness (cm)
A: cell area in (cm^2)

$$\sigma = \sigma_0 \exp\left(-\frac{E_A}{RT}\right) \tag{7.10}$$

σ: conductivity (Ω^{-1}cm^{-1})
σ_0: pre-exponential factor, a materials constant (Ω^{-1} cm^{-1})
T: temperature (K)
E_A: activation energy for the ionic conduction (J mol^{-1})
R: ideal gas constant: 8.314 J K^{-1} mol^{-1}

In the fuel-electrode-supported configuration, or the cathode in a CSC, this electrode is the thickest layer, while the electrolyte thickness is smaller than that of an ESC by factor of about two hundred. This significantly reduces the ohmic contribution of the electrolyte layer to the area specific cell resistance, in turn

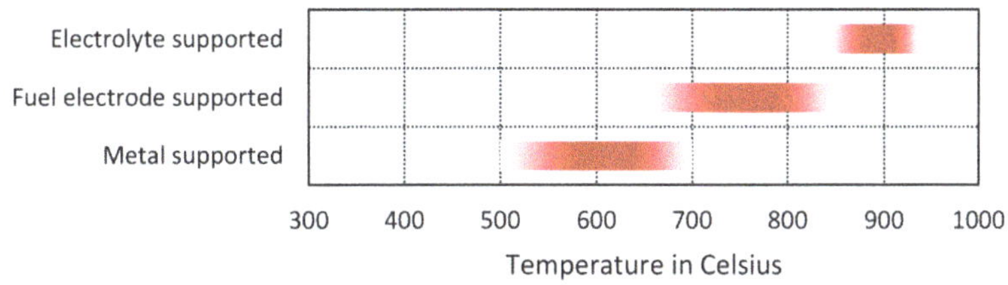

Figure 7.7. Typical operating temperatures for the three main cell configurations of SOECs.

allowing the operating temperature to be reduced down to ca. 650 °C (see figure 7.7) and increasing the importance of the electrode resistances.

In this temperature range, the activity of the electrodes, as determined by the materials compositions and structures, is largely responsible for the total ASR (see equation (7.8) and figure 7.7). This applies to the CSC and the MSC configurations as well. Typically, more active electrode materials are selected for the MSC configuration to allow for even lower operating temperatures and thus better protection of the support against corrosion.

Figure 7.8 illustrates how the individual resistances contribute to the overall ASR of a cell with ESC and CSC configurations. Apart from the previously mentioned shift of the major contribution from the electrolyte to the electrode resistances, the importance of the gas transport resistances also changes. Due to the thin electrode layers used in ESCs, their resistance due to gas transport is minor, while it becomes more significant in the fuel electrodes used for CSCs.

The cell configuration determines another important operational option, namely, to be able to work at the thermoneutral voltage (E_{tn}). This voltage is the minimum voltage at which a perfectly insulated electrolyzer would operate if there were no net inflow or outflow of heat. Thus, operating at the thermoneutral voltage does not require heating or cooling, and is interesting from a system point of view. The thermoneutral voltage is defined in equation (7.11).

$$E_{tn} = \frac{1}{nF}\Delta H_f \tag{7.11}$$

E_{tn}: thermoneutral voltage (V)
ΔH_f: molar enthalpy of the formation of H_2O (J mol^{-1})
n: number of electrons involved per reaction (two electrons)
F: Faraday's constant: 9.6485×10^4 As mol^{-1}.

In the case of steam (H_2O) electrolysis, the thermoneutral voltage is 1.48 V at 25 °C and 1.29 V at 850 °C. The system is endothermal when working below the thermoneutral voltage and exothermal when working above it. The region in which SOECs can operate depends on their iV characteristics, i.e. the relation between applied current density and voltage. The slope of the iV curve corresponds to the ASR of the cell and thus heavily depends on the materials and structures. However, overall trends can be deduced based on the cell configuration. These are illustrated in figure 7.9. Electrolyte-supported cells have a higher ASR, i.e. a steeper iV curve.

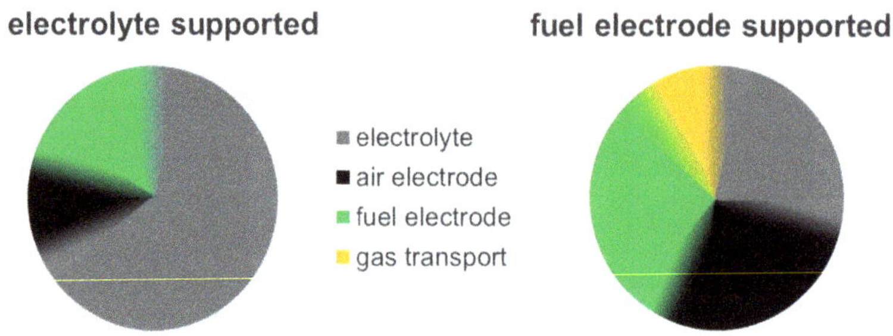

Figure 7.8. Typical distribution of resistance contributions to the total area specific cell resistance for ESCs and fuel electrode supported (CSC) (and MSC) configurations. The absolute numbers depend on the selected materials compositions and microstructures.

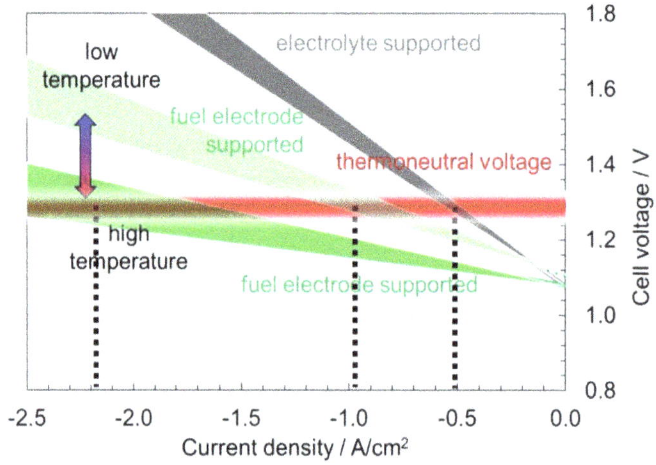

Figure 7.9. Dependence of the typical performance of an SOEC on the cell configuration—either electrolyte supported (gray area) or fuel electrode supported at the same temperature (green area) or at a lower temperature (patterned green area) in relation to the thermoneutral voltage (red horizontal line). The *iV* curves are representative for the two configurations.

Operating at the thermoneutral voltage is therefore possible with rather low current densities and this is therefore a realistic operating point. Fuel-electrode-supported cells typically have smaller ASR values. This flatter *iV* curve crosses the thermoneutral voltage at higher current densities (figure 7.9). State-of-the-art fuel-electrode-supported cells show increasing degradation rates with increasing current densities, which shorten their lifetimes. One solution is to decrease the operating temperature, thereby increasing the ASR towards levels where the thermoneutral voltage is reached at realistic current densities, leading to more stable operation (see figure 7.9).

7.4 Mechanical properties

The mechanical properties of an SOEC cell depend to a large degree on its support layer. Though strong and ductile metal-supported SOEC cells are being developed [3], the majority of the current SOEC designs are either fuel-electrode- or electrolyte-supported, and all rely on doped zirconia as the structural component. Pure zirconia has three polymorphs; from low to high temperature, these are monoclinic, tetragonal, and cubic. Due to the disruptive monoclinic ↔ tetragonal martensitic phase transformation, pure ZrO_2 is of little use. By doping zirconia with certain amounts of MgO, CaO, Sc_2O_3, Y_2O_3, or CeO_2, metastable tetragonal or cubic solid solutions can be maintained at room temperature, which are known as partially or fully stabilized zirconia. The mechanical properties of the materials most commonly used in fuel-electrode- and electrolyte-supported SOEC cells are presented in table 7.1.

Although it has inferior ionic conductivity, the superior strength of 3YSZ makes it a popular structural material in fuel-electrode-supported cells, in which the support layer is made of the Ni/3YSZ cermet. The superior strength of 3YSZ can be described using the so-called transformation toughening mechanism [4]. When a crack propagates in the cell, the stress at the crack tip initiates the phase transformation from the tetragonal to the monoclinic phase. As a consequence of this phase transformation, the energy of the propagating crack is dissipated. In addition, the volume expansion associated with the phase transformation creates compressive stresses, hindering the crack from growing and thus effectively enhancing the cell's strength. This is illustrated in figure 7.10.

In electrolyte-supported cells, 8YSZ and 10ScCeSZ, which have much higher ionic conductivity but inferior mechanical strength, have been used as electrolyte materials instead of 3YSZ. By adjusting the electrolyte layer thickness, a compromise can be achieved between the cell's mechanical strength and its ohmic resistance.

The mechanical strength of the support layer is furthermore dependent on its microstructure and temperature. This is especially true for Ni–3YSZ supports. A typical Ni–3YSZ support consists of Ni and 3YSZ in a volume ratio of 40/60 at 30%–35% porosity. The strength of the Ni(O)–3YSZ support decreases with increasing porosity [5]. Under oxidizing conditions, the mechanical strength of the NiO–3YSZ support is almost independent of temperature [6]. In the reduced state, the strength of Ni–3YSZ deteriorates with increasing temperature. It is, however, still higher than that of Ni–8YSZ at high temperature. Newly found stronger materials such as Ce and Y co-doped zirconia (CeYSZ) have enhanced the mechanical robustness of the support and its resistance to degradation.

Table 7.1. Typical mechanical properties of some common SOEC materials at room temperature (RT).

Material	Young's modulus (GPa)	Weibull strength (MPa)	Fracture toughness (MPa $m^{0.5}$)
3YSZ—$(Y_2O_3)_{0.03}(ZrO_2)_{0.97}$	214 [7]	423 [8]	5.0 [8]
8YSZ—$(Y_2O_3)_{0.08}(ZrO_2)_{0.92}$	221 [9]	232 [9]	1.79 [10]
10Sc1CeSZ	205 [8]	–	1.8 [8]
CGO10—$Ce_{0.9}Gd_{0.1}O_{1.85}$	218 [10]	134 [9]	1.47 [11]
LSM—$La_{1-x}Sr_xMnO_3$	108 [11]	52 [9]	–
NiO–3YSZ	183 [7]	412 [7]	3.05 [12]
NiO–8YSZ	150 [13]	118 [14]	1.8 [14]

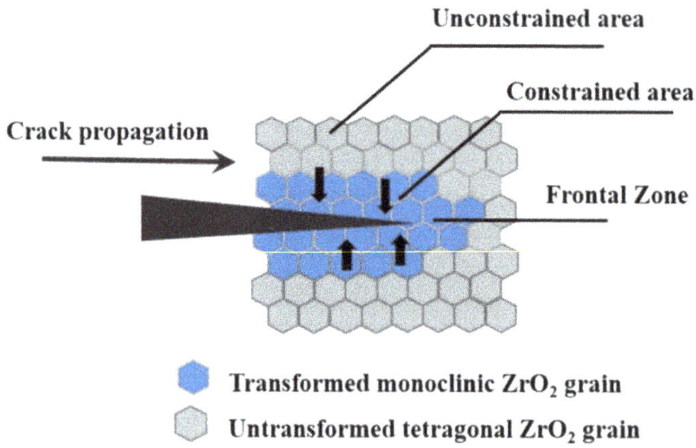

Figure 7.10. Schematic illustration of stress-induced transformation toughening. Reproduced from [14] with permission, © MDPI.

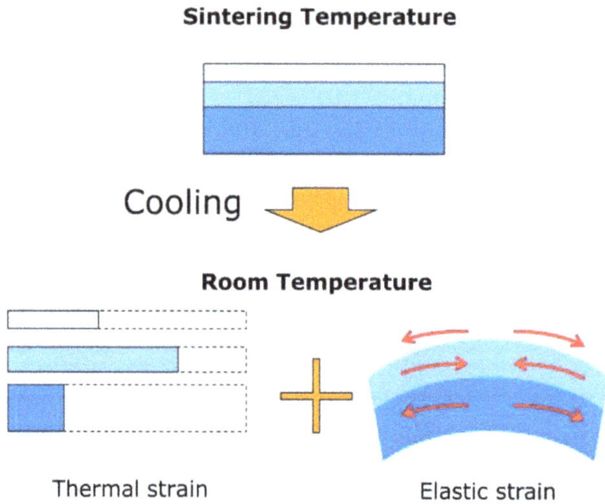

Figure 7.11. Schematic presentation of the buildup of residual stresses. Reproduced from [20] with permission, © Elsevier.

The mechanical reliability of an SOEC cell depends not only on the properties of its components, but also on its ability to withstand the mechanical stresses that arise during cell and stack production and operation. These stresses can have different origins, such as a mismatch of the thermal expansion coefficient (TEC) between different layers of the SOEC cell or between the cell and stack components (interconnects and seals), mechanical loading during stacking, chemical shrinkage or expansion during reduction or oxidation reactions, and uneven temperature distribution during operation [15]. For example, residual stresses build up after the cell is sintered due to a TEC mismatch (mismatches) between the different cell components. Compressive and tensile stresses are thus expected to form in components with lower and higher TECs, respectively, as illustrated in figure 7.11.

Table 7.2. Linear thermal expansion coefficients (TECs) of some common SOEC materials between RT and 1000 °C.

Material	$TEC_{RT\text{-}1000}$ (10^{-6} °C^{-1})	Reference
3YSZ—$(Y_2O_3)_{0.03}(ZrO_2)_{0.97}$	11.4	[16]
8YSZ—$(Y_2O_3)_{0.08}(ZrO_2)_{0.92}$	10.9	[17]
10ScCeSZ—$(ZrO_2)_{0.89}(Sc_2O_3)_{0.1}(CeO_2)_{0.01}$	10.6	[18]
CGO20—$Ce_{0.8}Gd_{0.2}O_{1.9}$	12.7	[18]
LSM—$La_{1-x}Sr_xMnO_3$	11–13	[18]
LSCF—$La_{0.6}Sr_{0.4}Co_{0.2}Fe_{0.8}O_{3-\delta}$	16.3 (RT-425 °C) and 24.5 (425 °C–1000 °C)	[19]
NiO–3YSZ	13.1	[20]
NiO–8YSZ	12.8	[20]
Ni–8YSZ	12.5	[18]

The TECs of common SOEC layer materials are listed in table 7.2. 10ScCeSZ and 8YSZ have the lowest TECs; thus, the electrolyte layer is very often exposed to compressive stress protecting the electrolyte layer if the stress level is properly controlled. The fuel electrode support and the active fuel electrode are exposed to tensile stress. The situation is less well defined in the case of the air electrode, which can be subjected to either compressive or tensile stress, depending on the materials (LSM–YSZ or LSCF–CGO) or the temperature [16].

The strength of the single- or multilayer components in SOEC cells can be measured using different methods, such as uniaxial tensile testing, four-point bending, ring-on-ring bending, ball-on-ring bending, etc while measurements of strain or residual stresses in multilayers can be done via high-temperature strain gages, X-ray diffraction (XRD), or indirectly by measuring the stress relaxation caused by the removal of material by hole drilling, a focused ion beam, or nanoindentation [20]. To evaluate the residual stresses and their influence on the mechanical reliability of multilayer samples, a statistical approach (Weibull distribution) is then used to assess the distribution of the strengths of brittle components under operational conditions [6].

7.5 Concluding remarks

SOECs have two major geometries: planar and tubular. The majority of mature SOECs adopt the planar geometry with an oxygen-ion-conducting electrolyte. The manufacture of planar cells and stacks is straightforward and suitable for large-scale production. Examples of tubular SOEC cells are rather limited; they are mainly used with proton-conducting electrolytes. The thickest layer (also called the support layer) in the cell determines the cell configuration. The three main configurations are electrolyte-, fuel electrode-, and metal-supported cells. The cell configuration in turn determines the range of applicable operating conditions and the overall cell properties, including the mechanical strength and the distribution of internal resistances. All three configurations have been adopted by industrial technology providers and are currently being commercialized.

References

[1] Lang M, Raab S, Lemcke M S, Bohn C and Pysik M 2020 Long-term behavior of a solid oxide electrolyzer (SOEC) stack *Fuel Cells* **20** 690–700

[2] Tucker M C 2020 Progress in metal-supported solid oxide electrolysis cells: a review *Int. J. Hydrogen Energy* **45** 24203–18

[3] Udomsilp D *et al* 2020 Metal-supported solid oxide fuel cells with exceptionally high power density for range extender systems *Cell Rep. Phys. Sci.* **1** 100072

[4] Chevalier J, Gremillard L, Virkar A V and Clarke D R 2009 The tetragonal-monoclinic transformation in zirconia: lessons learned and future trends *J. Am. Ceram. Soc.* **92** 1901–20

[5] Frandsen H L, Ramos T, Faes A, Pihlatie M and Brodersen K 2012 Optimization of the strength of SOFC anode supports *J. Eur. Ceram. Soc.* **32** 1041–52

[6] Ni D-W, Charlas B, Kwok K, Molla T T, Hendriksen P V and Frandsen H L 2016 Influence of temperature and atmosphere on the strength and elastic modulus of solid oxide fuel cell anode supports *J. Power Sources* **311** 1–12

[7] Fleischhauer F, Bermejo R, Danzer R, Mai A, Graule T and Kuebler J 2015 High temperature mechanical properties of zirconia tapes used for electrolyte supported solid oxide fuel cells *J. Power Sources* **273** 237–43

[8] Atkinson A and Selçuk A 2000 Mechanical behaviour of ceramic oxygen ion-conducting membranes *Solid State Ionics* **134** 59–66

[9] Selçuk A and Atkinson A 1997 Elastic properties of ceramic oxides used in solid oxide fuel cells (SOFC) *J. Eur. Ceram. Soc.* **17** 1523–32

[10] Morales M, Roa J J, Capdevila X G, Segarra M and Piñol S 2010 Mechanical properties at the nanometer scale of GDC and YSZ used as electrolytes for solid oxide fuel cells *Acta Mater.* **58** 2504–9

[11] Giraud S and Canel J 2008 Young's modulus of some SOFCs materials as a function of temperature *J. Eur. Ceram. Soc.* **28** 77–83

[12] Pećanac G, Wei J and Malzbender J 2016 Fracture toughness of solid oxide fuel cell anode substrates determined by a double-torsion technique *J. Power Sources* **327** 629–37

[13] Radovic M and Lara-Curzio E 2004 Mechanical properties of tape cast nickel-based anode materials for solid oxide fuel cells before and after reduction in hydrogen *Acta Mater.* **52** 5747–56

[14] Palmero P 2015 Structural ceramic nanocomposites: a review of properties and powders' synthesis methods *Nanomaterials* **5** 656–96

[15] Nakajo A, Kuebler J, Faes A, Vogt U F, Schindler H J, Chiang L-K, Modena S, Van herle J and Hocker T 2012 Compilation of mechanical properties for the structural analysis of solid oxide fuel cell stacks. constitutive materials of anode-supported cells *Ceramics Int.* **38** 3907–27

[16] Timakul P, Jinawath S and Aungkavattana P 2008 Fabrication of electrolyte materials for solid oxide fuel cells by tape-casting *Ceramics Int.* **34** 867–71

[17] Tietz F 1999 Thermal expansion of SOFC materials *Ionics* **5** 129–39

[18] Cai G, Liu R, Zhao C, Li L, Wang S and Wen T 2011 Anode performance of Mn-doped ceria–ScSZ for solid oxide fuel cell *J. Solid State Electrochem.* **15** 147–52

[19] Corbel G, Mestiri S and Lacorre P 2005 Physicochemical compatibility of CGO fluorite, LSM and LSCF perovskite electrode materials with $La_2Mo_2O_9$ fast oxide-ion conductor *Solid State Sci.* **7** 1216–24

[20] Charlas B, Frandsen H L, Brodersen K, Henriksen P V and Chen M 2015 Residual stresses and strength of multilayer tape cast solid oxide fuel and electrolysis half-cells *J. Power Sources* **288** 243–52

Chapter 8

Metal-supported cells

Martin Bram and Norbert H Menzler

Metal-supported solid oxide electrolysis cells (MS-SOECs) and metal-supported protonic ceramic electrolysis cells (MS-PCECs) form a promising subcategory of high-temperature electrolysis cells. The supporting component of the cells is made of a porous alloy, which is expected to decrease capital costs while improving thermal management and the robustness of the cells. Even though the development of this kind of cell is still in its early stages, this chapter reviews MS-SOEC and MS-PCEC technology with a focus on cell manufacture, including critical issues for the selection of alloy materials. The current status of MS-SOEC and MS-PCEC performance is summarized. Finally, the main issues affecting the performance and longevity of cells are addressed by considering the specific kinds of cell degradation mechanism introduced by metal supports and the strategies used to overcome them.

8.1 Background and motivation

The electrolysis of pure H_2O (steam electrolysis) and/or H_2O and CO_2 (co-electrolysis) are key technologies that support the change towards sustainable energy supplies [1, 2]. Solid oxide electrolysis cells (SOECs) and protonic ceramic electrolysis cells (PCECs) are the most efficient kinds of electrolysis cell from a thermodynamic perspective due to their high operating temperatures [1, 3–5]. One subcategory of this kind of cell is the metal-supported (MS) variant, consisting of MS-SOECs and MS-PCECs, in which a porous alloy acts as a supporting component that provides mechanical strength, electrical conduction, and improved heat distribution, while thin, porous, and dense ceramic layers ensure electro-catalytic function.

Development work on metal-supported electrochemical cells has been pursued for more than two decades, starting with the development of metal-supported solid oxide fuel cells (MS-SOFCs) [6]. The replacement of brittle ceramic or cermet supports with an inexpensive and mechanically robust porous metal enabled the

doi:10.1088/978-0-7503-3951-3ch8

Table 8.1. General advantages and challenges of MS electrochemical cells.

Advantages	Challenges
• Improved thermal management at high heating rates and under rapid thermal cycling • Thermal shock resistance • Decrease in cost when low-cost stainless steels are used • Mechanical robustness, e.g. when operating in electrolysis mode and applying electrochemical compression • Use of established metal joining techniques such as welding or brazing • Good redox tolerance	• Adaption of cell processing to the specific requirements of the metal substrate: (i) Reducing atmospheres are preferred, which are often incompatible with highly active oxide materials such as $La_{1-x}Sr_xCo_{1-y}Fe_yO_{3-\delta}$ (LSCF) and $BaZr_{1-x-y}Ce_xY_yO_{3-\delta}$ (BZCY); (ii) limitations imposed by the sintering temperature; (iii) chemical reactions at interfaces; (iv) adaption of the coefficient of thermal expansion (CTE) • Specific cell degradation issues caused by the metal substrate (e.g. breakaway oxidation, interdiffusion phenomena)

thickness of the electrochemically active layers to be reduced. This concept has already been successfully applied in anode-supported cell (ASC) concepts and provides significant advantages with respect to cost, manufacture, tolerance compensation, and operation, all of which are quite attractive for the commercialization of SOFC technology. Table 8.1 summarizes the general advantages and challenges of MS electrochemical cells.

The process and operational know-how gained during more than 20 years' work on MS-SOFCs can be directly transferred to the development of MS-SOECs and MS-PCECs. Nevertheless, the electrolysis mode raises some specific challenges for both concepts, which are addressed in more detail below.

Using concepts similar to those of reversible solid oxide cells (rSOCs), MS-SOFCs can be operated in reversible mode, so there is no need to develop completely new cell architectures. Furthermore, the usual operational temperatures in electrolysis mode are in a similar range of 650 °C to 900 °C [7]. Up until now, only a limited number of reports have been written describing the operation of MS-SOECs, which are summarized in section 8.3. Even if the data base is still limited, unacceptably high degradation rates during long-term operation form the general trend [7]. In addition to well-known SOEC degradation phenomena, such as Ni migration [8–11], there is a need to better understand and properly solve MS-SOEC-specific challenges like

- Severe oxidation of metal substrates in the case of high steam contents on the hydrogen electrode (cathode) or oxygen evolution on the counter electrode (anode).

- Risk of the delamination of functional layers in the case of high-differential-pressure operation based on electrochemical compression.
- Risk of Cr poisoning of active electrode materials such as LSCF.

In principle, MS-SOC concepts can also be transferred to MS-PCECs, and the same processing technologies can be applied. A remarkable advantage of MS-PCECs with respect to metal supports is that dry hydrogen is produced on the hydrogen electrode (cathode), which is expected to reduce high-temperature oxidation of the metal substrate to a minimum during operation. The preferred operational temperatures of PCECs are in the range of 500 °C–700 °C [7]. Nevertheless, up until now, almost no experimental data has been available that describes the operation of MS-PCECs in electrolysis mode, even on the lab scale. This is mainly caused by the fact that processing MS-PCFCs is much more challenging than processing MS-SOECs.

- Most proton-conducting electrolyte materials, such as $BaZr_{1-x-y}Ce_xY_yO_{3-\delta}$ (BZCY), $SrZr_{1-x-y}Ce_xY_yO_{3-\delta}$ (SZCY), or $La_{1-x}Ca_xNbO_{4-\delta}$ (LCN), require extremely high sintering temperatures in the range of 1300 °C–1700 °C. At these temperatures, there is a strong risk of Ba or Sr evaporation or reaction with the metal substrate [12–14]. Sintering activity can be increased by adding sintering aids such as LiF or MnO_2 [13, 14], but this further increases the risk of chemical interactions with the metal substrate.
- Up until now, materials for PCEC electrodes have not been standardized, so the functionality of most promising material combinations is still unclear.
- Gas-phase deposition of proton-conducting materials is challenging due to the change of chemical composition caused by the varying evaporation rates of the alloying elements [15].

Therefore, the development of MS-PCECs is still at an early stage. Nevertheless, some promising preliminary works are available, which are summarized in section 8.4.

8.2 The manufacture of metal-supported cells

8.2.1 Materials and processing of metal substrates

Ni, FeNi, NiCrAlY, and ferritic stainless steels have been investigated for use as substrate materials in the past (table 8.2). One of the key issues for mechanical compatibility between substrate and active layers is the adaption of the coefficient of thermal expansion (CTE) of the metal to that of the electrolyte. This adaption is essential to withstand thermal cycles. In addition to this, low material and manufacturing costs, as well as low oxidation rates, are also important issues.

Ni-based superalloys are highly attractive with respect to oxidation resistance, but their application is restricted due to their high costs and—for most electrolyte materials—unadapted CTE. Ni and Ni–Fe are promising alternatives, since established manufacturing technologies alleviate cell fabrication constraints. However, these materials have

Table 8.2. Properties of substrate materials for metal-supported cells [6, 16].

Metal	CTE (ppm K^{-1})	Costs	Oxidation resistance
NiCrAlY	15–16	Very expensive	Excellent
Hastelloy-X	15.5–16	Expensive	Excellent
Ni	16.5	Expensive	None[a]
Ni–Fe (1:1)	13.7	Fairly expensive	None[a]
Stainless steel, 300 series	18–20	Cheap	Poor
Stainless steel, 400 series	10–12	Cheap	Very good

Note that the CTEs of the established ceramic electrolytes (YSZ, CGO, and LSGM) are in the range of 10–12 ppm K^{-1}.
[a] Readily oxidizes in air or during cooling in a steam-containing atmosphere.

low oxidation resistance and relatively low mechanical strength [16]. It is well known from the literature that ferritic steels with high chromium contents >18 wt% and low specifically added manganese contents (~0.02–0.05 wt%) generally show sufficient long-term corrosion stability for use as dense interconnector plates in SOFCs [17–20]. However, the corrosion resistances of porous metallic structures with identical compositions are often significantly diminished. This behavior is mainly attributed to the much higher surface areas of such substrates. Since ferritic steels from the 400 series (e.g. SS430) and other similar alloys (e.g. Sandvik Sanergy, VDM Metals Crofer 22 APU, and Laves phase-strengthened Crofer 22 H) provide the best compromise for these requirements, they are preferentially used in current MS-SOFC concepts. Taking this principle a step further, porous structures made from the oxide-dispersion-strengthened (ODS) alloy Fe–26Cr–(Mo, Ti, Y_2O_3) developed by Plansee (trade name ITM) provided even better long-term corrosion stability [21, 22]. In this case, alloy-specific powder metallurgical processing leads to homogenous element distribution in the alloy and the formation of strongly adherent protective oxide scales during processing and cell operation. Furthermore, the ODS of the alloy by nano-sized Y_2O_3 particles increases the corrosion stability and creep resistance significantly, but this improved oxidation resistance has higher production costs.

Metal supports can be made either by perforating metal sheets, e.g. by laser drilling, or by powder metallurgical (PM) techniques such as pressing, tape casting, or extrusion. In the case of powder metallurgy, a subsequent sintering step is required for consolidation. Sintering is usually coupled with shrinkage, which can be advantageously used for further processing of the functional layers, e.g. by co-firing [6, 23]. In this context, it must be carefully considered that metal powders require reducing sintering conditions (i.e. low oxygen partial pressure) to avoid excessive oxidation during processing. Such sintering conditions are often not compatible with the functional materials, which are prone to decomposition under reducing conditions (e.g. LSCF, which is usually applied as the oxygen electrode [24], or BCZY, which is usually applied as proton-conducting electrolyte [7, 13]). The porosity and flow resistance of PM supports depend on a complex interplay of shaping technology, particle size distribution, particle shape, sintering conditions, and the optional application of pore formers, organic additives, etc.

8.2.2 The manufacture of metal-supported cells

In the manufacture of metal-supported electrochemical devices, there are a a few specific boundary conditions which must be carefully taken into account when selecting the preferred processing route. (i) The conventional sintering of functional layers in air at temperatures above 1000 °C—as established for the manufacture of SOFCs [25]—is not possible due to the strong oxidation of the metallic substrate during processing. (ii) The occurrence of interdiffusion phenomena at the interface (e.g. Cr and Fe diffusion into the Ni-containing electrodes) might deteriorate the electrochemical performance of the adjacent electrode [26]. (iii) The processing of dense electrolytes becomes challenging if prefabricated supports experience no or little sintering shrinkage during electrolyte processing. In this case, either the electrolyte material selected must be tolerant of constrained sintering, or thin-film deposition techniques have to be used for fabrication. (iv) The specific surface of the metal support must be kept as low as possible to reduce the risk of breakaway oxidation, even after short-term cell operation. Therefore, metal substrates usually require a coarse porous architecture [16], which makes the application of thin-film deposition technologies challenging or even impossible.

There are five main manufacturing routes, which were developed mainly for MS-SOFC applications during the last two decades [6]. All of these manufacturing routes can easily be adapted for the manufacture of MS-SOECs, since the same cell architecture and materials are often used. Some of the routes have also been adapted for the manufacture of MS-PCECs, but here other kinds of active material might require major changes to existing technologies or even the development of completely new processing routes.

8.2.2.1 Thermal spraying technologies

Thermal spraying technologies offer the possibility of producing metal-supported cells in a single step by the consecutive coating of all the functional layers onto the porous metal support without the need for additional post-sintering. These technologies are referred as 'low-temperature processes' since only a moderate thermal load is imposed on the metal support during layer deposition. Therefore, the risk of oxidation of the metal support, Ni coarsening, and interface reactions during processing is clearly reduced. Thermal spraying is based on melting a precursor powder in a spraying gun, followed by accelerating and directly depositing the molten droplets on the metal substrate. Compared to other film deposition techniques such as physical vapor deposition (PVD) or pulsed laser deposition (PLD), thermal spraying enables large areas to be coated at high deposition rates, making the process attractive for industrial production [27]. Furthermore, to a certain extent, it enables the direct coating of substrates with coarser pores. Of the many possible variants of thermal spraying, plasma spraying is the preferred method for the manufacture of metal-supported cells [28–31]. As an example, figure 8.1 shows the cross section of an MS-SOEC which was completely processed by plasma spraying [30].

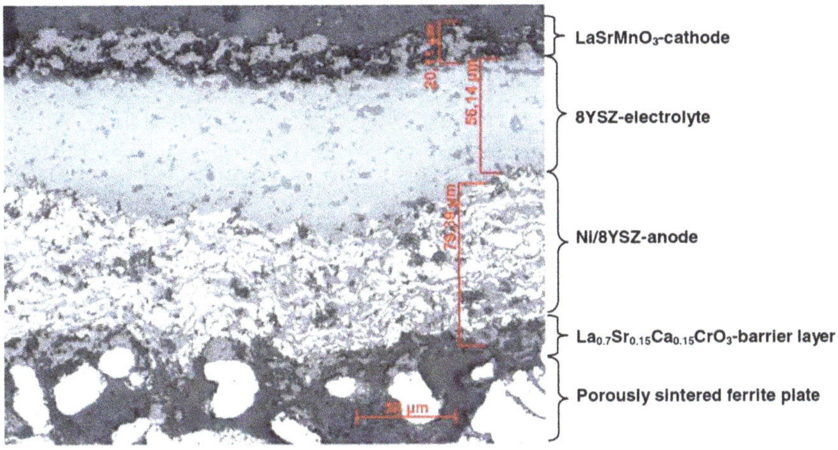

Figure 8.1. MS-SOEC manufactured by atmospheric plasma spraying (electrodes) and vacuum plasma spraying (electrolyte). Reproduced from [30] by permission of Springer.

The processing of highly porous electrodes in combination with a gastight electrolyte on a porous support by plasma spraying alone is challenging. For the processing of porous electrodes, atmospheric plasma spraying (APS) is usually applied [29, 30]. Even with optimized parameters, the porosity of such electrodes remains at <20 vol%, which may cause gas transport limitations. To overcome the low porosity of plasma-sprayed electrodes, screen-printed and sintered electrodes are applied as alternatives [26, 30]. For the manufacture of gastight electrolytes, APS with hot plasma conditions [29] or vacuum plasma spraying (VPS) [30] is used. In the latter case, higher particle velocities lead to higher densification of the deposited films. An electrolyte thickness of at least 30 μm is usually required to ensure sufficient gastightness. Furthermore, low oxygen partial pressure during deposition may lead to strain relaxation during post-thermal annealing, which is coupled with an inherent risk of microcrack formation. Furthermore, the capital and operational costs of VPS equipment are comparatively high.

Another promising variant of plasma spraying is the so-called plasma-spray physical vapor deposition (PS-PVD). Compared to conventional ceramic processing, PS-PVD allows deposition temperatures to be significantly lowered to the range of 800 °C to 1000 °C and enables rapid cooling rates [32]. PS-PVD is conducted under reducing conditions in a vacuum chamber with careful control of the residual oxygen partial pressure. Due to the low partial pressure, the starting powder is partly evaporated in addition to melting. These conditions enable the manufacturer to control the stoichiometry of the deposited layer and inhibit diffusion processes at the interface. Recently, metal-supported, proton-conducting $La_{6-x}WO_{12-\delta}$ (LWO) membranes with a thickness of around 50 μm were made directly on Crofer 22 APU substrates by PS-PVD. Promising hydrogen permeation rates in the temperature range of 650 °C–800 °C were reported [32, 33].

8.2.2.2 Wet chemical processing and electrophoretic deposition

Ceres Power—a spin-off company of Imperial College [34, 35]—established a manufacturing route for MS-SOFCs, which is optimized for automated mass production. In recent years, Ceres Power entered into a series of strategic partnerships with companies such as Bosch, Nissan, and Weichai Power with the objective of industrializing MS-SOFC technology for stationary and mobile applications. In parallel, Ceres Power aims to also use existing technology for MS-SOEC applications [36].

Ceres Power produces a planar-type, metal-supported cell, which is optimized for operational temperatures of less than 600 °C by the use of gadolinium-doped ceria (GDC) as the electrolyte [34, 37, 38]. A perforated Ti–Nb-stabilized 17% Cr-steel sheet is used for the metal support, which does not shrink during subsequent processing of the functional layers. The electrodes (Ni/GDC hydrogen electrode; doped lanthanum ferrite/GDC oxygen electrode) are deposited by conventional ceramic processing methods such as wet powder spraying or screen printing. Adapted thermal treatments under precisely controlled reducing and non-reducing conditions allow the support to be protected from excessive oxidation and avoid strong interdiffusion at the interfaces. Electrophoretic deposition with subsequent isostatic pressing is used to apply a 10–30 μm-thick GDC electrolyte [39]. Here, GDC particles are suspended in a liquid medium and migrate under the influence of an electric field to the metal support, which acts as an electrode during the coating process. The electric field enhances the packing density of the GDC particles on the support. After isostatic pressing, a sintering step at a temperature of less than 1000 °C is sufficient to achieve the required gastightness of the electrolyte [39]. Figure 8.2 shows a cross section of the Ceres Power cell concept [35].

8.2.2.3 Tape casting, lamination, and infiltration

Lawrence Berkeley National Laboratory (LBNL) and the Technical University of Denmark (DTU) developed multilayered metal-supported cell concepts which are

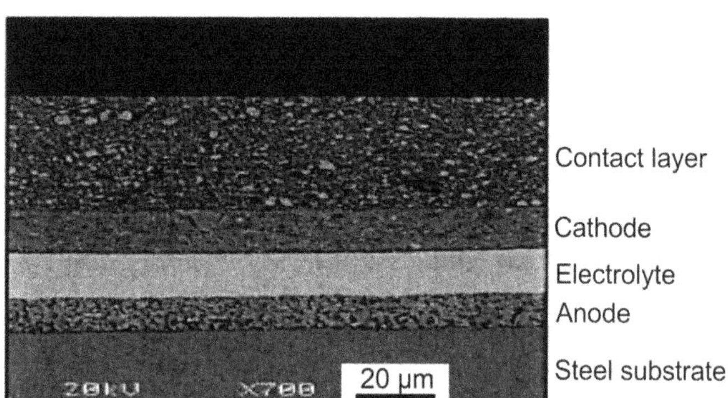

Figure 8.2. An MS-SOFC manufactured by electrophoretic deposition, isostatic pressing, and sintering. Adapted from [35] with permission from Springer.

based on the tape casting, lamination, and co-firing of freestanding stainless steel and yttria-stabilized zirconia (YSZ) tapes under reducing conditions [7]. In general, tape casting is a fully automated and established process in the ceramics industry. After lamination, electrochemically active catalysts are applied by infiltration and thermally treated at moderate temperatures. Since, according to this concept, catalysts are not structural elements of the cell, this results in improved mechanical integrity and tolerance against redox cycling. Furthermore, low-temperature treatment of catalysts reduces coarsening and interdiffusion at the interfaces to a minimum. Last but not least, coating the metal support with a catalyst layer and optionally using rare earth salts in the catalyst's precursor is expected to enhance the oxidation resistance of the metal support [40].

At LBNL, general proof of the infiltration approach was achieved using tubular cells [41, 42]. To increase the volume specific power density, the main results were then transferred to a planar cell concept with a symmetrical architecture [6, 43, 44]. As first step, a five-layered sandwich consisting of a scandia-stabilized zirconia (ScSZ) dense electrolyte layer (thickness 10 μm) enclosed by two porous ScSZ backbone layers (thickness 15 μm) and two porous stainless steel supports (thickness 180 μm) was laminated, isostatically pressed, and co-fired under reducing conditions in an Ar/H_2 atmosphere at a temperature of 1350 °C [44]. The concept of symmetric cells is an effective way to avoid bending due to a mismatch of CTEs or varying sintering activities of the different layers. To adjust the porosity of the ScSZ backbone layers, temporary pore formers (polymethyl methacrylate (PMMA) beads) were applied. It is important to mention that the ScSZ neither decomposed nor showed significant interfacial reactions with the adjacent metal supports (P434L stainless steel) during co-firing. After co-firing, electrode catalysts were introduced into the pores of the backbone layers by infiltrating the precursor solutions of nitrate salts. The cells were then thermally treated in air at a temperature between 600 °C and 850 °C to convert the nitrate salts into their respective oxides. Multiple infiltration–annealing cycles were required to achieve sufficient connectivity between the catalyst particles, which is disadvantageous with respect to time and energy consumption and leads to industrialization issues. Figure 8.3 shows the architecture of the MS-SOEC developed at LBNL [43].

Recently, Tucker's group transferred the symmetrical cell concept to MS-PCECs, investigating $Ba_3Ca_{1.18}Nb_{1.82}O_9$, $SrZr_{0.5}Ce_{0.4}Y_{0.1}O_3$, $BaZr_{0.7}Ce_{0.2}Y_{0.1}O_3$, and $BaCe_{0.4}Zr_{0.4}Y_{0.1}Yb_{0.1}O_{3-\delta}$ as electrolyte materials [12–14]. All these materials survived sintering in a reducing $Ar\text{-}H_2$ atmosphere, but the low sintering activity of the pure electrolyte materials, the evaporation of Ba and/or Sr, interfacial reactions with the Cr and Si of the metal support, and crack formation due to shrinkage and CTE mismatches remained challenging. Recently, Lau *et al* investigated the influence of sintering aids such as LiF and MnO_2 and varying particle size fractions for the metal substrate to better equalize the shrinkage during sintering [14]. To date, no electrochemical results of operating the LBNL MS-PCEC concept in electrolysis mode have been published.

A similar cell design was developed at DTU, which was based on a multilayered cell architecture fabricated by conventional ceramic processing techniques [23, 45].

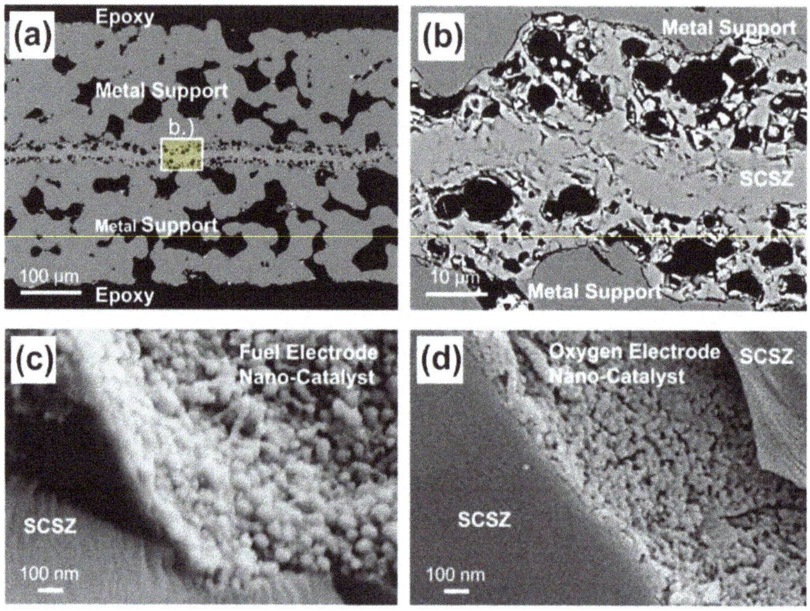

Figure 8.3. MS-SOEC manufactured by tape casting, lamination, co-firing, and infiltration of the electrodes. (a) Cross section, (b) electrolyte with porous backbone layers, (c) infiltrated hydrogen electrode, (d) infiltrated oxygen electrode. Adapted from [43] with permission from Wiley-VCH.

In the first step, the metal support and an electrode backbone layer were manufactured by individual tape casting, starting from Fe–22Cr powder for the substrate and a niobium-doped $SrTiO_3$/Fe–22Cr powder mixture for the anode backbone layer. These layers were laminated with the tape-cast scandia- and yttria-stabilized zirconia (ScYSZ) electrolyte. After debinding, all layers were co-fired at more than 1100 °C in a reducing Ar/H_2 atmosphere. After co-firing the electro-catalytically active hydrogen electrode, a mixture of Ni and GDC powders was infiltrated using the precursor solution technique. The electrolyte was PVD-coated with a 1 µm GDC diffusion barrier layer followed by the screen printing of a composite oxygen electrode consisting of an $La_{1-x}Sr_xCoO_{3-\delta}$ (LSC)/GDC10 mixture at a 50/50 ratio. On the top, a pure LSC current collector was applied by screen printing. The layers on the cathode side were sintered *in situ* during cell testing at 800 °C. To date, no results of operating this cell concept in electrolysis mode have been published.

8.2.2.4 Wet chemical processing and physical vapor deposition

All the cell concepts mentioned above have restrictions when used with electrolyte thicknesses of less than 10 µm. PVD technologies are used to overcome this problem. In general, PVD technologies such as electron-beam PVD (EB-PVD) or magnetron sputtering (MS) have been established for the automated production of thin films on an industrial scale, e.g. in the photovoltaic industry. A target of the respective material is evaporated by an electron beam or by noble gas ions accelerated to the target by applying an electric field. The substrate, which is placed opposite the target, is coated by the evaporated material. The PVD of complex materials might become challenging if the sputtering rates of the individual elements vary. This is, for example, the case for proton-conducting materials such as $BaZr_{0.7}Ce_{0.2}Y_{0.1}O_{3-\delta}$ [14].

Plansee SE developed an MS-SOFC concept based on wet chemical coating and physical vapor deposition in a combined processing route [46]. The manufacture of this cell concept was demonstrated on a pilot scale (cell size ~80 × 140 mm^2), but in 2018 Plansee SE decided to terminate their MS-SOFC activities. Nevertheless, the electrochemical results achieved with this specific cell concept in MS-SOFC mode recommend it as a very promising concept for the development of high-performance MS-SOECs and MS-PCECs in future.

The Plansee MS-SOFC concept is based on a coarse porous metal support with a thickness between 300 μm and 800 μm, which is manufactured by powder metallurgy starting from mechanically alloyed ODS powders (Plansee ITM). A thin GDC diffusion barrier layer (thickness ~1 μm) is applied by magnetron sputtering. A three-layered hydrogen electrode (Ni/YSZ and Ni/GDC) with an overall thickness of 40–60 μm and a stepwise reduction of pore size, porosity, and surface roughness is then deposited by screen printing. Accurate adaption of the Ni/GDC ratio and the particle size distribution were found to significantly enhance MS-SOFC performance [47]. Each layer was individually sintered in a hydrogen atmosphere using temperatures in the range of 1000 °C–1200 °C. A thin-film YSZ electrolyte with a thickness adjusted within the range of 2–6 μm was then deposited by innovative physical vapor deposition combined with gas flow sputtering (PVD-GFS) [48]. Here, residual stresses in the electrolyte led to bending of the half-cell, thereby aggravating subsequent cell assembly [47]. Another thin GDC diffusion barrier layer was deposited by magnetron sputtering. An oxygen electrode based on LSCF, LSC, or an LSCF/GDC composite was screen printed at a thickness between 20 μm and 40 μm and sintered/activated during the initial operation of the cell. Single-cell tests in MS-SOFC mode on the lab scale evidenced outstanding electrochemical performance, which competed with the best ASC results published so far [22]. Tests in MS-SOEC mode were not conducted using the Plansee cell concept. Figure 8.4 shows the cross section of the Plansee MS-SOFC.

Recently, Sata *et al* used a similar concept to manufacture an MS-PCEC with a $BaZr_{0.7}Ce_{0.2}Y_{0.1}O_{3-\delta}$ electrolyte [15]. To achieve an electrolyte thickness of around 1 μm on a Plansee ITM metal support, a $(La_{0.8}Sr_{0.2})_{0.95}MnO_{3-\delta}$ barrier layer and NiO-BZCY interlayers with a stepwise reduction of pore size were applied by screen printing followed by calcination in air at 950 °C. The authors achieved a promising low leakage rate of the $BaZr_{0.7}Ce_{0.2}Y_{0.1}O_{3-\delta}$ electrolyte in the range of 10^{-3} to 10^{-4} Pa m^3 s^{-1} cm^{-2}, but electrochemical performance data in electrolysis mode have not been reported so far.

8.2.2.5 Wet chemical processing and pulsed laser deposition

PLD is an alternative PVD method for the production of thin-film coatings at moderate temperatures. A high-energy pulsed laser beam hits a presintered target of

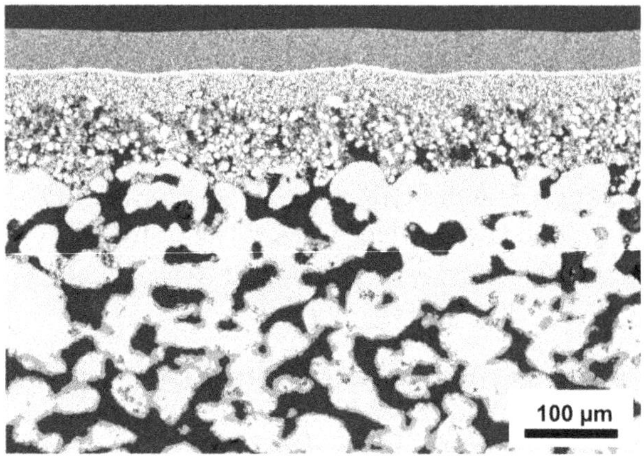

Figure 8.4. Microstructure of the Plansee MS-SOFC made by tape casting (metal substrate), screen printing (electrodes), and PVD (electrolyte).

the respective material. The material is evaporated from the target in a plasma plume, and a thin film is deposited on the substrate positioned at a fixed distance opposite the target. Preheating the substrate to temperatures up to 800 °C is an effective way of adjusting the resulting film morphology. For proton-conducting electrolytes, a column-like morphology is desirable in order to minimize grain boundary resistance to proton conduction. Optional, post-thermal treatments at temperatures below 1200 °C can be helpful to increase the crystallinity of the layer [49]. In general, the PLD process is conducted either in a vacuum or in a defined gas atmosphere. For the deposition of oxide layers, a well-defined oxygen partial pressure is usually applied to avoid partial reduction or decomposition of the deposited film. PLD is often used on the lab scale for the development of new functional materials, since targets can be easily produced by powder technology and the target stoichiometry can be reproduced with good accuracy, even in the case of complex oxides. On the other hand, PLD coatings are quite prone to cracking in the event of oxygen deficiency after coating, CTE mismatch, or insufficient adherence at the interface. Furthermore, uneven thickness distributions when scaling up the technology have been reported in the literature [16].

In metal-supported high-temperature electrolysis cells, PLD is preferentially applied for the deposition of proton-conducting materials such as $BaZr_{1-x}Y_xO_{3-\delta}$ (BZY), $BaCe_{1-x}Y_xO_{3-\delta}$ (BCY), $BaZr_{1-x}Yb_xO_{3-\delta}$ (BZYb), BZCY, $Ba_{1-x}Sr_xZr_{1-y-z}Ce_yY_zO_{3-\delta}$ (BSZCY) and others [15, 49–52], which are difficult to process by powder-based methods due their high sintering temperatures of more than 1400 °C [13, 15, 53]. BZCY-based materials are known for their excellent proton conductivity in the medium temperature range between 400 °C and 600 °C [54–57] and have therefore been preferentially used in MS-PCECs so far. The addition of Sr and Ce to BZY has been found to be an effective measure by which to increase the CTE, thus adapting it to other cell components such as the metal support and thereby reducing the risk of crack formation. To apply gastight PLD layers to coarse porous metal supports, micro- and nanoporous intermediate oxide layers are required for the stepwise reduction of pore

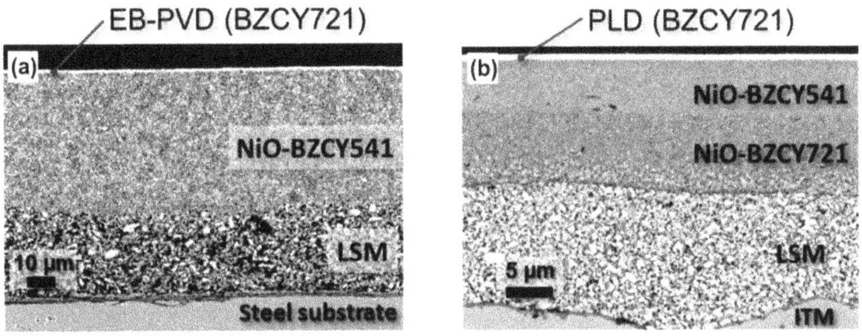

Figure 8.5. MS-PCECs manufactured by tape casting (substrate) and screen printing (diffusion barrier layer, hydrogen electrode): (a) electrolyte deposited by EB-PVD; (b) electrolyte deposited by PLD (BZCY721 = $BaZr_{0.7}Ce_{0.2}Y_{0.1}O_{3-\delta}$, BZCY541 = $BaZr_{0.5}Ce_{0.4}Y_{0.1}O_{3-\delta}$). Reproduced from [15] with permission from the Electrochemical Society.

size and surface roughness. $La_{0.5}Sr_{0.5}Ti_{0.75}Ni_{0.25}O_3$ is a promising material for this intermediate layer, which can be deposited by e.g. wet powder spraying with subsequent sintering under reducing conditions [49, 50]. Other groups have used screen printing to apply $La_{0.8}Sr_{0.2}MnO_{3-\delta}$ [15, 51], NiO-BZCY, [15] or NiO-GDC [52] as an intermediate layer. Figure 8.5 show cross sections of an MS-PCEC with a BZCY electrolyte deposited by PVD and PLD [15]. For electrochemical characterization, a screen-printed Pt layer can be provided [49].

8.3 Operational statuses of MS-SOECs

In most cases, established MS-SOFC concepts have been operated in the reversible mode to demonstrate the general potential of MS-SOECs. Nevertheless, only a limited amount of data are available for electrolysis mode, which makes it difficult to draw sound conclusions regarding the maximum cell performance, long-term stability, and specific degradation mechanisms of MS-SOECs. The following section summarizes the main results of the MS-SOEC technology currently reported in the literature in chronological order.

Schiller et al from DLR, Germany [21, 30] tested a ⌀125 mm plasma-sprayed MS-SOEC, which consisted of an ITM steel substrate, a $La_{0.7}Sr_{0.15}Ca_{0.15}CrO_3$ barrier layer (deposited by APS), a Ni–YSZ hydrogen electrode (deposited by APS), a YSZ electrolyte (deposited by VPS) and a LSCF oxygen electrode (deposited by APS). A single-cell test was performed at temperatures in the range of 750 °C–850 °C with varying steam contents. At a current density of 1 A cm^{-2}, cell voltages of 1.28 V (800 °C) and 1.4 V (850 °C) were measured. At 0.3 A cm^{-2}, the cell voltages decreased to 1.07 V (800 °C) and 1.04 V (850 °C). A 2027 h long-term test at 800 °C, a constant steam content of 43%, and a moderate current density of 0.3 A cm^{-2} was then conducted. An overall degradation of 3.2% was measured over a period of 1000 h (figure 8.6). Impedance measurements indicated an increase in the ohmic resistance, which was attributed to the formation of oxide scales in the metal substrate. A parallel increase of polarization resistance was attributed to the coarsening of Ni in the hydrogen electrode, while the

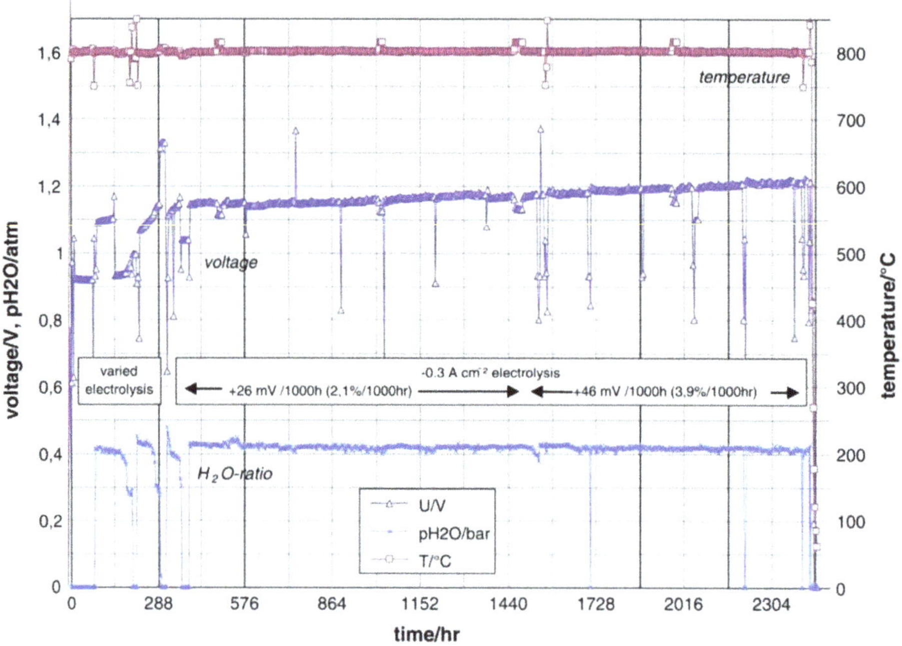

Figure 8.6. Long-term test of the plasma-sprayed MS-SOEC manufactured by DLR at 800 °C at a continuous current density of 0.3 A cm^{-2}. Reproduced from [30] with permission from Springer.

$La_{0.7}Sr_{0.15}Ca_{0.15}CrO_3$ barrier layer effectively suppressed the interdiffusion of Ni, Fe, and Cr at the interface between the metal support and the hydrogen electrode. Furthermore, the authors pointed out that contact between the cells in MS-SOEC stacks requires special attention, as it is an additional source of cell degradation.

Chen et al from the CAS Key Laboratory of Materials for Energy Conversion, Shanghai, China [58, 59] manufactured an MS-SOEC by laminating four tapes, including a 430L steel substrate, a 430L steel/YSZ composite backbone, a Sc- and Ce-doped ZrO_2 (SSZ) electrolyte, and a porous SSZ backbone layer, which were co-fired at 1300 °C for 4 h under a reducing N_2–5% H_2 atmosphere. Ni–Sm-doped CeO_2 and Nd_2O_3–Nd_2NiO_4 were then applied by infiltration as catalysts for the hydrogen electrode and oxygen electrode, respectively. The active area of the oxygen electrode was 35 mm^2. The catalysts were activated by an *in situ* thermal treatment at 800 °C for 2 h in air before the cell was operated. Single-cell tests in electrolysis mode were conducted at a steam content of 50%. At a cell voltage of 1.3 V, current densities of 2.05, 1.62, 1.06, and 0.60 A cm^{-2} were reported at 800 °C, 750 °C, 700 °C, and 650 °C, respectively. The current densities increased slightly with higher steam contents. In a mid-term test for 330 h at 600 °C, a steam content of 30%, and a current density of 0.166 A cm^{-2}, cell voltages between 1.1 V and 1.3 V were reported. An improvement in cell performance was observed during the first 100 h, but the study lacked a clear interpretation of this result. Furthermore, a detailed investigation of the microstructural changes was missing.

Wu et al from Institute of Nuclear Energy Research, Taoyuan, Taiwan [60, 61], manufactured an MS-SOEC on a 100 × 100 mm^2 NiMo substrate using APS. The cell configuration consisted of NiO-YSZ or NiO-GDC for the hydrogen electrode, La-doped ceria as a diffusion barrier layer, $La_{0.8}Sr_{0.2}Ga_{0.8}Mg_{0.2}O_{3-\delta}$ as the electrolyte, and $Sm_{0.15}Ce_{0.85}O_{3-\delta}$–$Sm_{0.5}Sr_{0.5}CoO_{3-\delta}$ as the oxygen electrode. At 800 °C, 0.35 A cm^{-2}, and a steam content of 90%, a cell voltage of 0.96 V was measured.

Nechache et al from DLR, Germany [62] operated an MS-SOEC with a Ni-catalyst-loaded $La_{0.1}Sr_{0.9}TiO_{3-\delta}$/GDC hydrogen electrode, a GDC/YSZ electrolyte, and a $La_{0.6}Sr_{0.4}Co_{0.2}Fe_{0.8}O_{3-\delta}$ oxygen electrode. Furthermore, $La_{0.7}Sr_{0.15}Ca_{0.15}CrO_3$ was applied to the metal support as a diffusion barrier layer. No information was given about the metal support and processing conditions. At 750 °C, 1.15 V, and a steam content of 20%, a current density of 0.75 A cm^{-2} was measured. In a 120 h mid-term test with a current density of 0.25 A cm^{-2}, the cell degradation was 13.7%/1000 h. A commercial ASC which was operated under the same conditions showed a lower degradation of 4%/1000 h.

Ishihara et al from Kyushu University, Japan [63] demonstrated CO_2 electrolysis using an MS-SOEC which was motivated by application in the steel industry. Their cell was based on a Ni–Fe metal support with a diameter of 20 mm, which was PLD coated with thin films of $Sm_{0.2}Ce_{0.8}O_2$ and $La_{0.9}Sr_{0.1}Ga_{0.8}Mg_{0.2}O_{3-\delta}$. After post-annealing in air at 800 °C for 2 h, a thin $Sm_{0.5}Sr_{0.5}CoO_{3-\delta}$ oxygen electrode was prepared by screen printing or PLD followed by a second firing at 800 °C for 30 min. The manufacture of the metal support started from a dense NiO-Fe_2O_3 compact. After layer deposition, the metal support was stepwise reduced to achieve open porosity and electrical conductivity while suppressing large shrinkage to retain the gastightness of the PLD electrolyte. Applying the oxygen electrode by PLD instead of screen printing increased the CO_2 electrolysis current at 700 °C, 1.5 V, and 50% CO_2 in Ar from 0.55 A cm^{-2} to 3 A cm^{-2}. The lower electrochemical performance of the screen-printed oxygen electrode was attributed to IR loss due to the post-sintering of the electrode.

Numata et al from the Tokyo Institute of Technology, Japan [31] demonstrated CO_2 electrolysis with an atmospheric plasma-sprayed cell. A 20 mm SUS 430 steel mesh was welded into an SUS 430 frame. The metal support was then coated with a $La_{0.6}Sr_{0.2}Ca_{0.2}CrO_3$ diffusion barrier layer, a Ni/YSZ CO_2 electrode, a YSZ electrolyte, and an LSCF oxygen electrode. At 800 °C, 4.82 V, and 50% CO_2 in N_2, a current density of 0.353 A cm^{-2} was measured. The low efficiency of CO_2 electrolysis of around 48% was justified by electrolyte leakage, which was mainly due to non-optimized processing conditions that resulted in high residual porosity.

The group of Tucker at LBNL, USA [43, 44] developed a metal-supported cell concept that was characterized by its symmetrical architecture, which clearly reduced the impact of varying shrinkage rates and CTE mismatch between the single layers (figure 8.3). Two metal supports made of P434L stainless steel with a thickness of 180 μm and a porosity of 30% embraced an assembly consisting of a gastight electrolyte with a porous backbone layer on both sides. All ceramic layers were made of Sc- and Ce-doped ZrO_2 (ScCeSZ). The porosity of the backbone layers

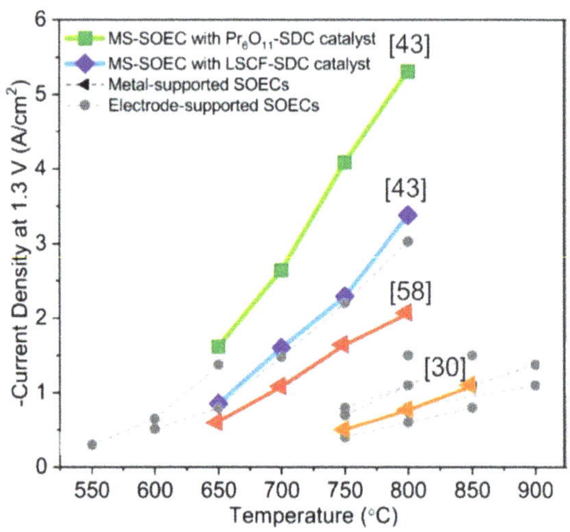

Figure 8.7. Benchmarks for the current densities of MS-SOECs produced at LBNL in the temperature range of 550 °C–900 °C (green and blue symbols). For details of electrode-supported SOECs, we refer the reader to the original work by Wang *et al* [43]. Adapted from [43] with permission from Wiley-VCH.

was 43%. To create the hydrogen electrode, one backbone layer was infiltrated with a mixture of Ni and SDC. The Ni content was varied between 20 vol%, 40 vol%, and 60 vol%. To create the oxygen electrode, the opposite backbone layer was infiltrated with pure LSM, LSCF, or Pr_6O_{11} or with a mixture of these materials with SDC. In single-cell tests, the best performances were achieved using Ni-SDC (at a ratio of 60/40 (SNDC40)) and Pr_6O_{11}/SDC electrodes [43]. At 1.3 V and 50% steam, current densities of 5.31, 4.09, 2.62, and 1.62 A cm^{-2} were measured at 800 °C, 750 °C, 700 °C, and 650 °C, respectively. When the steam contents were set to 3 vol%, 25 vol%, 50 vol%, and 75 vol% at 1.3 V and 700 °C, current densities of 0.63, 2.27, 2.64, and 3.48 A cm^{-2} resulted, respectively. This performance is among the highest reported for conventional SOECs and MS-SOECs (see figure 8.7).

Mid-term tests at 700 °C, 0.33 A cm^{-2}, and a steam content of 50% were subsequently conducted [44]. Before testing, infiltrated electrodes were pre-coarsened *in situ* at 750 °C to enhance the durability of their microstructure. MS-SOECs with SNDC40 and Pr_6O_{11}/SDC as well as with SNDC40 and LSCF/SDC electrodes were tested. When Pr_6O_{11}/SDC was used as the oxygen electrode, the cell test had to be interrupted after 146 h due to severe cell degradation of 54.1% in 100 h (figure 8.8). The main reason given for this behavior was Cr poisoning of the Pr_6O_{11}, followed by the formation of $PrCrO_4$. Two MS-SOECs with LSCF/SDC oxygen electrodes were operated for 860 h and 1000 h. The related degradation rates were 1.3% in 100 h and 1.6% in 100 h. For this kind of cell, the ohmic impedances in the sintered state, after pre-coarsening, and after long-term testing, were 0.27, 0.33, and 0.42 Ω cm^2, respectively. The respective polarization impedances were 0.48,

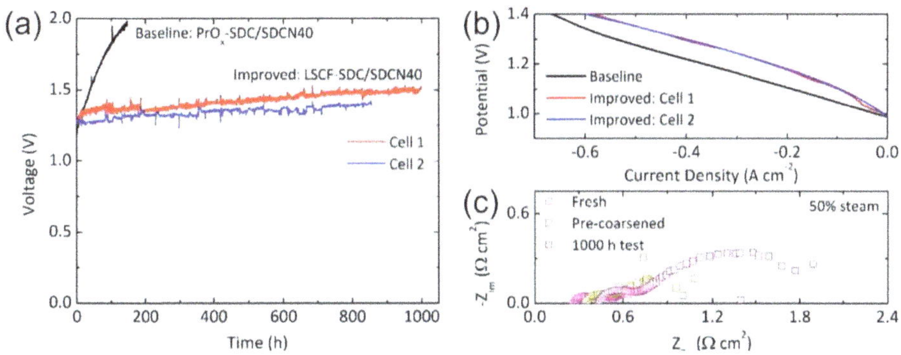

Figure 8.8. Long-term test of MS-SOECs produced at LBNL [44] with SDCN40 hydrogen electrodes and PrO_x-SDC or LSCF-SDC oxygen electrodes: (a) durability of the cells at 0.33 A cm^{-2}, 700 °C, and 50% H_2–50% H_2O; (b) Initial *I–V* curves before pre-coarsening. (c) Impedance spectra of Cell 1 at open-circuit voltage (OCV) with 50% steam. Adapted from [44] with permission from Elsevier.

0.69, and 1.51 Ω cm^2. Oxidation of the metal substrate, coarsening and agglomeration of the Ni catalyst in the hydrogen electrode, and Cr poisoning of the oxygen electrode were discussed as the main degradation mechanisms.

8.4 Operational statuses of MS-PCECs

MS-PCECs are an attractive alternative to MS-SOECs because they operate at lower temperatures (in the range of 500 °C–700 °C) and form dry, pure hydrogen at the hydrogen electrode [7]. These temperatures are a better fit to the temperatures of the steam or waste heat produced by industrial processes and might be advantageous with respect to the oxidation of the metal support, especially if the support only comes into contact with dry hydrogen. Up until now, the development of MS-PCECs has mainly been in its early stages, due to the severe challenges of processing proton-conducting materials such as barium zirconates in the presence of a metal support. Therefore, only a few electrochemical results on the lab scale and under open-circuit conditions have been made available so far.

A consortium of the University of Oslo and SINTEF (both in Norway) and DLR, Germany [15, 49–51, 64] has developed MS-PCECs based either on a coarse porous ITM substrate (⌀25 mm, pore size 10–30 µm) or a metal sheet with drilled holes (⌀25 mm, diameter 30–50 µm) made of Sanergy HT steel coated with Ce-Co [15]. Micro- and nanoporous intermediate oxide layers were deposited by wet chemical coating technologies (spraying, screen printing) with subsequent sintering under reducing Ar–4% H_2 conditions. These interlayers homogeneously cover the metal support, acting as a diffusion barrier layer and enabling the deposition of nanoporous hydrogen electrode and gastight electrolyte by PLD [15, 49–51, 64]. Recently, Sata *et al* introduced EB-PVD as alternative way of depositing a thin-film electrolyte [15]. LSTN [50] and LSM [15, 51] are preferred for the diffusion barrier layer. $BaZr_{0.85}Y_{0.15}O_{3-\delta}$–Ni or $BaZr_{0.9-x}Ce_xY_{0.1}O_{3-\delta}$–Ni composites were used for the hydrogen electrode. Amongst other materials, $Ba_{0.85}Sr_{0.15}Zr_{0.7}Ce_{0.1}Y_{0.2}O_3$ was used for the electrolyte. It was found that alloying barium zirconates with Sr and Ce increased the CTE of the electrolyte to $10–11\cdot10^{-6}\,K^{-1}$, adapting it better to the metal support and therefore avoiding cracking during thermal cycling. The systematic improvement made to the complete processing chain resulted in crack-free, smooth electrolyte coatings with promising leakage rates in the range of $10^{-3}–10^{-4}\,Pa\,m^3\,s^{-1}\,cm^{-2}$ [15], but no electrochemical results have been published so far.

The group of Tucker at LBNL, USA [7, 12–14] tried to transfer their symmetrical MS-SOEC concept to the manufacture of MS-PCECs with barium zirconate electrolytes by co-firing a metal support and electrolyte tapes. It was found that this approach became difficult because the optimum sintering conditions of most proton conductors are not compatible with the usual sintering conditions of stainless steels, which require temperatures in the range of 1300 °C–1450 °C and a reducing atmosphere. Some of the investigated electrolyte materials, such as $Ba_3Ca_{1.18}Nb_{1.82}O_9$, $SrZr_{0.5}Ce_{0.4}Y_{0.1}O_3$ and $BaZr_{0.7}Ce_{0.2}Y_{0.1}O_3$, were found to survive sintering at 1450 °C under reducing conditions, but the evaporation of Ba and Sr and the reaction of the electrolyte with Cr and Si from the steel remain critical issues. Furthermore, even at 1450 °C, sufficient gastightness of the electrolyte was not achieved. To lower the sintering temperature and enhance densification, alternative electrolyte materials such as $BaZr_{0.4}Ce_{0.4}Y_{0.1}Yb_{0.1}O_{3-\delta}$ and $BaZr_{0.1}Ce_{0.7}Y_{0.1}Yb_{0.1}O_{3-\delta}$ and different sintering aids such as Co_3O_4, ZnO, NiO, MnO_2, and LiF (in amounts of up to 2 wt%) have been tested [13, 14], but no electrochemical performance data have been published so far.

Miyamoto et al from Kyushu University, Japan [52] manufactured an MS-PCEC on a ZMG 232L stainless steel sheet (Ø 20 mm, thickness 300 μm) with laser-drilled holes. A Ni/GDC hydrogen electrode was applied by screen printing followed by sintering at 900 °C for 1 h in air. A proton-conducting $BaZr_{0.8}Yb_{0.2}O_3$ electrolyte was then deposited by PLD. Finally, an LSCF oxygen electrode was screen printed on the electrolyte and sintered at 900 °C in air. The MS-PCEC had an OCV too low for electrochemical characterization. The authors supposed that a CTE mismatch and an interface reaction between BZYb and GDC might have been the reason for this result.

8.5 Specific degradation issues of metal-supported cells

In general, MS-SOECs and MS-PCECs are subject to the same degradation mechanisms as their full ceramic counterparts, due to using the same electrochemically active materials. A literature review of general degradation issues in metal-supported high-temperature cells can be found in the excellent review by Zhou *et al* [65]. In the following, degradation mechanisms that are highly relevant to, or can only occur in, metal-supported cells are discussed in more detail.

8.5.1 Oxidation of the metal substrate

Degradation due to oxidation of the metal substrate is a fairly obvious mechanism for metal-supported cells. Oxide scales that form on the particle surface and—even more critically—those that form on the sintering necks between the particles directly increase the ohmic resistance of the cell. Furthermore, the growth of oxide scales in the pores of the metal substrate might limit or even block the gas transport. Another important aspect to consider is the occurrence of breakaway oxidation when the Cr concentration in the bulk falls below a critical value, usually 16% [19]. This effect is also influenced by the adherence of the oxide scale, which might be worse in the case of porous supports compared to dense and flat parts such as interconnects. When cells are operated in electrolysis mode, there are boundary conditions for the oxidation of the metal substrate which might be different from the conditions known from the fuel-cell mode.

Specific oxidation issues in MS-SOECs: in most MS-SOECs, the metal support is placed on the side of the hydrogen electrode. Here, higher steam contents are more usual than they are under MS-SOFC anode conditions, while the operational temperatures are in the same range of 650 °C–800 °C [7]. It is known from oxidation studies that higher steam contents generally accelerate oxidation [66], but the oxidation behavior appears manageable, since protective oxide scales are usually formed in this temperature range. For instance, Schiller *et al* [30] reported an only moderate increase in ohmic resistance due to the oxidation of the metal substrate after their 2000 h long-term test. In the symmetrical MS-SOEC concept of the Tucker group, a metal support is even placed on the oxygen electrode side. In MS-SOEC mode, dry oxygen is formed at this electrode. Ex-situ oxidation studies on pure metal substrates revealed a similar oxidation behavior to that of oxidation in air, with only a slight increase in oxidation rate [43, 44]. Furthermore, it is supposed that Cr evaporation is lowered in dry oxygen, which is said to be advantageous in retarding Cr poisoning, but proof has not yet been provided. In sum, Shen *et al* concluded that oxidation under MS-SOEC conditions is quite similar to the oxidation that takes place under MS-SOFC operational conditions [44].

Specific oxidation issues in MS-PCECs: a remarkable change of oxidation behavior is expected for MS-PCECs, which are usually operated in the temperature range of 500 °C–700 °C [7]. Here, dry hydrogen is formed on the hydrogen electrode, which is expected to be advantageous for the long-term operation of MS-PCECs, provided that the metal support is located on the side of the hydrogen electrode. The oxidation behavior is much more critical when a metal support is located on the oxygen electrode side, as is the case, for example, in the symmetrical concept of the Tucker group [43]. Wet air with high water contents in combination with intermediate operating temperatures at around 550 °C–650 °C makes the substrate quite sensitive to breakaway oxidation. Wang *et al* performed related oxidation studies using thin sheets ($d = 150$ µm, 15×15 mm^2) of 430 steel, 441 steel, and Crofer 22 APU in the temperature range of 450 °C–650 °C with a dwell time of 500 h [66]. Almost no oxidation was found in dry hydrogen (Ar–2.8% H_2). A steam content of 50% caused a significantly increased weight gain in an H_2 atmosphere as well as in air. The latter combination was found to be the most aggressive, leading to breakaway oxidation of all investigated steels at temperatures of 650 °C and 550 °C. The application of protective coatings was found to be an effective way to reduce the risk of breakaway oxidation. However, it must be ensured that such coatings completely cover all surfaces.

Breakaway oxidation at around 600 °C: when applying metal substrates for high-temperature electrolysis applications, it must be carefully considered that at temperatures in the range of 450 °C–650 °C, the formation of the protecting chromia scale might be reduced or even absent because Cr diffusion becomes less than the Cr consumption required for the formation of the oxide scale [7, 40, 66–68]. In this case, non-protective Fe oxide scales are formed on the outer surface, which are also known as breakaway oxidation. The Fe oxide scales are accompanied by Cr- and Mn-enriched inner oxide scales (figure 8.9). At lower temperatures, oxidation rates usually decrease, so diffusion might again become sufficient to form protective oxide scales, therefore avoiding breakaway oxidation [66, 68]. In general, the occurrence of the breakaway effect is quite sensitive to alloy composition, steam content, specific surface, and porosity [26, 40, 66]. Preoxidation at higher temperatures or the application of protecting coatings have been said to be effective measures to reduce the risk of breakaway oxidation [66].

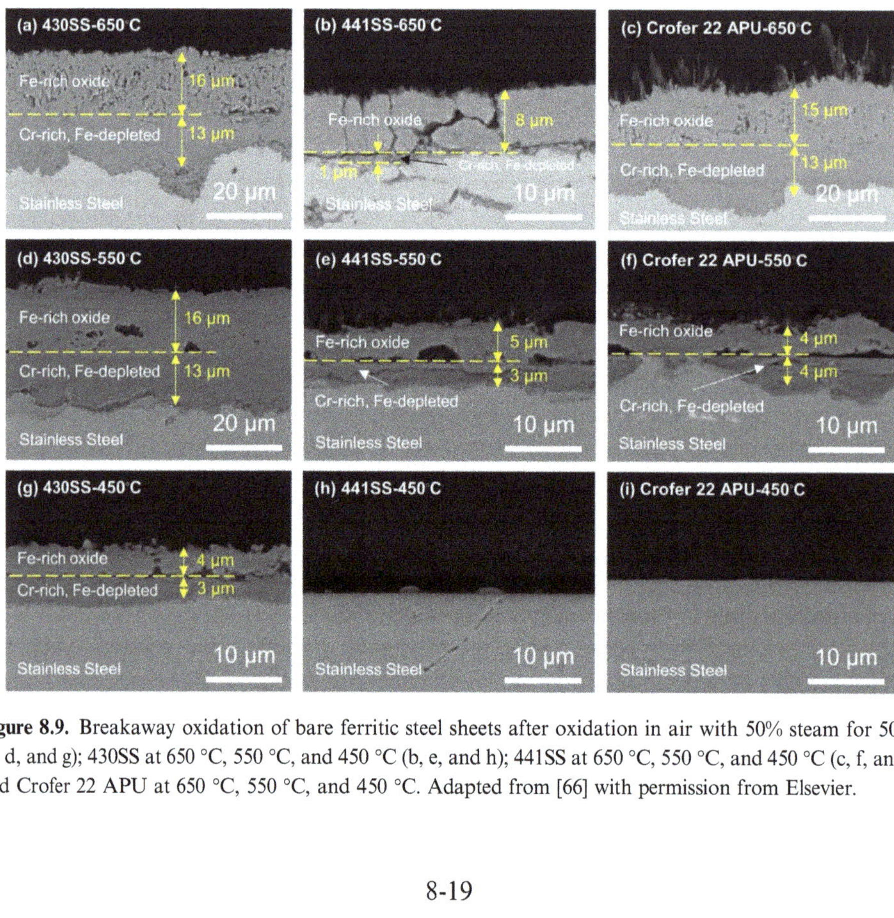

Figure 8.9. Breakaway oxidation of bare ferritic steel sheets after oxidation in air with 50% steam for 500 h (a, d, and g); 430SS at 650 °C, 550 °C, and 450 °C (b, e, and h); 441SS at 650 °C, 550 °C, and 450 °C (c, f, and i); and Crofer 22 APU at 650 °C, 550 °C, and 450 °C. Adapted from [66] with permission from Elsevier.

Development of protective coatings: the application of protective coatings is a well-known measure for improving the oxidation resistance of metal parts at high temperatures. In this context, it must be considered that homogeneous coatings might be difficult to apply in the case of porous metal substrates [69–71]. Several techniques for the deposition of such layers have been investigated, such as slurry coating, electrophoretic deposition, aerosol deposition, screen printing, PVD, electroplating, etc. [16, 66] but not all of them are suitable for porous substrates. For MS-PCEC application, Wang *et al* [66] investigated protective coatings based on Y_2O_3, $Ce_{0.02}Mn_{1.49}Co_{1.49}O_4$, $CuMn_{1.8}O_4$, and Ce/Co on 430 steel, 441 steel, and Crofer 22 APU steel sheets and demonstrated their function at 650 °C in air with 50% steam for selected material combinations. Stefan *et al* [71] applied a protective coating based on $La(Mn_{0.5}Co_{0.5})_{0.8}O_3$ for the same purpose.

Influence of geometrical issues: another important effect to consider for porous substrates is the influence of geometrical issues. It has been demonstrated in related oxidation studies that breakaway oxidation starts much earlier when the particle size or the thickness of the metal substrate falls below a critical value [26, 70, 72]. Large surface-to-volume ratios of small particles or thin sheets trigger breakaway oxidation due to accelerated Cr depletion. Larring *et al* [16] estimated that particle sizes larger than 80 μm are required for an alloy with 20 wt% chromium to form a stable 1.5 μm thick chromia scale without the risk of Cr depletion falling below the critical value.

8.5.2 Interdiffusion

Metal-supported cells are sensitive to interdiffusion at the interface between the metal support and adjacent functional layers [16]. Interdiffusion is often more pronounced during the processing of the cells than during operation due to higher temperatures, especially if sintering technologies are applied. In cells with Ni-containing electrodes, the interdiffusion of Fe, Cr, and Ni takes place, which might cause austenitization of the steel, resulting in a CTE mismatch with the other cell components. The diffusion of iron and chromium into Ni-based anodes may form passivating oxide scales on the nickel surface, thereby deteriorating electrochemical activity. Cell concepts based on the lamination of the electrolyte and the metal substrate with subsequent co-firing are prone to interfacial reactions at the interface between the metal substrate and the electrolyte. While ferritic steel and YSZ are less sensitive to interfacial reactions in MS-SOECs [6, 23], this completely changes for MS-PCECs. The co-firing of a P434 steel and BZCY electrolyte tapes at temperatures of more than 1300 °C led to the evaporation of Ba and an unacceptable reaction between the BZCY and the Cr and Si from the steel [12, 13]. The application of sintering aids such as LiF or MnO_2 might improve the sintering of BZCY, but further increases the risk of interfacial reactions.

Development of diffusion barrier layers: the application of a diffusion barrier layers with sufficient electrical conductivity between the metallic substrate and the adjacent cell component has been proposed in the literature to reliably avoid interdiffusion phenomena. For MS-SOECs, it has been demonstrated that

$La_{0.6}Sr_{0.2}Ca_{0.2}CrO_3$, $La_{1-x}Sr_xMnO_3$ (LSM), CeO_2, and $Ce_{0.8}Gd_{0.2}O_2$ are suitable candidates to act as effective diffusion barrier layers with acceptably low ohmic resistance [26, 73]. Recently, LSM was introduced as a diffusion barrier layer in the MS-PCEC concept of the consortium of University of Oslo, SINTEF, and DLR [15].

8.5.3 Ni migration

It is expected that MS-SOECs with sintered Ni/YSZ hydrogen electrodes are subject to the same Ni migration phenomena as those described for full ceramic SOECs [8–11]. Recent results hint that infiltrated Ni/YSZ electrodes might be less prone to Ni migration [44], but these results are still questionable due to the limited duration used for testing. An advantage of metal-supported cell concepts with PVD electrolytes could be the less difficult processing of Ni/GDC electrodes [47]. Ni/GDC electrodes are expected to be less sensitive to Ni migration than Ni/YSZ electrodes due to the electrochemical activity of the mixed-conduction GDC phase, but proof of this hypothesis is still lacking.

8.5.4 Chromium poisoning of the oxygen electrode

If the metal support in MS-PCECs is placed on the side of the oxygen electrode, there is an inherent risk of Cr poisoning due to the formation of volatile Cr species which might react with the oxygen electrode [44]. The risk of Cr poisoning increases with increasing steam content on the oxygen electrode. The application of Cr protection layers, as established in SOFC stacks in the air manifold, is promising, but applying a homogeneous and dense coating to a highly porous backbone remains quite challenging.

8.6 Concluding remarks

Metal-supported high-temperature electrolysis cells (MS-SOECs, MS-PCECs) are attractive due to their improved thermal management, mechanical robustness, and reduced capital costs when compared to full ceramic electrolysis cells. Up until now, only limited electrochemical results have been available, which were mostly achieved by reversing the operation of established MS-SOFCs. The first electrochemical performance data obtained for MS-SOECs are quite promising, but degradation in electrolysis mode was generally found to be higher than that of the fuel-cell mode. In the literature, several MS-SOEC-specific degradation mechanisms, such as oxidation of the metal substrate at high steam contents, have been discussed. Nevertheless, the limited number of available results, especially for long-term testing, prevents reliable conclusions from being drawn. The development of MS-PCECs has recently started, but it became obvious that processing this kind of cell is much more challenging than processing MS-SOECs. The reasons for this include the problem that the manufacture of gastight proton-conducting electrolytes on porous metal substrates by sintering technologies is limited by the low sintering activity of related electrolyte materials, the evaporation of Ba and Sr, and severe interface reactions in contact with the metal support. The deposition of proton-conducting electrolytes by PLD or PVD technologies appears to be more promising, but to date, no electrochemical results have been published for this kind of cell. For the successful implementation of metal-supported high-temperature electrolysis cells, further processing optimization, the implementation of high-performance materials, and a better understanding of specific degradation mechanisms are required.

Acknowledgments

The authors gratefully acknowledge funding provided by the Christian Doppler Laboratory for Interfaces in Metal-supported Electrochemical Energy Converters, the Austrian Federal Ministry for Digital and Economic Affairs (BMDW), and the Austrian National Foundation for Research, Technology and Development.

References

[1] Graves C, Ebbedesen S D, Mogensen M and Lackner K S 2011 Sustainable hydrocarbon fuels by recycling CO_2 and H_2O with renewable or nuclear energy *Renw. Sustain.* **15** 1–23

[2] Pöhlmann F and Jess A 2016 Influence of syngas composition on the kinetics of Fischer–Tropsch synthesis of using cobalt as catalyst *Energy Technol.* **4** 55–64

[3] Ebbesen S D and Mogensen M B 2009 Electrolysis of carbon dioxide in solid oxide electrolysis cells *J. Power Sources* **193** 349–58

[4] Ebbesen S D, Jensen S H, Hauch A and Mogensen M B 2014 High temperature electrolysis in alkaline cells, solid proton conducting cells, and solid oxide cells *Chem. Rev.* **114** 10697–734

[5] Tao G, Sridhar K R and Chan C L 2004 Study of carbon dioxide electrolysis at electrode/electrolyte interface: Part II. Pt-YSZ cermet/YSZ interface *Solid State Ionics* **175** 621–24

[6] Tucker M C 2010 Progress in metal-supported solid oxide fuel cells: a review *J. Power Sources* **195** 4570–82

[7] Tucker M C 2020 Progress in metal-supported solid oxide electrolysis cells: a review *Int. J. Hydrogen Energy* **45** 24203–18

[8] Ovtar S, Tong X F, Bentzen J J, Thyden K T S, Simonsen S B and Chen M 2019 Boosting the performance and durability of Ni/YSZ cathode for hydrogen production at high current densities via decoration with nano-sized electrocatalysts *Nanoscale* **11** 4394–406

[9] Mogensen M B, Hauch A, Sun X, Chen M, Tao Y, Ebbesen S D K V, Hansen K V and Hendriksen P V 2017 Relation between Ni particle shape change and Ni migration in Ni–YSZ electrodes—a hypothesis *Fuel Cells* **17** 434–41

[10] Hoerlein M P, Riegraf M and Costa R 2018 A parameter study of solid oxide electrolysis cell degradation: microstructural changes of the fuel electrode *Electrochim. Acta* **276** 162–75

[11] Frey C E, Fang Q, Sebold D, Blum L and Menzler N H 2018 A detailed post mortem analysis of solid oxide electrolyzer cells after long-term stack operation *J. Electrochem. Soc.* **165** F357–64

[12] Wang R, Byrne C and Tucker M C 2019 Assessment of co-sintering as a fabrication approach for metal-supported proton-conducting solid oxide cells *Solid State Ionics* **332** 25–33

[13] Wang R, Lau G Y, Ding D, Zhu T and Tucker M C 2019 Approaches for co-sintering metal-supported proton-conducting solid oxide cells with Ba(Zr,Ce,Y,Yb)$O_{3-\delta}$ electrolyte *Int. J. Hydrogen Energy* **44** 13768–76
[14] Lau Y and Tucker M C 2021 Development of metal-supported proton-conducting solid oxide cells via co-sintering *ECS Trans.* **103** 685–92
[15] Sata N, Han F, Zheng H, Dayaghi A M, Norby T, Stange M, Semerad R and Costa R 2021 Development of proton conducting ceramic cells in metal supported architecture *ECS Trans.* **103** 1779–89
[16] Larring Y and Fontaine M L 2013 Critical Issues of Metal-Supported Fuel Cell *Solid Oxide Fuels Cells: Facts and Figures* ed J T S Irvine and P Connor (Berlin: Springer) pp 71–93
[17] Franco T, Henne R, Lang M and Metzger P 2003 Metallic components for a plasma sprayed thin-film SOFC concept *Proc. of Solid Oxide Fuel Cell VIII (Paris)* ed S C Singhal and M Dokiya vol 2003-07 923–32
[18] Franco T and Ruckdäschel R 2006 Diffusion and protecting barrier layers in a substrate supported SOFC concept *7th European Solid Oxide Fuel Cell Forum Proc. (Oberrohrdorf, Switzerland)* P0802-051
[19] Molin S, Gazda M and Jasinski P 2010 High temperature oxidation of porous alloys for solid oxide fuel cell applications *Solid State Ionics* **181** 1214–20
[20] Fang Q, Blum L and Stolten D 2019 Electrochemical performance and degradation analysis of an SOFC short stack following operation of more than 100,000 hours *J. Electrochem. Soc.* **166** F1320–5
[21] Schiller G, Ansar A and Patz O 2010 High temperature water electrolysis using metal supported solid oxide electrolyser cells (SOEC) *Adv. Sci. Technol.* **72** 135–43
[22] Udomsilp D *et al* 2020 Development of metal-supported solid oxide fuel cells with exceptionally high power density for range extender systems *Cell Rep. Phys. Sci.* **1** 100072
[23] Blennow P, Hjelm J, Klemensø T, Persson Å H, Ramousse S and Mogensen M 2011 Manufacturing and characterization of metal-supported solid oxide fuel cells *Fuel Cells* **11** 661–68
[24] Udomsilp D, Thaler F, Menzler N H, Bischof C, de Haart L G J, Opitz A K, Guillon O and Bram M 2019 Dual-phase cathodes for metal-supported solid oxide fuel cells—processing, performance, durability *J. Electrochem. Soc.* **166** F506–10
[25] Menzler N H, Tietz F, Uhlenbruck S, Buchkremer H P and Stöver D 2010 Materials and manufacturing technologies for solid oxide fuel cells *J. Mater. Sci.* **45** 3109–35
[26] Brandner M, Bram M, Froitzheim J, Niewolak L, Buchkremer H P and Stöver D 2008 Electrically conductive diffusion barrier layers for metal supported SOFC's *Solid State Ionics* **179** 1501–6
[27] Hickey D, Alinger M, Shapiro A, Brown K, Striker T, Wang H, Gaunt S, Kinsey D and Hussaini I 2017 Stack development at GE–fuel cells *ECS Trans.* **78** 107–16
[28] Henne RH T, Franco T and Ruckdäschel R 2006 High-velocity DC-VPS for diffusion and protecting barrier layers in solid oxide fuel cells (SOFCs) *J. Thermal Spray Technol.* **15** 695–700
[29] Vaßen R, Hathiramani D, Mertens J, Haanappel V A C and Vinke I C 2007 Manufacturing of high performance solid oxide fuel cells (SOFCs) with atmospheric plasma spraying (APS) *Surf. Coat. Technol.* **202** 499–508
[30] Schiller G, Ansar A, Lang M and Patz O 2009 High temperature water electrolysis using metal supported solid oxide electrolyser cells (SOEC) *J. Appl. Electrochem.* **39** 293–301

[31] Numata Y, Nakajima K, Takasu H and Kato Y 2019 Carbon dioxide reduction on a metal-supported solid oxide electrolysis cell *ISIJ Int.* **59** 628–33

[32] Ivanova M E, Deibert W, Marcano D, Escolastico S, Mauer G, Meulenberg W A, Bram M, Serra J M, Vassen R and Guillon O 2019 Lanthanum tungstate membranes for H_2 extraction and CO_2 utilization: fabrication strategies based on sequential tape casting and plasma-spray physical vapor deposition *Sep. Purif. Technol.* **219** 100–12

[33] Marcano D, Ivanova M E, Mauer G, Sohn Y J, Schwedt A, Bram M, Menzler N H and Vassen R 2019 PS-PVD processing of single phase lanthanum tungstate layers for hydrogen related applications *J. Therm. Spray Technol.* **28** 1554–64

[34] Steele B C H 2000 Appraisal of $Ce_{1-y}Gd_yO_{2-y/2}$ electrolytes for IT-SOFC operation at 500 °C *Solid State Ionics* **129** 95–110

[35] Brandon N P *et al* 2013 Development of metal supported solid oxide fuel cells for operation at 500–600 °C *J Mat Eng Performance* **22** 2900–3

[36] www.spglobal.com/platts/en/market-insights/latest-news/electric-power/070121-ceres-targets-solid-oxide-electrolyzers-for-hydrogen-in-industrial-processes-e-fuels retrieved on 16 November 2021

[37] Bance P, Brandon N P, Girvan B, Holbeche P, O'Dea S and Steele B C H 2004 Spinning-out a fuel cell company from a UK university – 2 years of progress at Ceres Power *J. Power Sources* **131** 86–90

[38] Leah R T, Brandon N P and Aguiar P 2005 Modelling of cells, stacks and systems based around metal-supported planar IT-SOFC cells with CGO electrolytes operating at 500–600 °C *J. Power Sources* **145** 336–52

[39] Steele B C H, Lewis G, Oishi N and Selcuk A 2013 Densification of ceria based electrolytes, Patent EP 1608605

[40] Reisert M, Berova V, Aphale A, Singh P and Tucker M C 2020 Oxidation of porous stainless steel supports for metal-supported solid oxide fuel cells *Int. J. Hydrogen Energy* **45** 30882–97

[41] Tucker M C, Lau G Y, Jacobson C P, DeJonghe L C and Visco S J 2007 Performance of metal-supported SOFCs with infiltrated electrodes *J. Power Sources* **171** 477–82

[42] Tucker M C, Lau G Y, Jacobson C P, DeJonghe L C and Visco S J 2008 Stability and robustness of metal-supported SOFCs *J. Power Sources* **175** 447–51

[43] Wang R, Dogdibegovic E, Lau G Y and Tucker M C 2019 Metal supported solid oxide electrolysis cell with significantly enhanced catalysis *Energy Technol.* **7** 1801154

[44] Shen F, Wang R and Tucker M C 2020 Long term durability test and post mortem for metal-supported solid oxide electrolysis cells *J. Power Sources* **474** 228618

[45] Nielsen J, Persson Å H, Thanh Muhl T and Brodersen K 2018 Towards high power density metal supported solid oxide fuel cell for mobile applications *J. Electrochem. Soc.* **165** F90–6

[46] Haydn M *et al* 2014 Multi-layer thin-film electrolytes for metal supported solid oxide fuel cells *J. Power Sources* **256** 52–60

[47] Bischof C, Nenning A, Malleier A, Martetschläger L, Gladbach A, Schafbauer W, Opitz A K and Bram M 2019 Microstructure optimization of nickel/gadolinium-doped ceria anodes as key to significantly increasing power density of metal-supported solid oxide fuel cells *Int. J. Hydrogen Energy* **44** 31475–87

[48] Tang S and Schulz U 2009 Gas flow sputtering—an aproach to coat complex geometries and non-line of sight areas *Surf. Coat. Technol.* **204** 1087–91

[49] Stange M A, Dayaghi A M, Denonville C, Larring Y, Rørvik P M P, Haugsrud R and Norby T 2019 Fabrication of metal-supported proton-conducting electrolysers with thin film Sr- and Ce-doped BZY electrolyte *ECS Trans.* **91** 941–49

[50] Stefan E, Stange M, Denonville C, Larring Y, Hildenbrand N, Norby T and Haugsrud R 2017 Layered microstructures based on $BaZr_{0.85}Y_{0.15}O_{3-\delta}$ by pulsed laser deposition for metal supported proton ceramic electrolyzer cells *J. Mater. Sci.* **52** 6486–97

[51] Han F, Zhou X, Dayaghi A M, Norby T, Stange M, Sata N and Costa R 2019 Development of metal supported cells using $BaZrO_3$-based proton conducting ceramics *ECS Trans.* **91** 1035–45

[52] Miyamoto K, Kawabata T, Tachikawa Y, Matsuda J, Shunsuke Taniguchi S, Matsuzaki Y, Hayashi A and Sasaki K 2021 Preparation of model SOFCs with proton-conducting electrolytes on metal supports using pulsed laser deposition *ECS Trans.* **103** 2033–40

[53] Ricote S, Bonanos N, Manerbino A and Coors W G 2012 Conductivity study of dense $BaCexZr_{(0.9-x)}Y_{0.1}O_{(3-\delta)}$ prepared by solid state reactive sintering at 1500 °C *Int. J. Hydrogen Energy* **37** 7954–61

[54] Kreuer K D 1996 Proton conductivity: materials and applications *Chem. Mater.* **8** 610–41

[55] Norby T 1999 Solid-state protonic conductors: principles, properties, progress and prospects *Solid State Ionics* **125** 1–11

[56] Ishihara T, Koga M, Matsumoto H and Yoshio M 2007 Electrochemical intercalation of hexafluorophosphate anion into various carbons for cathode of dual-carbon rechargeable battery *Electrochem. Solid-State Lett.* **10** A74–6

[57] Matsuzaki Y *et al* 2019 Leakage current and chemical potential profile in proton-conducting bi-layered solid oxide electrolyte with BZY and hole-blocking layers *ECS Trans.* **91** 1009–18

[58] Chen T, Zhou Y, Liu M, Yuan C, Ye X, Zhan Z and Wang S 2015 High performance solid oxide electrolysis cell with impregnated electrodes *Electrochem. Commun.* **54** 23–7

[59] Zhou Y, Xin X, Li J, Ye X, Xia C, Wang S and Zhan Z 2014 Performance and degradation of metal-supported solid oxide fuel cells with impregnated electrodes *Int. J. Hydrogen Energy* **39** 2279–85

[60] Wu S H, Lin J K, Shiu W H, Liu C K, Lin T N and Lee R Y *et al* 2019 Performance test for anode-supported and metal-supported solid oxide electrolysis cell under different current densities *Proc. of the 42nd Int. Conf. on Advanced Ceramics and Composites, Ceramic Engineering and Science Proc.* 39 ed J Salem *et al* (New York: Wiley) 13 139–48

[61] Hwang C S, Tsai C H, Hwang T J, Chang C L, Yang S F and Lin J K 2016 Novel metal substrates for high power metal-supported solid oxide fuel cells *Fuel Cells* **16** 244–51

[62] Nechache A, Han F, Semerad R, Schiller G and Costa R 2017 Evaluation of performance and degradation profiles of a metal supported solid oxide fuel cell under electrolysis operation *ECS Trans.* **78** 3039–47

[63] Ishihara T, Kusaba H, Kim H H and Kang B S 2019 Preparation of $La_{0.9}Sr_{0.1}Ga_{0.8}Mg_{0.2}O_3$ film by pulse laser deposition (PLD) method on porous Ni–Fe metal substrate for CO_2 electrolysis *ISIJ Int.* **4** 613–8

[64] Stange M, Stefan E, Denonville C, Larring Y, Rørvik P M and Haugsrud R 2017 Development of novel metal-supported proton ceramic electrolyser cell with thin film BZY15–Ni electrode and BZY15 electrolyte *Int. J. Hydrogen Energy* **42** 13454–62

[65] Zhou Z, Karthik Nadimpalli V, Pedersen D B and Esposito V 2021 Degradation mechanisms of metal-supported solid oxide cells and countermeasures: a review *Materials* **14** 3139

[66] Wang R, Sun Z, Choi J-P, Basu S N, Stevenson J W and Tucker M C 2019 Ferritic stainless steel interconnects for protonic ceramic electrochemical cell stacks: oxidation behavior and protective coatings *Int. J. Hydrogen Energy* **44** 25297–309

[67] Bram M, Niewolak L, Shah N, Sebold D and Buchkremer H P 2011 Interaction of sealing material mica with interconnect steel for solid oxide fuel cells application at 600 °C *J. Power Sources* **196** 5889–96

[68] Alnegren P, Sattari M, Svensson J-E and Froitzheim J 2018 Temperature dependence of corrosion of ferritic stainless steel in dual atmosphere at 600–800 °C *J. Power Sources* **392** 129–38

[69] Molin S, Gazda M and Jasinski P 2011 Coatings for improvement of high temperature corrosion resistance of porous alloys *J. Eur. Ceram. Soc.* **31** 2707–10

[70] Karczewski J, Brylewski T, Miruszewski T, Andersen K B, Jasinski P Z and Molin S 2019 High-temperature kinetics study of 430L steel powder oxidized in air at 600–850 °C *Corros. Sci.* **149** 100–7

[71] Stefan E, Denonville C, Larring Y, Stange M and Haugsrud R 2020 Oxidation study of porous metal substrates for metal supported proton ceramic electrolyzer cells *Corros. Sci.* **164** 108335

[72] Young D J, Chyrkin A and Quadakkers W J 2012 A simple expression for predicting the oxidation limited life of thin components manufactured from FCC high temperature alloys *Oxid. Met.* **77** 253–64

[73] Franco T, Schilbinger K, Ilhan Z, Schiller G and Venskutonis A 2007 Ceramic diffusion barrier layers for metal supported SOFCs *ECS Trans.* **7** 771–80

IOP Publishing

High-Temperature Electrolysis
From fundamentals to applications
Werner Sitte and Rotraut Merkle

Chapter 9

Advanced data analysis

Dino Klotz, Sebastian Dierickx, Jochen Joos and André Weber

This chapter introduces measurement, characterization, and advanced data-analysis procedures. Those comprise *in situ* electrochemical measurements and *ex situ* microstructural reconstruction. Combining the results obtained by these two approaches promises to untangle in the best possible detail the physicochemical effects that govern the electrochemical reaction(s) of fuel cell/electrolyzer electrodes. To the best of our knowledge, the descriptions and examples gathered here represent the current state of the art.

In essence, we will summarize the approaches used, from impedance measurement to modeling and parameter identification. We start with a brief introduction to the essentials of electrochemical measurements, focusing on brevity and referencing more detailed tutorials and introductory articles. We then present current approaches for electrode modeling, commenting on the effort required, their accuracy, and their ease of handling. In order to obtain true material parameters, geometrical information about the exact microstructure is required. Therefore, we will introduce microstructural reconstruction as a means of relating the electrochemical information to physical parameters by considering the detailed geometry of the respective structure.

An important prerequisite for advanced data analysis is advanced and reliable data. This has to be collected using a reliable test setup equipped with appropriate devices and tested/double-checked for consistency. Therefore, we suggest best-practice guidelines for impedance measurements and microstructural reconstruction.

9.1 Introduction

As the previous chapters in this book have shown, high-temperature electrolysis using SOECs is a promising technique, but it also requires optimized materials, microstructures, and cell designs to deliver its full potential with respect to performance and efficiency. It is quite challenging to obtain a deep understanding of all the electrochemical processes and material characteristics involved and how they contribute to performance and durability in concert with the operating conditions.

doi:10.1088/978-0-7503-3951-3ch9

For the characterization, development, and diagnosis of the electrodes and cells used for high-temperature electrolysis, it is paramount to identify the performance-limiting factors, quantify the physical parameters that describe the respective processes, and correlate them with material and microstructural properties. The main figure of merit for a given electrode is the area specific resistance (ASR), which provides a measure of the operational losses. However, the ASR of an electrode is only an effective value—which also depends on the temperature and the operating point (mainly determined by the supplied gases and polarization)—that shows a complex dependency on the key parameters of the electrode, such as the surface exchange coefficient k_{chem} (catalytic activity), electronic conductivity, and ionic conductivity (bulk diffusion). The deconvolution and identification tasks become even more complex for technical electrode structures which are porous and/or multiphase, such as the standard electrodes in SOECs.

We require advanced analysis in order to understand the behavior of technical electrode structures and to design better-performing electrode and cell structures by choosing material(s), composition, and microstructure. In addition, a proper understanding of the behavior of pristine samples can help to identify the physics responsible for degradation modes, such as poisoning of the active area and intrinsic degradation due to surface segregation and interphase formation. Finding the right models and the governing parameters for the most common electrode materials has been the subject of many studies in recent decades.

In this chapter, we briefly review the setup and equipment required for the reliable characterization of SOEC single cells, with a focus on providing an appropriate environment for electrolysis tests as opposed to fuel cell operation, as these have been much less covered in the available literature. Here, we focus on the characterization of incremental cells (with a $1 \cdot 1\ \mathrm{cm}^2$ active area, if not stated otherwise). Such cell areas can be assumed to be free of lateral gradients under operational conditions, making it possible to access individual electrode processes, which are blurred if large cells or even stacks are tested under nominal operating conditions, i.e. nonzero current densities resulting in severe temperature (100 K to 200 K) and gas concentration gradients (with gas conversion of up to 90% along the cell).

We then introduce electrochemical impedance spectroscopy (EIS), an operando characterization technique which provides the most detailed information that is attainable by purely electrical means [1]. This introduction provides a reference for the detailed basics of EIS and focuses on the more advanced techniques used to develop adequate physically motivated electrode models. Some peculiarities for electrolysis are pointed out as well.

For a comprehensive analysis, we also require microstructural data for the device under test. An introduction, overview, and guidelines for microstructure reconstruction and analysis techniques are the subjects of section 9.3. An advanced microstructural reconstruction is required which can be linked to the electrochemical data to identify concrete material properties. Today's microstructures feature pore and particle sizes in the range of tens of nanometers to several micrometers, and focused ion beam tomography (FIB/SEM) is by far the most appropriate and most applied technique for SOEC electrodes. For larger structures and whole cells, X-ray tomography (μCT) has recently gained increased interest. We compare both methods, provide practical advice, and present some of the most valuable image-processing, segmentation, and reconstruction tools.

For the comprehensive data analysis, we then explain how the validation and preidentification of impedance data can be performed using the Kramers–Kronig test and the distribution of relaxation times (DRT), respectively [2–5]. The latter is a powerful data-analysis method that increases the obtainable/accessible information in a spectrum; it has been applied to solid-oxide fuel cells (SOFCs) for almost 20 years now. DRT analysis recently gained a boost in interest, as reflected by the increased number of publications and tools available for DRT calculation, which has increased accessibility but also caused some confusion about which tool to choose. Therefore, we will provide a brief overview of the most important aspects of a reliable DRT calculation.

On the basis of electrochemical and microstructural data, precise electrode models can be constructed that are able to predict the performance of an electrode based on its material and microstructural properties, including the electrochemical features of model electrodes. The last part of this chapter will introduce those models, describe their characteristics and individual simplifications, and use real-life examples to demonstrate the state of the art and its capacity for electrode modeling.

9.2 Electrochemical characterization of SOECs

9.2.1 SOEC testing in general

Incremental single cells with an active electrode area of about 1 cm^2 have been the most successful in investigations of the electrochemical characteristics of an electrode. Smaller cells exhibit serious edge effects (such as 'elephant's foot'), and in such cells, the part of the electrode area where the thickness can be assumed to be homogenous is smaller than in larger cell areas. Larger cells, however, show additional contributions due to lateral gradients caused by gas conversion [6] and heat generation [7], which makes the overall analysis of single characteristics more challenging. Gas conversion happens alongside the gas channels as the feed gases are consumed and depleted towards the exhaust ends of the channels. This process is different from gas diffusion through the electrode structure itself (i.e. through the thickness of the electrode). The longer the gas channel and the thicker and less porous the electrode structure, the larger the influence of gas conversion and diffusion on cell performance, and the more difficult it becomes to discern the two different effects. Special model electrodes are a good way to investigate the basic electrochemical reactions and their parameters (or the material parameters) at clearly defined interfaces. However, the geometrical expansion of the reaction zone has to be considered [8]. Technical electrodes exhibit porous structures with microstructural features in the submicrometer to micrometer scale, which potentially have a non-negligible influence on the reaction paths and the electrochemically active regions in the electrode, such that the material potentially behaves differently in the different situations (model electrode versus technical electrode). Furthermore, the purity of the material, including any potential secondary phases, might be different due to different manufacturing protocols and annealing temperatures. Therefore, results obtained using model electrodes have to be carefully cross-checked and corroborated against the results gained from technical electrodes in order to make predictions for technical cells, stacks, and systems. This chapter therefore tries to describe the current status of data analysis for technical electrodes and single cells.

A comprehensive summary of electrochemical measurement techniques for SOFCs is provided in [9]; together with an advanced data analysis that has been constantly refined in recent years, there is now quite a knowledge base that can be used to set up an environment to understand the electrochemistry of SOECs.

For the electrochemical testing of incremental cells, there are commercial test rigs available and also some helpful descriptions of ways to build customized test setups. Figure 9.1 displays a complete setup including the gas lines and the cell housing mounted inside the furnace. The details of this experimental setup and common testing procedures are given in [9].

For SOECs, the steam (and/or CO_2) content in the gas supplied to the fuel electrode is an important issue. It is necessary to thoroughly check whether the desired humidification level can be achieved and whether the humidification actually reaches the electrode. In case of single-cell testing, steam can be generated in different ways. While gas bubbling and membrane humidifiers require a carrier gas and thus are limited in the steam contents they can generate, steam evaporators and combustion chambers can produce reactants with up to 100 vol% steam. In any case, the presence of a steady state has to be confirmed, for example by checking the stability of the open-circuit voltage (OCV). A common problem is condensation, which occurs when parts of the piping providing the humidified gas are placed

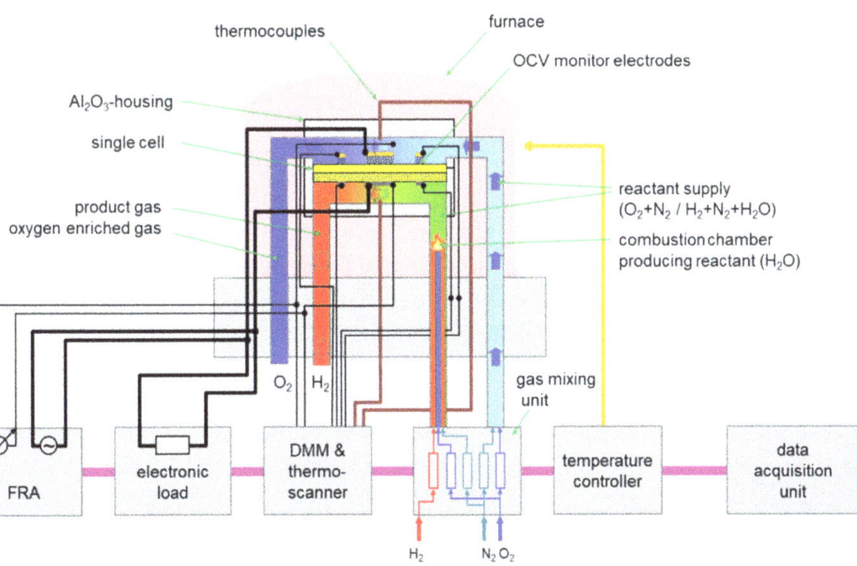

Figure 9.1. Test setup for SOECs with a 1 cm^2 active electrode area (adapted from [10]).

outside the hot zone at temperatures below 100°C. Any cold part of the setup causes water vapor to condense upstream of the cell, decreases the steam content delivered to the cell, and, in the long run, clogs pipes upstream and downstream of the cell. In the case of SOECs, it might be thought that this problem is less relevant downstream of the cell, but as some tests are performed at the OCV, at which no steam is consumed, the exhaust gas lines have to be designed appropriately. Another common issue is the provision of a reliable cell temperature. Significant changes in the cell temperature often arise when the humidification level is changed due to the different resultant temperature of the incoming feed gas. Therefore, it becomes even more important to monitor the cell temperature and OCV over time than it is in the case of SOFC tests.

Considering today's testing infrastructure, which is commercially available for cell sizes from button cells up to large stacks, huge amounts of data can be easily generated, and the focus changes to appropriate data validation and analysis. The most important and meaningful data used to quantify performance and losses at all operating points is provided by the *J–V* characteristics of a cell. But with such stationary DC techniques, the deconvolution of the individual contributions to performance loss is impeded; even a meaningful assessment of the ASRs of the respective cell components remains elusive for DC measurement methods.

A detailed description of the main three electrochemical measurement methods relevant for SOEC testing, i.e. the OCV, *J–V* characterization, and EIS can be found in [9], which mainly targets SOFC measurements but can readily be adapted for SOECs.

OCV measurements are an important part of the testing; even though not too much information can be gained from them, they are an essential indicator of cell conditions, and therefore an important control measurement for checking the validity of the obtained data. Again, this is described in more detail with an example of failure/setup problems in [9]. In the case of common solid-oxide cells (SOCs), a deviation of the OCV from the expected thermodynamic value might be related to gas leakages in the sealant or the electrolyte, a short circuit due to the test bench, a mixed-conduction electrolyte, or a non-equilibrated fuel gas (for carbon-containing fuels). In the case of symmetrical cells, any measurable OCV $\neq 0$ is an error indication.

The standard technique for rating the performance of a cell is to examine the *J–V* characteristics. It is a very easy technique to apply and makes it possible to check whether the cell has any problem apart from leakage. The correct power density is attainable over time only if the gas distribution works flawlessly, if the membrane–electrode assembly (MEA) is reliable, and if the temperature control is accurate. If the power density is far too low as compared to the published values, further characterization of the cell is useful only for troubleshooting and diagnosing what went wrong. A further investigation of the electrochemical processes is not helpful for such a problematic cell.

9.2.2 Electrochemical impedance spectroscopy

Electrochemical impedance spectroscopy is a powerful and versatile measurement technique. This short chapter cannot provide a comprehensive introduction;

however, the theory and many practical hints are discussed in the books of Barsoukov and Macdonald [11], Orazem and Tribollet [12], and Lasia [13], which are recommended for a good overview of EIS. A more basic introduction is provided in [14].

EIS is the measurement method that provides the most attainable information about an electrochemical system at the operating point. As a frequency-domain technique, EIS is also a quite robust technique for the experimental designs. Due to its sinusoidal excitation signal, its signal energy is concentrated at one measurement frequency (lock-in technique, frequency-domain measurement), which efficiently mitigates the influence of measurement noise. Errors can be detected via validation methods (Kramers–Kronig, see section 9.4.1).

By measuring at different frequencies, EIS enables us to specifically look at different contributions to the electrochemical behavior, as all processes occur with a certain 'speed,' which means that each process has its characteristic frequency or time constant. By separating these processes in the frequency domain, one can deconvolute the electrochemical behavior into a number of individual processes. A lot of work has been invested into measurement techniques, instruments, and ways to perform this deconvolution.

In SOC research, the Nyquist plot (shown in the example in figure 9.2) is the most popular graphical representation. An ideal electrochemical process shows up as a semicircle in this type of plot. As much as this graphical separation of processes is helpful and intuitive, identifying actual numbers from impedance spectra is a more difficult task. For this quantification, a so-called equivalent circuit model (ECM) is often used, which consists of electrical equivalents to the electrochemical behavior that have actual parameter values.

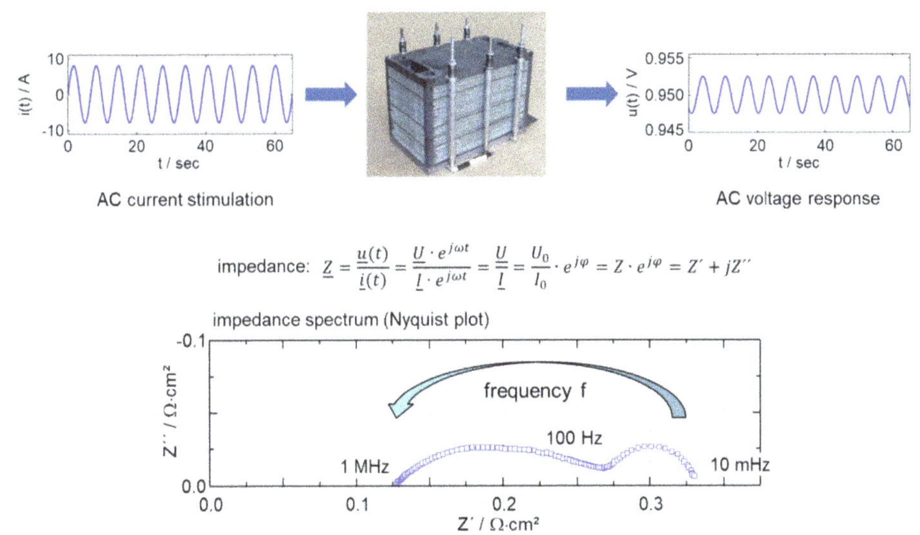

Figure 9.2. The use of impedance spectroscopy for the analysis of low-resistance devices—such as fuel cells and electrolyzers—the cell is usually stimulated by a small AC current; the AC voltage response divided by the current provides the impedance at the related frequency (photo courtesy of Forschungszentrum Jülich GmbH, Jülich, Germany).

Figure 9.2 illustrates the concept of impedance spectroscopy for the analysis of low-resistance devices, such as fuel cells and batteries. In case of such low-impedance samples ($|\underline{Z}| < 1\ \Omega$), the excitation signal is preferably galvanostatic, i.e. a fixed AC current is used as the excitation signal[1]. The amplitude selected should be as small as possible to ensure linearity, but sufficiently high to obtain an acceptable signal-to-noise ratio. In case of a test bench with a low noise level, an AC voltage response amplitude of 10 mV to 20 mV provides high-quality spectra. The details of how to perform impedance spectroscopy on single SOFCs have been summarized in [9].

For a proper analysis of impedance spectra (including the DRT, as discussed in section 9.4), any kind of error in the spectra should be carefully avoided. There are a number of error sources, such as the wiring of samples and the selection of measurement parameters, which are obvious to electrical engineers but often neglected in the literature.

9.3 Microstructural analysis and reconstruction

As mentioned above, detailed quantitative knowledge about the electrode's microstructure is mandatory for the evaluation of electrochemical data. Technical electrodes in SOECs are porous multiphase compounds with a random arrangement of particles and pores; thus, any 2D microstructural analysis performed on a qualitative basis over the last few decades has been limited. Techniques such as optical microscopy, scanning electron microscopy, and transmission electron microscopy are routinely applied to analyze the microstructure of electrodes in R&D as well as for quality assurance. A disadvantage of this approach is the lack of quantification. The 2D gray scale images obtained can hardly be used to extract viable microstructural parameters. In related publications, one can often find statements such as 'the electrode with the finer microstructure exhibited a 43.7% lower polarization resistance.' Whereas the polarization resistance can be measured to a high accuracy, the 'finer microstructure' is only visible in the SEM images. This rating of the microstructure is based on the experience of the person performing the analysis and is thus subjective. Any quantitative parameters such as grain sizes, triple-phase boundary (TPB) densities, porosities, or tortuosities, which are required in homogenized models and are thus directly related to measurable performance parameters, are often missing.

The use of effective microstructural parameters obtained from 3D reconstructions is an appropriate way to include the microstructure in electrochemical data analysis. There are several approaches to reconstructing the microstructure of a porous electrode and generating a 3D image that represents its spatial material distribution. Depending on the scale (i.e. nanometer to micrometer), different methods are applicable.

9.3.1 FIB-SEM and μCT

For the 3D reconstruction of microscale electrodes with particle and pore sizes in the submicrometer and micrometer range, FIB/SEM became the method of choice for SOC [15–19] in the second half of the last decade. In the case of nanoscale electrodes,

[1] Depending on the equipment manufacturer and model, other modes might be preferable.

other methods, such as high-angle annular dark-field (HAADF) scanning transmission electron microscopy (STEM) tomography, which are able to provide a much higher resolution, are required [20]. Coarser structures such as metal supports in SOFCs (particle and pore sizes in the 10 μm to 100 μm range) require the reconstruction of much larger volumes, at which point FIB/SEM tomography becomes expensive and methods such as μCT are preferable [21, 22].

In any kind of 3D reconstruction it is essential to analyze a volume that is sufficiently large to be representative of the microstructure [23]. Such a representative volume element (RVE) has to be acquired at a high resolution so that it is possible to access microstructural details, enable the correct assignment of phases, and determine microstructural parameters such as tortuosity and inner surface area correctly.

A 3D reconstruction requires a number of consecutive tasks to be performed: sample preparation, image acquisition, image preprocessing, segmentation, and the reconstruction of the 3D image (figure 9.3). When the work on the microstructural reconstruction of SOC electrodes started in 2006, no software was available for most of the these tasks, the number of publications in the field of FIB/SEM microstructural reconstruction of fuel cell and battery electrodes was limited [15], and it was necessary to establish suitable procedures and algorithms for the reconstruction [22–26].

For the analysis of SOC electrodes, a sufficiently small (usually several mm^3) region of interest has to be cut out of a larger sample (i.e. a single cell). FIB analysis requires an infiltration of the pores to avoid the redeposition of ablated material in the pore volume and to obtain a defined cutting surface. It is essential to achieve sufficient contrast between the infiltrate and the electrode materials. The sample is usually infiltrated by an epoxy resin using a vacuum infiltration procedure. This provides good contrast with respect to SOC electrode materials such as metal oxides [24], nickel, and zirconia [27].

In case of μCT, which is basically a non-destructive method, such preprocessing of the sample can be avoided. It is possible to reconstruct a complete SOC with dimensions in the cm range at the expense of spatial resolution. If micrometer-scale structures are of interest, i.e. porous electrodes with pore and grain sizes in the range of 1 to 100 μm, a suitable piece has to be cut out of the cell to reduce absorption by the surrounding material and increase the spatial resolution.

Most SOC electrodes show particle and pore sizes in the range of several 100 nm to values slightly above 1 μm and thus enable the reconstruction of a sufficiently large RVE by FIB/SEM with sufficient contrast between all the phases involved. However, if this is not the case, the method must be extended to make all phases visible: to visualize the spatial distribution of primary and secondary phases at the interface of $La_{0.58}Sr_{0.4}Co_{0.2}Fe_{0.8}O_{3-\delta}$ cathode/10 mol% gadolinia-doped ceria/8 mol% yttria-stabilized zirconia electrolyte, it is necessary to combine data sets from different detectors [28] in a so-called 'correlative tomography' approach. Another example in which correlative tomography approaches are useful is the case of electrodes with particle sizes at different length scales. FIB/SEM then even reveals information about small particles and details, while μCT gives access to the distribution and size of the agglomerates in the electrode layer. Such correlative tomography has recently gained

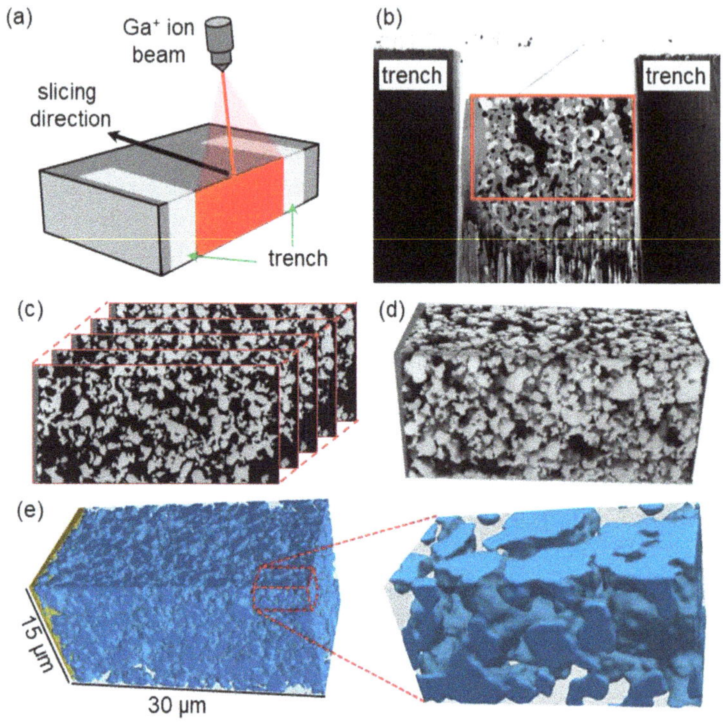

Figure 9.3. (a) Bulk milling technique and (b) exemplary SEM image of a prepared sample at the beginning of sectioning (the red box indicates the area which can be used for reconstruction); (c) by stacking and aligning the 2D images in 3D space and (d) expanding the pixels in the slicing direction, a 3D reconstruction consisting of voxels is derived; (e) by assigning each voxel of the structure to its corresponding phase, a 3D material distribution of the sample is obtained.

increased interest for use with fuel cells [29] and batteries [30]. In any case, the reconstruction procedure provides a huge amount of image data that has to be transformed into a 3D reconstruction that provides detailed information about the spatial distributions of different materials and pores.

9.3.2 Image processing, segmentation, and reconstruction

The reconstruction procedure should provide a 3D data set consisting of voxels (volumetric pixels) representing the microstructure of the electrode. It is essential for the quality of the final reconstruction to perform image processing of the raw data. In most cases, the 3D data set is available as a stack of consecutive 2D images. Consecutive images obtained from FIB-SEM tomography usually have to first be aligned with each other, as the imaged area can move due to mechanical drifts and sample charging during FIB-SEM sectioning. This image alignment can be done using the public domain image-processing software ImageJ (plugin StackReg [31]), or commercial software tools such as GeoDict™ (Math2Market GmbH) or Avizo (Thermo Fisher Scientific).

Depending on the tomography method and data quality, the application of different filters is necessary in the next step. Image quality determines the number and sequence of the filtering steps. Typically, each set of images requires a unique filtering procedure determined by manual experimentation. From our experience, anisotropic diffusion or non-local means filters are useful in order to reduce noise and promote intraregion smoothing. However, the goal of the filtering procedure is to reduce the high-frequency noise while retaining the particle boundaries.

After filtering, all areas in the recorded data must be unambiguously assignable to their actual phase (e.g. electrode materials and pore space). This segmentation procedure is one of the most critical steps. It is useful to compare different segmentation methods for a reconstruction, such as the global threshold [24] and local threshold methods [32]. In general, all phases should be distinguishable by their brightness due to the SEM detector settings and the infiltration method used [25, 32]. The most common method for segmentation is based on selecting global threshold values. Finding the correct threshold values for each entire image data set is usually very challenging, and an incorrect value would jeopardize the calculated microstructural parameters [24].

In high-quality image data, the peaks in the histograms do not (or only slightly) overlap; the different phases can be distinguished by their brightness values. However, brightness gradients are often present, as reported by many groups, which typically results in wider peaks that overlap. There can be multiple reasons for brightness gradients in the raw data. The most common reason is the shadow effect, which occurs when parts of the sample next to the reconstructed area partially block the path from the reconstructed area to the detector. This effect typically becomes more pronounced with increasing milling time, since the milled area reaches deeper into the sample and redeposited material can form additional blockers. The shadow effect can be minimized by milling large trenches along the sides of the volume to be reconstructed [25, 32]. Other prominent reasons are charging of the sample and the imaged area or slight instabilities of the system (e.g. the SEM emitter and/or detector) during image acquisition. The application of the local threshold method has improved the quality of the segmentation in such cases (detailed in reference [32]).

However, the benefit of reconstruction has so far been limited to nice, colorful pictures and movies that provide a qualitative insight into the electrode's microstructure (figure 9.4). Even though differences are clearly visible, a quantitative comparison of the different microstructures is impossible at this point. It is thus necessary to homogenize the reconstruction and describe its properties using global parameters, such as the volume fractions and tortuosities of the different phases, the particle and pore size distributions, the volume-specific interfacial and surface areas, the volume-specific TPB length and others. Methods for quantifying the microstructural parameters from the reconstruction have been developed; details of the methods can be found e.g. in [19, 21–27]. Today, such tasks can be performed using commercial software such as GeoDict™ (Math2Market GmbH), which provides effective microstructural parameters. These parameters are correlated with the electrode's performance and thus enable a direct comparison of different microstructures [22, 27].

In the case of optimized electrodes, a number of loss mechanisms quite often limit the performance simultaneously. In this case, a direct comparison of the

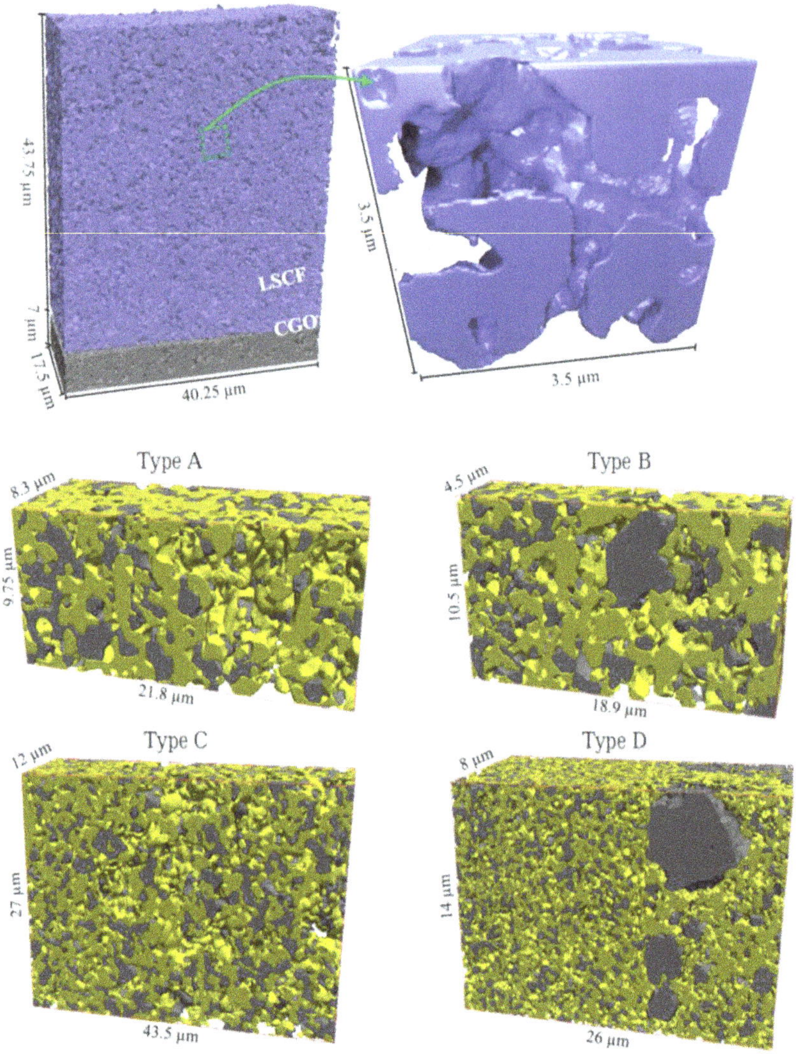

Figure 9.4. Three dimensional reconstructions of porous SOC electrodes obtained using FIB-SEM: $La_{1-x}Sr_xCo_{1-y}Fe_yO_{3-\delta}$ (LSCF) cathode (top) and different Ni/YSZ cermet anodes (bottom).

microstructural parameters is not constructive and it is necessary to set up an appropriate electrode model that considers the coupling of the different loss mechanisms [33]. This approach not only enables a comparison of existing electrodes but also facilitates model-based electrode design [34].

9.4 Impedance data analysis

The microstructural parameters obtained from the microstructural reconstruction provide the geometrical part of an electrode model. For the electrochemical part, EIS can be applied. For a meaningful impedance data analysis, which should provide quantitative information about the electrochemical processes in the investigated cell, valid spectra are required. Only if the validity of the spectra is confirmed is further analysis useful. Hence, we will first discuss the validation of impedance data and introduce the Kramers–Kronig relation, which is used to ensure the validity of the measured spectra, before discussing approaches to the subsequent impedance data analysis. Equivalent circuit models will be introduced, which have represented the state of the art for impedance data analysis and the quantification of electrochemical parameters for some decades. The advantages of model-free deconvolution approaches such as differential impedance analysis and the distribution of relaxation times will be discussed. Furthermore, it will be shown that equivalent circuit modeling approaches are not only suitable for the correlation of a measured impedance spectrum with the material and microstructural features of the investigated cell, but also enable performance predictions to be made for future electrode and cell concepts.

9.4.1 Validity of impedance data

Judging the quality and validity of impedance spectra is important, but difficult, as no fully automatable procedure covers all the possible errors. The shape of the spectrum and its noise level are insufficient criteria. A powerful and independent test for the validity of a spectrum is the Kramers–Kronig test (KKT). The KKT checks for causality, linearity, and time invariance using the Kramers–Kronig relations [2, 35]. Algorithms based on the Kramers–Kronig relation enable the detection of errors in the spectra; appropriate software tools are available for this task [36]. Typical error sources detectable by the KKT are: (i) a drift in cell performance or operating conditions that affects the time invariance; (ii) the influence of the line frequency or other oscillating operating parameters that violate causality; or (iii) an overlarge perturbation amplitude, resulting in a nonlinear response.

Furthermore, there are additional error sources, such as impedance contributions from the wiring and other electrical devices on the test bench, which affect the measured impedance but cannot be detected by the KKT if their impedance contributions also fulfill the Kramers–Kronig relations. To detect these error sources, it is useful to make reference measurements under short- and open-circuit conditions as well as measurements of reference impedances based on electronic components that are similar to the cell.

9.4.2 Equivalent circuit modeling

The most common way to analyze the measured impedance spectra of an electrochemical system is to set up an ECM, consisting of linear equivalent circuit elements such as resistors, capacitors and inductances. Such a linear model approach is permissible because the highly nonlinear current/voltage behavior of the cell is linearized in the impedance spectrum. In addition to these basic elements, more complex elements representing dispersed physicochemical processes in porous electrodes, such as Warburg [37] or Gerischer impedances [38], are often included. These transmission-line models (TLMs) represent the coupling of transport processes by spatially dispersed electrochemical reactions and capacitive storage [39]. An ECM is advantageous because the complex coupling of highly nonlinear electrochemical reactions and diffusive transport processes in a multiphase electrode structure is (i) homogenized, and (ii) linearized, resulting in a much simpler model structure with a limited number of discrete equivalent circuit elements as free fitting parameters. Thus, the complex nonlinear least-squares (CNLS) fitting of the model to a measured spectrum is simplified.

It should be noted that there are different ECM approaches: (i) black-box models that only reproduce the measured spectrum but do not contain any relation to physics or chemistry [40]; (ii) gray-box models that commonly represent a physicochemical process by a resistor R in parallel with a constant-phase element Q (an RQ element) but do not consider further details such as material or microstructural properties; and (iii) white-box models, whose parameters are physicochemically meaningful [34]. Apart from linear ECMs, more complex nonlinear models can be applied as well, but their fitting is aggravated as the computation time is drastically increased and the nonlinear information is missing from the impedance spectrum.

A general disadvantage of fitting approaches is the ambiguity of an impedance spectrum. There is always more than one model that is able to represent the spectrum [41]. Furthermore, with an increasing number of equivalent circuit elements, the growing degrees of freedom enable more accurate fitting. Thus, it is difficult to select the correct equivalent circuit model. An excellent agreement between the measured values and an equivalent circuit model in no way guarantees the correctness of the model. To develop a physicochemically meaningful model, the following points should be taken into account:

- The physicochemical processes expected in the cell should appear in the measured spectrum.
- All equivalent circuit elements in the model should correspond to a physicochemical process in the cell.
- The same model structure should be applicable to any set of operating parameters, and the evaluated model parameters should exhibit meaningful operating parameter dependencies.

Quite often, it is difficult to detect all the expected physicochemical processes in the measured spectra. The most prominent reason for this is a strong overlap of two or more processes in the frequency domain. To solve this problem, methods that increase the resolution of the spectrum and enable the deconvolution of different processes with quite close relaxation frequencies are required.

9.4.3 Impedance data deconvolution approaches

There are different ways in which information about a cell and its electrodes can be extracted from an impedance spectrum. A straightforward approach suggested by Bernard Boukamp 45 years ago is to fit and subsequently subtract prevailing elements corresponding to the dominant electrochemical processes in the cell and then analyze the residual spectrum independently. This method was implemented in his nonlinear least squares fitting program, EQIVCT [42, 43], and is still a useful approach. Its application might be critical if the spectrum contains two processes with quite similar time constants that are considered to be one in the fit.

The analysis of differences in impedance spectra (ADIS) is a powerful tool for analyzing a series of impedance spectra measured under selectively varied operating conditions or corresponding to different health states of a cell [44]. The difference of impedance spectra measured before and after durability testing provides direct access to the degenerated electrochemical processes in the cell [45].

One prominent solution for the deconvolution of the processes in an impedance spectrum is the distribution of relaxation times. When a spectrum is transformed into a related DRT, an RC element (R||C) in the spectrum is converted into a Dirac pulse at the related relaxation time (respectively, relaxation frequency) in the DRT. Thus, at least theoretically, two processes with a minor difference in their relaxation frequencies can be separated in the DRT, whereas in conventional ways of displaying the spectrum, they strongly overlap and thus appear as one process. Even though the concept of distributed relaxation times was already introduced in 1907 by Schweidler [46], its application in the analysis of impedance spectra took several decades [47–55]. In 2002, Schichlein [3] applied the DRT to analyze the impedance spectra of SOCs for the first time.

The relation between the impedance spectrum $Z(\omega)$ and the distribution of relaxation times $\gamma(\tau)$, which is most commonly displayed as a distribution $g(f)$ versus the relaxation frequencies, is given by equation (9.1).

$$Z(\omega) = R_0 + Z_{\text{pol}}(\omega) = R_0 + \int_0^\infty \frac{\gamma(\tau)}{1 + j\omega\tau} d\tau \tag{9.1}$$

This relation is valid for any kind of system that fulfills the fundamental demands of impedance spectroscopy—linearity (at least within the stimulation amplitude), causality, and time invariance, and is not limited to capacitive systems as derived in [56]. Different approaches are feasible for calculating the DRT of a spectrum, such as discrete Fourier transformation [3] or the Tikhonov regularization [57–59]. While the discrete Fourier transform method requires extrapolation and filtering with a number of parameters and settings that have to be properly chosen, the Tikhonov regularization only requires one important parameter, the regularization parameter λ, which determines the smoothness of the DRT, but also the width of the peaks. In the case of excellent data quality or synthetic data, a small regularization parameter can be used. For noisy data the regularization parameter has to be increased to suppress false peaks and oscillations in the DRT. Today, a number of commercial and freeware software tools for calculating the DRT are available, such as Ftikreg [60, 61] and others [62–70].

The increased resolution of the DRT is visualized in figure 9.5 for the impedance spectrum of a SOC. The DRT clearly resolves the individual processes in the spectrum. The area underneath each peak corresponds to the related losses. Noise in the spectrum is eliminated by the Tikhonov regularization (the impact of errors and noise on the DRT is discussed in [4]). In addition to the DRT of the measured spectra, this figure also shows the DRT of the CNLS fit of a physicochemical ECM (see section 9.4.4) as well as the DRTs of the individual ECM components.

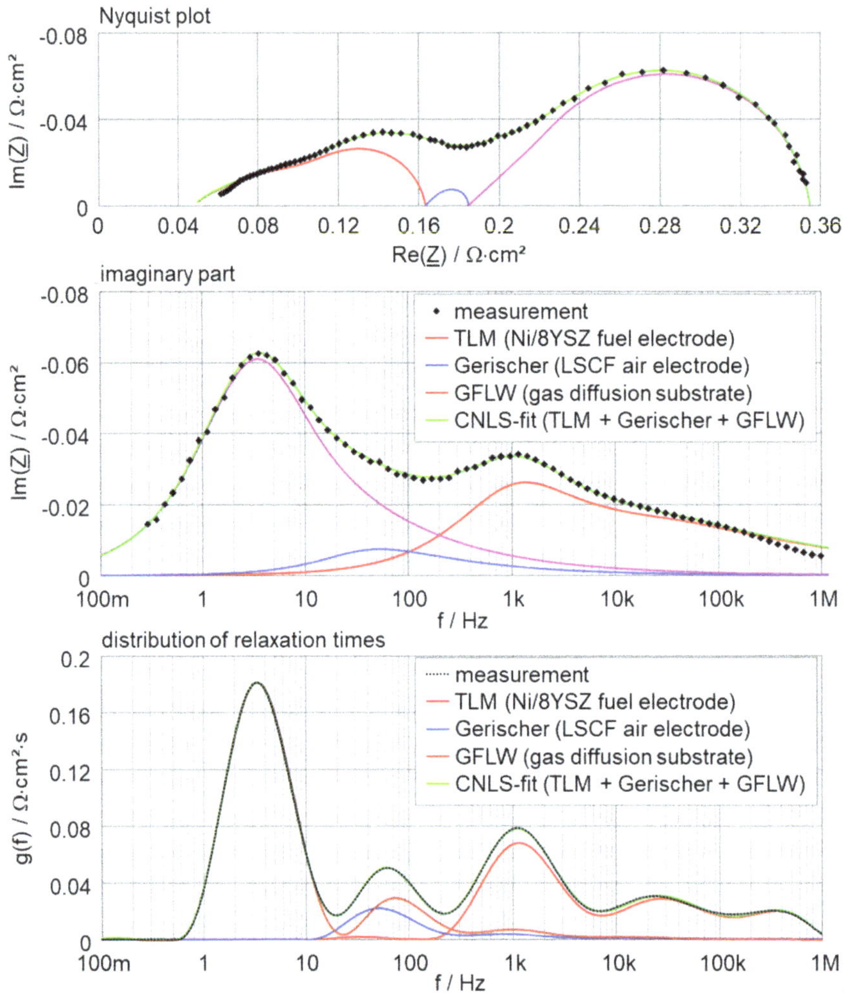

Figure 9.5. Impedance spectrum in a Nyquist plot (top), imaginary part of the spectrum plotted versus the frequency (center), and DRT plotted versus the relaxation frequencies (bottom) of a fuel-electrode-supported cell. A comparison of the imaginary part and the DRT reveals the higher resolution of the DRT. The plotted lines show both measurements and equivalent circuit modeling results [71]. The cell impedance can be properly described by a transmission-line model (TLM), a Gerischer element, and a generalized finite-length Warburg element (GFLW), which account for the fuel electrode, the mixed-conduction air electrode, and the gas diffusion in the porous substrate, respectively.

9.4.4 DRT-based equivalent circuit modeling and simulation

Physicochemically meaningful ECMs are useful to analyze the loss mechanisms and degradation phenomena in a cell. To establish such a model, the deconvolution of the processes in the spectra, the assignment of their physicochemical backgrounds, and the quantification of model parameters can be performed by a combined DRT/CNLS-fitting approach [72]. The results of such a fit are displayed in figure 9.5.

To deconvolute the different physicochemical processes in the impedance spectrum of an electrochemical cell, their different dependencies on the operating parameters have to be considered. Such dependencies often can hardly be identified in the spectrum, but become resolvable in the DRT.

In the case of an SOC, it is possible to vary a number of operating parameters such as current density, temperature, and gas composition. These parametric variations should not be limited to technically meaningful operating conditions; the operating parameters should be selected in such a way that the process of interest becomes accessible. In [73], this approach of operating parameter variation and DRT analysis was systematically implemented to develop an ECM for anode-supported SOFCs for the first time. Impedance spectra were measured during stepwise variations of individual operating parameters. The DRT analysis visualized the operational parameter dependencies of the different polarization processes. Based on this preidentification by the DRT, a physicochemically meaningful ECM was developed and parameterized by a CNLS fit. By applying appropriate operating conditions during the EIS measurements, even minor processes can be resolved, such as the aging of an $La_{1-x}Sr_xCo_{1-y}Fe_yO_{3-\delta}$ (LSCF) SOFC cathode [74], gas diffusion in thin $La_{1-x}Sr_xMnO_3$ (LSM) [75], or Ni/ceria fuel electrodes [76]. Moreover, the selective poisoning of electrodes or variations in electrode microstructure and thickness are possible measures that can be used to correlate the physicochemical processes in the cell with the peaks in the DRT.

9.4.5 Correlation of impedance and physicochemically meaningful parameters

The electrochemical behavior of porous SOC electrodes is directly linked to (i) electrode geometry, (ii) electrode microstructure, and (iii) electrode materials—specifically, their kinetic and catalytic properties. In this context, a number of electrode models have been established for different electrode configurations, which facilitate a direct correlation between the abovementioned properties and their contributions to the impedance of the electrodes.

The impedance behavior of a single electrode layer with thickness L can be modeled with a two-channel TLM, also known as Bisquert-element, which can be analytically described by the following general impedance function [11, 77]:

$$Z_{\mathrm{TLM}}(\omega)=\frac{\chi_1\chi_2}{\chi_1+\chi_2}\cdot\left(\mathrm{L}+\frac{2\cdot\kappa}{\sinh(L/\kappa)}\right)+\kappa\cdot\frac{\chi_1^2+\chi_2^2}{\chi_1+\chi_2}\cdot\coth\left(\frac{L}{\kappa}\right), \tag{9.2}$$

and

$$\kappa=\sqrt{\frac{\zeta}{\chi_1+\chi_2}}. \tag{9.3}$$

In essence, such a TLM approach defines the electrode impedance by considering the resistances of the ionic and the electronic transport paths (χ_1 and χ_2 in Ωm^{-1}) and the coupling charge-transfer reaction (ζ in Ωm). Depending on the electrode configuration, this approach can either be simplified or must be extended to model the electrochemical behavior of all SOC electrode types. Here, we show a range of approaches that comprehensively describe electrode behavior for some relevant SOC electrode structures.

One-phase single-layer electrode—One important application of TLMs in SOC research is the Gerischer element (see figure 9.6(a)), which is applied in the Adler–Lane–Steele (ALS) model to describe the impedance behavior of mixed ionic–electronic conduction (MIEC) air electrodes [78, 79]:

$$Z_{\text{Gerischer}}(\omega) = \frac{R_{\text{chem}}}{\sqrt{1 + j\omega \cdot \tau_{\text{chem}}}}. \tag{9.4}$$

The Gerischer element can be derived from equation (9.2) by assuming an infinite electronic conductivity ($\chi_1 \gg \chi_2$) and an electrode thickness significantly larger than the penetration depth λ (see figure 9.6(a)) of the electrochemical reaction ($L>3\cdot\lambda \rightarrow$ $\coth(L/\lambda)\approx 1$). In this model, the impedance response is directly correlated to the microstructural and material parameters of a porous mixed-conduction electrode. R_{chem} corresponds to the DC resistance of the electrode, which depends on the oxygen diffusion and surface exchange coefficient of the mixed-conduction bulk material and its surface, respectively. τ_{chem} is the related time constant determined by the surface exchange coefficient and the chemical capacity of the bulk material. In addition to material properties related to the stoichiometry and oxygen exchange kinetics of an MIEC, microstructural features can also affect performance.

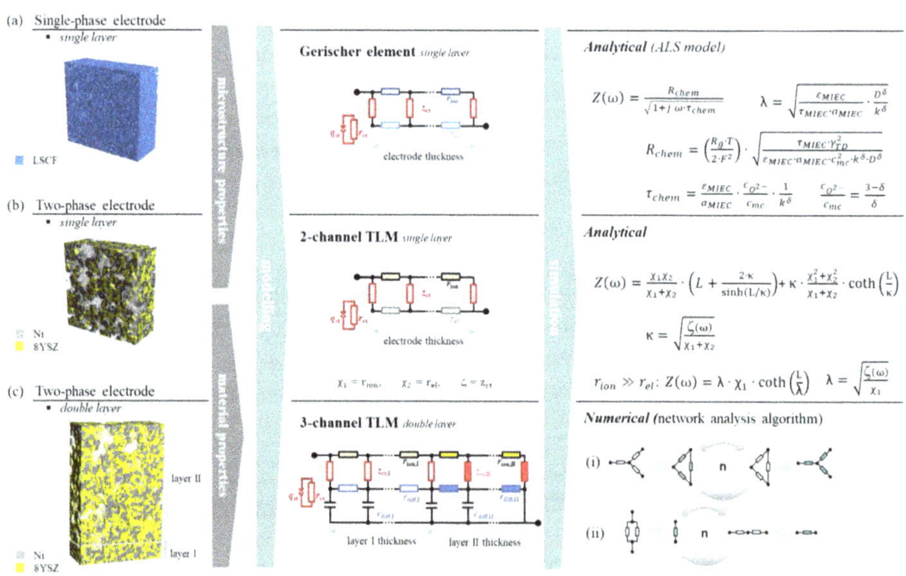

Figure 9.6. Illustration of different electrode configurations, the corresponding physically meaningful impedance models, and their implementation: (a) Gerischer element for single-layer one-phase MIEC electrodes described by the ALS model, (b) two-channel TLM for single-layer two-phase electrodes modeled via the Bisquert approach, and (c) three-channel TLM for double-layer two-phase electrodes calculated by numerical algorithms based on network analysis.

In porous electrode structures, the impact of intrinsic material parameters is always weighted by effective microstructural parameters such as the porosity, tortuosity, and inner surface area.

Two-phase single-layer electrode—As long as the two-phase electrode is formed from only one electrochemically active single layer, equations (9.2) and (9.3) are directly applicable (see figure 9.6(b)). For electrodes with a highly electronically conductive phase (e.g. metals such as nickel), the loss contribution of the electronic path is negligible, thus simplifying equations (9.2) and (9.3) to:

$$Z_{\text{TLM}} = \lambda \cdot \chi_1 \cdot \coth\left(\frac{L}{\lambda}\right), \tag{9.5}$$

The penetration depth λ provides information about the active thickness of the electrode.

$$\lambda = \sqrt{\frac{\zeta}{\chi_1}} \tag{9.6}$$

In the last decade, the applicability of such a TLM model to porous composite electrodes was demonstrated for several SOC material compositions, such as Ni/CGO-infiltrated cermet anodes [80], LSM/YSZ cathodes [81], Ni/YSZ anodes [34, 82, 83], and Ni/ScYSZ anodes [84]. However, this model has limitations in describing multiphase materials with more than one layer and makes certain assumptions that are not always fulfilled, e.g. that the restriction of gas diffusion to the reaction zone is negligible.

Two-phase double-layer electrode—A popular approach to the consideration of additional electrode processes is the separate modeling of the electrochemical processes and the transport by two distinct, serially connected elements (e.g. a Gerischer element + an RQ element that account for an MIEC-cathode and an MIEC/electrolyte interface, respectively, or two RQ elements + a GFLW or a TLM + a GFLW that account for a gas phase, an ionic transport, and their coupling by the charge-transfer reaction) [34, 39, 73]. However, as soon as the electrochemical reaction zone extends over more than one electrode layer and/or the gas transport in the electrode layers is no longer negligible, serial modeling approaches fail [71]. Therefore, in order to model multilayer electrodes in a physically meaningful way, (i) coupling of multiple electrode layers (coupling of the TLMs) as well as (ii) an extension consisting of a 'third TLM channel' representing the gas transport path are required (see figure 9.6(c)). Due to the complexity of the resulting TLM, a simple analytical expression is no longer possible.

With great effort, a mathematical solution of a three-channel TLM can be obtained by assembling a system of ordinary differential equations (ODEs), but so far, this has only been demonstrated for a single-layer electrode [85]. Another workflow results from the direct simulation of the electrical circuit in a graphical programming environment (e.g. OrCAD®, PSpice®, Designer, or MathWorks MATLAB™ Simulink). The solution of the ODE system is then embedded in the software, but this approach requires time-consuming preparatory work (assembling and predefining the electrical circuit elements) and high computational costs (to solve the complex ODE system) [71].

Nevertheless, a cost-effective solution for such a complex circuit can be numerically derived using a simulation approach based on calculation steps taken from network analysis [71]. This approach allows the individual parametrization of each circuit element while requiring a simulation duration of less than 1 s, thus enabling the simulation and even the fitting of impedance spectra of gradient and multilayer electrodes. This algorithm has been successfully employed for double-layered Ni/YSZ fuel electrodes and can be adapted from [71].

Parametrization—Most TLM approaches in the literature are commonly parameterized by the CNLS fitting of impedance spectra. Here, it should be considered that the correlation between ECM parameterization and an impedance spectrum is generally ambiguous, i.e. an impedance spectrum simulated via a TLM model can equally be represented by different parameter sets [34]. To ensure an unambiguously correct and physically meaningful parametrization of the individual transport and reaction processes, all the required parameters have to be determined by different experimental methods, such as conductivity measurements, model electrodes [86, 87], and FIB/SEM tomography [24]. A complete parameterization has been successfully performed for Ni/YSZ electrodes [34, 71]; an example of this type of impedance data analysis for a fuel-electrode-supported cell is shown in figure 9.5.

9.5 Concluding remarks

As we have seen in this chapter, relating the behavior of an SOEC electrode to its fundamental and characteristic physical parameters requires a variety of advanced measurement, data-analysis, and evaluation tools. However, recent literature has provided some guidelines and a framework for advanced electrode modeling that future work can build upon.

It has been pointed out that precision is decisive, both for the electrochemical measurements and for the microstructural analysis. Advanced data sets are required and have to be corroborated against published results. The good news is that this involved analysis has become necessary because of the strong improvement in SOECs (high-performing electrodes with high efficiency and low losses). Thus, it has become even harder to identify and quantify the remaining issues. The developed models can be used to compare material choices and microstructures and to monitor degradation, while being able to particularly attribute changes to specific failure modes.

Advanced measurements and subsequent data analysis allow physicochemically meaningful ECMs to be developed that are not restricted to existing electrodes and cells. The impact of any desired material and/or microstructural optimization can be analyzed by simulation in advance. In [35], it was shown that microstructural (tortuosity, triple-phase boundary length) as well as material properties (e.g. the conductivity of the electrolyte in a composite electrode) strongly affect the ASR. Considering the technically relevant materials and processing and stability limitations, an optimum can only be achieved if electrode materials and microstructures are considered. In future simulation studies it will be essential that parameters are varied in meaningful ranges and that the interactions between the different micro-structural parameters [88] are considered.

References

[1] Macdonald D D 2006 Reflections on the history of electrochemical impedance spectroscopy *Electrochim. Acta* **51** 1376–88

[2] Schönleber M, Klotz D and Ivers-Tiffée E 2014 A method for improving the robustness of linear Kramers–Kronig validity tests *Electrochim. Acta* **131** 20–7

[3] Schichlein H, Müller A C, Voigts M, Krügel A and Ivers-Tiffée E 2002 Deconvolution of electrochemical impedance spectra for the identification of electrode reaction mechanisms in solid oxide fuel cells *J. Appl. Electrochem.* **32** 875–82

[4] Ivers-Tiffée E and Weber A 2017 Evaluation of electrochemical impedance spectra by the distribution of relaxation times *J. Ceram. Soc. Japan* **125** 193–201

[5] Dierickx S, Weber A and Ivers-Tiffée E 2020 How the distribution of relaxation times enhances complex equivalent circuit models for fuel cells *Electrochim. Acta* **355** 136764

[6] Primdahl S and Mogensen M 1999 Gas diffusion impedance in characterization of solid oxide fuel cell anodes *J. Electrochem. Soc.* **146** 2827–33

[7] Njodzefon J-C, Klotz D, Kromp A, Weber A and Ivers-Tiffée E 2013 Electrochemical modeling of the current-voltage characteristics of an SOFC in fuel cell and electrolyzer operation modes *J. Electrochem. Soc.* **160** F313–23

[8] Doppler M C, Fleig J, Bram M and Opitz A K 2019 Comparison of electrochemical hydrogen oxidation on different metal/ceramic model anodes and mechanistic implications *J. Phys.: Energy* **1** 035001

[9] Klotz D, Weber A and Ivers-Tiffée E 2017 Practical guidelines for reliable electrochemical characterization of solid oxide fuel cells *Electrochim. Acta* **227** 110–26

[10] Weber A 2002 *Entwicklung von Kathodenstrukturen für die Hochtemperatur-Brennstoffzelle SOFC* Universität Karlsruhe (TH) https://doi.org/10.5445/IR/13662002

[11] Barsoukov E and Macdonald R (ed) 2018 *Impedance Spectroscopy: Theory, Experiment, and Applications* 3 (Hoboken, NJ: Wiley)

[12] Orazem M E and Tribollet B 2008 *Electrochemical Impedance Spectroscopy* (Hoboken, NJ: Wiley)

[13] Lasia A 2014 *Electrochemical Impedance Spectroscopy and its Applications* (New York: Springer)

[14] Klotz D 2012 Characterization and modeling of electrochemical energy conversion systems by impedance techniques *Dissertation* Karlsruher Institut für Technologie

[15] Wilson J R, Kobsiriphat W, Mendoza R, Chen H Y, Hiller J M, Miller D J, Thornton K, Voorhees P W, Adler S B and Barnett S A 2006 Three-dimensional reconstruction of a solid-oxide fuel-cell anode *Nat. Mater.* **5** 541–4

[16] Gostovic D, Smith J R, Kundinger D P, Jones K S and Wachsman E D 2007 Three-dimensional reconstruction of porous LSCF cathodes *Electrochem. Solid-State Lett.* **10** B214

[17] Shearing P R, Golbert J, Chater R J and Brandon N P 2009 3D reconstruction of SOFC anodes using a focused ion beam lift-out technique *Chem. Eng. Sci.* **64** 3928–33

[18] Iwai H *et al* 2010 Quantification of SOFC anode microstructure based on dual beam FIB-SEM technique *J. Power Sources* **195** 955–61

[19] Rüger B, Joos J, Weber A, Carraro T and Ivers-Tiffée E 2019 3D electrode microstructure reconstruction and modelling *ECS Trans.* **25** 1211–20

[20] Dieterle L, Bockstaller P, Gerthsen D, Hayd J, Ivers-Tiffée E and Guntow U 2011 Microstructure of nanoscaled $La_{0.6}Sr_{0.4}CoO_{3-\delta}$ cathodes for intermediate-temperature solid oxide fuel cells *Adv. Energy Mater.* **1** 249–58

[21] Reiss G, Frandsen H L, Brandstätter W and Weber A 2015 Numerical evaluation of micro-structural parameters of porous supports in metal-supported solid oxide fuel cells *J. Power Sources* **273** 1006–15

[22] Ender M, Joos J, Weber A and Ivers-Tiffée E 2014 Anode microstructures from high-energy and high-power lithium-ion cylindrical cells obtained by X-ray nano-tomography *J. Power Sources* **269** 912–9

[23] Joos J, Ender M, Carraro T, Weber A and Ivers-Tiffée E 2012 Representative volume element size for accurate solid oxide fuel cell cathode reconstructions from focused ion beam tomography data *Electrochim. Acta* **82** 268–76

[24] Joos J, Carraro T, Weber A and Ivers-Tiffée E 2011 Reconstruction of porous electrodes by FIB/SEM for detailed microstructure modeling *J. Power Sources* **196** 7302–7

[25] Ender M, Joos J, Carraro T and Ivers-Tiffée E 2011 Three-dimensional reconstruction of a composite cathode for lithium-ion cells *Electrochem. Commun.* **13** 166–8

[26] Joos J, Rüger B, Carraro T, Weber A and Ivers-Tiffée E 2019 Electrode reconstruction by FIB/SEM and microstructure modeling *ECS Trans.* **28** 81–91

[27] Joos J, Ender M, Rotscholl I, Menzler N H and Ivers-Tiffée E 2014 Quantification of double-layer Ni/YSZ fuel cell anodes from focused ion beam tomography data *J. Power Sources* **246** 819–30

[28] Wankmüller F, Szász J, Joos J, Wilde V, Störmer H, Gerthsen D and Ivers-Tiffée E 2017 Correlative tomography at the cathode/electrolyte interfaces of solid oxide fuel cells *J. Power Sources* **360** 399–408

[29] Burnett T L *et al* 2015 Correlative tomography *Sci. Rep.* **4** 4711

[30] Moroni R, Börner M, Zielke L, Schroeder M, Nowak S, Winter M, Manke I, Zengerle R and Thiele S 2016 Multi-scale correlative tomography of a Li-ion battery composite cathode *Sci. Rep.* **6** 30109

[31] Thevenaz P, Ruttimann U E and Unser M 1998 A pyramid approach to subpixel registration based on intensity *IEEE Trans. Image Process.* **7** 27–41

[32] Ender M, Joos J, Carraro T and Ivers-Tiffée E 2012 Quantitative characterization of $LiFePO_4$ cathodes reconstructed by FIB/SEM tomography *J. Electrochem. Soc.* **159** A972–80

[33] Illig J, Ender M, Weber A and Ivers-Tiffée E 2015 Modeling graphite anodes with serial and transmission line models *J. Power Sources* **282** 335–47

[34] Dierickx S, Joos J, Weber A and Ivers-Tiffée E 2018 Advanced impedance modelling of Ni/8YSZ cermet anodes *Electrochim. Acta* **265** 736–50

[35] Boukamp B A 1995 A linear Kronig–Kramers transform test for immittance data validation *J. Electrochem. Soc.* **142** 1885–94

[36] Schönleber M 2015 https://iam.kit.edu/et/english/Lin-KK.php

[37] Warburg E 1899 Ueber das Verhalten sogenannter unpolarisirbarer Elektroden gegen Wechselstrom *Ann. Phys. Chem.* **303** 493–9

[38] Gerischer H 1951 Wechselstrompolarisation von Elektroden mit einem potentialbestimmenden Schritt beim Gleichgewichtspotential I *Z. Phys. Chem.* **198** 286–313

[39] Weber A 2021 Impedance analysis of porous electrode structures in batteries and fuel cells *tm - Tech. Mess.* **88** 1–16

[40] Schmidt J P, Berg P, Schönleber M, Weber A and Ivers-Tiffée E 2013 The distribution of relaxation times as basis for generalized time-domain models for Li-ion batteries *J. Power Sources* **221** 70–7

[41] Macdonald J R and Garber J A 1977 Analysis of impedance and admittance data for solids and liquids *J. Electrochem. Soc.* **124** 1022–30

[42] Boukamp B A, Hildenbrand N, Nammensma P and Blank D H A 2011 The impedance of thin dense oxide cathodes *Solid State Ionics* **192** 404–8

[43] Boukamp B A 1986 A nonlinear least squares fit procedure for analysis of immittance data of electrochemical systems *Solid State Ionics* **20** 31–44

[44] Jensen S H, Hauch A, Hendriksen P V, Mogensen M, Bonanos N and Jacobsen T 2007 A method to separate process contributions in impedance spectra by variation of test conditions *J. Electrochem. Soc.* **154** B1325

[45] Blennow P, Hjelm J, Klemensø T, Ramousse S, Kromp A, Leonide A and Weber A 2011 Manufacturing and characterization of metal-supported solid oxide fuel cells *J. Power Sources* **196** 7117–25

[46] v Schweidler E R 1907 Studien über die Anomalien im Verhalten der Dielektrika *Ann. Phys.* **329** 711–70

[47] Cole K S and Cole R H 1941 Dispersion and absorption in dielectrics I. Alternating current characteristics *J. Chem. Phys.* **9** 341–51

[48] Fuoss R M and Kirkwood J G 1941 Electrical properties of solids. VIII. Dipole moments in polyvinyl chloride-diphenyl systems *J. Am. Chem. Soc.* **63** 385–94

[49] Misell D L and Sheppard R J 1973 The application of deconvolution techniques to dielectric data *J. Phys. D: Appl. Phys.* **6** 303

[50] Giese K 1973 On the numerical evaluation of the dielectric relaxation time distribution function from permittivity data *Adv. Mol. Relax. Process.* **5** 363–73

[51] Colonomos P and Gordon R G 1979 Bounded error analysis of experimental distributions of relaxation times *J. Chem. Phys.* **71** 1159–66

[52] Salefran J L and Dutuit Y 1981 The use of a discriminative window in deconvolution method applied to dielectric data *J. Chem. Phys.* **74** 3056–63

[53] Franklin A D and De Bruin H J 1983 The fourier analysis of impedance spectra for electroded solid electrolytes *Phys. Status Solidi* **75** 647–56

[54] Morgan F D and Lesmes D P 1994 Inversion for dielectric relaxation spectra *J. Chem. Phys.* **100** 671–81

[55] Paulson K S, Jouravleva S and McLeod C N 2000 Dielectric relaxation time spectroscopy *IEEE Trans. Biomed. Eng.* **47** 1510–7

[56] Schönleber M and Ivers-Tiffée E 2015 Approximability of impedance spectra by RC elements and implications for impedance analysis *Electrochem. Commun.* **58** 15–9

[57] Groetsch C W 1984 *The Theory of Thikonov Regularization for Fredholm Equations of the First Kind* (Boston, MA: Pitman) p 104

[58] Tikhonov A N, Goncharsky A V, Stepanov V V and Yagola A G 1995 *Numerical Methods for the Solution of Ill-Posed Problems* (Dordrecht: Springer Netherlands)

[59] Sonn V, Leonide A and Ivers-Tiffée E 2008 Combined deconvolution and CNLS fitting approach applied on the impedance response of technical Ni/8YSZ cermet electrodes *J. Electrochem. Soc.* **155** B675

[60] Weese J 1992 A reliable and fast method for the solution of Fredhol integral equations of the first kind based on Tikhonov regularization *Comput. Phys. Commun.* **69** 99–111
[61] Weese J 1992 http://cpc.cs.qub.ac.uk/summaries/ACGH_v1_0.html
[62] Graves C, Ebbesen S D and Mogensen M 2011 Co-electrolysis of CO_2 and H_2O in solid oxide cells: performance and durability *Solid State Ionics* **192** 398–403
[63] Wuillemin Z, Antonetti Y, Beetschen C, Millioud O, Ceschini S, Madi H and Van herle J 2013 Local activation and degradation of electrochemical processes in a SOFC *ECS Trans.* **57** 561–70
[64] Sumi H, Yamaguchi T, Hamamoto K, Suzuki T and Fujishiro Y 2014 Electrochemical analysis for anode-supported microtubular solid oxide fuel cells in partial reducing and oxidizing conditions *Solid State Ionics* **262** 407–10
[65] Ciucci F and Chen C 2015 Analysis of electrochemical impedance spectroscopy data using the distribution of relaxation times: a Bayesian and hierarchical Bayesian approach *Electrochim. Acta* **167** 439–54
[66] Zhang Y, Chen Y, Li M, Yan M, Ni M and Xia C 2016 A high-precision approach to reconstruct distribution of relaxation times from electrochemical impedance spectroscopy *J. Power Sources* **308** 1–6
[67] Subotić V, Schluckner C, Strasser J, Lawlor V, Mathe J, Rechberger J, Schroettner H and Hochenauer C 2016 *In-situ* electrochemical characterization methods for industrial-sized planar solid oxide fuel cells. Part I: methodology, qualification and detection of carbon deposition *Electrochim. Acta* **207** 224–36
[68] Nechache A, Mansuy A, Petitjean M, Mougin J, Mauvy F, Boukamp B A, Cassir M and Ringuedé A 2016 Diagnosis of a cathode-supported solid oxide electrolysis cell by electrochemical impedance spectroscopy *Electrochim. Acta* **210** 596–605
[69] Kobayashi K, Sakka Y and Suzuki T S 2016 Development of an electrochemical impedance analysis program based on the expanded measurement model *J. Ceram. Soc. Japan* **124** 943–9
[70] Katayama S, Nakai H and Kita A 2016 Electrochemical impedance analysis of $LiCoO_2$/graphite lithium ion batteries by distribution of relaxation times method *ECS Meet. Abstr.*
[71] Dierickx S, Mundloch T, Weber A and Ivers-Tiffée E 2019 Advanced impedance model for double-layered solid oxide fuel cell cermet anodes *J. Power Sources* **415** 69–82
[72] Klotz D, Schmidt J P, Kromp A, Weber A and Ivers-Tiffée E 2019 The distribution of relaxation times as beneficial tool for equivalent circuit modeling of fuel cells and batteries *ECS Trans.* **41** 25–33
[73] Leonide A, Sonn V, Weber A and Ivers-Tiffée E 2008 Evaluation and modeling of the cell resistance in anode-supported solid oxide fuel cells *J. Electrochem. Soc.* **155** B36
[74] Endler-Schuck C, Leonide A, Weber A, Uhlenbruck S, Tietz F and Ivers-Tiffée E 2011 Performance analysis of mixed ionic–electronic conducting cathodes in anode supported cells *J. Power Sources* **196** 7257–62
[75] Kornely M, Leonide A, Weber A and Ivers-Tiffée E 2011 Performance limiting factors in anode-supported cells originating from metallic interconnector design *J. Power Sources* **196** 7209–16
[76] Grosselindemann C, Russner N, Dierickx S, Wankmueller F and Weber A 2021 Deconvolution of gas diffusion polarization in Ni/gadolinium-doped ceria fuel electrodes *J. Electrochem. Soc.*
[77] Bisquert J 2000 Influence of the boundaries in the impedance of porous film electrodes *Phys. Chem. Chem. Phys.* **2** 4185–92

[78] Adler S B 1998 Mechanism and kinetics of oxygen reduction on porous $La_{1-x}Sr_xCoO_{3-\delta}$ electrodes *Solid State Ionics* **111** 125–34
[79] Adler S B, Lane J A and Steele B C H 1996 Electrode kinetics of porous mixed-conducting oxygen electrodes *J. Electrochem. Soc.* **143** 3554–64
[80] Nielsen J, Klemensø T and Blennow P 2012 Detailed impedance characterization of a well performing and durable Ni:CGO infiltrated cermet anode for metal-supported solid oxide fuel cells *J. Power Sources* **219** 305–16
[81] Nielsen J and Hjelm J 2014 Impedance of SOFC electrodes: a review and a comprehensive case study on the impedance of LSM:YSZ cathodes *Electrochim. Acta* **115** 31–45
[82] Shin E-C, Ahn P-A, Seo H-H, Nguyen D T, Kim S-D, Woo S-K, Yu J H and Lee J-S 2015 Pinning-down polarization losses and electrode kinetics in cermet-supported LSM solid oxide cells in reversible operation *Solid State Ionics* **277** 1–10
[83] Shin E-C, Ma J, Ahn P-A, Seo H-H, Nguyen D T and Lee J S 2016 Deconvolution of four transmission-line-model impedances in Ni-YSZ/YSZ/LSM Solid oxide cells and mechanistic insights *Electrochim. Acta* **188** 240–53
[84] Mohammadi R, Søgaard M, Ramos T, Ghassemi M and Mogensen M B 2014 Electrochemical impedance modeling of a solid oxide fuel cell anode *Fuel Cells* **14** 645–59
[85] Bisquert J, Grätzel M, Wang Q and Fabregat-Santiago F 2006 Three-channel transmission line impedance model for mesoscopic oxide electrodes functionalized with a conductive coating *J. Phys. Chem.* B **110** 11284–90
[86] Utz A, Leonide A, Weber A and Ivers-Tiffée E 2011 Studying the CO–CO_2 characteristics of SOFC anodes by means of patterned Ni anodes *J. Power Sources* **196** 7217–24
[87] Utz A, Störmer H, Leonide A, Weber A and Ivers-Tiffée E 2010 Degradation and relaxation effects of Ni patterned anodes in H_2–H_2O atmosphere *J. Electrochem. Soc.* **157** B920
[88] Joos J, Geisler H, Weber A and Ivers-Tiffée E 2016 Designing porous cathode structures for SOFCs *Proceedings of 12th European SOFC & SOE Forum 2016* ed N Brandon (Lucerne: European Fuel Cell Forum AG) 8 B0803: 29–35 https://www.efcf.com/efcf-library/proceedings-free/2016

Chapter 10

Long-term stack tests

Qingping Fang and Norbert H Menzler

Although long-term stack tests are time- and cost-intensive, they are important and necessary to understand the degradation behavior and predict the lifetime of solid oxide cell stacks, as degradation mechanisms/rates can change within the operating period, depending on the conditions. Therefore, short-term operation or accelerated degradation tests may easily mislead results and conclusions. In this chapter, long-term stack tests (more than 3000 h) of planar-type stacks from the last two decades are summarized and discussed, with a focus on the degradation behaviors of the stacks and the major degradation mechanisms observed during steam electrolysis, for both electrolyte- and fuel-electrode-supported cells. The complexity and difficulties of long-term stack tests, which may affect experimental results, are also discussed.

10.1 Introduction

A solid oxide cell stack is the core unit of a solid oxide cell (SOC) system. The stack consists of multiple single cells connected either in parallel or in series by means of a metallic or ceramic interconnect material. Solid oxide cells and stacks can generally be grouped into tubular and planar designs. The first development of solid oxide fuel cell (SOFC) and solid oxide electrolysis cell (SOEC) technologies can be traced back to the late 1950s and the 1960s, respectively [1–4]. Due to the difficulty of sealing and preparing planar cells with large areas and thin layers at high temperatures, tubular designs were preferred, and led the early development. With the ongoing R&D activities and achievements in thin-film technology and materials optimization, the working temperature could be decreased from over 1000 °C down to 800 °C and even lower, which tremendously decreased the technical hurdles for sealant and metallic interconnects. In particular, driven by the U.S. Department of Energy's (DOE's) Solid State Energy Conversion Alliance (SECA) program, dating from 2000 and aimed at developing a low-cost, large SOFC system [5], more attention returned to the planar design due to its high power density and diminishing manufacturing costs. Currently, with the exception

doi:10.1088/978-0-7503-3951-3ch10

of a few firms in the USA and Japan, most of the world's stack manufacturers are focusing on the planar design.

Although SOEC stack tests in both planar and tubular designs had already been demonstrated before 1980 [4, 6, 7], the major R&D activities of SOC technology focused on SOFCs, which aimed for high-efficiency power generation using carbon-containing fuels. It was only during the last two decades that the interest and R&D work on SOECs grew rapidly, as a result of increasing renewable power availability and the goals of reducing CO_2 emissions and dependency on fossil resources. Both the material and stack designs of SOECs have profited from the long-term development of SOFC technology. For these reasons, most SOFC stack components and designs can be directly used for SOEC operation. It is notable that SOCs are the only fuel-cell/electrolysis technology (compared to, e.g. alkaline or polymer electrolyte systems) that can be operated in both modes in a single system. On the one hand, many degradation mechanisms relating to chemical reactions inside the stacks have been extensively investigated for SOFCs and can be further applied to the SOEC mode. On the other hand, even though the stacks can be operated in the SOEC mode or in reversible mode, the existing design for SOFCs may not be particularly optimal for SOEC operation with regard to the different requirements for airflow and current density, for instance. Other degradation effects may arise for electrolytic operation that do not exist, or only exist to a very low degree, in fuel-cell mode. Currently, the state-of-the-art stacks may reach a voltage degradation of 0.3% kh^{-1} or even less in SOFC mode, but in the SOEC mode, degradation is generally greater than 0.4% kh^{-1} or 0.5% kh^{-1}.

In this chapter, the discussion of long-term stack tests focuses on planar stacks. Most of these were initially designed for SOFC operation.

10.2 General overview of the degradation tests of SOEC stacks

The degradation of SOCs is a complex scientific and engineering topic that incorporates chemistry, electrochemistry, materials science, thermal mechanics, mechanical engineering, etc. After decades of extensive investigations by researchers around the world, many degradation mechanisms have been identified and intensively studied [8–18]. In tables 10.1 and 10.2, general overviews of the major degradation mechanisms in SOC stacks with the most common components are summarized for both fuel-cell and electrolysis modes, including some concerns from the test periphery, which may in turn distort the test results. Depending on the test conditions and durations, different degradation mechanisms may dominate the stack behavior. Therefore, each degradation mechanism is important, and may play a critical role under certain circumstances. The ratings related to importance and probability are based on the experience gained from long-term stack tests conducted under a relatively wide range of working conditions, seeking to provide a rough and qualitative comparison of the mechanisms subjected to the testing conditions and testing periods thus far. Depending on the stack designs and testing conditions, the ratings can differ.

Most of the listed degradation phenomena can take place in both fuel-cell and electrolysis modes. For those with less dependence on electrical current and potential, the degradation behavior should be similar and comparably independent of the

Table 10.1. Overview of the degradation mechanisms in SOC stacks under normal/expected working conditions.

	Mechanism	Consequence	Dependency	Importance
Ni-based fuel electrode	Ni coarsening in the functional layer and substrate	Loss of triple-phase boundary	Temperature, steam content (also dependent on current density and utilization/ conversion), microstructure	+
	Ni migration at the interface between the electrolyte and the functional layer	Increase of ohmic resistance in functional layer	Temperature, steam content (also dependent on current density and utilization/ conversion), overpotential, microstructure, and composition	++
Zr-based electrolyte	Phase stability	Decrease in ionic conductivity	Temperature, local oxygen partial pressure	–
	Microstructural evolution (e.g., void formation along grain boundaries in SOECs)	Possible decrease in conductivity and mechanical stability	Temperature, current density	o
Ce-based barrier layer	Interdiffusion with electrolyte [19]	Interphase formation, increase in ohmic resistance	Sintering temperature	o
	Interdiffusion with air electrode elements [19]	Secondary phase formation, increase in polarization resistance	Sintering temperature, air electrode composition and stoichiometry, density of barrier layer	–
	Porous structure [19]	Formation of $SrZrO_4$—mainly during sintering	Thin-film preparation, sintering temperature	–

(*Continued*)

Table 10.1. (*Continued*)

	Mechanism	Consequence	Dependency	Importance
$La_{1-x}Sr_xCo_{1-y}Fe_yO_{3-\delta}$ (LSCF)/$La_{1-x}Sr_xCoO_{3-\delta}$ (LSC) electrode	Sr, Co segregation	Decrease in conductivity and phase stability, formation of $SrZrO_4$, $SrCrO_4$, and Co_3O_4	Temperature, humidity in air, microstructure, and composition	+
	Cr poisoning	Loss of triple-phase boundary and phase stability	Temperature, humidity in air, local current density and oxygen partial pressure, microstructure, composition, periphery	+
	Delamination in SOEC mode	Loss of contact with electrolyte	Temperature, local current density, microstructure, and composition	o
Ceramic contact layer	Cr getter	Enhanced contact (ohmic) resistance	Temperature, humidity in air, microstructure, and composition	–
	Contact loss with cell or interconnect	Power loss and risk of hot spot	Temperature, temperature gradient, preparation technique, microstructure, and composition	o
Protective coating on interconnect	Porous structure (Cr diffusivity) [20, 21]	High oxidation rate of steel and more Cr evaporation	Temperature, humidity in air, thickness, density, preparation technique, microstructure, and composition	+
	Delamination	High oxidation rate of steel and more Cr evaporation	Temperature, steel thickness and density, preparation technique, microstructure, and composition	–

Metallic interconnect	Oxidation on both fuel and air sides [22]	Increase in ohmic resistance and Cr evaporation	Temperature, humidity in air, steam content in fuel, steel thickness, surface treatment, coating	+
	Creep	Decrease in stability and risk of short circuit and contact loss	Temperature, temperature gradient, steel thickness	o
	Interdiffusion with Ni	Formation of austenite	Temperature, steel	–
	Evaporation of minor constituents	Poisoning of fuel electrode	Temperature, humidity, oxygen partial pressure, overpotential, interconnect composition*	o/+*
Glass sealant	Interaction with coatings	Risk of short circuit with decreased resistivity	Temperature, coating composition, glass composition	o
	Cracking	Leakage	Temperature, temperature gradient, thermal cycling, stack design, glass strength	o
Mica sealant	Interaction with steel	Accelerated corrosion	Temperature, steel	–
	Cracking	Leakage	Thermal cycle, clamping weight, and stack design	o

Table 10.2. Overview of the degradation mechanisms in SOC stacks under abnormal/unexpected working conditions.

	Mechanism	Consequence	Dependency	Probability
Ni-based fuel electrode	Poisoning (S, Si, Cl, Mn, …)	Loss of triple-phase boundary	Temperature, gas composition, gas/steam cleanness, periphery	++
	Carbon deposition	Loss of triple-phase boundary, blocking of gas transport	Temperature, steam/carbon ratio, local current density and utilization, electrode composition	+
	Redox	Loss of triple-phase boundary, microcracking of functional layer, cell fracture	Temperature, local current density and utilization, microstructure, and composition	+
Zr-based electrolyte	Mechanical imperfection	Increased risk of mechanical failure due to pinholes or microcracks	Production and assembly	–
$La_{1-x}Sr_xCo_{1-y}Fe_yO_{3-\delta}$ (LSCF)/ $La_{1-x}Sr_xCo_{1-y}O_3$ (LSC) electrode	Poisoning (Si, S, …)	Loss of triple-phase boundary and phase stability	Temperature, humidity in air, local current density and oxygen partial pressure, microstructure and composition, periphery	+
	Reduction (in case of leakage)	Electrode delamination	Temperature, local current density, microstructure, and composition	–
Contact between stack components	Contact loss or inhomogeneous contact	Risk of hot spot and wrong degradation calculation	Temperature gradient, production tolerance, stack concept	+

Testing periphery	Impurities (Cr, Si, Mn, …)	Poisoning of electrodes and possible clogged gas or water paths	Temperature, materials, media cleanness	+
	Unstable fuel or steam supply	Local starvation	Temperature, high utilization/ conversion, inhomogeneous contact	+
	Insufficient trace heating	Condensate in piping and unstable fuel or steam supply	High utilization/conversion, inhomogeneous contact	–
	Global leakage	Deviation of desired gas composition	High utilization/conversion rate, maintenance, sealing concept	–

operating mode (whether SOFC or SOEC). For others that have a strong dependency, the degradation behavior can substantially differ. For instance, although Ni relocation was recently observed following SOFC operation [23], it had never previously been considered or observed to be a noticeable degradation effect in SOFCs. However, Ni relocation, especially Ni depletion from the interface of the electrolyte and the functional layer, has proven to be a major degradation mechanism in SOECs for fuel-electrode-supported cells (FSCs) [24–26]. Another example is the poisoning effect of Si on the fuel electrode. Si was detected in the electrode after stationary SOFC stack operation [27]. It is also believed that Si deposits may degrade the charge-transfer and gas diffusion properties [28]. However, the poisoning effect is not comparable to that in SOEC mode, where severe degradation behavior can be attributed to Si impurities from either glass sealants or the testing periphery [29–31]. Nevertheless, the relatively mutual understanding of the degradation mechanisms in SOFC mode established a solid basis that facilitated the degradation analysis in SOEC mode.

As noted above, degradation is a complex topic. Each individual degradation mechanism might be simulated and easily investigated with a single cell measurement. However, the total degradation of a stack is not simply the mathematical sum of all the individual effects under simulated conditions, as the degradation may change when other degradation mechanisms or processes occur simultaneously, and the different processes may then compete with one another. For this reason, even though the long-term stack tests are generally time- and cost-intensive, they are always necessary for degradation analysis, especially when different degradation mechanisms may predominate under different conditions. A successful long-term stack test not only places a high demand on stack quality but also requires high reliability from the test facilities and infrastructures. Undesired interruptions from the testing environment may easily deteriorate the stack and cause misleading results. Skafte *et al* [13] and Küngas [32] summarized the long-term SOEC stack tests published in the last decade, in which it can be seen that most of the tests were carried out with limited stack types by a small group of institutions. The stack suppliers include the Commissariat à l'énergie atomique et aux énergies alternatives (CEA), Ceramtec, Forschungszentrum Jülich GmbH (FZJ), Fraunhofer Institute for Ceramic Technologies and Systems (IKTS), Ningbo Institute of Materials Technology and Engineering (NIMTE), Topsoe, Materials and Systems Research Inc. (MSRI), SOLIDPower, Sunfire, and Versa Power, whereas the testing institutions include the CEA, German Aerospace Center (DLR), Danish Technical University (DTU), European Institute for Energy Research (EIFER), Ecole Polytechnique fédérale de Lausanne (EPFL), FZJ, IKTS, NIMTE, Topsoe, Idaho National Laboratory (INL), Sunfire, Versa Power, and Technical Research Centre of Finland Ltd (VTT). Accordingly, all stack tests in steam electrolysis mode with an operating time of more than 3000 h are summarized in table 10.3. Each of the listed stacks is of the planar design. Clearly, Europe has been in the leading position in the field of SOECs for a decade as a result of the continuously increasing demand and capacity of renewable energy sources. Amongst others, Topsoe and DTU from Denmark made the most important contribution in the 2000s, and are still leading. In Germany, EIFER also started SOEC investigations in the 2000s, and carried out many valuable long-term tests for nearly all of the industrial partners (e.g., Topsoe,

Table 10.3. Overview of long-term stack tests in steam electrolysis mode with an operating time of more than 3000 h (based on the data of Küngas [32], with small corrections).

Stack	Cell type	Active area (cm^2/layer)	No. of layers	Testing period (kh)	Voltage degradation rate (% kh^{-1})	Year of publication	Reference
Sunfire (DE)	ESC	128	30	3.4	0.5	2019	[33]
IKTS (DE)	ESC	127	30	3.6[a]	0.5	2017	[34]
CEA (FR)	ESC	100	1[b]	3.6	1.8	2012	[35]
Topsoe (DK)	ESC	100	5	4.1	5.6	2011	[36]
Sunfire (DE)	ESC	128	30	4.8	0.7	2019	[37]
Sunfire (DE)	ESC	128	5	7.0	1.5	2016	[38]
Sunfire (DE)	ESC	128	30	8.5	0.4~0.8	2021	[39]
CEA (FR)	FSC	100	5	3.0	8	2011	[40]
FZJ (DE)	FSC	80	2	3.0	0~1.5	2013	[41]
Versa Power (US)	FSC	120	28	5.0	–	2012	[42]
Versa Power (US)	FSC	120	6	5.8	3.1	2012	[42]
Topsoe (DK)	FSC	88	25	9.0	2.2	2015	[43]
SOLIDpower (IT)	FSC	48	6	10.7	0.5	2015	[44, 45]
FZJ (DE)	FSC	80	2	20.0	0.6	2018	[25, 46, 47]

[a] including 1500 h of co-electrolysis.
[b] single cell with stack repeating unit.

SOLIDPower, and Sunfire). The major stack developers in Germany (e.g. the FZJ, IKTS, and Sunfire) focused only on SOFCs before they became involved with SOECs in the 2010s. Note that only steam electrolysis is specified in table 10.3, without considering co-electrolysis and pure CO_2 electrolysis. Therefore, table 10.3 cannot be used to fully reflect the current activities of the listed institutions.

Both electrolyte-supported cell (ESC) stacks and FSC stacks are presented in table 10.3. In the next section, the performance data and major findings of the selected tests are summarized and discussed. Note the calculated voltage degradation rates are only listed in table 10.3 to give a rough impression of the degradation behavior and should not be used for comparisons of different stack designs, since the voltage degradation rate is strongly dependent on the operating parameters.

10.3 Long-term SOEC stack tests

10.3.1 Stacks of ESC cells

One of the earliest stack tests with ESC cells was conducted by EIFER using a five-layer Topsoe stack with coated Crofer22 APU as the interconnect [36], as shown in figure 10.1. The cells were fabricated by ECN (NL) and consisted of a 3mol% yttria-stabilized zirconia (3YSZ) electrolyte, a Ni/ceria gadolinia oxide (GDC) fuel electrode, and an $La_{1-x}Sr_xCo_{1-y}Fe_yO_{3-\delta}$ (LSCF) oxygen electrode. The stack was tested in a furnace at about 820 °C using a feed gas of 90% H_2O in H_2 with a current density of -0.4 A cm^{-2} and a steam conversion rate of 39%. The cell voltages increased from 1.2 V to ~1.6 V after 4055 h, resulting in an average voltage degradation rate of 5.6%

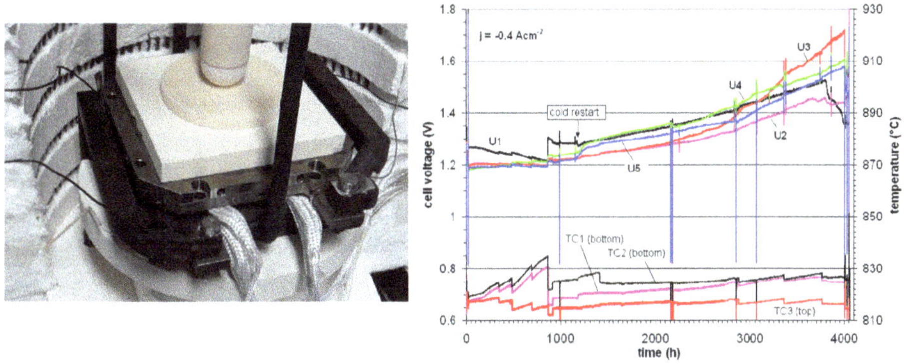

Figure 10.1. Five-layer Topsoe stack tested at EIFER (reproduced from [36] with permission, copyright The Electrochemical Society 2005).

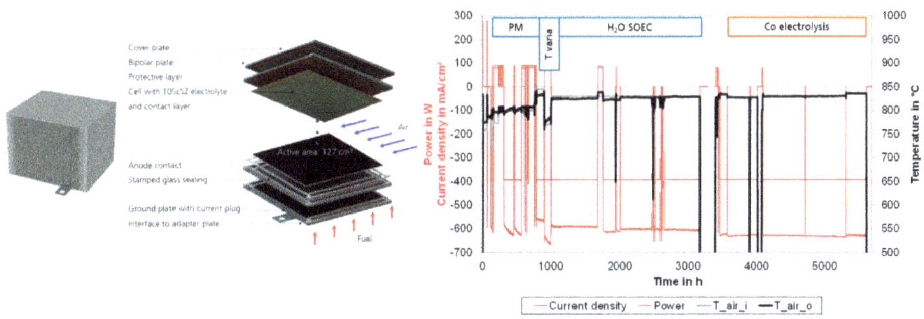

Figure 10.2. MK352 stack design by IKTS and a long-term test with a ten-layer MK352 stack (reproduced from [34] with permission, copyright The Electrochemical Society 2005).

kh^{-1}. However, the degradation rate was not constant and increased with time. Impedance measurements during the last 1000 h of operation indicated that the degradation was dominated by an increase in ohmic resistance. The post-mortem analysis was difficult due to the unexpected dry operation under the electrolysis current, which led to a cell voltage of more than 2.3 V and a reduction in the 3YSZ electrolyte. However, even under such extreme conditions, the adhesion between the LSCF and 3YSZ was found to be intact.

IKTS performed a long-term test using a ten-layer MK352 stack in both steam- and co-electrolysis modes [34]. The stack consisted of a coated CFY interconnect and 10ScSZ cells with IKTSG3 electrodes (i.e. LSMM′/ScSZ and Ni/gadolinium-doped ceria (GDC) electrodes [48], where LSMM′ stands for lanthanum strontium manganite with an additional transition metal at the B site). The stack was tested in a furnace at about 820 °C using 80% steam in H_2 with a current density of -0.39 A cm^{-2} and a steam conversion rate of 75%. After 2100 h of steam electrolysis, the feed gas was changed to $H_2O/CO_2 = 2$ for another 1500 h, as shown in figure 10.2. In the given constant-current mode, the power degradation rate was 0.5% kh^{-1} for both steam- and co-electrolysis operation, which corresponds to 17 $m\Omega\,cm^2\,kh^{-1}$. This degradation rate was comparable to that of a 30-layer stack after 20 000 h of SOFC operation in a hotbox (20 mΩ cm^2 kh^{-1}) [49], which means that no major degradation difference was observed between the fuel-cell and electrolysis modes within the given testing period.

Sunfire's stacks generally consist of coated Crofer22 APU interconnects and 3YSZ cells with Ni/GDC and LSCF electrodes. Several long-term tests have been conducted in recent years by different institutions [33, 37–39, 50], which have shown stable and reproducible performance in both SOFC and SOEC modes. An ongoing (status as of 13/10/2021) SOFC test with a 30-layer SMK-B230-type stack, operating with H_2 at a current density of –0.226 A cm^{-2} and a fuel utilization of 75%, has been operating in a hotbox for more than 23 000 h. The linear area specific resistance (ASR) degradation rate was ~17 mΩ cm^2 kh^{-1} for 16 000 h and has been ~11 mΩ cm^2 kh^{-1} during the last 7000 h of operation [39]. A 30-layer stack of the same type was tested at EIFER in a furnace at ~820 °C using 90% steam in H_2 at a current density of –0.5 A cm^{-2} and a steam conversion rate of 70%. The linear degradation rate was 18 mΩ cm^2 kh^{-1} after 4800 h of operation [37], as shown in figure 10.3. A 30-layer stack of the latest SMK-B240 design was tested at Sunfire in a hotbox [39]. The evolution of the average ASR is shown in figure 10.4. The stack was tested under different current densities ranging from –0.25 to –0.65A cm^{-2} (from endothermic to exothermic), while maintaining a steam conversion rate of 70%. Hot-standby, thermal cycles, and the redox cycle were also performed to simulate realistic working conditions. The linear degradation rate of ~21 mΩ cm^2 kh^{-1} for the total operation of 8500 h was slightly higher than the previous results, which was assumed to be due to the small amount of Si deposition in the fuel electrode, as confirmed by the post-mortem analysis. A similar degradation rate was also observed in co-electrolysis mode [39].

Although Topsoe has worked with ESCs, their main focus is FSC technology. IKTS and Sunfire are considered to be the state-of-the-art European ESC stack developers. According to the available long-term test results (up to 23 000 h for SOFCs and 8500 h for SOECs), stacks of ESCs did not exhibit noticeable differences in degradation behavior between the SOFC and SOEC modes. As the thick electrolyte

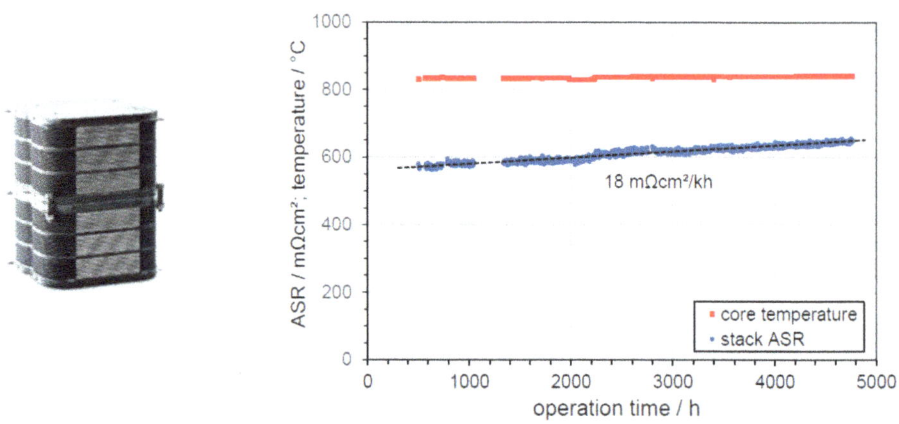

Figure 10.3. SMK-B230-type stacks from Sunfire and a long-term test with a 30-layer SMK-B230-type stack in a furnace at EIFER (reproduced from [37] with permission, copyright The Electrochemical Society 2005).

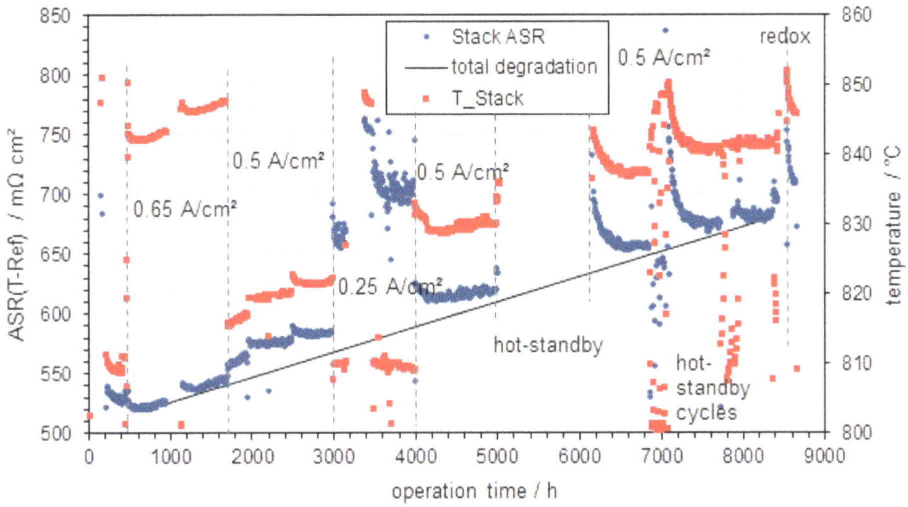

Figure 10.4. Long-term test with a 30-layer SMK-B240-type stack in a hotbox at Sunfire (reproduced from [39] with permission, copyright Sunfire, Dresden, Germany, personal communication).

makes the major contribution to the total ASR, any possible differences in the degradation behaviors of the electrodes may not be as visible, as in the case of FSCs.

10.3.2 Stacks of FSC cells

Versa Power Systems (VPS) has been developing SOFCs since 1997. They also developed cells and stacks for reversible solid oxide cells (rSOCs) in the 2000s. Within the framework of the DOE project 'Advanced Materials for RSOFC Dual Operation with Low Degradation' (2009–11), VPS presented several long-term stack tests [42]. Amongst others, six- and a 28-layer KW-class stacks were tested at ~750 °C with 50% H_2O in 50% purge gas (4% H_2 in N_2) for more than 5000 h. The current density and conversion rate were mainly kept at −0.39 A cm^{-2} and 50%, respectively. No details regarding the cell and stack components were noted. Both stacks exhibited a voltage degradation rate of ~3% kh^{-1}. VPS was taken over by FuelCell Energy in 2012. In the DOE project, 'Solid Oxide Based Electrolysis and Stack Technology with Ultra-High Electrolysis Current Density (>3 A cm^{-2}) and Efficiency' (2015–17), a 20-cell stack with an active area of 22.3 cm^2, consisting of high-power-density cells, was tested at 750 °C for 1265 h at −2 A cm^2 and 50% steam utilization. The degradation rate was less than 0.7% kh^{-1} [51].

In Europe, one of the earliest long-term stack tests with FSC cells was again carried out at EIFER using a five-layer stack made by Topsoe [52]. The cells consisted of an 8mol% yttria-stabilized zirconia (8YSZ) electrolyte with Ni/yttria-stabilized zirconia (YSZ) and $La_{1-x}Sr_xMnO_3$ (LSM)/YSZ electrodes. The stack could be tested for 2600 h at different current densities between −0.4 and −0.8 A cm^{-2} with steam conversion rates between 39% and 58%. However, due to the inconsistent temperature variations caused by the unstable end-plate contacts, a reliable degradation rate was difficult to calculate.

CEA presented a stack test that took place over more than 3000 h using a five-layer stack with a 10 mm-thick Crofer22 APU interconnect and a mixed sealant of Macor® and glass [40]. The cells consisted of an 8YSZ electrolyte with Ni/YSZ and Yttria-doped ceria (YDC)/LSCF electrodes. The endurance test was carried out at 800 °C in a furnace with 90% H_2O in H_2. The current density and conversion rate were kept at -0.5 A cm^{-2} and 25%, respectively. A high average voltage degradation rate of ~10% kh^{-1} was measured. Both the ohmic and polarization losses increased, possibly indicating a growing contact issue during operation.

Using a 25-layer stack from Topsoe, EIFER performed another stationary test for more than 9000 h [43]. Figure 10.5 shows the stack and the evolution of its voltage. The cells consisted of an Ni/YSZ fuel electrode, a YSZ electrolyte, and a $La_{1-x}Sr_xCo_{1-y}O_3$ (LSC) oxygen electrode, each of which had an active area of 87.7 cm^2. The stack was tested in a furnace at 750 °C with 90% H_2O in H_2 and a constant steam conversion rate of 50%. The current density was increased from -0.57 A cm^{-2} to -0.72 A cm^{-2} after an initial 2175 h of operation. Due to an interruption in the DC power supply, the stack was kept at the open-circuit voltage (OCV) for roughly 500 h before electrolysis could be restarted. The overall operational hours at the two current densities were 2528 h and 5640 h, respectively. The average linear voltage degradation rate was ~2% kh^{-1} for both current densities. No further discussion regarding the degradation mechanisms was presented. However, this test is an invaluable example that demonstrates the complexities and difficulties of performing long-term stack tests and degradation analysis.

Stacks, especially large ones, feature intrinsic inhomogeneity in their performance due to inhomogeneous distributions of temperature and gas flows. This inhomogeneity is undesirable, but can be tolerated by choosing appropriate operating parameters. The inhomogeneity and uncertainty of the contact within the stack is, in most cases, critical to stack performance and stability, as the layers with the worst or unstable contacts generally determine the performance of the entire stack. For instance, as is shown in figure 10.5, the stack temperature increased rapidly during the first 2000 h of operation, but was relatively stable afterwards. This was due to a

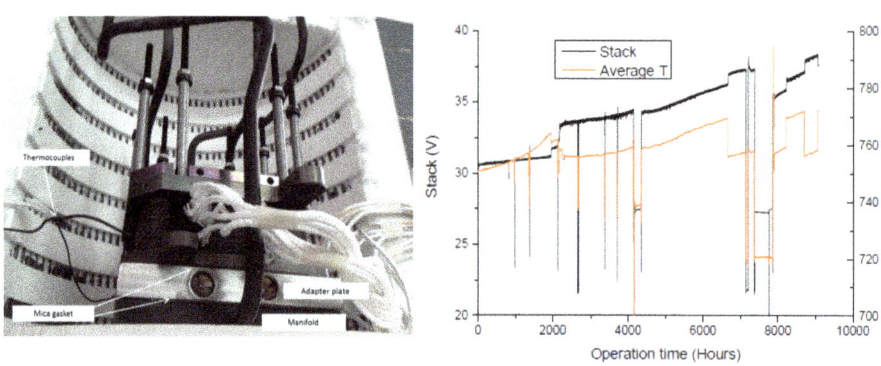

Figure 10.5. A 25-layer Topsoe stack was tested at EIFER for over 9000 h (reproduced from [43] with permission, copyright The Electrochemical Society 2015).

contact issue with the stack end plate. The increased temperature had an influence on the adjacent layers and in turn on the overall degradation behavior.

Due to different factors, the stack's degradation rate may not be constant during operation. Short-term fluctuations or variations in the stack's performance can distort the degradation rate if the testing period is insufficiently long. Note that the increase in the stack voltage accelerates after ~3500 h in figure 10.5. Long-term tests can be interrupted by various unexpected events, including:

- Malfunctions or defects in the test facilities, including electronics, valves, MFCs, pumps, evaporators, heaters, trace heating, etc.
- Interruption of the infrastructure, including electricity, gas, deionized water, cooling water, exhaust, safety monitoring, etc.
  Unstable wateror steam supply, especially at high conversion rates
- Maloperation or programming failures

Another long-term test was performed at EIFER using a six-layer stack made by SOLIDpower [44, 45]. The cells had an active area of 48 cm^2 and the typical YSZ electrolyte (Ni/YSZ), a GDC compatibility layer, and an LSC electrode. The stack was tested in a furnace with 90% H_2O in H_2. The air outlet temperature was kept between 710 °C and 730 °C. The evolution of the cell voltages, as well as some other parameters, are displayed in figure 10.6. The test began at a current density of −0.6 A cm^{-2} and a steam conversion rate of 50%. After a water supply interruption at 3250 h, the current density and steam conversion rate were reduced to −0.5 A cm^{-2}

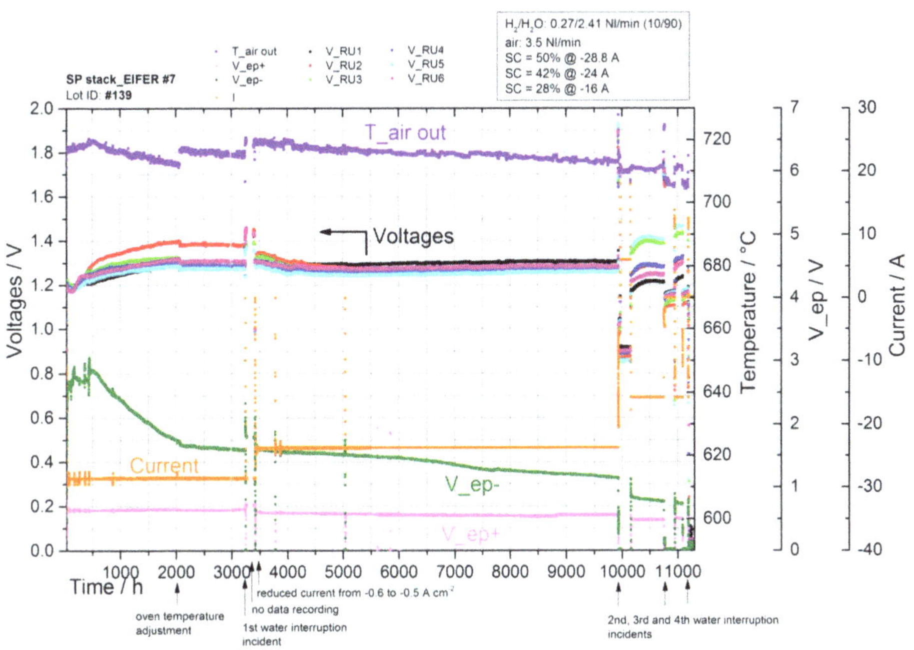

Figure 10.6. A six-layer SOLIDpower stack tested at EIFER for over 10 000 h (reproduced from [44, 45] with permission, copyright 2017 John Wiley & Sons).

and 42%, respectively. Water supply interruptions occurred three further times after 10 000 h, which caused remarkable degradation, especially in repeating units 3 and 5. The stack indeed exhibited good performance and stability and all layers showed similar voltages. The voltage degradation rate between 5000 h and 10 000 h was less than 0.5% kh^{-1}, and could have been even lower had the temperature been kept constant. On the other hand, the stack behavior was affected by the unstable contact of the negative end plate, which improved throughout the entire operation, as can be seen from the voltage V_ep− line in figure 10.6. This was also thought to be the reason for the decreasing temperature at the air outlet. Following the test, the cells from repeating units 1 and 4 were intensively analyzed. The major findings of this analysis can be summarized as follows:

- Ni depletion had occurred in the electroactive region of the fuel-electrode functional layer (to a depth of 5 μm)
- Void formation had occurred in the electrolyte along and inside the YSZ grain boundaries, especially in proximity of the GDC
- A dense layer filled with $SrZrO_3$ had formed between the YSZ and the GDC
- Si had accumulated at the interface between the Ni/YSZ electrode and the YSZ
- Delamination during the operation could not be confirmed

The longest SOEC test with a planar stack was conducted in a furnace at the FZJ using a two-layer F10-design stack [25]. The stack consisted of Crofer22 APU interconnects with an $MnCo_{1.9}Fe_{0.1}O_4$ (MCF) coating applied by atmospheric plasma spraying (APS). The cells featured the general components of FSCs, including an 8YSZ electrolyte (Ni/YSZ), LSCF electrodes, and a GDC barrier layer. In order to monitor the stack temperature, two thermocouples were inserted into the thick interconnect between the two layers, as shown in figure 10.7. The stack was tested with 50% H_2O in H_2 at a current density of −0.5 A cm^{-2} and a steam conversion rate of 50%. During the first 8000 h of operation, the stack was operated in electrolysis mode at three temperatures (namely 800 °C, 700 °C, and 750 °C),

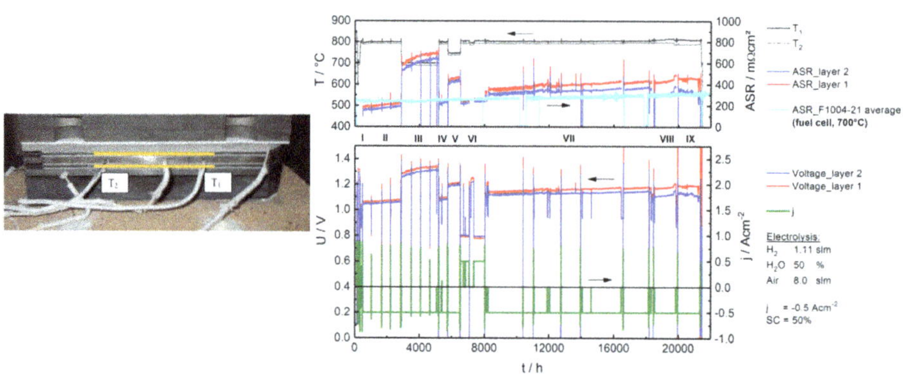

Figure 10.7. A two-layer F10-design stack was tested at the FZJ for over 20 000 h (reproduced from [25, 46] with permission, copyright The Electrochemical Society 2015/2018).

followed by a short period of operation in fuel-cell mode with the same current density and fuel utilization. Starting from 8000 h, the stack was kept in electrolysis mode at 800 °C. The evolution of the cell voltages and ASRs is shown in figure 10.7.

The voltage degradation rate during the first 2400 h of operation at 800 °C was 0.7% kh^{-1}. At 700 °C, a progressive increase in both voltages and ASRs could be noticed for more than 1000 h. The linear voltage degradation rate over 2300 h at 700 °C was 1.9% kh^{-1}, while the ASR degradation rates at both temperatures were in the same range of 9–10% kh^{-1}. Repeated operation at 800 °C directly after that at 700 °C confirmed that there was no accelerated degradation at lower temperatures within the same testing period. The short operation at 750 °C exhibited similar behavior to that at 700 °C. However, the period of progressive increase in voltages and ASRs seemed to be shorter. During 1500 h of fuel-cell operation between 6000 h and 8000 h, little degradation was observed. However, small voltage losses were noticed as a result of some unplanned interruptions. Starting from ~18 000 h, layer two suffered from an internal leak, which led to local overheating on the air outlet side of the stack and finally blocked the air channels. This was confirmed by the post-mortem analysis, and explained the temperature rise after 18 000 h and the large diffusion polarization on the air side observed by the distribution of relaxation times (DRT) analysis. The average linear degradation rates of the stack for the operation period of ~18 000 h before the severe leakage in layer two started, including the losses during all unplanned load cycles and the intermediate fuel-cell operation, were 0.6% kh^{-1} and 8.2% kh^{-1} for the voltage and ASR degradations, respectively, at 800 °C. Impedance measurements near the OCV condition were performed regularly throughout the entire operation. It can be seen from the Nyquist plots that the degradation up until 20 000 h was dominated by a continuous increase in the ohmic resistance, whereas the electrode polarization increased only slightly. According to the DRT analysis, this slight growth in the electrode polarization could be attributed to the air electrode. In figure 10.7, the evolution of the ASRs is also compared to that of a stack with the same type of cell but operated in SOFC mode at a furnace temperature of 700 °C for 34 000 h [53]. In SOFC mode, the stack clearly exhibited lower degradation rates (0.3% kh^{-1} and 1% kh^{-1} for the voltage and ASR degradation rates, respectively), indicating different dominant degradation mechanisms for SOFC versus SOEC operations. The precise reason for the increased leakage in layer two after 18 000 h will be discussed in the next section, but is difficult to confirm. However, several unplanned interruptions at ~18 000 h made the performance of layer two even worse. The reasons for the interruptions during the entire operation include:

- Defects in the evaporator
- Defects in the trace heating
- Defects in the computer
- Defects in the electrical load
- Defects in the pump for deionized water
- Defects in the temperature display with the safety function
- Emergency shutdown of the exhaust system
- An accident in the cooling water supply
- Maloperation

Amongst other problems, the evaporator had to be exchanged and cleaned nearly every half year due to choking. It is highly likely that this choking was due to impurities in the water, including silica. For this reason, it seemed that the stack did not suffer from Si poisoning, which generally leads to a high degradation rate in the fuel electrode.

10.4 Degradation mechanisms

Depending on the stack and cell components and designs, as well as the operating conditions, the long-term performance of stacks can be affected by different degradation phenomena and mechanisms. Even for the same stack, the major degradation mechanism can differ during different operating periods. Based mainly on the long-term test results, the major degradation issues under 'normal' (i.e. no failure or malfunction of balance-of-plant (BoP) components) operating conditions are summarized and briefly discussed below.

10.4.1 Oxidation of the interconnect

Unfortunately, oxidation of the metallic interconnect cannot be avoided at working temperatures, especially under dual atmospheres. Until now, the major effects aiming on Cr retention by applying a protective layer to the air side of the chromia-forming alloys such as Crofer22 APU, CFY, ITM, Sanergy®HT, SS441, etc. With an MCF coating on Crofer22 APU, the voltage degradation rate of Jülich's F10-design short stacks could be reduced from 1% kh^{-1} to 0.3% kh^{-1} at a current density of 0.5 A cm^{-2} in a temperature range of 700 ~750 °C for at least ~30 000 h (in SOFC mode). Correspondingly, the ASR degradation rate was reduced from ~20 mΩ cm^2 kh^{-1} to less than 5 mΩ cm^2 kh^{-1} [53–56]. The protective function of the MCF coating was also proven by an 18-layer F20-design stack with an active area of 360 cm^2 [57]. With other types of coating and stack design, metal oxidation and increases in interfacial resistance due to the growth of oxide scales can effectively be reduced [34, 58–60]. A stack test for 100 000 h demonstrated a clear reduction in the degradation rate after 40 000 h, which may suggest that oxide scale growth can reach a kinetic equilibrium between Cr oxidation and Cr evaporation [56].

Up until now, no evidence has been amassed indicating that the oxidation of the metallic interconnects on the air side behaves differently in SOEC mode. An increased oxygen partial pressure can, in principle, promote Cr evaporation. However, this effect is as yet not noticeable in stack tests, as air is still used for flushing in most cases. On the fuel side, oxidation can be accelerated under high humidity and higher oxygen partial pressures, and this may play a more important role with increasing operating time, especially when the degradation of the cells is further reduced in the future. Considering the possible release of Mn and thus the risk of fatal consequences for cell performance [23, 47, 61–63], a protective coating on the fuel side of the interconnect would be important and necessary to enable a long lifetime of more than 40 000 h.

10.4.2 Degradation of the YSZ electrolyte

A single FSC from the FZJ was tested at EIFER at −1 A cm^{-2} for 90 000 h [64]. A detailed post-mortem analysis revealed an enhanced fracture mode along the grain boundaries due to void formation. Horizontally structured pores at the nanometer scale could be observed over the entire electrolyte at the GDC/YSZ interface [65], as shown in figure 10.8. A similar phenomenon was also observed in the stack after 10 700 h of operation (figure 10.6) [45] and tests at high current densities [66].

The increased porosity may decrease the mechanical stability and ionic conductivity of the YSZ. The influence of degradation on the ionic conductivity cannot yet be separated from general contact loss in the stack and the effect of Ni depletion. However, this effect can be further investigated using a single cell measurement and Ni-free electrodes. Weakness in the mechanical stability can hardly be realized and characterized during operation. However, the risk of mechanical failure grows with increasing operating times. Sudden failure could be more easily induced by unplanned

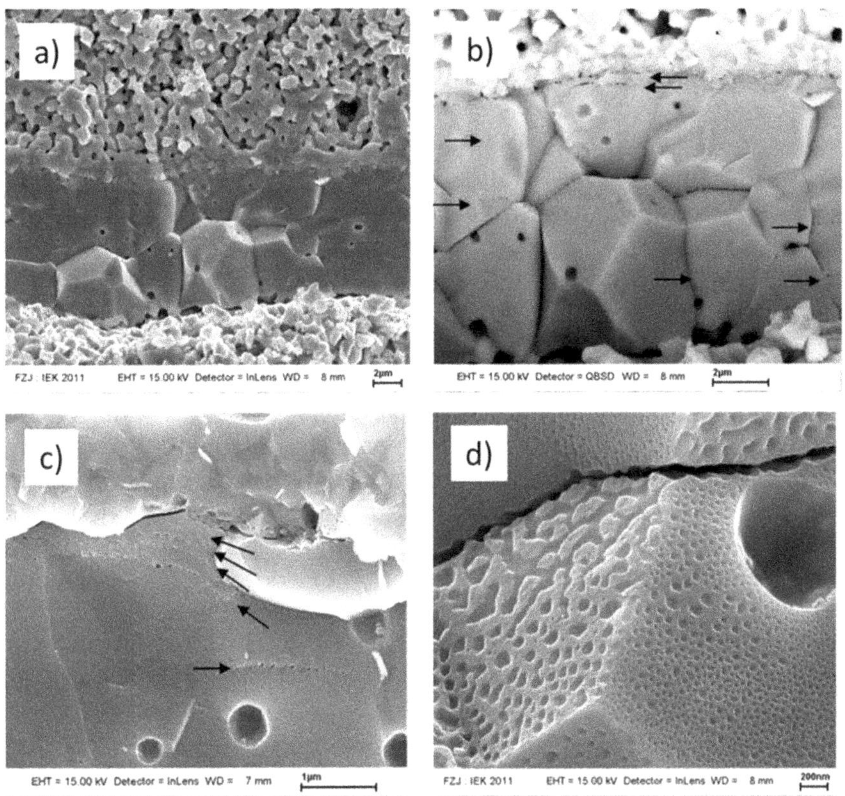

Figure 10.8. Images of fracture surfaces of the electrolyte layer after 9000 h of the single cell test. (a) Area of inter- and intragranular fracture; (b) the same region in back-scattering mode shows the horizontally structured pores (arrows) more clearly over the entire electrolyte layer; and (c) at the GDC/YSZ interface; and (d) a highly enlarged 8YSZ grain surface with a fractured grain boundary (reproduced from [65] with permission, copyright *Journal of Power Sources*, Elsevier 2013).

impacts and interruptions, which are generally coupled with fast variations in temperature and pressure gradients within the stack. For this reason, operation near the thermoneutral voltage may have the advantage of reducing the mechanical impact in the case of interruptions, especially after long operation. It should also be mentioned here that the cell suffered from an interrupted steam supply at the end of operation [64], which may have had an impact on the observed microstructure and defects.

Another possible degradation mechanism related to Mn diffusion and oxidation in the YSZ electrolyte has not attracted much attention thus far, as this can only happen in stacks after extremely long periods of operation. Mn was found at the interface of the fuel electrode and electrolyte after long-term SOFC stack operation, and was assumed to be the reason for the observed progressive degradation after ~23 000 h of operation in two stacks [62, 63]. Tiny amounts of Mn were also identified at the interface after 100 000 h of operation, but without a noticeable degradation effect [23]. The main reason for the latter is that the amount of Mn in the interconnect used for this stack is only about one tenth of the amount found in the usual materials. Under fuel-cell conditions, Mn can diffuse though the gas phase to the YSZ/fuel-electrode interface and form an oxide, leading to internal stress, microcracking, and possibly, finally, electrolyte cracking. In the stack used for the 20 000 h SOEC test, some pore features could be found in the air electrode. Further analysis of the pores indicated the presence of MnOx deep within the YSZ and on its surface. MnOx was also detected in the crack of the electrolyte [47]. As noted in the previous section, layer two of the stack started to exhibit increasing leakage rates and degradation after ~18 000 h. Any correlation between the observed degradation and Mn diffusion in YSZ is only speculation and cannot be proven yet. The overall interaction between the Mn and the fuel-side cell materials must be clarified in the future.

10.4.3 Degradation of the LSC(F) air electrode

Most long-term tests are assumed to have been conducted with synthetic or compressed air, which contains fewer impurities and humidity than ambient air. The possible poisoning effects on the air electrode (except for the effect of Cr) have therefore not been remarkable thus far. It is known that LSC(F) electrodes can also be poisoned in SOFCs by S, B, and Si [67, 68]. Purging with CO_2 instead of air has led to higher degradation in the electrode in SOEC mode [69]. Delamination of the air electrode during electrolysis is well known in LSM electrodes [70]. With increased mixed-conduction and ceria-based interdiffusion layers, the LSC(F)-based electrode exhibited much less of an issue with delamination [71]. However, delamination of the air electrode could still be observed after long-term operation [25, 45, 47]. A possible weakness near the GDC/YSZ interface due to void formation along the grain boundaries in YSZ, which could lead to delamination, has been reported after single cell and stack tests [24, 45, 65]. In this case, it may look like delamination of the air electrode, as the GDC layer is very thin.

The high activity of Sr in LSC(F) is one of the causes of degradation, such as the formation of $SrCrO_4$ and $SrZrO_4$. It has been proposed that the depletion of oxygen vacancies under electrolytic polarization could drive Sr release from the LSCF structure [72], which could facilitate the reaction with Cr on the electrode surface and Zr below the (porous) GDC layer during operation. The effect of the formation of $SrCrO_4$ and $SrZrO_4$ during electrolysis on ohmic resistance may not be significant within the current testing periods compared to other contributions. Changes in stoichiometry and structure due to Sr and Co segregation may decelerate the electrode reactions. As can be seen from the DRT analysis for the stack with 20 000 h of operation, the air electrode polarization continuously increased over time [25].

10.4.4 Degradation of the Ni-based electrode

10.4.4.1 Ni depletion

A comparison between stacked fuel electrodes after 20 000 h of SOEC operation at 800 °C [25] and those of the same type of cell after 34 000 h of SOFC operation at 700 °C [53] showed obviously different porosities in the fuel electrode's functional layer close to the YSZ. Energy-dispersive X-ray spectroscopy (EDS) mapping confirmed nearly complete depletion of Ni in this region, leaving a porous YSZ structure behind [47]. The Ni depletion zone was thicker on the fuel inlet side than on the outlet side. Similar phenomena were also observed in most of the long-term tests that used FSCs [24, 26, 45] (figure 10.9).

Although Ni depletion has been widely accepted as one of the major degradation mechanisms for FSCs in electrolysis mode, Ni migration in the opposite direction has also been reported in both electrolysis [73] and fuel-cell modes [23]. Therefore, it could be more suitable to refer to this type of degradation as Ni migration rather than depletion. However, Ni enrichment near the electrolyte was only reported at higher temperatures of more than 900 °C. For state-of-the-art FSCs with LSC- and LSCF-type electrodes, the working temperatures are generally below 900 °C or 850 °C. For this reason, the term 'Ni depletion' is still used in this chapter instead of 'Ni migration.'

There are different theories for Ni depletion, including the formation of $Ni(OH)_2$ [74], the contact angle and interface free energy between Ni and YSZ [75], the

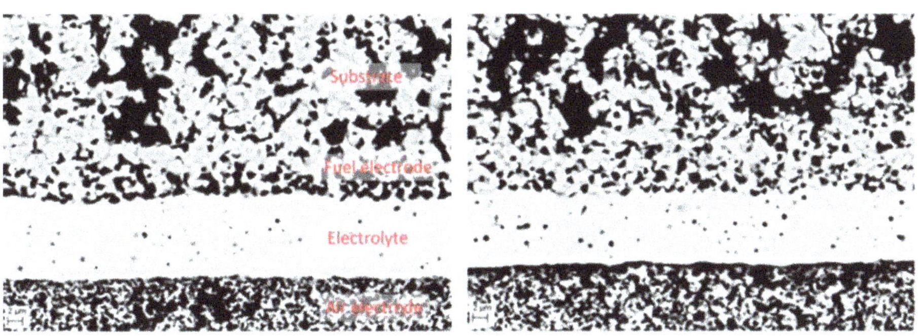

Figure 10.9. SEM cross-sections of (left) cell 1 after 20 000 h of SOEC operation at 800 °C [25]; and (right) the same type of cell after ~34 000 h of fuel-cell operation at 700 °C [53], which display obvious differences in the region's porosity at the electrolyte/fuel-electrode interface (reproduced from [25] with permission, copyright 2018 The Electrochemical Society).

accumulation of vacancies at high cathodic overpotential [76], etc. Recently, Mogensen *et al* [17, 77] conducted extensive reviews of Ni migration, seeking to develop a unified migration mechanism to explain different Ni migration behaviors under different conditions. Their hypothesis asserts that Ni migration is, through the movement of $Ni(OH)_x$ or other similar compounds with Ni, in a positive oxidation state (e.g. NiOH dominates in the SOEC mode; alternatively, $Ni(OH)_2$ dominates in the SOFC mode) driven by the local electrical potential difference caused by the loss of contact between Ni–Ni (due to Ni coarsening) and Ni–YSZ (due to a strong cathodic overpotential [75]). According to this hypothesis, maintaining contact is essential for preventing Ni migration, regardless of the working mode and conditions. Gas-phase transport alone is thermodynamically unlikely, or cannot explain the high depletion rate over the operating time.

10.4.4.2 Impurities

Impurities in water are not only harmful for the electrochemical performance of the cell but also critical for testing facilities such as pumps and evaporators. Amongst other factors, Si poisoning has been recognized as one of the major degradation mechanisms which is more severe than the similar poisoning effect that occurs in SOFC mode. This can probably be explained by reference to the equilibrium between $Si(OH)_4$ and SiO_2. In the case of SOFC operation, the generated H_2O can promote the formation of gaseous $Si(OH)_4$ [17, 29]. The sources of Si-containing impurities may include raw materials and Si-containing gaskets [29]. Recently, Schäfer *et al* [31] investigated the purities of water and steam at different positions from a water purification system to the stack inlet, and found the following:

- Species such as silicates and borates may easily pass through mixed-bed ion exchangers. The background concentrations can be tens of $\mu g\ L^{-1}$.
- The carryover of impurities in boilers/evaporators takes place through vaporous and mechanical carryover.
- An overloaded steam generator greatly increases the mechanical carryover.
- A used steam generator containing deposits can transfer vast amounts of contaminants (>500 $\mu g\ L^{-1}$).
- Moderately acidic feed water combined with unstable operation can cause significant releases of metals from the evaporator.
- Exponential increases of Si, As, Mo, B, and Sb concentrations in commercial gas preheaters can occur at high temperatures of more than 650 °C. The poisoning effect can be seen in a ten-layer F20-design stack operated at higher fuel gas preheater temperatures, as shown in figure 10.10. This phenomenon could be reproduced with two other stacks. What must still be investigated is (1) whether Si is the only critical impurity and (2) the mechanism that lies behind the (partially) reversible poisoning effect.

10.4.4.3 Void formation and delamination

As noted in sections 10.4.2 and 10.4.3, void formation has been reported in the grain boundaries of the YSZ electrolyte near the GDC/YSZ interface, which may

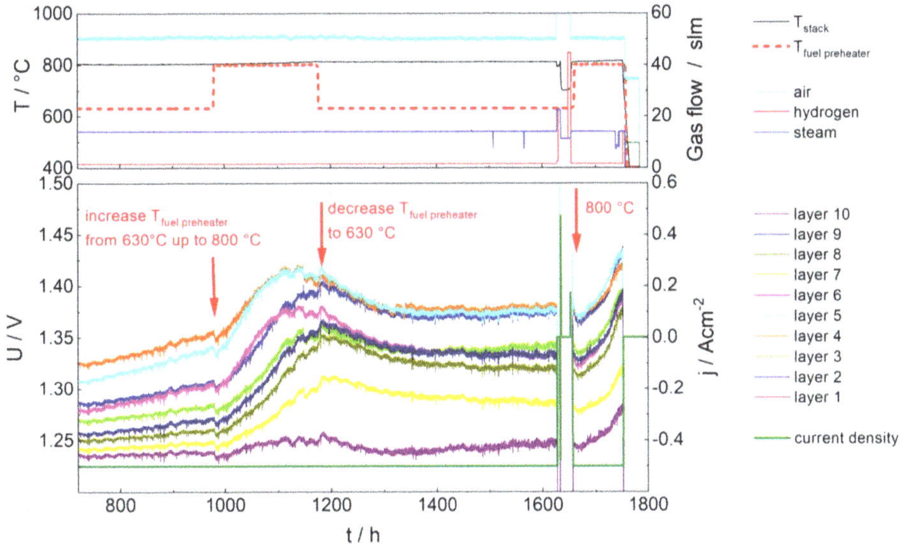

Figure 10.10. Behavior of a ten-layer F20-design stack, showing the degradation dependence on the temperature of the fuel gas preheater (reproduced from [31] with permission, copyright 2020 European Fuel Cell Forum EFCF).

deteriorate conductivity and mechanical stability. Delamination of the air electrode with or without GDC may occur as a result of the weakened bonding. On the fuel side, the main degradation mechanism is centered on Ni depletion or migration. In the case of Ni depletion from the functional layer, the porous YSZ structure behind it can increase not only the electrical resistance but also the instability at the interface. Poor contact at the electrolyte–fuel-electrode interface was noted in the cell after 9000 h of operation [65]. Delamination of the fuel electrode following electrolysis has not been reported thus far. However, a post-mortem analysis of more of the stacks operated at the FZJ revealed poor bonding at this interface in the cells with higher degradation rates. One example of a cell after ~4000 h of co-electrolysis operation under relatively mild conditions at 800 °C ($|J| \leqslant 0.7$ A cm^{-2}, H_2O:CO_2:H_2 = 6:3:1, conversion rate: $\leqslant$70%) is shown in figure 10.11. Of course, the possibility of such delamination occurring during sample preparation cannot be excluded entirely. Nevertheless, the potential risk and possibility of fuel-electrode delamination should also be kept in mind for the degradation analysis.

10.4.5 Degradation due to contact in stacks

The influence of the contact situation inside the stack on stack performance was also discussed in section 10.3.2. The stack tests shown in figures 10.5 and 10.6 were affected by unstable contact with the end plate. In many other cases, the influence of the contact is not as visible, as displayed in the figures. Therefore, this can lead to the derivation of the wrong degradation rate or incorrect conclusions about the stack. If the contact area is smaller than designed or expected for any reason, the cell and

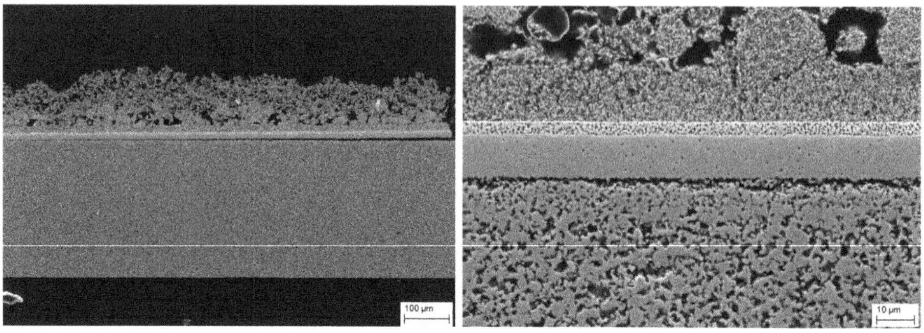

Figure 10.11. One example of possible poor bonding between the electrolyte and fuel electrode of a cell after ~4000 h of SOEC operation at 800 °C at different magnifications.

stack may exhibit a higher voltage degradation rate in the constant-current operating mode, as they are, in fact, being operated at a higher current density. With such a higher current density, the voltage drop as well as the calculated degradation rate will be larger, despite the fact that the degradation is not really accelerated. If the total contact area is correct, but has an inhomogeneous distribution, then the local current density and conversion rate can also be higher. Note that the 'real' contact area (porous air electrode/porous contact material) and 'obvious' contact area (simply the visible area size) differ drastically. Depending on the actual local conditions, the consequences of local gas starvation may have different impacts on the cell performance and stability. One example of this type of degradation was demonstrated with a four-layer F10-design stack at the FZJ [78]. The stack was tested in SOFC mode at different current densities and fuel utilizations for more than 10 000 h. During the test, it exhibited a slightly higher degradation rate and an unexpectedly strong dependence on fuel utilization. A post-mortem analysis confirmed that the degradation behavior was influenced by inhomogeneous contact between the cells and the interconnect, especially on the air side.

10.5 Concluding remarks

Stack tests in a furnace or hotbox are important and necessary in order to understand the degradation and lifetime of solid oxide electrolysis cells and stacks, as the realistic operating environment of a stack cannot be fully simulated in a test bench for single or button cells. Long-term stack tests are generally time-consuming and costly, but still difficult to replace, as there are neither convincing ways of modeling, nor reliable methods for accelerated testing. The degradation rate of SOEC stacks is not constant and changes with the operating time, depending on the conditions. Calculating degradation and predicting lifetimes by extrapolating the measured curves is not always meaningful and may easily result in misleading conclusions. For this reason, long-term tests of several thousands of hours are necessary in order to avoid large distortions in the results. In this chapter, our discussion of the degradation behavior focused mainly on the published stack results with measurement times of more than 3000 h.

The ESC stacks with both 10ScSZ- and 3YSZ-based cells showed the same ASR degradation rates in SOEC and SOFC modes under the testing conditions used. Except for Si poisoning of the Ni/GDC electrode, no other degradation mechanism was apparent compared to SOFC operation. Degradation could mainly be observed as an increase in the ohmic resistance, which was dominated by the electrolytic and interfacial resistance between the metallic interconnect and the air electrode. Compared to the change in ohmic resistance, degradation of the electrodes did not yet play an important role.

FSC stacks with 8YSZ-based cells have been investigated more intensively. The major degradation mechanisms in SOEC mode include Ni migration (mainly depletion), Si poisoning, void formation along YSZ grain boundaries, air electrode delamination, Sr/Co segregation, and possible weakened bonding at the electrolyte and fuel-electrode interface. The effect of Mn deposition at the electrolyte–fuel-electrode interface and its diffusion in the electrolyte requires further attention.

Long-term stack tests impose high demands on stack quality and testing facilities. A homogeneous and stable contact within the stack is the precondition for a reliable and reproducible test. Possible impurities from the infrastructure and test bench should be examined and excluded in order to avoid misleading stack performance and degradation analyses.

References

[1] Möbius H-H 1997 On the history of solid electrolyte fuel cells *J. Solid State Electrochem.* **1** 2–16

[2] Behling N H 2012 History of solid oxide fuel cells *Fuel Cells* ed N H Behling 1 (Amsterdam: Elsevier) pp 223–421

[3] Spacil H S and Tedmon C S 1969 Electrochemical dissociation of water vapor in solid oxide electrolyte cells *J. Electrochem. Soc.* **116** 1618

[4] Inami S, Mc Cullough C, Ring S, Smart W and Weissbart J 1969 Development of a carbon dioxide-water solid oxide electrolyte electrolysis system Annual Report, 29 March 1968–29 May 1969

[5] Surdoval W A *et al* 2001 The Solid State Energy Conversion Alliance (SECA) - A U. S. Department of Energy Initiative to Promote the Development of Mass Customized Solid Oxide Fuel Cells for Low-Cost Power *Seventh International Symposium on Solid Oxide Fuel Cells (SOFC-VII) Proc. Vol.* **2001–16** 53

[6] Doenitz W, Schmidberger R, Steinheil E and Streicher R 1980 Hydrogen production by high temperature electrolysis of water vapour *Int. J Hydrogen Energy* **5** 55–63

[7] Isenberg A O 1981 Energy conversion via solid oxide electrolyte electrochemical cells at high temperatures *Solid State Ionics* **3–4** 431–7

[8] Huczkowski P, Shemet V, Piron-Abellan J, Singheiser L, Quadakkers W J and Christiansen N 2004 Oxidation limited life times of chromia forming ferritic steels *Mater. Corros.* **55** 825–30

[9] Yokokawa H, Tu H, Iwanschitz B and Mai A 2008 Fundamental mechanisms limiting solid oxide fuel cell durability *J. Power Sources* **182** 400–12

[10] de Haart L G J, Mougin J, Posdziech O, Kiviaho J and Menzler N H 2009 Stack degradation in dependence of operation parameters; the real-SOFC sensitivity analysis *Fuel Cells* **9** 794–804

[11] Yokokawa H, Horita T, Yamaji K, Kishimoto H and Brito M 2012 Degradation of SOFC cell/stack performance in relation to materials deterioration *J. Korean Ceram. Soc.* **49** 11–8

[12] Moçoteguy P and Brisse A 2013 A review and comprehensive analysis of degradation mechanisms of solid oxide electrolysis cells *Int. J Hydrogen Energy* **38** 15887–902

[13] Skafte T L, Hjelm J, Blennow P and Graves C 2016 Quantitative review of degradation and lifetime of solid oxide cells and stacks *12th European SOFC & SOE Forum* (Lucerne: European Fuel Cell Forum AG) 8–27 https://www.efcf.com/efcf-library/proceedings-free/2016

[14] Hauch A, Brodersen K, Chen M, Graves C, Jensen S H and Jørgensen P S *et al* 2017 A decade of solid oxide electrolysis improvements at DTU energy *ECS Trans.* **75** 3–14

[15] Wang F, Kishimoto H, Ishiyama T, Develos-Bagarinao K, Yamaji K and Horita T *et al* 2020 A review of sulfur poisoning of solid oxide fuel cell cathode materials for solid oxide fuel cells *J. Power Sources* **478** 228763

[16] Horita T 2021 Chromium poisoning for prolonged lifetime of electrodes in solid oxide fuel cells – review *Ceram. Int.* **47** 7293–306

[17] Mogensen M B, Chen M, Frandsen H L, Graves C, Hauch A and Hendriksen P V *et al* 2021 Ni migration in solid oxide cell electrodes: review and revised hypothesis *Fuel Cells* **21** 415–29

[18] Niewolak L, Tietz F and Quadakkers W J 2016 Interconnects *High-Temperature Solid Oxide Fuel Cells for the 21st Century* (Amsterdam: Elsevier) pp 195–254

[19] Szász J, Wankmüller F, Wilde V, Störmer H, Gerthsen D and Menzler N H *et al* 2018 Nature and functionality of $La_{0.58}Sr_{0.4}Co_{0.2}Fe_{0.8}O_{3-\delta}/Gd_{0.2}Ce_{0.8}O_{2-\delta}/Y_{0.16}Zr_{0.84}O_{2-\delta}$ interfaces in SOFCs. *J. Electrochem. Soc.* **165** F898

[20] Vaßen R, Grünwald N, Marcano D, Menzler N H, Mücke R and Sebold D *et al* 2016 Aging of atmospherically plasma sprayed chromium evaporation barriers *Surf. Coat. Technol.* **291** 115–22

[21] Grünwald N, Sohn Y J, Yin X, Menzler N H, Guillon O and Vaßen R 2019 Microstructure and phase evolution of atmospheric plasma sprayed Mn–Co–Fe oxide protection layers for solid oxide fuel cells *J. Eur. Ceram. Soc.* **39** 449–60

[22] Zurek J, Margaritis N, Naumenko D, Menzler N H and Quadakkers W J 2019 Behaviour of metallic materials in simulated service environments of CO_2/H_2O co-electrolysis systems for Power-to-X application *Oxid. Met.* **92** 353–77

[23] Menzler N H, Sebold D, Sohn Y J and Zischke S 2020 Post-test characterization of a solid oxide fuel cell after more than 10 years of stack testing *J. Power Sources* **478** 228770

[24] The D, Grieshammer S, Schroeder M, Martin M, Al Daroukh M and Tietz F *et al* 2015 Microstructural comparison of solid oxide electrolyser cells operated for 6100 h and 9000 h *J. Power Sources* **275** 901–11

[25] Fang Q P, Frey C E, Menzler N H and Blum L 2018 Electrochemical performance and preliminary post-mortem analysis of a solid oxide cell stack with 20,000 h of operation *J. Electrochem. Soc.* **165** F38–45

[26] Trini M, Hauch A, De Angelis S, Tong X, Hendriksen P V and Chen M 2020 Comparison of microstructural evolution of fuel electrodes in solid oxide fuel cells and electrolysis cells *J. Power Sources* **450** 227599

[27] Horita T, Kishimoto H, Yamaji K, Brito M E and Yokokawa H 2010 Effects of impurities on the degradation of SOFC stacks *ECS Trans.* **26** 349–56

[28] Madi H, Lanzini A, Diethelm S, Papurello D, Van herle J and Lualdi M *et al* 2015 Solid oxide fuel cell anode degradation by the effect of siloxanes *J. Power Sources* **279** 460–71
[29] Hauch A, Jensen S H, Bilde-Sørensen J B and Mogensen M 2007 Silica segregation in the Ni/YSZ electrode *J. Electrochem. Soc.* **154** A619
[30] Kiebach R, Norrman K, Chatzichristodoulou C, Chen M, Sun X and Ebbesen S D *et al* 2014 TOF-SIMS characterization of impurity enrichment and redistribution in solid oxide electrolysis cells during operation *Dalton Trans.* **43** 14949–58
[31] Dominik S L Q, Qingping F and Ludger B 2020 SOE stack deterioration caused by reactant contaminations in realistic operation conditions *14th European SOFC & SOE Forum 2020* (Lucerne: European Fuel Cell Forum AG) https://www.efcf.com/efcf-library/proceedings-free/2020
[32] Küngas R 2020 Solid oxide electrolysis stack tests 2010–2019. https://figshare.com/articles/dataset/Solid_oxide_electrolysis_stack_tests_2010-2019/12137112
[33] Lang M, Raab S, Lemcke M S, Bohn C and Pysik M 2019 Long term behavior of solid oxide electrolyser (SOEC) stacks *ECS Trans.* **91** 2713–25
[34] Megel S, Dosch C, Rothe S, Folgner C, Trofimenko N and Rost A *et al* 2017 Co-electrolysis with CFY-stacks *ECS Trans.* **78** 3089
[35] Mougin J, Chatroux A, Couturier K, Petitjean M, Reytier M and Gousseau G *et al* 2012 High temperature steam electrolysis stack with enhanced performance and durability *Energy Procedia* **29** 445–54
[36] Schefold J, Brisse A, Zahid M, Ouweltjes J P and Nielsen J U 2011 Long term testing of short stacks with solid oxide cells for water electrolysis *ECS Trans.* **35** 2915–27
[37] Geipel C, Hauptmeier K, Herbrig K, Mittmann F, Münch M and Pötschke M *et al* 2019 Stack development and industrial scale-up *ECS Trans.* **91** 123–32
[38] Brisse A, Schefold J and Dailly J 2016 20 000 hours steam electrolysis with solid oxide cell technology *12th European SOFC & SOE Forum* (Lucerne: European Fuel Cell Forum AG) https://www.efcf.com/efcf-library/proceedings-free/2016
[39] Christian Walter K S, Boltze M, Herbrig K and Surrey A 2020 Status of stack and system development at sunfire *14th European SOFC & SOE Forum 2020* (Lucerne: European Fuel Cell Forum AG) https://www.efcf.com/efcf-library/proceedings-free/2020
[40] Petitjean M, Reytier M, Chatroux A, Bruguière L, Mansuy A and Sassoulas H *et al* 2011 Performance and durability of high temperature steam electrolysis: from the single cell to short-stack scale *ECS Trans.* **35** 2905–13
[41] Nguyen V N, Fang Q P, Packbier U and Blum L 2013 Long-term tests of a julich planar short stack with reversible solid oxide cells in both fuel cell and electrolysis modes *Int. J. Hydrogen Energy* **38** 4281–90
[42] Tang E, Wood T, Benhaddad S, Brown C, He H and Nelson J *et al* 2012 Advanced materials for RSOFC dual operation with low degradation. Report for United States Department of Energy. https://www.osti.gov/servlets/purl/1058912
[43] Corre G and Brisse A 2015 9000 Hours operation of a 25 solid oxide cells stack in steam electrolysis mode *ECS Trans.* **68** 3481–90
[44] Bertoldi M, Bucheli O and Ravagni A 2015 Development, manufacturing and deployment of SOFC-based products at SOLID power *ECS Trans.* **68** 117–23
[45] Rinaldi G, Diethelm S, Oveisi E, Burdet P, Van herle J and Montinaro D *et al* 2017 Post-test analysis on a solid oxide cell stack operated for 10,700 hours in steam electrolysis mode *Fuel Cells* **17** 541–9
[46] Fang Q, Blum L and Menzler N H 2015 Performance and degradation of solid oxide electrolysis cells in stack *J. Electrochem. Soc.* **162** F907–12

[47] Frey C E, Fang Q P, Sebold D, Blum L and Menzler N H 2018 A detailed post mortem analysis of solid oxide electrolyzer cells after long-term stack operation *J. Electrochem. Soc.* **165** F357–64

[48] Trofimenko N, Kusnezoff M and Michaelis A 2017 Optimization of ESC performance for co-electrolysis operation *ECS Trans.* **78** 3025–37

[49] Bienert C, Brandner M, Skrabs S, Venskutonis A, Sigl L S and Megel S *et al* 2015 CFY-stack technology: the next design. *ECS Trans.* **68** 2159–68

[50] Lang M, Raab S, Lemcke M S, Bohn C and Pysik M 2020 Long-term behavior of a solid oxide electrolyzer (SOEC) stack *Fuel Cells* **20** 690–700

[51] Eric Tang T W, Brown C, Casteel M, Pastula M, Richards M and Petri. R 2018 Solid oxide based electrolysis and stack technology with ultra-high electrolysis current density (>3 A/cm^2) and efficiency https://www.osti.gov/servlets/purl/1513461

[52] Schefold J, Brisse A and Zahid M 2010 Long term testing of solid oxide fuel cell stacks with yttria stabilized zirconia electrolyte in the H2O electrolysis mode *ECS Trans.* **28** 357–67

[53] Gross-Barsnick S M, Fang Q, Batfalsky P, Niewolak L, Blum L and Quadakkers W J 2019 Post-test characterization of metallic materials and adjacent components in an SOFC stack after 34,000 h operation at 700 °C *Fuel Cells* **19** 84–95

[54] Blum L, Fang Q, de Haart L, Malzbender J, Margaritis N and Menzler N H *et al* 2019 Forschungszentrum Jülich – progress in SOC development *ECS Trans.* **91** 2443–53

[55] Blum L, Fang Q, Groß-Barsnick S M, de Haart L G J, Malzbender J and Menzler N H *et al* 2020 Long-term operation of solid oxide fuel cells and preliminary findings on accelerated testing *Int. J. Hydrogen Energy* **45** 8955–64

[56] Fang Q P, Blum L and Stolten D 2019 Electrochemical performance and degradation analysis of an SOFC short stack following operation of more than 100,000 hours *J. Electrochem. Soc.* **166** F1320–5

[57] Fang Q P, Blum L, Batfalsky P, Menzler N H, Packbier U and Stolten D 2013 Durability test and degradation behavior of a 2.5 kW SOFC stack with internal reforming of LNG *Int. J. Hydrogen Energy* **38** 16344–53

[58] Brandner M, Bienert C, Megel S, Kusnezoff M, Trofimenko N and Sauchuk V *et al* 2013 Long term performance of stacks with chromium-based interconnects (CFY) *ECS Trans.* **57** 2235–44

[59] Sauchuk V, Megel S, Girdauskaite E, Trofimenko N, Kusnezoff M and Michaelis A 2011 Influence of protective layers on SOFC operation *Russ. J. Electrochem* **47** 522–30

[60] Fang Q P, Heinrich M and Wunderlich C 2011 Crofer22 APU in real SOFC stacks *Advances in Solid Oxide Fuel Cells VII Ceramic Engineering and Science Proceedings* ed N P Bansal *et al* 32 (New York: Wiley) pp 101–14

[61] Menzler N H, Beez A, Grunwald N, Sebold D, Fang Q P and Vassen R 2017 Diffusion-related SOFC stack degradation *Solid Oxide Fuel Cells 15 (Sofc-XV) ECS Trans.* **78** 2223

[62] Menzler N H, Sebold D and Guillon O 2018 Post-test characterization of a solid oxide fuel cell stack operated for more than 30,000 hours: the cell *J. Power Sources* **374** 69–76

[63] Menzler N H, Sebold D and Zischke S 2019 SOC degradation: long-term and small-scale effects *ECS Trans.* **91** 719–29

[64] Schefold J, Brisse A and Tietz F 2011 Nine thousand hours of operation of a solid oxide cell in steam electrolysis mode *J. Electrochem. Soc.* **159** A137–44

[65] Tietz F, Sebold D, Brisse A and Schefold J 2013 Degradation phenomena in a solid oxide electrolysis cell after 9000 h of operation *J. Power Sources* **223** 129–35

[66] Knibbe R, Traulsen M L, Hauch A, Ebbesen S D and Mogensen M 2010 Solid oxide electrolysis cells: degradation at high current densities *J. Electrochem. Soc.* **157** B1209

[67] Zhao L, Hyodo J, Chen K, Ai N, Amarasinghe S and Ishihara T *et al* 2013 Effect of boron deposition and poisoning on the surface exchange properties of LSCF electrode materials of solid oxide fuel cells *J. Electrochem. Soc.* **160** F682–F6

[68] Jiang S P 2019 Development of lanthanum strontium cobalt ferrite perovskite electrodes of solid oxide fuel cells—a review *Int. J. Hydrogen Energy* **44** 7448–93

[69] Sun X, Hendriksen P V, Hauch A, Clausen A K, Lehtinen T and Noponen M 2021 Effect of anode side purge gas on the degradation of solid oxide electrolysis cells *ECS Trans.* **103** 1083–91

[70] Chen K and Jiang S P 2016 Review—materials degradation of solid oxide electrolysis cells *J. Electrochem. Soc.* **163** F3070–83

[71] Hjalmarsson P, Sun X, Liu Y-L and Chen M 2013 Influence of the oxygen electrode and inter-diffusion barrier on the degradation of solid oxide electrolysis cells *J. Power Sources* **223** 349–57

[72] Laurencin J, Hubert M, Sanchez D F, Pylypko S, Morales M and Morata A *et al* 2017 Degradation mechanism of $La_{0.6}Sr_{0.4}Co_{0.2}Fe_{0.8}O_{3-\delta}/Gd_{0.1}Ce_{0.9}O_{2-\delta}$ composite electrode operated under solid oxide electrolysis and fuel cell conditions. *Electrochim. Acta* **241** 459–76

[73] Hauch A, Ebbesen S D, Jensen S H and Mogensen M 2008 Solid oxide electrolysis cells: microstructure and degradation of the Ni/yttria-stabilized zirconia electrode *J. Electrochem. Soc.* **155** B1184

[74] Gubner A, Landes H, Metzger J, Seeg H and Stübner R 1997 Investigations into the Degradation of the Cermet Anode of a Solid Oxide Fuel Cell *Fifth International Symposium on Solid Oxide Fuel Cells (SOFC-V) ECS Proc. Volumes* **1997-40** 844–50

[75] Jiao Z, Busso E P and Shikazono N 2020 Influence of polarization on the morphological changes of nickel in fuel electrodes of solid oxide cells *J. Electrochem. Soc.* **167** 024516

[76] Monaco F, Hubert M, Vulliet J, Ouweltjes J P, Montinaro D and Cloetens P *et al* 2019 Degradation of Ni-YSZ electrodes in solid oxide cells: impact of polarization and initial microstructure on the Ni evolution *J. Electrochem. Soc.* **166** F1229–42

[77] Mogensen M B, Hauch A, Sun X, Chen M, Tao Y and Ebbesen S D *et al* 2017 Relation between Ni particle shape change and Ni migration in Ni-YSZ electrodes – a hypothesis *Fuel Cells* **17** 434–41

[78] Fang Q, de Haart U, Schäfer D, Thaler F, Rangel-Hernandez V and Peters R *et al* 2020 Degradation analysis of an SOFC short stack subject to 10,000 h of operation *J. Electrochem. Soc.* **167** 144508

IOP Publishing

High-Temperature Electrolysis
From fundamentals to applications
Werner Sitte and Rotraut Merkle

Chapter 11

Proton and mixed proton/hole-conducting materials for protonic ceramic electrolysis cells

Rotraut Merkle

Protonic ceramic electrolytes have higher ionic conductivity at intermediate temperatures than oxide ion conductors, opening the possibility of decreasing the operating temperature. However, for good performance, positrode materials require 'triple conduction' by protons, oxygen vacancies, and electron holes. This chapter presents the defect chemistry and transport properties of proton-conducting electrolyte and triple-conducting positrode materials for protonic ceramic electrolysis cells (PCECs). A short overview of the most important materials is given, and some aspects of the surface reaction kinetics are discussed.

11.1 Introduction

Proton conductivity can be found in a number of chemically rather different families of solids, see e.g. [1–5]. Prominent examples are:

(i) proton-conducting polymers (sulfonic acid groups attached to a fluorinated or unfluorinated polymer backbone) and, in recent years, hydroxide-ion-conducting polymers (quarternary amine groups). Their operating temperatures are typically limited to less than 120 °C, as they rely on the presence of water in the hydrophilic domains (max. $T \approx 200$ °C in the case of polybenzimidazole).

(ii) Hydroxides such as NaOH; however, their proton conductivities are rather low unless they approach the molten state.

(iii) Hydrogen phosphates such as CsH_2PO_4 (whose typical operating temperatures are higher than 230 °C).

(iv) Hydrogen sulfates such as $CsHSO_4$ (typical operating temperatures are 140 °C–170 °C).

(v) Phosphate-based glasses (typical temperatures are 400 °C–600 °C).

doi:10.1088/978-0-7503-3951-3ch11

(vi) Oxides that dissociatively incorporate water and thus develop a perceptible proton conductivity, typically at 300 °C–700 °C. The most prominent examples are acceptor-doped $Ba(Zr,Ce)O_3$ perovskites, which are discussed in more detail below.

(vii) At temperatures of less than about 100 °C, significant proton conductivity may develop in the surface layers of oxides due to surface OH groups and/or adsorbed water layers; such conductivity may be further enhanced if capillary condensation of water occurs. However, owing to the limited temperature range, this will not be further considered here.

Since this book is dedicated to high-temperature electrolysis, we will from now on focus on proton-conducting oxides in PCECs. These oxides exhibit the required mechanical stability, and properly chosen material compositions also exhibit long-term chemical stability. Specific materials are discussed in section 11.2. A general feature of proton-conducting oxides is that the proton migration barrier is typically much lower than the migration barrier for oxide ions/oxide ion vacancies in oxide ion conductors. This leads to a lower activation energy of the ionic conductivity, as shown in figure 11.1(a). In particular, in an intermediate-temperature range of 200 °C–500 °C, this yields significantly higher proton conductivity σ_{H^+} compared to oxide ion conductivity $\sigma_{O^{2-}}$. The use of a proton-conducting electrolyte further offers the advantage that water formation/water oxidation occurs on the oxygen side. In fuel-cell mode (as protonic ceramic fuel cells (PCFCs)) this means that the fuel is not diluted by produced water, which facilitates high fuel utilization. In electrolysis mode, compressed (relatively) dry hydrogen can be obtained directly (more details in chapter 13).

An electrochemical cell based on a protonic ceramic electrolyte also has specific requirements for the electrode material at the oxygen side (the cathode in fuel-cell operation, the anode in electrolysis mode, recently often termed the 'positrode' to comprise both cases). At the positrode, oxygen is reduced to water by reaction with protons and electrons (fuel-cell mode), or vice versa (electrolysis mode). Figure 11.1(b) indicates that this reaction can always occur at the three-phase-boundary (TPB) where

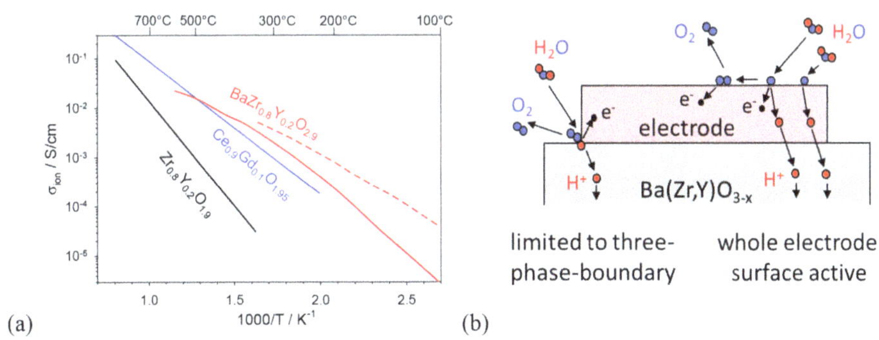

Figure 11.1. (a) Comparison of ceramic oxide ion and proton conductors: total conductivities of sintered polycrystalline samples; dashed red line = bulk conductivity of $BaZr_{0.8}Y_{0.2}O_{2.9}$. Data taken from [6, 7]. (b) Locations of the water oxidation/oxygen reduction reaction at the positrode. Only if the positrode has a certain proton conductivity can the reaction extend beyond the three-phase boundary.

the gas phase, the electrolyte, and the positrode meet. However, even in a composite electrode with an optimized morphology, the very narrow TPB zone accounts only for a small fraction of the overall electrode surface area. Thus, it is desirable to extend this reactive zone to a larger part of the electrode. To achieve this, protons must be able to migrate from the electrolyte through the grain interior (or 'bulk') of the electrode material (right part of figure 11.1(b)). Fortunately, lower σ_{H^+} values than those of the electrolyte materials suffice (the required transport length in the positrode is typically smaller than the electrolyte thickness). Positrode materials and further details are discussed in section 11.3.

As is the case for SOFCs/SOECs, the PCFC/PCEC fuel-side electrode ('negatrode') often consists of a composite of the electrolyte material and nickel. The hydrogen oxidation/reduction reaction is typically kinetically easier than the oxygen reaction, thus the overall performance of PCECs is instead limited by the electrolyte and positrode performances. Some specific aspects of the negatrode reaction and materials, for example, related to direct CO_2 conversion or ammonia synthesis are discussed in chapter 14.

Compared to SOFCs/SOECs based on oxide ion conductors, protonic ceramic cells are in an earlier stage of technological optimization. In recent years, their performance has improved drastically (see e.g. [8–12]), and experience has been gained in the field of stack assembly (see e.g. [13]). Regarding cost, it is expected that protonic ceramic cells can become competitive with oxide-ion-electrolyte-based cells upon upscaling [14]. More details of the present stage and ongoing developments of PCEC applications are given in chapters 13 and 14.

11.2 Proton-conducting oxides

11.2.1 Proton incorporation reaction and thermodynamics

In order to be able to dissociatively incorporate water, the respective oxides need a large concentration of oxygen vacancies $V_O^{\bullet\bullet}$ (Kröger–Vink notation, see chapter 2), which are typically introduced by acceptor doping (e.g. replacement of Zr^{4+} by Y^{3+}, forming a negative defect Y'_{Zr}). According to the electroneutrality condition $[Y'_{Zr}] = 2[V_O^{\bullet\bullet}]$, these acceptors are then compensated by doubly positively charged $V_O^{\bullet\bullet}$. In a humid atmosphere, dissociative water incorporation occurs according to the following reaction (11.1):

$$H_2O_{gas} + V_O^{\bullet\bullet} + O_O^{x} \rightleftharpoons 2OH_O^{\bullet} \qquad K_{hydrat} = \frac{[OH_O^{\bullet}]^2}{pH_2O[V_O^{\bullet\bullet}][O_O^{x}]}. \tag{11.1}$$

Square brackets indicate concentrations. The microscopic process is illustrated in figure 11.2(a): one proton from the water molecule is attached to an oxide ion, and the remaining hydroxide ion is incorporated into a $V_O^{\bullet\bullet}$. Both processes form a 'protonic defect,' $OH_O^{\bullet}$, i.e. a hydroxide ion located on an oxygen ion site. Since the O–H covalent bond length of approximately 1 Å is shorter that the oxide ion radius (1.4 Å), the proton actually resides within the oxygen's electron cloud, and the effective radius of $OH_O^{\bullet}$ is actually slightly smaller than that of a regular oxide ion.

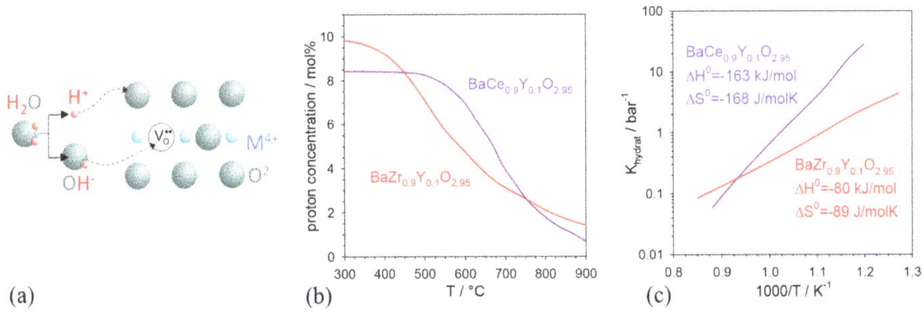

Figure 11.2. (a) Dissociative water incorporation into an oxygen vacancy. (b) Proton concentration of $BaCe_{0.9}Y_{0.1}O_{2.95}$ and $BaZr_{0.9}Y_{0.1}O_{2.95}$ perovskites as a function of temperature (pH_2O = 20 mbar). (c) van 't Hoff plots. Data for (b) and (c) was taken from [4].

Figure 11.2(b) shows exemplary proton concentrations for $BaCe_{0.9}Y_{0.1}O_{2.95}$ and $BaZr_{0.9}Y_{0.1}O_{2.95}$. At temperatures below about 300 °C at 20 mbar pH_2O, all oxygen vacancies are filled by the hydration reaction, leading to a plateau of proton concentration at $[Y'_{Zr}]=[OH^{\bullet}_{O}]$ (the fact that the maximum $[OH^{\bullet}_{O}]$ of $BaCe_{0.9}Y_{0.1}O_{2.95}$ in figure 11.2(b) is less than 10% suggests that the effective acceptor concentration of this sample was a bit lower than the nominal value). The hydration reaction can be quantified by the mass action constant K_{hydrat} in equation (11.1). The van 't Hoff plot of $\ln(K_{hydrat})$ versus $1/T$ yields the standard enthalpy ΔH^0_{hydrat} and entropy ΔS^0_{hydrat} of the hydration reaction from the slope and the vertical axis intercept of the linear regression line, respectively (figure 11.2(c)). Within the perovskite family, the hydration enthalpy is typically more exothermic for materials with larger basicity, see e.g. [4]. This may appear surprising at first glance, because the hydration reaction comprises the protonation of O (which is more favorable for a more basic material) as well as the filling of an oxygen vacancy (which is more favorable for a more acidic material). However, density functional theory (DFT) calculations of proton and hydroxide affinities show that the latter varies less with material composition [15], i.e. the variation of the proton affinity also determines ΔH^0_{hydrat}. This explains the empirical correlation with the material's basicity, e.g. the more negative hydration enthalpy of $BaCeO_3$ compared to that of $BaZrO_3$. Symmetry aspects also play a role in the hydration entropy [4], leading to more negative ΔS^0_{hydrat} values for more distorted perovskites such as $BaCeO_3$.

The hydration reaction (11.1) leads to significant so-called chemical expansion, e.g. a linear expansion of 0.4% for full hydration of Ba $Zr_{0.85}Y_{0.15}O_{2.925}$ [4]. This can become critical in the case of abrupt dehydration, causing severe crack formation. Some cell architectures are being explored that avoid using materials with such a chemical expansion as the mechanical support (metal-supported cells, cells on non-hydrating oxidic support).

11.2.2 Proton transport

The mechanism of proton transport in perovskites is illustrated in figure 11.3, which is based on molecular dynamics simulations [4, 16]. The proton can easily and

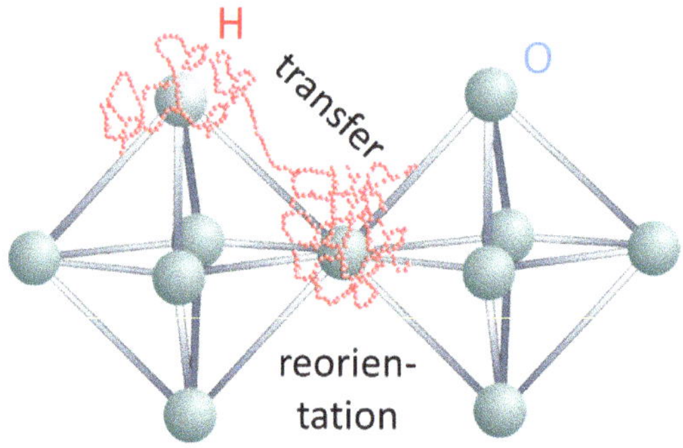

Figure 11.3. Proton trajectory (red trace) in $BaCeO_3$ perovskite obtained from a molecular dynamics simulation. Figure adapted from [4] with permission, copyright Annual Reviews (2003).

quickly reorient around the O, to which it is bound by a short covalent O–H bond. On a moderately slower timescale, the proton is transferred to a neighboring oxide ion. This process is assisted by phonons, which approach the two involved O ions to a suitable distance for a sufficiently strong hydrogen bond to be formed to the final oxygen ion. The lower migration barrier of protons (0.44–0.6 eV [4]) compared to oxide ions (0.8–1 eV for $Ba(Zr,Ce)O_3$ [17]) results from two facts: (i) the lower charge of the mobile ion (less charged ions typically have smaller migration barriers); (ii) the specific interaction of the proton with neighboring oxide ions by hydrogen bonds, which are weaker than the covalent O–H bonds but nevertheless important for stabilizing the transition state.

It has been observed that the effective proton mobility depends on the nature and the concentration of the acceptor dopant, see e.g. [4, 18], and the highest mobilities are observed for large acceptors such as Y^{3+} and high acceptor concentrations. This indicates attractive interactions between protons and acceptors ('trapping,' most pronounced at low T) caused by a combination of electrostatic and elastic effects. Recent DFT and kinetic Monte Carlo simulations give a detailed microscopic picture [19].

At temperatures above approx. 300 °C, the materials start to partially dehydrate (figure 11.2(b)). The decreasing proton concentration then flattens the slope in the proton conductivity Arrhenius plot (figure 11.1(a)). It is important to note that at elevated temperatures, the conductivity gradually changes from protonic to $V_O^{\bullet\bullet}$ transport. Thus, very high temperatures and/or low humidity gradually change a proton-conducting ceramic into an oxide ion conductor (at 700 °C in 50 mbar H_2O, the $V_O^{\bullet\bullet}$ conductivity of $BaZr_{0.8-x}Ce_xY_{0.2}O_{2.9}$ may exceed the proton conductivity [20]). The high pH_2O typically applied in electrolysis mode favors the predominance of σ_{H^+} at comparably high T. The mobility of the $V_O^{\bullet\bullet}$ is also important for the hydration kinetics of thick samples; for a high degree of hydration, the water chemical diffusion coefficient $D_{H_2O}^{\delta}$ approaches the proton diffusivity $D_{OH_O^\bullet}$ [21].

11.2.3 Electronic defects in proton-conducting materials

Perovskites such as acceptor-doped $BaZr_{1-x}Ce_xO_3$ are typically treated as protonic electrolyte materials, implying that they have negligible electronic conductivity.

The large bandgaps of zirconate and cerate perovskites prevent n-type conductivity even under strongly reducing conditions. However, under oxidizing conditions a small but nonnegligible concentration of electron holes $h^{\bullet}$ forms according to reaction (11.2):

$$0.5O_2 + V_O^{\bullet\bullet} \rightleftharpoons O_O^x + 2h^{\bullet} \qquad K_{ox} = \frac{[O_O^x][h^{\bullet}]^2}{[V_O^{\bullet\bullet}]\sqrt{pO_2}}. \tag{11.2}$$

Due to the larger mobility of holes compared to protons, this leads to perceptible p-type electronic conductivity at high pO_2 and temperatures of more than ≈500 °C, see e.g. [20, 22, 23]. The electronic transference number can exceed 0.2 at 600 °C; a high pH_2O and a high Ce content help to decrease it. This undesired p-type electronic conductivity is relevant in the following three situations:

(i) Half-cell measurements of positrode materials typically performed under oxidizing conditions, i.e. the whole electrolyte is exposed to high oxygen activity. The electronic conductivity of a $BaZr_{1-x}Ce_xO_3$ electrolyte can (partially) short circuit the oxygen surface reaction resistance. The larger the reaction resistance compared to the electrolyte's electronic resistance, the stronger the effect (i.e. it is more critical for pore-free microelectrodes than for porous electrodes with large active surface areas) [24, 25].

(ii) Protonic ceramic fuel cells. The electrolyte is in contact with air at the positrode and subjected to strongly reducing conditions at the negatrode. An oxygen chemical potential gradient μ_O exists over the thickness of the electrolyte, such that the hole concentration and thus electronic conductivity is suppressed close to the fuel side. This allows the cell to actually function; however, a small decrease in the open-circuit voltage/faradaic efficiency cannot be avoided [22].

(iii) Protonic ceramic electrolysis cells. Here, the hole conductivity is also decreased close to the fuel side, but the chemical potential gradient that develops under high currents strongly decreases this hole suppression [22]. Severe losses of faradaic efficiency (the ratio of the H_2 flux produced to the applied electrical current) can occur, and gas-phase analysis is highly recommended to monitor it [26, 27]. Minimization of the air electrode resistance helps to improve this situation [28].

Cases (ii) and (iii) are discussed in more detail in chapter 12.

11.2.4 Grain-boundary properties and processing issues

Protonic ceramic electrolytes are typically used in PCECs in the form of sintered polycrystalline layers. It has been found that large-bandgap proton and oxide ion conductors typically exhibit a more or less pronounced blocking behavior for ion transport at the grain boundaries (GBs); see [29] and the references therein. This is evidenced by an additional semicircle in the impedance spectra at frequencies lower than the bulk response but higher than the electrode frequency range (see the example in figure 11.4(a)). The capacitance caused by the less conductive layers at the grain boundaries is typically two to four orders of magnitude larger than the bulk capacitance (according to the 'brick layer model,' see e.g. [30]); this is an important diagnostic for identifying the GB response. The GB semicircle is usually

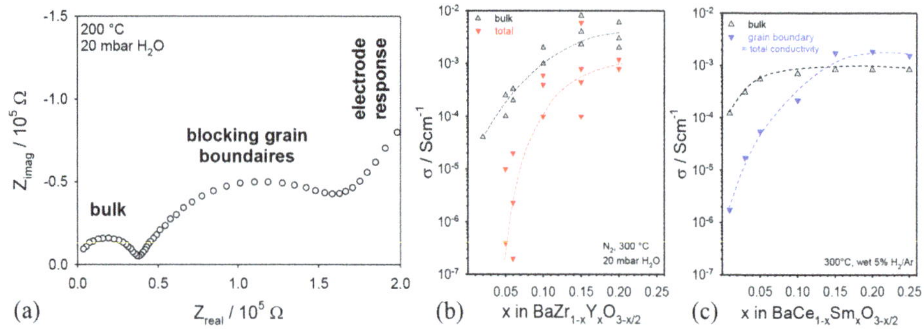

Figure 11.4. (a) Exemplary impedance spectrum of a polycrystalline $BaZr_{0.8}Ce_{0.2}Y_{0.2}O_{2.9}$ sample. (b, c) Bulk and total proton conductivities of $BaZr_{1-x}Y_xO_{3-x/2}$ and $BaCe_{1-x}Sm_xO_{3-x/2}$ ceramics as a function of acceptor concentration. Reproduced with permission from [29], copyright Elsevier (2017).

most visible at comparatively low temperatures (100 °C–300 °C); its relative magnitude decreases at higher temperatures because the GB conductivity typically has a larger activation energy than the bulk proton transport (figure 11.1(a)).

The blocking character of GBs in proton-conducting perovskites is largely caused by space charge depletion effects. Transmission electron microscopy (TEM) investigations show ‘clean’ grain boundaries, i.e. an absence of blocking secondary phases. The structurally distorted grain-boundary core carries an excess positive charge caused by $V_O^{\bullet\bullet}$ and/or proton segregation [31], and the driving force for such segregation is larger in $BaZrO_3$ than in $BaCeO_3$ [32]. This positive core charge is then compensated by the depletion of positive $V_O^{\bullet\bullet}$ and $OH_O^{\bullet}$ carriers in adjacent space charge zones. Experimental evidence for this space charge depletion has been obtained from different experiments, e.g. [33–35]. One characteristic feature of this depletion is that the magnitude of blocking decreases with increasing dopant concentration (figure 11.4(b)), because at higher acceptor and $OH_O^{\bullet}$ concentrations, a thinner depletion zone is sufficient to compensate the GB core charge (which is approximately constant).

Barium zirconate ceramics require very high sintering temperatures (typically 1600 °C–1700 °C, see e.g. [7]). The grain size remains small, i.e. the number of blocking grain boundaries is high. This is related to solute drag; for grain growth, acceptors that accumulate close to the positively charged GB cores have to migrate together with the grain boundary. The sintering temperature can be reduced by partial Ce substitution. Sintering aids such as ZnO and NiO have been used [36, 37], which foster grain growth by forming a transient liquid phase. However, this typically decreases proton uptake and bulk conductivity (e.g. [38, 39]) because the $(Ba,Ni,Y)O_x$ transient liquid phase causes a detrimental Ba deficiency in the grain interior [40]. To scale up from lab-scale button cells to larger tubular or planar cells (see chapters 13 and 14), established ceramic processing techniques such as extrusion or tape casting are employed. A thin $Ba(Zr,Ce,Acceptor)O_3$ electrolyte layer is typically co-sintered with a thicker NiO/electrolyte composite layer that forms the fuel electrode. NiO from the composite can act as a sintering aid for the electrolyte layer, but also introduces the drawback of decreased proton conductivity (the σ_{H^+} value extracted from cell data is typically significantly lower than the value expected for the respective material sintered in the absence of NiO).

11.2.5 Material examples

11.2.5.1 $BaZrO_3$–$BaCeO_3$ perovskites

The vast majority of electrolyte materials used in PCFCs and PCECs stem from solid solutions of acceptor-doped barium zirconates and cerates, see e.g. [4, 41, 42]. $BaZrO_3$ is a cubic perovskite with a Goldschmidt tolerance factor of $t = 1.004$, i.e. close to unity. Acceptor dopants such as Y on the Zr site have a solubility of up to about 20%; higher concentrations lead to non-cubic distortions and/or secondary phase formation (only for Sc has a high dopant concentration of 70% been successful [43]). The advantage of $BaZrO_3$-based materials is their good long-term chemical stability against decomposition by CO_2 and water, and their high bulk proton mobility in particular when large acceptors such as Y are used, which minimize trapping effects. Yb-doping or Y,Yb co-doping increases the proton conductivity only slightly [44]. The disadvantages of $BaZrO_3$ materials are that they are harder to sinter (which also results in smaller grain sizes) and the character of their grain boundaries is more blocking than those of $BaCeO_3$ materials.

Because of its Goldschmidt factor of 0.937, $BaCeO_3$ acquires a slight orthorhombic distortion (i.e. tilting of the CeO_6 octahedra). For temperatures up to 700 °C, $BaCe_{0.9}Y_{0.1}O_{2.9}$ has a higher degree of hydration than $BaZr_{0.9}Y_{0.1}O_{2.9}$ (figure 11.2(b)), which is beneficial for suppressing *p*-type electronic conductivity [20]. However, the non-cubic distortion leading to nonequivalent O and proton sites slightly decreases the bulk proton mobility. A worse problem is that $BaCeO_3$ is thermodynamically unstable at intermediate temperatures with respect to decomposition in the presence of CO_2 or water [4, 41]. Barium cerate–zirconate solid solutions have been explored with the aim of achieving a compromise between improved sintering and sufficient chemical stability. Perovskites with 0.5 Ce per formula unit are found to be stable in the long term [41]; occasionally, compositions with 0.7 Ce are also used, but these have to be regarded with some caution. Combining Y,Yb co-doping and Zr,Ce yields compositions such as $BaCe_{0.7}Zr_{0.1}Y_{0.1}Yb_{0.1}O_{2.9}$ (BCZYYb7111) and $BaCe_{0.4}Zr_{0.4}Y_{0.1}Yb_{0.1}O_{2.9}$ (BCZYYb4411), which are used rather frequently (see figure 11.5(a)) and have slightly higher conductivity than their respective singly doped Y or Yb compositions [44, 45]. The use of slightly smaller Yb^{3+} instead of Y^{3+} has been found to decrease the issues resulting from transient liquid-phase formation that arise when sintering takes place in contact with NiO [46]. It has been shown that partial Sr and Ce substitution, as in $Ba_{0.85}Sr_{0.15}Zr_{0.7}Ce_{0.1}Y_{0.2}O_{2.9}$, can increase the thermal expansion coefficient, e.g. in order to better match that of a metal support [47].

11.2.5.2 Other perovskites

Perovskites that mainly have Sr on the A site have also been intensively investigated for proton transport. However, with Goldschmidt factors of 0.947 and 0.885

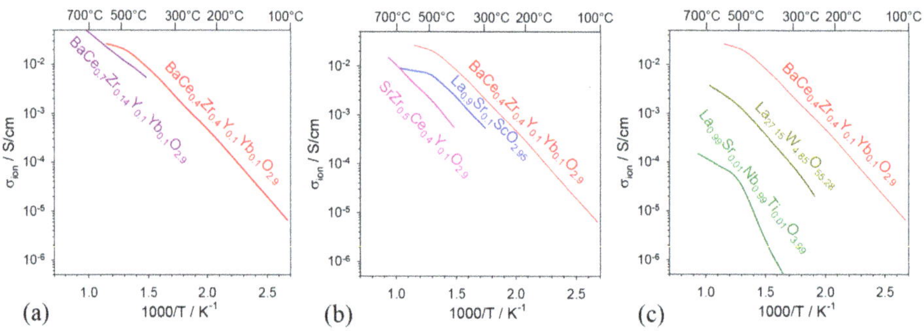

Figure 11.5. Total conductivities of various polycrystalline proton conductors in a humid N_2, Ar, or dilute H_2 atmosphere: (a) $BaCe_{0.7}Zr_{0.1}Y_{0.1}Yb_{0.1}O_{2.9}$ (BCZYYb7111) and $BaCe_{0.4}Zr_{0.4}Y_{0.1}Yb_{0.1}O_{2.9}$ (BCZYYb4411). (b) $SrZr_{0.5}Ce_{0.4}Y_{0.1}O_{2.95}$ and $La_{0.9}Sr_{0.1}ScO_{2.95}$. (c) $La_{0.99}Ca_{0.01}Nb_{0.99}Ti_{0.01}O_{3.99}$ and $La_{27.15}W_{4.85}O_{55.28}$. Data taken from [44, 45, 48, 51, 54, 55].

respectively, $SrZrO_3$ and $SrCeO_3$ are strongly orthorhombically distorted. Inequivalent oxygen positions lead to a levelling-off of the proton uptake far below the [Acceptor′] = $[OH_O^{\bullet}]$ limit [49]. The decreased symmetry also lowers the bulk proton mobility (with increased proton migration barriers ≈0.6 eV [49]), resulting overall in a significantly lower conductivity than for the respective barium perovskites (figure 11.5(b)). $SrCeO_3$ has stability issues with CO_2 and H_2O comparable to those observed for $BaCeO_3$. Mixed compositions such as $SrZr_{0.5}Ce_{0.4}Y_{0.1}O_{2.95}$ exhibit sufficient stability. They have the advantage of a sintering onset at lower T than the respective barium perovskites, and have been successfully used for the electrolyte/NiO composite layer, as their increased sintering activity helps to densify the thinner $BaZr_{0.44}Ce_{0.36}Y_{0.2}O_{2.9}$ top electrolyte layer [50].

$A^{3+}B^{3+}O_3$ perovskites can also incorporate protons; the most prominent example is $LaScO_3$, see e.g. [51, 52]. In these materials the acceptor doping used to create oxygen vacancies can be applied to the A site and/or the B site. Typically, it is applied to the A site, as this has a lower tendency to trap protons than acceptors on the B site. The advantage of these perovskites is their low or zero content of alkali-earth cations, which improves CO_2 and H_2O stability. However, their proton conductivity does not fully reach that of $Ba(Zr,Ce,Y)O_{3-z}$ (figure 11.5(b)).

Several other perovskites and related materials develop proton conductivity upon the dissociative hydration of oxygen vacancies, for example, $Ba_3Ca_{1.18}Nb_{1.82}O_{8.73}$, $Ba_2In_2O_5$, etc. which are currently less considered for actual application in protonic ceramic electrochemical cells. An overview can be found in [53].

11.2.5.3 LaNbO$_4$

The structure of $LaNbO_4$ consists of NbO_4 tetrahedra connected by LaO_8 polyhedra (figure 11.6(a)). Oxygen vacancies can be introduced by acceptor doping (e.g. Ca on the La site, Ti on the Nb site), but the dopant solubility is much lower than in perovskites. Typical sintering temperatures are about 1500 °C, resulting in comparatively large grains (approx. 10 μm). These materials hydrate and develop protonic conductivity with a proton migration barrier of 0.55 eV [54]. However,

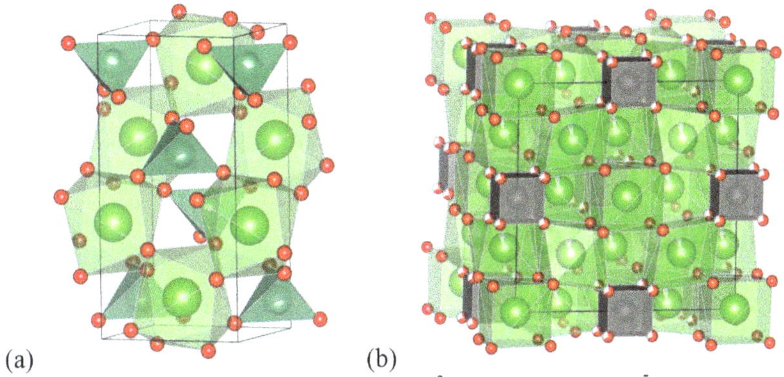

Figure 11.6. Crystal structures of: (a) $LaNbO_4$, La^{3+} = light green, Nb^{5+} = dark green; (b) 'La_6WO_{12},' La^{3+} = light green, W^{6+} = gray.

the proton conductivity remains significantly lower than that of $Ba(Zr,Ce,Y)O_{3-z}$ (figure 11.5(c)). These materials may be of interest for use as hydrogen permeation membranes (due to their excellent chemical stability and the possibility of introducing *n*-type electronic conductivity) or as hole-blocking layers for bilayer electrolyte architectures (due to their low hole transference number even at high *T*).

11.2.5.4 La_6WO_{12}

The structure of 'La_6WO_{12}' oxides is loosely related to a distorted fluorite structure, and has W in sixfold and La largely in eightfold coordination (figure 11.6(b)). The correct formula is actually $La_{28-x}W_{4+x}O_{54+1.5x}v_{2-1.5x}$. It includes (2–1.5$x$) unoccupied oxygen sites ('oxygen vacancies' present even in the absence of dopants), which means that some W are substituted into La sites [55]. The La/W ratio can be tuned in a certain range ($0.78 \leqslant x \leqslant 1.08$), which yields higher ionic conductivity for smaller x. La can be replaced by other rare-earth cations, and W can be partially replaced by Mo to increase its reducibility. Typical sintering temperatures are about 1500 °C, resulting in comparatively large grains (approx. 10 μm). As in the case of $LaNbO_4$, these materials have only moderate proton conductivity (figure 11.5(c)) and might more be of interest for use as hydrogen permeation membranes.

11.3 Mixed proton/hole-conducting materials

Materials used for the oxygen electrode (positrode) have to combine electronic (hole) and proton conductivity; thus, they typically have a large fraction of mixed-valent (redox-active) transition metals. This also implies a variable oxygen stoichiometry. Since the mobility of holes is much larger in perovskite (related) oxides than in other crystal structures, the PCEC positrode materials being investigated and applied today all belong to this family of materials; see e.g. [56].

11.3.1 Proton incorporation reactions and thermodynamics, defect interactions

The triple-conducting perovskite (related) materials typically contain oxygen vacancies, which can be hydrated by reaction (11.1) similarly to the protonic electrolyte materials. Depending on the cation composition as well as T and pO_2, the $V_O^{\bullet\bullet}$ concentration may even be higher than in $Ba(Zr,Ce,Y)O_{3-z}$ electrolytes. In addition, the oxygen incorporation reaction (11.2) has to be considered when redox-active cations are present. Reaction (11.2) can equivalently be formulated for the formation of holes $h^{\bullet}$, or it can include transition metals (TM) in different oxidation states TM_{TM}^{x}, $TM_{TM}^{\bullet}$:

$$0.5O_{2,\,gas} + V_O^{\bullet\bullet} + 2TM_{TM}^{x} \rightleftharpoons O_O^{x} + 2TM_{TM}^{\bullet} \qquad K'_{ox} = \frac{[TM_{TM}^{\bullet}]^2}{\sqrt{pO_2}\,[V_O^{\bullet\bullet}][O_O^{x}][TM_{TM}^{x}]^2}. \tag{11.3}$$

The combination of reactions (11.1) and (11.2) yields proton incorporation by a redox reaction at the expense of holes:

$$H_2O_{gas} + 2O_O^{x} + 2h^{\bullet} \rightleftharpoons 2OH_O^{\bullet} + 0.5O_2 \qquad K_H = \frac{\sqrt{pO_2}\,[OH_O^{\bullet}]^2}{pH_2O[O_O^{x}]^2[h^{\bullet}]^2}, \tag{11.4}$$

which is, however, not an independent reaction. From a general defect chemical model (assuming ideally dilute behavior) one can derive that for the condition $2[V_O^{\bullet\bullet}] > [h^{\bullet}]$ reaction (11.1) dominates after an increase of pH_2O, while otherwise proton uptake occurs largely as described by reaction (11.4) [57].

The proton concentration can be determined from thermogravimetric measurements. It is advisable to first sinter the sample into a pellet and then crush and sieve it to a particle size of approximately 100 μm to decrease the surface area and thus avoid complications caused by water surface adsorption. Since the relative weight changes are small, large sample weights (in the gram range) are beneficial. Isothermal measurements made while changing pH_2O from dry to humid and back are the most reliable mode, as the reversibility can be checked. To ensure $2[V_O^{\bullet\bullet}] > [h^{\bullet}]$ (and thus obtain the conversion factor from weight change to proton concentration according to reaction (11.1)), the measurement in figure 11.7(a) is taken with the sample particles

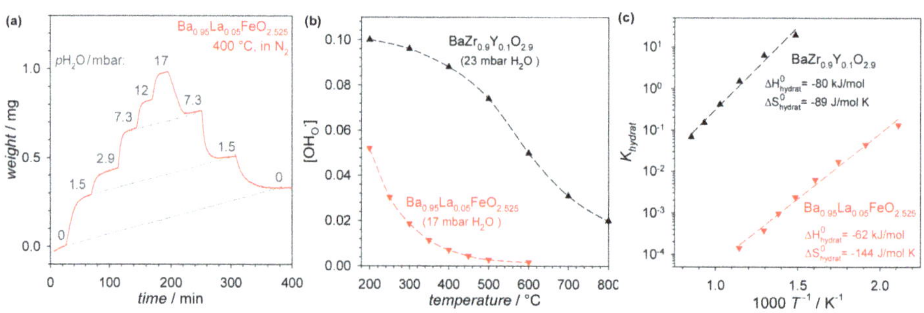

Figure 11.7. (a) Exemplary thermogravimetric measurement of isothermal hydration according to reaction (11.1) on coarse particles of $Ba_{0.95}La_{0.05}FeO_{2.525}$ quenched from 700 °C in N_2. (b) Proton concentrations of the electrolyte $BaZr_{0.9}Y_{0.1}O_{2.95}$ and the triple conductor $Ba_{0.95}La_{0.05}FeO_{2.525}$. (c) Corresponding van 't Hoff plots. Reproduced with permission from [56], copyright Annual Reviews (2021).

quenched from 700 °C in N_2. This freezes the kinetically difficult oxygen exchange reaction (11.3), while the much easier hydration reaction (11.1) still proceeds at the lower measurement temperatures (the sloping baseline in figure 11.4(a) is caused by slow oxidation with trace O_2 but can be corrected). The alternative approach of comparing thermogravimetry runs in dry and in humid atmospheres at fixed pO_2 may lead to larger uncertainties: (i) some materials exhibit large weight changes from the T-dependent oxidation reaction (11.3), thus the water uptake is the small difference of two large numbers. (ii) if the sample exhibits a large proton uptake, the corresponding $[V_O^{\bullet\bullet}]$ decrease may in turn affect the oxidation reaction (11.3). The mass action constant K_{hydrat} is obtained from the measured proton concentration (figure 11.7(b)) according to equation (11.1), and the van 't Hoff plot (figure 11.7(c)) yields the standard enthalpy ΔH^0_{hydrat} and entropy ΔS^0_{hydrat} of hydration.

The exemplary data in figure 11.7 demonstrate the characteristics observed for triple-conducting oxides: despite a high $[V_O^{\bullet\bullet}]$, the degree of hydration remains lower than for electrolytes (the saturation observed for $Ba(Ce,Zr,Y)O_{3-z}$ at temperatures of less than ≈300 °C is never achieved). This is caused by a combination of a less negative ΔH^0_{hydrat} and a more negative ΔS^0_{hydrat}.

General chemical trends (decreasing cation radius and increasing element electronegativity when moving to the right in the transition-metal rows of the periodic table), as well as DFT calculations (e.g. [58], showing strong TM*3d*–O*2p* orbital hybridization) indicate that (Fe,Co,Ni)–O bonds in the positrode materials have a higher degree of covalency than (Ce,Zr)–O bonds in the electrolyte materials. This decreases the negative charge density of oxygen. Correspondingly, its protonation is less favorable, and the overall hydration reaction also becomes less exothermic (cf. the correlation between proton affinity and hydration enthalpy [15]). The reasons for the systematically more negative ΔS^0_{hydrat} of positrode materials are still under investigation (and require quite demanding phonon calculations).

The comparatively high covalency of the (Fe,Co,Ni)–O bonds has further consequences. It means that when the transition-metal cation is oxidized, the electron hole formed is, to a significant degree, transferred to the oxygen ion. Thus, it affects a larger volume (the complete TMO_6 octahedron or even more) than the simplistic expression in reaction (11.3) in which the hole is localized at the transition metal. This leads to deviations from an ideally dilute defect chemical model. Since holes perceive each other already at relatively low concentrations, the higher the hole concentration, the less negative the enthalpy of the oxidation reaction (11.3) becomes; see e.g. [59]. In addition to this 'hole–hole interaction,' an increased hole concentration decreases the basicity of the oxygen ions, which disfavors their protonation ('hole–proton interaction' [56, 60]). These defect interactions have also been confirmed by DFT calculations for $BaFeO_{3-\delta}$ [58]. In fuel-cell operation, the oxygen chemical potential within the positrode particles is decreased relative to the ambient pO_2 (see e.g. [61, 62]), which favors proton uptake. However, in electrolysis mode it is increased, and proton uptake becomes less favorable.

11.3.2 Proton transport in triple-conducting perovskites

Proton migration in triple conductors follows the same mechanism as in perovskite electrolytes (section 11.2.2). However, some differences need to be considered. On one hand, the larger local lattice distortions related to the higher $V_O^{\bullet\bullet}$ concentration (figure 11.8(a)) are expected to increase proton migration barriers. On the other hand, the lower charge of Fe^{3+} compared to $Ce^{4+}$and Zr^{4+} decreases the repulsion of the migrating proton in the transition state. Trapping is expected to be less pronounced, because the high electronic conductivity can effectively screen electrostatic interactions.

Direct measurements of proton conductivity or proton mobility in triple conductors are challenging because the total conductivity is dominated by the electronic contribution, and the ionic part also comprises mobile $V_O^{\bullet\bullet}$. The proton diffusion coefficients of $Ba_{0.5}Sr_{0.5}Fe_{0.8}Zn_{0.2}O_{3-\delta}$ have been estimated using chemical diffusion experiments on dense samples after a pH_2O step (figure 11.8(b) [63]). The values are in the same range as those of $Ba(Ce,Zr,Y)O_{3-z}$ electrolytes. For a few triple conductors, proton migration barriers in the range of 0.23–0.62 eV have been determined using DFT calculations (see e.g. [64–66]). However, a systematic investigation of their dependence on material parameters, such as the lattice parameter, B cation charge, $V_O^{\bullet\bullet}$ concentration, or local distortions, has not been published so far.

While in electrolyte materials the transport across grain boundaries is significantly impeded for protons and oxygen vacancies (section 11.2.4), the blocking behavior largely vanishes for redox-active positrode materials. Excess oxygen vacancies probably still accumulate in the grain-boundary cores, but the respective charge can be compensated by valence changes of the transition-metal cations, and the respective space charge potentials remain small.

Since the oxygen vacancies in triple conductors are typically hydrated only to a small degree, these materials also exhibit $V_O^{\bullet\bullet}$ ionic conductivity. The $V_O^{\bullet\bullet}$ conductivity may become relevant if the cell is operated under conditions in which the electrolyte becomes a combined $V_O^{\bullet\bullet}$ and proton conductor due to its dehydration (typically at $T > 650$ °C). Interestingly, the $V_O^{\bullet\bullet}$ defect diffusion coefficients (proportional to the

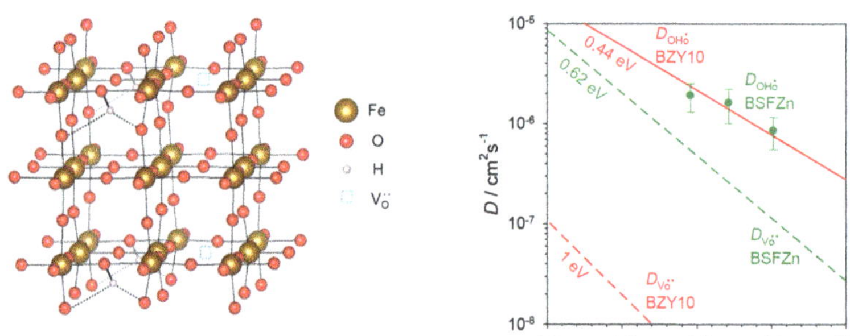

(a) (b)

Figure 11.8. (a) Exemplary structure of a $Ba_8Fe_8O_{23}H$ 2 × 2 × 2 supercell containing one proton and one $V_O^{\bullet\bullet}$ according to DFT calculations (data from [58]). (b) Proton ($D_{OH_O^\bullet}$) and oxygen vacancy ($D_{V_O^{\bullet\bullet}}$) diffusion coefficients of $BaZr_{0.9}Y_{0.1}O_{2.95}$ (BZY10) and $Ba_{0.5}Sr_{0.5}Fe_{0.8}Zn_{0.2}O_{3-\delta}$ (BSFZn). Data taken from [4, 17, 63]

$V_O^{\bullet\bullet}$ mobility) in Ba-rich triple conductors are higher than in the $Ba(Zr,Y)O_{3-z}$ electrolyte (figure 11.8(b)). This might be related to their similarity to $Ba_{0.5}Sr_{0.5}Co_{0.8}Fe_{0.2}O_{3-\delta}$, which was initially developed as an SOFC cathode, and shows one of the highest $V_O^{\bullet\bullet}$ mobilities in perovskites.

11.3.3 Electronic conductivity, conflicting trends

It is instructive to explore the relations between the different properties which are desirable for good positrode materials. While actual data are still scarce, many Ba-rich triple conductors show proton concentrations in the few-percent range, which most probably suffices to make large parts of the positrode active for oxygen reduction/water oxidation.

Figure 11.9(a) compiles the proton concentrations and electronic conductivities σ_{eon} of triple conductors. For each of the material families (ferrates, Zn-doped ferrates, Co-containing materials), an anticorrelation between the two quantities is observed. This is in line with the discussion in section 11.3.1, which states that increased orbital hybridization between a transition metal and oxygen—which is beneficial for σ_{eon}—decreases proton uptake. A number of $(Ba,Sr)(Fe,Co,Acc)O_{3-\delta}$ perovskites successfully used in PCFCs have comparatively low σ_{eon} values in the range of $\leqslant 10$ S cm^{-1} (e.g. $BaCo_{0.4}Fe_{0.4}Zr_{0.1}Y_{0.1}O_{3-\delta}$ [67]). Values of σ_{eon} that are too low might also be compensated by adjusting the cell architecture or using an additional current-collector layer.

Other properties are also in mutual conflict (figure 11.9(b)). For example, high Ba contents increase proton uptake and catalytic activity, but decrease stability versus carbonate formation and transformation into a hexagonal perovskite structure. Partial Co substitution is typically beneficial for catalytic activity and σ_{eon}, but detrimental for proton uptake and stability. Thus, rather than maximizing a single property, a good balance is needed. Partial replacement of the transition metal by Zr, Y, or Zn is beneficial for proton uptake and stability (section 11.3.5.1); further approaches beyond simple perovskites are discussed in sections 11.3.5.2–11.3.5.4.

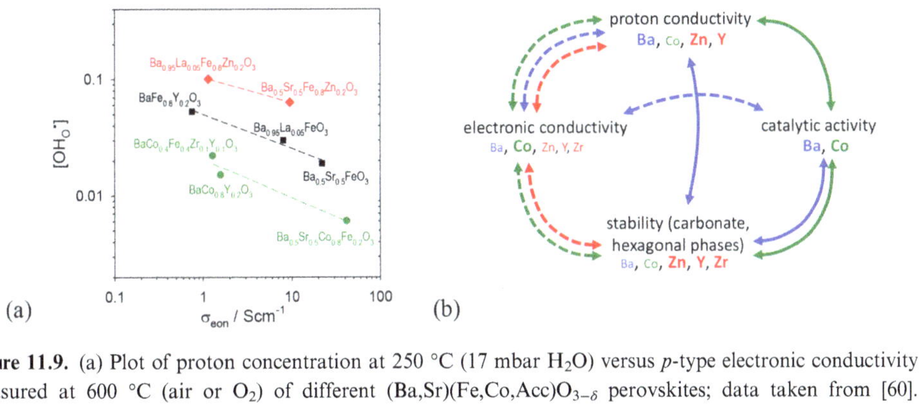

Figure 11.9. (a) Plot of proton concentration at 250 °C (17 mbar H_2O) versus *p*-type electronic conductivity measured at 600 °C (air or O_2) of different $(Ba,Sr)(Fe,Co,Acc)O_{3-\delta}$ perovskites; data taken from [60]. (b) Conflicts between different PCFC/PCEC positrode material requirements; elements which are beneficial for the respective property are shown in bold type. Adapted with permission from [56], copyright Annual Reviews (2021).

11.3.4 Surface oxygen exchange kinetics and mechanism

For a few materials, such as $PrBa_{0.5}Sr_{0.5}Co_{1.5}Fe_{0.5}O_{5+\delta}$ and $BaGd_{0.3}La_{0.7}Co_2O_{5+\delta}$, it has been directly demonstrated using dense microelectrodes that the whole surface area is active, i.e. their proton conductivity suffices to transfer protons from the electrolyte to the positrode/air interface [44, 68]. Based on measured proton concentrations examples in section 11.3.5 and estimated proton mobilities (section 11.3.2), perceptible proton conductivities and thus an extension of the reactive zone beyond the air/electrolyte/positrode three-phase boundary are also expected for many other Ba-rich perovskites and related materials.

Regarding the mechanism, let us first summarize results for oxygen exchange kinetics at SOFC cathodes, which has been investigated in detail (also using pore-free microelectrodes, which make it possible to decouple morphological effects from the intrinsic material properties). For many perovskites it has been found that oxygen vacancies appear before or in the rate-determining step (rds) and thus determine or co-determine the oxygen exchange rate (see e.g. [69, 70]). The barrier for breaking the O=O double bond is strongly decreased when intermediate O_2^- (superoxo) or O_2^{2-} (peroxo) species are incorporated in a surface oxygen vacancy.

For the oxygen surface reaction at PCFC/PCEC positrodes (figure 11.1(b)), i.e.

$$O_{2,\,\mathrm{gas}} + 4OH_O^{\bullet} + 4TM_{TM}^{x} \rightleftharpoons 2H_2O + 4TM_{TM}^{\bullet}, \tag{11.5}$$

oxygen vacancies are, in principle, unnecessary, because the reduced O is desorbed as H_2O and not incorporated as O_O^x as it is in SOFCs. Nevertheless, oxygen vacancies, as catalytically active centers, might facilitate O=O splitting; cf. DFT results for $BaZr_{0.75}Co_{0.25}O_{3-\delta}$ which shows a very small dissociation barrier when intermediate O_2^{y-} species are adsorbed inside an oxygen vacancy [71]. Studies using dense micro-electrodes are rare for protonic positrode materials. Thus, the details of the surface reaction mechanism and its dependence on material properties, such as electronic structure and defect concentrations, have largely remained elusive so far. As long as the deviation from equilibrium is not too large, the principle of microscopic reversibility implies that the same rds applies to both forward and backward reactions, i.e. it also suggests that the same overall activation energy and pO_2, pH_2O dependencies apply (oxygen reduction is often considered, as it appears more intuitive than water oxidation, but the final equilibrium rate expression is identical).

Possible reaction pathways for the oxygen reduction/water oxidation reaction are shown in figure 11.10(a) [72]. Related mechanisms are suggested in e.g. [42, 73, 74], but it should be noted that the dissociation of O_2 into neutral adsorbed O is not very probable—some early electron transfer is expected. Scenarios with and without assistance of an oxygen vacancy for the O=O dissociation are given. If oxygen vacancies are involved, they have to be regenerated in a final water desorption step, as they are only a catalyst, not a stoichiometric reaction partner. The various branches differ by the point at which the intermediate oxygen species become protonated. Adsorbed O_2^- or O_2^{2-} are expected to have a comparatively low basicity, thus their protonation by surface hydroxide ions or bulk $OH_O^{\bullet}$ defects is less probable than the protonation of adsorbed O^-.

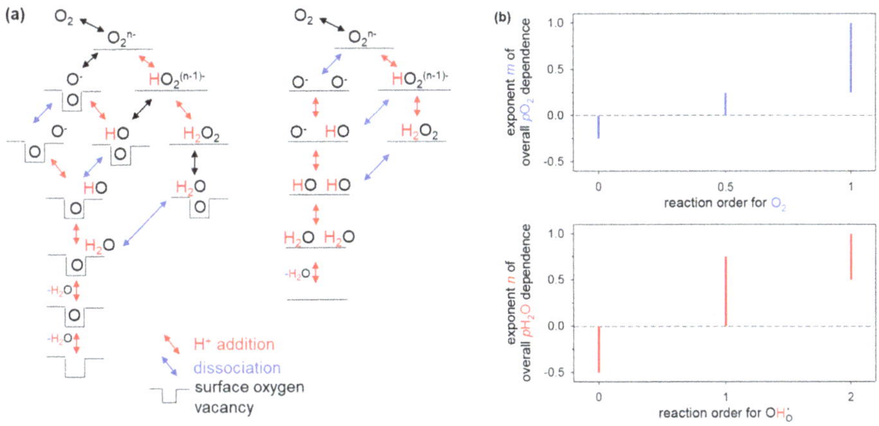

Figure 11.10. (a) Reaction pathways for oxygen reduction/water oxidation on protonic positrode materials, with and without surface $V_O^{••}$ assistance for the dissociation. (b) The range of exponents *m*,*n* of the overall pO_2 and pH_2O dependences of the equilibrium exchange rate $\mathfrak{R}$ derived for the pathways shown in (a); data taken from [72].

The area specific resistance of a positrode is inversely proportional to the equilibrium exchange rate $\mathfrak{R}$ of the surface reaction, which depends on pO_2 and potentially also on pH_2O. Figure 11.10(b) shows the range of exponents *m*,*n* of the overall pO_2 and pH_2O dependences of $\mathfrak{R} \propto (pO_2)^m$, $\mathfrak{R} \propto (pH_2O)^n$ extracted from the pathways in figure 11.10(a) [72]. It is important to note that these overall dependencies comprise the direct contribution of the involved point defects as well as their dependence on pO_2 and pH_2O. Thus a path which contains molecular oxygen species in the rds with a direct contribution from $(pO_2)^1$ may have an overall exponent *m* as low as 1/4. The effect of the involved protons on *m* is small, the negative pO_2 dependence of the proton concentration slightly decreases *m*. The exponent *n* of the overall pH_2O dependence contains a direct contribution $(pH_2O)^1$ if two protons, resulting from the incorporation of one H_2O, appear before/in the rds. In addition, it comprises the pH_2O dependences of other defects (e.g. $[V_O^{••}]$ with a slightly negative pH_2O dependence) participating before or in the rds. The latter term may even lead to an overall negative exponent $n = -1/2$ when no protons appear before/in the rds. In the cases for which the pH_2O dependence of the surface reaction resistance was measured, no pronounced acceleration of the reaction rate was observed with increasing pH_2O. This means that there is no strong evidence for the appearance of protons before/in the rds of the oxygen reduction reaction. Occasionally, increasing pH_2O has even been found to decelerate the O reduction rate.

One has to consider that surface defect concentrations differ, in principle, from the respective bulk concentrations. For $V_O^{••}$ and $h^•$, it may be a reasonable approximation to also use the bulk pO_2 and pH_2O dependencies for surface defects. However, while bulk proton concentrations are low, the concentration of surface OH groups is typically high, potentially approaching a saturation limit. It vanishes only at high temperatures (see e.g. [75]); perceptible surface OH coverage has even been observed on oxides such as ceria, which do not show bulk proton incorporation [76].

This means that surface OH groups might partially block the reactive sites required for oxygen reduction, such as $V_O^{\bullet\bullet}$ or undercoordinated cations; see [77].

An interesting comparison is given in [78], in which $BaCo_{0.4}Fe_{0.4}Zr_{0.1}Y_{0.1}O_{3-\delta}$ and $BaCo_{0.38}Fe_{0.38}Zr_{0.095}Y_{0.095}MO_{3-\delta}$ (M=Mn, Cu, Ni, Zn) perovskites are deposited to form a porous positrode on either oxide-ion-conducting $Ce_{0.8}Sm_{0.2}O_{1.9}$ or $BaCe_{0.7}Zr_{0.1}Y_{0.1}Yb_{0.1}O_{2.9}$ electrolytes (applying identical annealing temperatures to obtain comparable microstructures). The area specific resistances on the protonic electrolyte are systematically higher (factor 5–10) than on $Ce_{0.8}Sm_{0.2}O_{1.9}$. In this context it is important to remember that the part of a porous electrode which actually participates in the oxygen reduction reaction (the 'penetration depth') depends on the ionic (for PCFC/PCEC: protonic) conductivity of the electrode material. A higher ionic conductivity makes a larger part of the electrode active. Thus, the electrode resistance not only depends on specific surface area and the surface rate constant k but also on ionic/protonic conductivity (or equivalently, the tracer diffusivity D^*) [79]. For PCFC/PCEC positrode materials with limited protonic conductivity that are not used as composites with a protonic electrolyte, this might lead to a larger resistance.

11.3.5 Materials examples

PCEC positrode materials are closely related to PCFC positrode materials; thus, the following example materials stem from PCEC as well as PCFC research. A more detailed overview of PCFC cathode materials is given e.g. in [11, 42, 56, 78, 80–82] and chapter 14. One has to keep in mind that overall device performances depend on many cell components; even the area specific resistance of the positrode still depends on its microstructure (see chapter 9). However, some requirements differ for PCFCs. In electrolysis mode the positrode is usually exposed to higher pH_2O (for pressurized operation, pH_2O can even exceed 1 atm) than in PCFC operation. This requires better stability in high humidity. Since a high positrode overpotential increases the issues with electronic transport in the electrolyte (chapter 12), a high electrocatalytic activity achieved by optimizing material composition as well as electrode morphology is even more important than in fuel-cell mode. Another important difference is that in electrolysis cell operation, the presence of an overpotential at the positrode corresponds to an increased oxygen chemical potential in the material (a high hole concentration), which tends to decrease the proton uptake (section 11.3.1).

11.3.5.1 Simple perovskites

The largest systematic investigation of proton uptake as function of cation composition was performed for simple perovskites [56]; the trends found there might to some degree also be transferable to double perovskites and Ruddlesden–Popper phases. Starting from $BaFeO_{3-\delta}$ (stabilized by 5% La doping of the cubic perovskite structure), figure 11.11 shows that the replacement of Ba^{2+} by Sr^{2+} or larger amounts of La^{3+} decreases the proton uptake, in parallel with a decrease in the overall basicity of the materials. Partial replacement of Fe by Co increases the TM–O covalency and also decreases proton uptake, in line with the discussion in

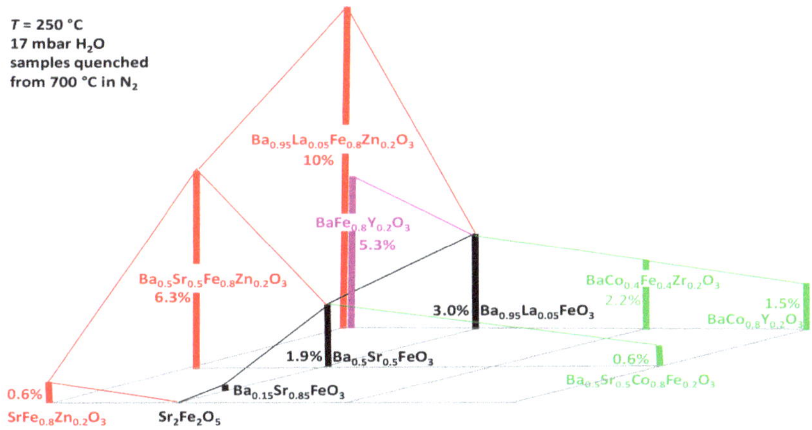

Figure 11.11. Proton uptakes (mol%) of simple perovskites measured by thermogravimetry at 250 °C in 17 mbar H_2O (coarse particles, quenched from 700 °C in N_2). Reproduced with permission from [56], copyright Annual Reviews (2021).

section 11.3.1. Interestingly, partial substitution with redox-inactive oversized cations such as Zn^{2+} and Y^{3+} significantly increases the proton uptake. This substitution causes local lattice distortions and in particular a bending of Fe–O–Fe connections [83], which decreases the Fe–O covalency and increases the basicity of the oxygen ions (see section 11.3.1).

As mentioned, systematic investigations of the surface kinetics on pore-free model electrodes are lacking. By analogy to the results for SOFC positrodes [69], one might expect that the presence of at least some cobalt on the perovskite's B site is beneficial for the oxygen reduction/water oxidation kinetics on PCFC/PCEC. $BaCo_{0.4}Fe_{0.4}Zr_{0.1}Y_{0.1}O_{3-\delta}$ and $Ba_{0.5}Sr_{0.5}Co_{0.8}Fe_{0.2}O_{3-\delta}$ have been successfully employed in performant lab-scale PCFCs [67, 84] and PCECs [26].

11.3.5.2 Double perovskites

Double perovskites have an expanded unit cell, caused by the ordering of the A- or B-site cations. Two families have been investigated in more detail for use as PCFC/PCEC positrode materials:

(i) Materials such as $NdBa_{0.5}Sr_{0.5}Co_{1.5}Fe_{0.5}O_{5+\delta}$ [85], $PrBa_{0.5}Sr_{0.5}Co_{1.5}Fe_{0.5}O_{5+\delta}$ [27, 44], $PrBa_{0.8}Ca_{0.2}Co_2O_{5+\delta}$ [86], and $BaGd_{0.8}La_{0.2}Co_2O_{5+\delta}$ [25], which show a layering of divalent (Ba^{2+}, Sr^{2+}, Ca^{2+}) and trivalent (La^{3+}, Pr^{3+}, Nd^{3+}, Gd^{3+}) A-site cations, and which are typically Co-rich. They have been employed in performant lab-scale PCFCs [44, 85, 86] and PCECs [27, 86, 87]. Although they have a lower concentration of alkali-earth cations per transition metal than the simple perovskites described in section 11.3.5.1, a comparably high proton uptake has been reported (3% for $PrBa_{0.5}Sr_{0.5}Co_{1.5}Fe_{0.5}O_{5+\delta}$ at 250 °C in air [44], 3% for $BaGd_{0.8}La_{0.2}Co_2O_{5+\delta}$ at 400 °C in 0.0004 bar O_2 [25]).

(ii) Materials such as $Sr_2Fe_{1.5}Mo_{0.5}O_{6-\delta}$, in which the pure Fe and mixed (Fe/Mo) sites show an alternating arrangement, and the high $Mo^{6+/5+}$ oxidation state pushes iron into lower $Fe^{2+/3+}$ states. A hydration enthalpy of −0.3 eV and a proton migration barrier of 0.5 eV have been obtained using DFT [88]. $Sr_2Fe_{1.5}Mo_{0.5}O_{6-\delta}$ has been employed as lab-scale PCEC positrode; however, its performance was moderate due to the high ohmic resistance of the protonic electrolyte [89].

11.3.5.3 Ruddlesden–Popper phases

A Ruddlesden–Popper phase of order p can be regarded as a stack of perovskite-type $A_pB_pO_{3p}$ layers which alternate with rock salt-type AO layers. Such a phase can exhibit oxygen deficiency or oxygen excess relative to the perfect structure (cf. chapter 3); oxygen excess is realized by O_i'' interstitials in the AO layer. If the constituent materials contain oxygen vacancies, hydration occurs according to reaction (11.1). If no oxygen vacancies are available, hydration can occur via the formation of interstitial hydroxide ions:

$$H_2O_{gas} + O_O^x \rightleftharpoons OH_O^{\bullet} + OH_i' \qquad K_{RP} = \frac{[OH_O^{\bullet}][OH_i']}{pH_2O[O_O^x]}. \qquad (11.6)$$

However, this reaction is expected to be energetically less favorable than the vacancy-filling (11.1). Ruddlesden–Popper phases have two chemically distinct types of oxygen ions in the AO and inner part of perovskite-type layers, which are connected to only one or to two TM cations. Correspondingly, oxygen in the AO layers, which has a more basic character, offers energetically more favorable binding sites for protons (see e.g. the DFT results in [90]). This spatial decoupling of proton-accepting AO layers with a more ionic character and more covalent perovskite layers supplying electronic conductivity may be advantageous for PCFC/PCEC positrode materials. However, to some degree this advantage is lost due to an increase in proton migration barriers from the AO to perovskite layers [90]. Two families have been investigated in more detail for use as PCFC/PCEC positrode materials:

(i) First-order Ruddlesden–Popper phases that typically have nickel as the B-site cation, such as $(Pr_{1-x}Sr_x)_2NiO_{4\pm\delta}$ [91]. Materials for which $x \leqslant 0.25$ are in the oxygen-excess regime and proton uptake is correspondingly low. Nevertheless, $Pr_2NiO_{4+\delta}$ showed an area specific resistance comparable to that of $PrBaCo_2O_{5+\delta}$ [74]. An overview of the more recently investigated (and rather strongly varying) performance of $Ln_2NiO_{4\pm\delta}$-based positrodes is given in [92].

(ii) Higher-order Ruddlesden–Popper phases that typically have iron as the B-site cation, such as $Sr_3Fe_2O_{7-\delta}$, for which good lab-scale PCFC performance has been reported [90]. The exclusive (or at least predominant) occupation of the A site by 2+ alkali-earth ions ensures the presence of oxygen vacancies that can be hydrated.

11.3.5.4 Complex morphologies

The use of a composite formed by mixing positrode and electrolyte powders is well-established for SOFCs/SOECs. The material combinations must be free of detrimental reactions under sintering conditions, and a microstructure must be fabricated in which the gas phase, electrolyte particles, and positrode particles each form a percolating network. The benefits are a better match of thermal expansion between the electrolyte and the composite electrode, and in the case of positrode materials with comparatively low ionic conductivity, another benefit is improved ionic transport within the electrode layer. The same strategy is frequently used for PCFCs/PCECs as it allows some of the conflicting properties to be distributed (figure 11.9(b)) to different phases. For positrode materials with negligible proton conductivity, such as $La_{0.6}Sr_{0.4}Co_{0.2}Fe_{0.8}O_{3-\delta}$, a composite with the protonic electrolyte drastically increased the device performance [93]. Even for $BaCo_{0.4}Fe_{0.4}Zr_{0.1}Y_{0.1}O_{3-\delta}$, which has some proton conductivity, a composite with the electrolyte material was beneficial [67]. Recently, a modified approach was suggested in which a composite forms by a self-assembly process from a common (amorphous) precursor, e.g. from $BaCo_{0.7}Ce_{0.24}Y_{0.06}O_{3-\delta}$ or $Sr_{0.9}Ce_{0.1}Fe_{0.8}Ni_{0.2}O_{3-\delta}$ [94, 95]. Furthermore, infiltration or decoration with additional catalysts on a moderately active mixed-conduction backbone is possible. Interestingly, catalytically active noble-metal clusters can be exsolved from $BaCo_{0.4}Fe_{0.4}Zr_{0.1}Y_{0.1}O_{3-\delta}$ in oxidizing humid atmospheres when proton incorporation at the expense of holes (reaction (11.4)) leads to a decrease in the average transition-metal oxidation state [96].

11.4 Concluding remarks

In terms of electrolyte materials, mainly perovskites are employed in PCECs. Their fundamental defect and transport properties are well understood. The detailed optimization of composition (Ce content, Y/Yb co-doping) and sintering processes for large-area cells/tubes is in progress. Potential issues caused by hole transport in PCEC mode are being recognized and mitigated. In the field of positrode materials, the range of materials is larger, comprising simple and double perovskites and occasionally also Ruddlesden–Popper phases. The perovskites have high Ba and/or Sr contents, which favor hydration but may cause stability issues. The defect chemistry of such materials, which have protons, oxygen vacancies, and holes, is complex (comprising also pronounced mutual defect interactions) and still developing. The proton uptake is lower than for electrolyte materials, but for many Ba/Sr-rich materials, it is expected to be adequate to extend the reactive zone beyond the three-phase boundary. More investigations are required to decouple the overall surface kinetics into morphological aspects and intrinsic materials properties (preferably on pore-free model electrodes), and elucidate the rds of oxygen reduction to water in PCEC operation. Overall, both PCECs and PCFCs are in an earlier stage of technological development than SOECs/SOFCs, but are making rapid progress. They have reached a stage at which not only the individual component materials, but also their interplay in cells and stacks, has become relevant (see chapters 12–14). PCFCs/PCECs have a strong potential to become competitive, in particular regarding their specific advantages of decreased operating temperature and direct dry hydrogen production.

Acknowledgments

The author wishes to thank C Berger, M F Hoedl, Y Huang, K D Kreuer, J Maier, D Poetzsch, G Raimondi, M Shirpour, R Zohourian (MPI Stuttgart) and

D Gryaznov, E A. Kotomin (University of Riga) for very fruitful collaboration and numerous valuable discussions.

References

[1] Iwahara H, Esaka T, Uchida H and Maeda N 1981 Proton conduction in sintered oxides and its application to steam electrolysis for hydrogen production *Solid State Ionics* **3** 359–63
[2] Kreuer K D 1996 Proton conductivity: materials and applications *Chem. Rev.* **8** 610–41
[3] Norby T 1999 Solid-state protonic conductors: principles, properties, progress and prospects *Solid State Ionics* **125** 1–11
[4] Kreuer K D 2003 Proton-conducting oxides *Annu. Rev. Mater. Sci.* **33** 333–59
[5] Kreuer K D 2014 Ion conducting membranes for fuel cells and other electrochemical devices *Chem. Mater.* **26** 361–80
[6] Steele B C H 2001 Material science and engineering: the enabling technology for the commercialization of fuel cell systems *J. Mater. Sci.* **36** 1053–68
[7] Yamazaki Y, Hernandez-Sanchez R and Haile S M 2009 High total proton conductivity in large-grained yttrium-doped barium zirconate *Chem. Mater.* **21** 2755–62
[8] Lei L B, Zhang J H, Yuan Z H, Liu J P, Ni M and Chen F L 2019 Progress report on proton conducting solid oxide electrolysis cells *Adv. Funct. Mater.* **29** 1903805
[9] Duan C, Huang J, Sullivan N and O'Hayre R 2020 Proton-conducting oxides for energy conversion and storage *Appl. Phys. Rev.* **7** 011314
[10] Ji H I, Lee J H, Son J W, Yoon K J, Yang S and Kim B K 2020 Protonic ceramic electrolysis cells for fuel production: a brief review *J. Kor. Ceram. Soc.* **57** 480–94
[11] Medvedev D 2019 Trends in research and development of protonic ceramic electrolysis cells *Int. J. Hydr. Energy* **44** 26711–40
[12] Su H and Hu Y H 2022 Degradation issues and stabilization strategies of protonic ceramic electrolysis cells for steam electrolysis *Energy Sci. Eng.* **10** 1706–25
[13] Le L Q, Hernandez C H, Rodriguez M H, Zhu L, Duan C, Ding H, O'Hayre R P and Sullivan N P 2021 Proton-conducting ceramic fuel cells: scale up and stack integration *J. Power Sources* **482** 228868
[14] Dubois A, Ricote S and Braun R J 2017 Benchmarking the expected stack manufacturing cost of next generation, intermediate-temperature protonic ceramic fuel cells with solid oxide fuel cell technology *J. Power Sources* **369** 65–77
[15] Bjorheim T S, Hoedl M F, Merkle R, Kotomin E A and Maier J 2020 Proton, hydroxide ion, and oxide ion affinities of closed-shell oxides: importance for the hydration reaction and correlation to electronic structure *J. Phys. Chem.* C **124** 1277–84
[16] Munch W, Seifert G, Kreuer K D and Maier J 1996 A quantum molecular dynamics study of proton conduction phenomena in $BaCeO_3$ *Solid State Ionics* **86** 647–52
[17] Bohn H G and Schober T 2000 Electrical conductivity of the high-temperature proton conductor $BaZr_{0.9}Y_{0.1}O_{2.95}$ *J. Am. Ceram. Soc.* **83** 768–72
[18] Kreuer K D, Adams S, Munch W, Fuchs A, Klock U and Maier J 2001 Proton conducting alkaline earth zirconates and titanates for high drain electrochemical applications *Solid State Ionics* **145** 295–306
[19] Draber F M, Ader C, Arnold J P, Eisele S, Grieshammer S, Yamaguchi S and Martin M 2020 Nanoscale percolation in doped $BaZrO_3$ for high proton mobility *Nat. Mater.* **19** 338–46

[20] Han D, Liu X, Bjorheim T S and Uda T 2021 Yttrium-doped barium zirconate-cerate solid solution as proton conducting electrolyte: why higher cerium concentration leads to better performance for fuel cells and electrolysis cells *Adv. Energ. Mater.* **11** 2003149

[21] Kreuer K D, Schonherr E and Maier J 1994 Proton and oxygen diffusion in $BaCeO_3$ based compounds – a combined thermal gravimetric analysis and conductivity study *Solid State Ionics* **70** 278–84

[22] Zhu H and Kee R J 2016 Membrane polarization in mixed-conducting ceramic fuel cells and electrolyzers *Int. J. Hydr. Energy* **41** 2931–43

[23] Nomura K and Kageyama H 2007 Transport properties of $Ba(Zr_{0.8}Y_{0.2})O_{3-\delta}$ perovskite *Solid State Ionics* **178** 661–5

[24] Poetzsch D, Merkle R and Maier J 2013 Investigation of oxygen exchange kinetics in proton-conducting ceramic fuel cells: effect of electronic leakage current using symmetric cells *J. Power Sources* **242** 784–9

[25] Strandbakke R, Cherepanov V A, Zuev A Y, Tsvetkov D S, Argirusis C, Sourkouni G, Prunte S and Norby T 2015 Gd- and Pr-based double perovskite cobaltites as oxygen electrodes for proton ceramic fuel cells and electrolyser cells *Solid State Ionics* **278** 120–32

[26] Duan C, Kee R, Zhu H, Sullivan N, Zhu L, Bian L, Jennings D and O'Hayre R 2019 Highly efficient reversible protonic ceramic electrochemical cells for power generation and fuel production *Nat. Energy* **4** 230–40

[27] Choi S, Davenport T C and Haile S M 2019 Protonic ceramic electrochemical cells for hydrogen production and electricity generation: exceptional reversibility, stability, and demonstrated faradaic efficiency *Energy Environ. Sci.* **12** 206–15

[28] Shimada H, Yamaguchi T, Sumi H, Yamaguchi Y, Nomura K, Mizutani Y and Fujishiro Y 2019 A key for achieving higher open-circuit voltage in protonic ceramic fuel cells: lowering interfacial electrode polarization *ACS Appl. Energy Mater.* **2** 587–97

[29] Gregori G, Merkle R and Maier J 2017 Ion conduction and redistribution at grain boundaries in oxide systems *Prog. Mater. Sci.* **89** 252–305

[30] Haile S M, West D L and Campbell J 1998 The role of microstructure and processing on the proton conduction properties of gadolinium-doped barium cerate *J. Mater. Res.* **13** 1576–95

[31] Lindman A, Bjorheim T S and Wahnstrom G 2017 Defect segregation to grain boundaries in $BaZrO_3$ from first-principles free energy calculations *J. Mater. Chem.* A **5** 13421–9

[32] Lindman A, Helgee E E and Wahnstrom G 2017 Comparison of space-charge formation at grain boundaries in proton-conducting $BaZrO_3$ and $BaCeO_3$ *Chem. Mater.* **29** 7931–41

[33] Kjolseth C, Fjeld H, Prytz O, Dahl P I, Estournes C, Haugsrud R and Norby T 2010 Space-charge theory applied to the grain boundary impedance of proton conducting $BaZr_{0.9}Y_{0.1}O_{3-\delta}$ *Solid State Ionics* **181** 268–75

[34] Iguchi F, Chen C T, Yugami H and Kim S 2011 Direct evidence of potential barriers at grain boundaries in Y-doped $BaZrO_3$ from dc-bias dependence measurements *J. Mater. Chem.* **21** 16517–23

[35] Shirpour M, Rahmati B, Sigle W, van Aken P A, Merkle R and Maier J 2012 Dopant segregation and space charge effects in proton-conducting $BaZrO_3$ perovskites *J. Phys. Chem.* C **116** 2453–61

[36] Babilo P and Haile S M 2005 Enhanced sintering of yttrium-doped barium zirconate by addition of ZnO *J. Am. Ceram. Soc.* **88** 2362–8

[37] Tong J, Clark D, Bernau L, Sanders M and O'Hayre R 2010 Solid-state reactive sintering mechanism for large-grained yttrium-doped barium zirconate proton conducting ceramics *J. Mater. Chem.* **20** 6333–41
[38] Han D, Uemura S, Hiraiwa C, Majima M and Uda T 2018 Detrimental effect of sintering additives on conducting ceramics: yttrium-doped barium zirconate *ChemSusChem* **11** 4102–13
[39] Loureiro F J A, Nasani N, Reddy G S, Munirathnam N R and Fagg D P 2019 A review on sintering technology of proton conducting $BaCeO_3$–$BaZrO_3$ perovskite oxide materials for protonic ceramic fuel cells *J. Power Sources* **438** 226991
[40] Huang Y, Merkle R and Maier J 2021 Effects of NiO addition on sintering and proton uptake of $Ba(Zr,Ce,Y)O_{3-\delta}$ *J. Mater. Chem.* A **9** 14775–85
[41] Ryu K H and Haile S M 1999 Chemical stability and proton conductivity of doped $BaCeO_3$-$BaZrO_3$ solid solutions *Solid State Ionics* **125** 355–67
[42] Wang W, Medvedev D and Shao Z 2018 Gas humidification impact on the properties and performance of perovskite-type functional materials in proton-conducting solid oxide cells *Adv. Funct. Mater.* **28** 1802592
[43] Hyodo J, Kitabayashi K, Hoshino K, Okuyama Y and Yamazaki Y 2020 Fast and stable proton conduction in heavily scandium-doped polycrystalline barium zirconate at intermediate temperatures *Adv. Energ. Mater.* **10** 2000213
[44] Choi S, Kucharczyk C J, Liang Y, Zhang X, Takeuchi I, Ji H I and Haile S M 2018 Exceptional power density and stability at intermediate temperatures in protonic ceramic fuel cells *Nat. Energy* **3** 202–10
[45] Yang L, Wang S, Blinn K, Liu M, Liu Z and Liu M L 2009 Enhanced sulfur and coking tolerance of a mixed Ion conductor for SOFCs: $BaZr_{0.1}Ce_{0.7}Y_{0.2-x}Yb_xO_{3-\delta}$ *Science* **326** 126–9
[46] Han D, Jiang L and Zhong P 2021 Improving phase compatibility between doped $BaZrO_3$ and NiO in protonic ceramic cells via tuning composition and dopant *Int. J. Hydrogen Energy* **46** 8767–77
[47] Dayaghi A M, Haugsrud R, Stange M, Larring Y, Strandbakke R and Norby T 2021 Increasing the thermal expansion of proton conducting Y-doped $BaZrO_3$ by Sr and Ce substitution *Solid State Ionics* **359** 115534
[48] Leonard K, Lee Y S, Okuyama Y, Miyazaki K and Matsumoto H 2017 Influence of dopant levels on the hydration properties of SZCY and BZCY proton conducting ceramics for hydrogen production *Int. J. Hydrogen Energy* **42** 3926–37
[49] Kreuer K D 1999 Aspects of the formation and mobility of protonic charge carriers and the stability of perovskite-type oxides *Solid State Ionics* **125** 285–302
[50] Leonard K, Deibert W, Ivanova M E, Meulenberg W A, Ishihara T and Matsumoto H 2020 Processing ceramic proton conductor membranes for use in steam electrolysis *Membranes* **10** 339
[51] Nomura K, Takeuchi T, Kamo S, Kageyama H and Miyazaki Y 2004 Proton conduction in doped $LaScO_3$ perovskites *Solid State Ionics* **175** 553–5
[52] Farlenkov A S *et al* 2017 Water uptake, ionic and hole transport in $La_{0.9}Sr_{0.1}ScO_{3-\delta}$ *Solid State Ionics* **306** 126–36
[53] Kochetova N, Animitsa I, Medvedev D, Demin A and Tsiakaras P 2016 Recent activity in the development of proton-conducting oxides for high-temperature applications *RSC Adv.* **6** 73222–68
[54] Huse M and Haugsrud R 2012 Effects of A and B site acceptor doping on hydration and proton mobility of $LaNbO_4$ *Int. J. Hydrogen Energy* **37** 8004–16

[55] Magraso A and Haugsrud R 2014 Effects of the La/W ratio and doping on the structure, defect structure, stability and functional properties of proton-conducting lanthanum tungstate $La_{28-x}W_{4+x}O_{54+\delta}$. A review *J. Mater. Chem.* A **2** 12630–41

[56] Merkle R, Hoedl M F, Raimondi G, Zohourian R and Maier J 2021 Oxides with mixed protonic and electronic conductivity *Ann. Rev. Mater. Res.* **51** 461–93

[57] Poetzsch D, Merkle R and Maier J 2015 Stoichiometry variation in materials with three mobile carriers-thermodynamics and transport kinetics exemplified for protons, oxygen vacancies, and holes *Adv. Funct. Mater.* **25** 1542–57

[58] Hoedl M F, Gryaznov D, Merkle R, Kotomin E A and Maier J 2020 Interdependence of oxygenation and hydration in mixed-conducting (Ba,Sr)$FeO_{3-\delta}$ perovskites studied by density functional theory *J. Phys. Chem.* C **124** 11780–9

[59] Mizusaki J, Mima Y, Yamauchi S, Fueki K and Tagawa H 1989 Nonstoichiometry pf the perovskite-type oxides $La_{1-x}Sr_xCoO_{3-\delta}$ *J. Solid State Chem.* **80** 102–11

[60] Zohourian R, Merkle R, Raimondi R and Maier J 2018 Mixed-conducting perovskites as cathode materials for protonic ceramic fuel cells: understanding the trends in proton uptake *Adv. Funct. Mater.* **28** 1801241

[61] Kawada T, Suzuki J, Sase M, Kaimai A, Nigara Y, Mizusaki J, Kawamura K and Yugami H 2002 Determination of oxygen vacancy concentration in a thin film of $La_{0.6}Sr_{0.4}CoO_{3-\delta}$ by an electrochemical method *J. Electrochem. Soc.* **149** E252–9

[62] Nakamura T, Oike R, Kimura Y, Tamenori Y, Kawada T and Amezawa K 2017 Operando Soft X-ray absorption spectroscopic study on a solid oxide fuel cell cathode during electrochemical oxygen reduction *ChemSusChem* **10** 2008–14

[63] Poetzsch D, Merkle R and Maier J 2015 Proton uptake in the H^+-SOFC cathode material Ba0.5$Sr_{0.5}Fe_{0.8}Zn_{0.2}O_{3-\delta}$: transition from hydration to hydrogenation with increasing oxygen partial pressure *Farad. Discuss.* **182** 129–43

[64] Munoz-Garcia A B and Pavone M 2016 From oxide to proton conduction: a quantum-chemical perspective on the versatility of $Sr_2Fe_{1.5}Mo_{0.5}O_{6-\delta}$-based materials *Int. J. Quant. Chem.* **116** 1501–6

[65] Xu X, Wang H, Fronzi M, Wang X, Bi L and Traversa E 2019 Tailoring cations in a perovskite cathode for proton-conducting solid oxide fuel cells with high performance *J. Mater. Chem.* A **7** 20624–32

[66] Zhu K, Yang Y, Huan D, Hu X, Shi N, Xie Y, Li X, Xia C, Peng R and Li Y 2021 Theoretical and experimental investigations on K-doped $SrCo_{0.9}Nb_{0.1}O_{3-\delta}$ as a promising cathode for proton-conducting solid oxide fuel cells *ChemSusChem* **14** 3876–86

[67] Duan C, Tong J, Shang M, Nikodemski S, Sanders M, Ricote S, Almansoori A and O'Hayre R 2015 Readily processed protonic ceramic fuel cells with high performance at low temperatures *Science* **349** 1321–6

[68] Amezawa K 2021 Keynote talk KN46 at SSPC-20, (online) How Do We Investigate Reaction Mechanism in Oxide Electrode on Proton-Conducting Electrolyte?

[69] Kuklja M M, Kotomin E A, Merkle R, Mastrikov Y A and Maier J 2013 Combined theoretical and experimental analysis of processes determining cathode performance in solid oxide fuel cells *Phys. Chem. Chem. Phys.* **15** 5443–71

[70] Siebenhofer M, Riedl C, Schmid A, Limbeck A, Opitz A K, Fleig J and Kubicek M 2021 Investigating oxygen reduction pathways on pristine SOFC cathode surfaces by *in situ* PLD impedance spectroscopy *J. Mater. Chem.* A **10** 2305–19

[71] Wang Z, Yang W, Zhu Z, Peng R, Wu X, Xia C and Lu Y 2014 First-principles study of O_2 reduction on $BaZr_{1-x}Co_xO_3$ cathodes in protonic-solid oxide fuel cells *J. Mater. Chem.* A **2** 16707–14

[72] Poetzsch D, Merkle R and Maier J 2015 Oxygen reduction at dense thin-film microelectrodes on a proton-conducting electrolyte I. Considerations on reaction mechanism and electronic leakage effects *J. Electrochem. Soc.* **162** F939–50

[73] Peng R, Wu T, Liu W, Liu X and Meng G 2010 Cathode processes and materials for solid oxide fuel cells with proton conductors as electrolytes *J. Mater. Chem.* **20** 6218–25

[74] Grimaud A, Mauvy F, Bassat J M, Fourcade S, Rocheron L, Marrony M and Grenier J C 2012 Hydration properties and rate determining steps of the oxygen reduction reaction of perovskite-related oxides as H^+-SOFC cathodes *J. Electrochem. Soc.* **159** B683–94

[75] Stoerzinger K A, Comes R, Spurgeon S R, Thevuthasan S, Ihm K, Crumlin E J and Chambers S A 2017 Influence of $LaFeO_3$ surface termination on water reactivity *J. Phys. Chem. Lett.* **8** 1038–43

[76] Feng Z A, El Gabaly F, Ye X, Shen Z X and Chueh W C 2014 Fast vacancy-mediated oxygen ion incorporation across the ceria-gas electrochemical interface *Nat. Comm.* **5** 4374

[77] Huang Y L, Pellegrinelli C and Wachsman E D 2016 Fundamental impact of humidity on SOFC cathode ORR *J. Electrochem. Soc.* F171–82

[78] Liang M, He F, Zhou C, Chen Y, Yang G, Zhou W and Shao Z 2021 Nickel-doped $BaCo_{0.4}Fe_{0.4}Zr_{0.1}Y_{0.1}O_{3-\delta}$ as a new high-performance cathode for both oxygen-ion and proton conducting fuel cells *Chem. Eng. J.* **420** 127717

[79] Adler S B, Lane J A and Steele B C H 1996 Electrode kinetics of porous mixed-conducting oxygen electrodes *J. Electrochem. Soc.* **143** 3554–64

[80] Papac M, Stevanovic V, Zakutayev A and O'Hayre R 2021 Triple ionic-electronic conducting oxides for next-generation electrochemical devices *Nat. Mater.* **20** 301–13

[81] Mather G C, Muñoz-Gil D, Zamudio-Garcia J, Porras-Vázquez J M, Marrero-López D and Pérez-Coll D 2021 Perspectives on cathodes for protonic ceramic fuel cells *Appl. Sci.* **11** 5363

[82] Tang C, Akimoto K, Wang N, Fadillah L, Kitano S, Habazaki H and Aoki Y 2021 The effect of an anode functional layer on the steam electrolysis performances of protonic solid oxide cells *J. Mater. Chem.* A **9** 14032–42

[83] Raimondi G, Giannici F, Longo A, Merkle R, Chiara A, Hoedl M F, Martorana A and Maier J 2020 X-ray spectroscopy of $(Ba,Sr,La)(Fe,Zn,Y)O_{3-\delta}$ identifies structural and electronic features favoring proton uptake *Chem. Mater.* **32** 8502–11

[84] An H, Lee H W, Kim B K, Son J W, Yoon K J, Kim H, Shin D, Ji H I and Lee J H 2018 A 5 × 5 cm^2 protonic ceramic fuel cell with a power density of 1.3 W cm^{-2} at 600 °C *Nat. Energy* **3** 870–5

[85] Kim J, Sengodan S, Kwon G, Dong D, Shin J, Liu M L and Kim G Triple-conducting layered perovskites as cathode materials for proton-conducting solid oxide fuel cells *ChemSusChem* **7** 2811–5

[86] Zhou Y *et al* 2021 An active and robust air electrode for reversible protonic ceramic electrochemical cells *ACS Energy Lett.* **6** 1511–20

[87] Vollestad E, Strandbakke R, Tarach M, Catalan-Martinez D, Fontaine M L, Beeaff D, Clark D R, Serra J M and Norby T 2019 Mixed proton and electron conducting double perovskite anodes for stable and efficient tubular proton ceramic electrolysers *Nat. Mater.* **18** 752–9

[88] Munoz-Garcia A B and Pavone M 2016 First-principles design of new electrodes for proton-conducting solid-oxide electrochemical cells: a-site doped $Sr_2Fe_{1.5}Mo_{0.5}O_{6-\delta}$ perovskite *Chem. Mater.* **28** 490–500

[89] Lei L, Tao Z, Wang X, Lemmon J P and Fang C L 2017 Intermediate-temperature solid oxide electrolysis cells with thin proton-conducting electrolyte and a robust air electrode *J. Mater. Chem.* A **5** 22945–51

[90] Wang Z, Yang W, Shafi S P, Bi L, Wang Z, Peng R, Xia C, Liu W and Lu Y 2015 A high performance cathode for proton conducting solid oxide fuel cells *J. Mater. Chem.* A **3** 8405–12

[91] Grimaud A, Mauvy F, Bassat J M, Fourcade S, Marrony M and Grenier J C 2012 Hydration and transport properties of the $Pr_{2-x}Sr_xNiO_{4+\delta}$ compounds as H^+-SOFC cathodes *J. Mater. Chem.* **22** 16017–25

[92] Tarutin A P, Lyageva J G, Medvedev D A, Bi L and Yaremchenko A A 2021 Recent advances in layered $Ln_2NiO_{4+\delta}$ nickelates: fundamentals and prospects of their applications in protonic ceramic fuel and electrolysis cells *J. Mater. Chem.* A **9** 154–95

[93] Yang L, Liu Z, Wang S, Choi Y M, Zuo C and Liu M L 2010 A mixed proton, oxygen ion, and electron conducting cathode for SOFCs based on oxide proton conductors *J. Power Sources* **195** 471–4

[94] Song Y, Chen Y, Wang W, Zhou C, Zhong Y, Yang G, Zhou W, Liu M and Shao Z P 2019 Self-assembled triple-conducting nanocomposite as a superior protonic ceramic fuel cell cathode *Joule* **3** 2842–53

[95] Song Y, Liu J, Wang Y, Guan D, Seong A, Liang M, Robson MJ, Xiong X, Zhang Z, Kim G, Shao ZP and Ciucci F 2021 Nanocomposites: a new opportunity for developing highly active and durable bifunctional air electrodes for reversible protonic ceramic cells. *Adv. Energ. Mater.* **11** 2101899

[96] Kim H J, Hong J, Lim D K, Ahn S, Kim J, Kim J K, Oh D, Jeon S H, Song S J and Jung W C 2022 Water as a hole-predatory instrument to create metal nanoparticles on triple-conducting oxides *Energy Environ. Sci.* **15** 1097–105

Chapter 12

Thermodynamics, transport, and electrochemistry in protonic ceramic electrolysis cells

Huayang Zhu, Sandrine Ricote and Robert J Kee

This chapter discusses the theory and performance of electrolysis cells based upon proton-conducting ceramics (protonic ceramic electrolysis cells (PCECs)). To put PCECs in perspective in relation to other ceramic-based electrochemical cells, the discussion initially compares oxygen-ion-conducting membranes as well as fuel-cell performance. Although the proton-conducting membrane is at the heart of the cell, its electrolytic performance also significantly depends on electrode performance, including porous-media transport, charge transfer, and defect-incorporation chemistry. Proton-conducting ceramics (e.g. doped barium zirconates or cerates) are typically mixed ionic–electronic conductors (MIECs). The electronic conduction, in the form of positively charged small polarons or electron holes, leads to 'electronic leakage.' In an ideal steam electrolysis cell, one gas-phase H_2 molecule is produced from every two electrons delivered by an external power source. However, the electronic flux associated with MIEC membranes contributes to reduced faradaic efficiency. The chapter discusses and applies models that can be used to interpret and predict the performance of these cells. The illustrative results consider the behavior of a cell based upon a $BaCe_{0.7}Zr_{0.1}Y_{0.1}Yb_{0.1}O_{3-\delta}$ (BCZYYb) electrolyte membrane.

12.1 Introduction

Ceramic ion-conducting electrolyte membranes will likely play central roles in technologies including fuel cells, electrolysis, and separation membranes. This chapter is primarily concerned with electrolysis using ceramic proton-conducting membranes such as doped barium zirconates and cerates. Although the present focus is on protonic ceramic electrolysis, the introductory writing draws on examples that compare alternative oxygen-ion-conducting membranes and the reverse polarization used for fuel-cell operation.

Although ion transport (e.g. oxygen ions or protons) is most relevant for this technology, the materials are often, but not always, MIECs. The present chapter

doi:10.1088/978-0-7503-3951-3ch12

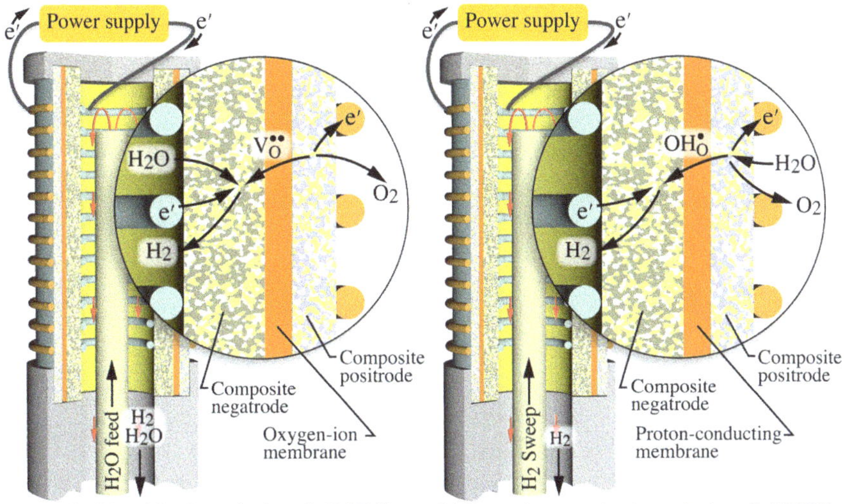

Figure 12.1. Comparison of the SOEC and PCEC principles in tubular configurations [1]. Reproduced by permission, copyright IOP Publishing.

briefly compares, contrasts, and discusses the functional membrane behaviors of alternative and widely used electrolyte materials. Yttria-stabilized zirconia (YSZ) and gadolinium-doped ceria (GDC) are primarily oxygen ion conductors. Yttrium-doped barium zirconate (BZY) and yttrium- and ytterbium-doped solid solutions of barium cerate and zirconate (e.g. BCZYYb) are primarily proton conductors. While the underpinning theory is much more general, the examples throughout this chapter focus particularly on the BCZYYb electrolyte. Figure 12.1 illustrates aspects of the operating principles for solid oxide electrolysis cells (SOECs) and PCECs. The operating principles for the two alternatives are quite different. Electrolysis cells based on oxygen ion conduction (SOECs) represent the more mature technology. However, PCEC technology offers some potentially significant advantages, including lower-temperature operation and the production of nearly dry H_2.

12.1.1 SOEC function

Consider first the SOEC behavior (figure 12.1(a)). Steam is introduced into the tube's interior (the hydrogen-collection side) and air is present on the tube's exterior (the oxygen-production side). Within the porous composite negatrode (electrolysis cathode; typically, Ni-YSZ or Ni-GDC), hydrogen is produced via electrochemical charge transfer as follows (in Kröger–Vink notation):

$$H_2O(g) + V_O^{\bullet\bullet}(el) + 2e'(ed) \rightleftharpoons H_2(g) + O_O^{\times}(el). \quad (12.1)$$

The parenthetical (g), (ed) and (el) represent the gas, electrode, and electrolyte phases, respectively. On the positrode (electrolysis anode) side, the charge-transfer chemistry may be written as

$$O_O^{\times}(el) \rightleftharpoons V_O^{\bullet\bullet}(el) + 2e'(ed) + \frac{1}{2}O_2(g). \quad (12.2)$$

The composite electrodes typically comprise three phases—an ion-conducting phase, an electron-conducting phase, and a percolating pore phase that hosts the gas phase. Within YSZ, the oxygen vacancies $V_O^{\bullet\bullet}$ are mobile and driven primarily by electrostatic potential gradients (i.e. migration). The gases are mobile within the composite electrode pore volumes. The electrons come from and are released to the external circuit power supply. In SOECs, the H_2O is introduced on the negatrode side of the cell. The desired H_2 product, which is produced on the negatrode side, is mixed with H_2O. Thus, subsequent separation is needed to deliver dry H_2.

12.1.2 PCEC function

Protonic ceramic cells are based on electrolyte membranes that are predominantly proton conductors. However, as discussed subsequently, they are in fact mixed ionic–electronic conductors. As illustrated in figure 12.1(b), H_2O is introduced on the positrode (electrolysis anode) side, where it reacts electrochemically to produce O_2 and protons, as follows:

$$H_2O(g) + 2O_O^{\times}(el) \rightleftharpoons \frac{1}{2}O_2(g) + 2OH_O^{\bullet}(el) + 2e'(ed), \qquad (12.3)$$

On the negatrode (electrolysis cathode) side, the protons recombine electrochemically to produce the desired H_2, as follows:

$$2OH_O^{\bullet}(el) + 2e'(ed) \rightleftharpoons H_2(g) + 2O_O^{\times}(el). \qquad (12.4)$$

Note that equation (12.4) is written in the cathodic direction, meaning that electrons are consumed. This correctly captures the charge-transfer process that produces H_2. However, as discussed subsequently, the forward (anodic) reaction direction has important implications for evaluations of the reaction rate of progress using the Butler–Volmer formulation.

12.1.3 Practical tradeoffs

PCECs can be competitive with SOECs in terms of performance while operating at lower temperatures (about 500 °C). In contrast with SOECs, in which oxygen ions are transported through the electrolyte, PCECs are based on proton-conducting ceramic electrolytes that have lower activation energies (0.4 eV compared to 0.6–0.7 eV) [2]. The decrease in operating temperature is expected to enable the use of less expensive materials for the balance-of-plant and interconnect components, resulting in cost reductions. The water-splitting electrode and its chemical function in an SOEC are quite different than they are for a PCEC. Hydrogen in an SOEC is produced on the electrode that splits steam, resulting in the production of humidified hydrogen. High steam partial pressures on the cermet electrode may contribute to degradation. By contrast, water splitting in a PCEC occurs on the positrode (an oxidizing environment). The protons produced are transported through the electrolyte, where they recombine to form nominally dry hydrogen on the negatrode (reducing environment) side.

12.2 Electrolyte and electrode compositions

The most studied high-temperature proton conductors crystallize in the perovskite structure, in which the A site is occupied by a divalent cation (e.g. Ba^{2+}, Sr^{2+}) and the B site is occupied by tetravalent cations (e.g. Zr^{4+}, Ce^{4+}) and trivalent cation dopants (Y^{3+}, Yb^{3+}) [2–6]. Other non-perovskite materials, including pyrochlores [7, 8], niobates [9], and tungstates [10], can conduct protons [11]. However, their conductivities are substantially lower than those of the perovskites.

The proton uptake in the perovskite $(Ba,Sr)(Ce,Zr,Y,Yb)O_{3-\delta}$ materials depends on the basicity of the oxygen. Kreuer [12] reported that the enthalpy of hydration becomes more exothermic as the electronegativity of the cations interacting with the lattice oxygen decreases (i.e. with a decrease in the Brønsted basicity of the oxide). Norby *et al* [13] proposed, based on a large set of experimental data, a correlation between the hydration enthalpy $\Delta H^{\circ}_{H_2O}$ (kJ mol^{-1}) and the difference in electronegativity between the cations populating the perovskite's B sites ($\chi_B^{AR} = \sum_i x_{B,i}\chi_i^{AR}$) and A sites ($\chi_A^{AR} = \sum_i x_{A,i}\chi_i^{AR}$), as follows:

$$\Delta H^{\circ}_{H_2O} = 400\left(\chi_B^{AR} - \chi_A^{AR}\right) - 180, \quad (12.5)$$

where $\chi_{A,i}$ and $\chi_{B,i}$ are the chemical compositions at the A and B sites. Table 12.1 lists the Allred–Rochow electronegativities χ_i^{AR} for the cations in $(Ba,Sr)(Ce,Zr,Y,Yb)O_{3-\delta}$. Using these electronegativities, table 12.1 also lists estimated hydration enthalpies for several common compositions.

The electronegativity analysis shows that dehydration temperatures increase as the hydration enthalpies (i.e. $\Delta H^{\circ}_{H_2O}$ for $H_2O + V_O^{\bullet\bullet} + O_O^{\times} \rightleftharpoons 2OH_O^{\bullet}$) become more negative. This trend was confirmed by Han *et al* [14], who studied $BaCe_{0.8-x}Zr_xY_{0.2}O_{3-\delta}$ in 3.1% moist oxygen with temperatures ranging from 375°C to 620 °C. Using high-temperature X-ray diffraction (HT-XRD) measurements while increasing the cerium content, they reported increasing dehydration temperatures. A more negative hydration enthalpy is desirable because the target operating temperatures are around 500 °C–550 °C. However, more negative hydration enthalpies can adversely affect electrolyte stability. Doped barium cerates exhibit the greatest proton uptake at temperatures from 500 °C to 600 °C, but they are the most unstable [15, 16]. Therefore, a practical compromise is needed between performance and stability. Leonard *et al* [17] studied the influence of dopant

Table 12.1. Allred–Rochow elemental electronegativities and compound hydration enthalpies evaluated using equation (12.5).

Element	χ_i^{AR}	Compound	$\Delta H^{\circ}_{H_2O}$ (kJ mol^{-1})
Ba	0.97	$BaZr_{0.8}Y_{0.2}O_{3-\delta}$	−88.80
Ce	1.07	$BaCe_{0.2}Zr_{0.7}Y_{0.1}O_{3-\delta}$	−96.40
Zr	1.22	$BaCe_{0.4}Zr_{0.4}Y_{0.1}Yb_{0.1}O_{3-\delta}$	−114.80
Y	1.11	$BaCe_{0.35}Zr_{0.44}Y_{0.2}O_{3-\delta}$	−114.68
Yb	1.06	$BaCe_{0.7}Zr_{0.1}Y_{0.1}Yb_{0.1}O_{3-\delta}$	−132.80
		$BaCe_{0.8}Y_{0.2}O_{3-\delta}$	−136.80

levels on the hydration properties of $SrZr_{1-x-y}Ce_xY_yO_{3-\delta}$ (SZCY) and $BaZr_{1-x-y}Ce_xY_yO_{3-\delta}$ (BZCY) proton-conducting ceramics. They found that $BaCe_{0.35}Zr_{0.44}Y_{0.2}O_{3-\delta}$ had a maximum proton concentration of 17.1 mol%, compared to only 6.03 mol% for $SrCe_{0.35}Zr_{0.44}Y_{0.2}O_{3-\delta}$, and retained the highest proton conductivity of 1.44×10^{-2} S cm^{-1} in wet 1% H_2 at 600 °C. Sufficient chemical stability was achieved under 80% steam for 200 h.

The high-temperature proton conductors are complex mixed ionic–electronic conductors. For oxygen partial pressures of less than approximately 10^{-6} atm, the perovskite (Ba,Sr)(Ce,Zr,Y,Yb)$O_{3-\delta}$ materials are effectively pure ion conductors. In addition to the protonic contribution to the conductivity, oxygen ion conduction becomes noticeable at temperatures above 600 °C. The concentration of O-site polaron conductivity increases according to $p_{O_2}^{1/4}$ [18–20]. At higher oxygen partial pressures, O-site polarons contribute significantly to the conductivity. For cerium-containing perovskites, some electronic conductivity under very reducing conditions has been observed at very high temperatures ($T > 900$ °C) [18]. However, because PCECs are typically not operated at such high temperatures, this effect is usually negligible. However, under polarization, evidence of cerium reduction has been reported [21].

To achieve high faradaic efficiency (see section 12.3), the conductivity of O-site polarons needs to be minimized and that of the proton conductivity maximized. It is known that electronic conductivity gradually diminishes with the increase of p_{H_2O} because there is competition to occupy the oxygen vacancies (hydration versus oxidation) [2, 6, 18–20]. Additionally, the transference number (i.e. partial conductivity fraction) of O-site polarons decreases with increasing Y doping from 10 to 20 mol%, but it does not further decrease when the Y content exceeds 20 mol% [22]. Furthermore, both Ba deficiency and Ba excess lead to an increase in the O-site polaron transference number. It has also been found that the partial replacement of Ba by Ca or Sr can increase the proton concentration and hence its transference number [6].

An interesting way to suppress the electronic leakage through the electrolyte is to add an electron-blocking layer between the electrolyte and the steam/air electrode. The ideal candidate for this electron-blocking layer would be a pure proton conductor. Li *et al* [23] used a $La_2Zr_2O_7$ pyrochlore layer between a BZY20 electrolyte and a $Pr_2NiO_{4+\delta}$ electrode. In addition to blocking the electrons, such a barrier layer can be used to prevent the degradation of high-cerium-content materials in high steam environments.

The foregoing discussion focused on the electrolyte material. However, cell performance also depends on electrode performance. The negatrodes (reducing environment) are typically porous ceramic–metallic composites (cermets) that use Ni as the electron-conducting phase and a BCZY as the ion-conducting phase. The steam/air electrode (positrode) has the largest effect on the durability and performance of PCECs. Positrode materials must be chemically stable in moist oxidizing environments and should also be chemically and physically compatible with the common proton-conducting electrolyte materials. Moreover, to achieve good performance and minimize polarization resistances, positrode materials should possess sufficient electronic and protonic conductivities and good catalytic activity for H_2O oxidation. He *et al* [24] concluded that the rate-limiting steps for the PCEC air electrode electrochemical process are the transfer of protons from decomposed water to the three-phase boundaries (TPBs) and the transport of protons from TPBs to the dense electrolyte membrane. Therefore, increasing the proton conductivity instead of the oxygen ion conductivity in the air electrode is crucial for improving the performance of PCECs. The most common positrode materials for PCEC applications are:

- cobalt-doped triple-conducting perovskites (e.g. $BaCo_{0.4}Fe_{0.4}Zr_{0.2-x}Y_xO_{3-\delta}$, $(Ba,Sr,La)(Fe,Co,Zn,Y)O_{3-\delta}$) [25–27] and
- double perovskites (e.g. $BaGd_{0.8}La_{0.2}Co_2O_{6-\delta}$, $PrBa_{0.5}Sr_{0.5}Co_{1.5}Fe_{0.5}O_{5+\delta}$) [28, 29] and
- Ruddlesden–Popper structured oxides (e.g. $(La,Pr)_2NiO_{4+\delta}$) [23, 30].

Cell performance depends on design considerations such as electrolyte membrane thickness, although these are not material properties per se. For example, thinner membranes beneficially reduce ohmic resistance. However, depending on the material, electrolyte membranes that are thin can adversely affect faradaic efficiency [31]. In all cases, the membranes must be pinhole free and adhere well to the adjoining electrode structures. The review by Winiarz *et al* [32] summarizes alternative techniques for fabricating thin BCZY electrolytes.

12.3 Faradaic and energy efficiencies

Faradaic efficiency characterizes the relationship between hydrogen production/consumption and the supplied/delivered electric current. In electrolysis mode, faradaic efficiency is defined as the ratio of the H_2 production flux J_{H_2} and the imposed electric current density i_e, as follows:

$$\eta_{F,EC} = \frac{2FJ_{H_2}}{i_e}. \tag{12.6}$$

In fuel-cell mode, faradaic efficiency is defined to be the ratio of the produced current density output i_e and the H_2 consumption flux J_{H_2}, as follows:

$$\eta_{F,FC} = \frac{i_e}{2FJ_{H_2}}. \tag{12.7}$$

Energy efficiency, which is related to faradaic efficiency, serves as a measure of the economic effectiveness of electrochemical cells. In electrolysis mode, energy efficiency can be evaluated as follows:

$$\eta_{E,EC} = \frac{-\Delta H_{H_2} J_{H_2}}{i_e E_{cell}} = \frac{-\Delta H_{H_2}}{2FE_{cell}} \frac{2FJ_{H_2}}{i_e} = \frac{E_h^\circ}{E_{cell}} \eta_{F,EC}, \tag{12.8}$$

Table 12.2. Thermodynamics of defect reactions for BCZYYb. The values of $\Delta H°$ and $\Delta S°$ are taken from [19].

Reactions	$\Delta H°$(kJ mol^{-1})	$\Delta S°$(J mol^{-1} K^{-1})	K_p (600 °C)	K_p (700 °C)	K_p (800 °C)
$\frac{1}{2}H_2 + O_O^{\bullet} \rightleftharpoons OH_O^{\bullet}$	−243.23	−54.87	4.85×10^{11}	1.55×10^{10}	9.41×10^{08}
$\frac{1}{2}O_2 + O_O^{X} + V_O^{\bullet\bullet} \rightleftharpoons 2O_O^{\bullet}$	+114.88	−60.30	9.49×10^{-11}	4.82×10^{-10}	1.81×10^{-09}
$H_2O + V_O^{\bullet\bullet} + O_O^{X} \rightleftharpoons 2OH_O^{\bullet}$	−130.00	−126.47	1.48×10^{01}	2.35×10^{00}	5.27×10^{-01}
$X_B' + O_O^{\bullet} \rightleftharpoons (X_B' - O_O^{\bullet})$	−90.00	−14.41	4.28×10^{04}	1.20×10^{04}	4.25×10^{03}
$H_2 + \frac{1}{2}O_2 \rightleftharpoons H_2O$	−248.11	−55.48	8.81×10^{11}	2.63×10^{10}	1.51×10^{09}

where ΔH_{H_2} is the enthalpic change of oxidation of the produced H_2 to steam, and $E_h^\circ = -\Delta H_{H_2}/2F$ is the thermoneutral voltage, at which there is no heat exchange with the surroundings. In the fuel-cell mode, the energy efficiency is evaluated as follows:

$$\eta_{E,FC} = \frac{i_e E_{cell}}{-\Delta H_{H_2} J_{H_2}} = \frac{2FE_{cell}}{-\Delta H_{H_2}} \frac{i_e}{2FJ_{H_2}} = \frac{E_{cell}}{E_h^\circ} \eta_{F,FC}. \tag{12.9}$$

Equations (12.8) and (12.9) show that the faradaic and energy efficiencies are related through the ratio of the cell operating voltage E_{cell} and the thermoneutral voltage E_h°. The open-circuit (Nernst) potential can be evaluated using $E_{OCV} = -\Delta G_{H_2}/2F$ for the overall reaction (i.e. $H_2 + \frac{1}{2}O_2 \rightleftharpoons H_2O$). Because $\Delta G = \Delta H - T\Delta S$, $E_{OCV} = E_h^\circ + T\Delta S/2F$. The overall hydrogen oxidation is exothermic, with $\Delta S_{H_2}^\circ \approx -55.48$ J mol^{-1} K^{-1} < 0 (see table 12.2). Thus, $E_{OCV} < E_h^\circ$ for H_2 electrochemical oxidation. If a cell is operated as a fuel cell, the operating cell potential E_{cell} must be lower than the open-circuit potential E_{OCV}. Therefore $E_{cell} < E_{OCV} < E_h^\circ$, and equation (12.9) indicates that $\eta_{E,FC} < \eta_{F,FC}$. If a cell is operated as an electrolysis cell, the operating cell potential E_{cell} must be higher than the open-circuit potential E_{OCV} (i.e. $E_{cell} > E_{OCV}$). If the cell voltage is lower than the thermoneutral voltage (i.e. $E_{OCV} < E_{cell} < E_h^\circ$), equation (12.8) indicates that $\eta_{E,EC} > \eta_{F,EC}$. However, if the cell operating voltage exceeds the thermoneutral voltage (i.e. $E_{cell} > E_h^\circ$), equation (12.8) indicates that $\eta_{E,EC} < \eta_{F,EC}$.

12.4 Electrolyte membrane performance

Yttrium and ytterbium co-doped barium zirconate cerate (e.g. BCZYYb) has a higher conductivity at temperatures below 750 °C than that of YSZ, GDC, and $BaZr_{0.8}Y_{0.2}O_{3-\delta}$ (BZY20). Thus, BCZYYb is considered to be a promising proton-conducting ceramic electrolyte membrane for intermediate-temperature electrochemical applications, including PCFCs, PCECs, H_2 pumps and electrochemical compressors, gas separation membranes, and membrane reactors.

12.4.1 BCZYYb equilibrium defect chemistry

Table 12.2 lists defect-incorporation reactions for H_2, O_2, and H_2O at gas–membrane interfaces together with the relevant thermodynamic parameters (ΔH° and ΔS°) as well as the equilibrium constants evaluated at 600 °C, 700 °C, and 800 °C [19]. The three *mobile* defects are protons $OH_O^\bullet$, oxygen vacancies $V_O^{\bullet\bullet}$, and O-site polarons $O_O^\bullet$. The three *immobile* defects are lattice oxygen $O_O^\times$, trapped polaron complexes $(X'_B - O_O^\bullet)$, and untrapped dopants $X'_B = (Y'_B, Yb'_B)$. The 'polaron trapping' reaction applies within the membrane as well as at the gas interface.

The equilibrium relationships for the defect reactions may be expressed in terms of the gas-phase partial pressures p_k and formula-unit defect concentrations $[X_k]_L$ as follows:

$$K_{p,H_2} = \frac{[OH_O^\bullet]_L}{p_{H_2}^{1/2}[O_O^\bullet]_L}, \tag{12.10}$$

$$K_{p,O_2} = \frac{[O_O^\bullet]_L^2}{p_{O_2}^{1/2}[O_O^\times]_L[V_O^{\bullet\bullet}]_L}, \tag{12.11}$$

$$K_{p,H_2O} = \frac{[OH_O^\bullet]_L^2}{p_{H_2O}[O_O^\times]_L[V_O^{\bullet\bullet}]_L}, \tag{12.12}$$

$$K_{p,Trap} = \frac{[(X'_B - O_O^\bullet)]_L}{[X'_B]_L[O_O^\bullet]_L}. \tag{12.13}$$

The defect formula-unit concentrations $[X_k]_L$ are related to their molar concentrations $[X_k]$ through the lattice molar volume V_m according to $[X_k]_L = [X_k]V_m$, where $V_m = 5.132\,3 \times 10^{-5}$ m^3 mol^{-1} for BCZYYb. The equilibrium constants K_p (i.e. K_{p,H_2}, K_{p,O_2}, K_{p,H_2O}, and $K_{p,Trap}$) can be evaluated from the reactions' Gibbs free energy changes ($\Delta G^\circ = \Delta H^\circ - T\Delta S^\circ$) as follows:

$$K_p = \exp\left(-\frac{\Delta G^\circ}{RT}\right). \tag{12.14}$$

On the electrolyte surface and within the bulk of the BCZYYb, site constraints and electroneutrality must be preserved. The number of oxygen sites in a perovskite lattice must be maintained at three, as follows:

$$[V_O^{\bullet\bullet}]_L + [OH_O^\bullet]_L + [O_O^\bullet]_L + [O_O^\times]_L + [(X'_B - O_O^\bullet)]_L = 3. \tag{12.15}$$

The total yttrium and ytterbium concentrations remain at the dopant level of $[X'_B]_L^\circ = 0.2$, as follows:

$$[X'_B]_L + [(X'_B - O_O^\bullet)]_L = [X'_B]_L^\circ. \tag{12.16}$$

Charge neutrality requires that

$$2[V_O^{\bullet\bullet}]_L + [OH_O^{\bullet}]_L + [O_O^{\bullet}]_L - [X_B']_L = 0. \tag{12.17}$$

12.4.2 Defect and charge transport

Considering the three mobile charge-carrying defects (i.e., $OH_O^{\bullet}$, $O_O^{\bullet}$, and $V_O^{\bullet\bullet}$), the charged-defect transport through the electrolyte membrane is generally driven by the gradients of the defect electrochemical potentials $\tilde{\mu}_k$. The mobile defect molar fluxes $\mathbf{J}_k$ can be expressed as

$$\mathbf{J}_k = -B_k[X_k]\nabla\tilde{\mu}_k, \tag{12.18}$$

where $[X_k]$ and B_k are the defect molar concentrations and mobilities, respectively. The mobilities are related to the defect diffusion coefficients D_k and conductivities σ_k through the Nernst–Einstein relationship, as follows:

$$\sigma_k = z_k^2 F^2 B_k[X_k] = \frac{z_k^2 F^2}{RT} D_k[X_k], \tag{12.19}$$

where z_k are the defect charges, R is the gas constant, and F is the Faraday constant. The total conductivity within the electrolyte membrane σ_{el} is the sum of the individual defect conductivities (i.e. $\sigma_{el} = \sum_k \sigma_k$) and the defect transference numbers are defined as $t_k = \sigma_k/\sigma_{el}$.

The defect electrochemical potentials $\tilde{\mu}_k$ can be expressed as

$$\tilde{\mu}_k = \mu_k^\circ + RT \ln \gamma_k[X_k] + z_k F\Phi_{el}, \tag{12.20}$$

where μ_k° are standard-state chemical potentials, γ_k are activity coefficients, and Φ_{el} is the electrostatic potential. Under the dilute-defect-interaction assumption ($\gamma_k \approx 1$), the defect transport fluxes $\mathbf{J}_k$ can be represented as functions of the defect concentration gradients (diffusion) and electrostatic potential gradients (migration), as follows:

$$\mathbf{J}_k = -D_k\nabla[X_k] - \frac{z_k F}{RT} D_k[X_k]\nabla\Phi_{el}. \tag{12.21}$$

The defect diffusivities D_k may be represented in Arrhenius form as

$$D_k = D_k^\circ \exp\left(-\frac{E_k}{RT}\right), \tag{12.22}$$

where D_k° are the pre-exponential factors and E_k are the activation energies. Table 12.3 lists the fitted parameters D_k° and E_k for the diffusion coefficients $OH_O^{\bullet}$, $O_O^{\bullet}$, and $V_O^{\bullet\bullet}$ [19].

Figure 12.2 illustrates how the BCZYYb equilibrium conductivity and mobile defect transference number depend on oxygen partial pressure and temperature. The oxygen is balanced with 3% H_2O and N_2. The total conductivity increases as p_{O_2} and temperature increase. Under highly reducing environments, the O-site polaron

Table 12.3. Diffusion coefficients of mobile defects for BCZYYb. The values of D_k° and E_k are taken from [19].

Defect	D_k° ($m^2\ s^{-1}$)	E_k ($kJ\ mol^{-1}$)	D_k (600 °C) ($m^2\ s^{-1}$)	D_k (700 °C) ($m^2\ s^{-1}$)	D_k (800 °C) ($m^2\ s^{-1}$)
$OH_O^\bullet$	1.02×10^{-07}	42.65	2.88×10^{-10}	5.26×10^{-10}	8.60×10^{-10}
$V_O^{\bullet\bullet}$	1.73×10^{-07}	59.70	4.64×10^{-11}	1.08×10^{-10}	2.15×10^{-10}
$O_O^\bullet$	2.41×10^{-05}	8.40	7.56×10^{-06}	8.52×10^{-06}	9.38×10^{-06}

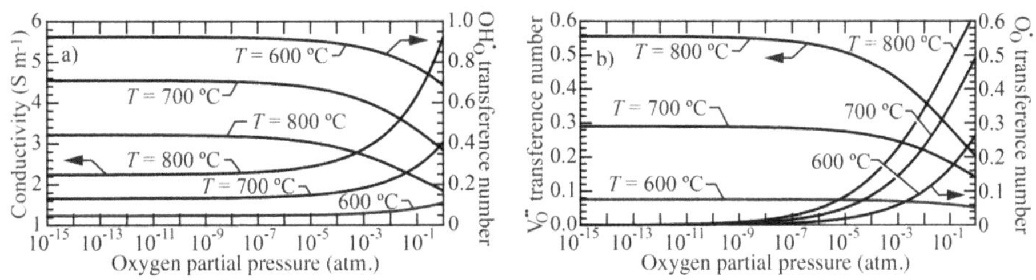

Figure 12.2. Effects of temperature and oxygen partial pressure on defect conductivities for BCZYYb. (a) Equilibrium total conductivity and transference number for $OH_O^\bullet$. (b) Transference numbers of $V_O^{\bullet\bullet}$ and $O_O^\bullet$. Three temperatures are considered. The gas compositions contain 3% H_2O balanced with N_2.

transference number is small, and the proton and oxygen vacancy transport dominate. However, polaron transport becomes significant in oxidizing environments and at high temperatures. The proton transference number $t_{OH_O^\bullet}$ decreases as the temperature increases. By contrast, both $t_{V_O^{\bullet\bullet}}$ and $t_{O_O^\bullet}$ increase with increasing temperature. In reducing environments, the transference numbers depend very weakly on the oxygen partial pressure.

The temporal and spatial variations of the defect concentrations $[X_k]$ within the electrolyte membrane can be described by defect mass conservation as follows (the Nernst–Planck problem) [33–36]:

$$\frac{\partial [X_k]}{\partial t} + \nabla \cdot \mathbf{J}_k = \dot{r}_k, \tag{12.23}$$

where $\dot{r}_k$ are the molar production rates of the defects *within* the bulk of the electrolyte membrane (i.e. the polaron-trapping reaction).

The current density resulting from the mobile charged-defect transport can be expressed as

$$\mathbf{i}_{el} = \sum_k z_k F \mathbf{J}_k = -\sigma_{el} \nabla \Phi_{el} + \mathbf{i}_c, \tag{12.24}$$

where $\mathbf{i}_c$ is the charge-transport flux due to the charged-species concentration gradients

$$\mathbf{i}_c = -\sum_k z_k F D_k \nabla [X_k]. \tag{12.25}$$

Considering that the bulk defect reactions inherently preserve the charge balance (i.e. $\sum_k z_k F \dot{r}_k = 0$), the local electrostatic potential profile Φ_{el} can be obtained by imposing strict local electroneutrality, i.e. $\sum_k z_k F[X_k] = 0$, which leads to a charge-conservation equation within the electrolyte membrane, as follows:

$$\nabla \cdot \mathbf{i}_{el} = 0. \tag{12.26}$$

If $\Phi_{el,n}$ denotes the electrostatic potential Φ_{el} at the negatrode–electrolyte interface and $\Phi_{el,p}$ at the positrode–electrolyte interface, the ohmic overpotential loss across the electrolyte membrane can be evaluated as follows: $\eta_{ohm} = \Delta\Phi_{el} = \Phi_{el,p} - \Phi_{el,n}$.

12.4.3 Half-cell reversible potential and cell voltage

An equilibrated polaron–electron annihilation reaction at the electrode–electrolyte interfaces can be represented as [36–39]:

$$\mathrm{O_O^\times(el) = O_O^\bullet(el) + e'(ed)}. \tag{12.27}$$

The half-cell reversible potential $E^{eq} = \Phi_{ed}^{eq} - \Phi_{el}^{eq}$ can be estimated as follows:

$$E^{eq} = E^\circ + \frac{RT}{F}\ln\frac{[\mathrm{O_O^\bullet(el)}]^{eq}}{[\mathrm{O_O^\times(el)}]^{eq}}, \tag{12.28}$$

where $E^\circ = (\mu^\circ_{\mathrm{O_O^\bullet(el)}} + \mu^\circ_{\mathrm{e'(ed)}} - \mu^\circ_{\mathrm{O_O^\times(el)}})/F$, and $[\mathrm{O_O^\bullet}]^{eq}$ and $[\mathrm{O_O^\times}]^{eq}$ are the equilibrated molar concentrations of $\mathrm{O_O^\bullet}$ and $\mathrm{O_O^\times}$ at the electrode–electrolyte interface.

The operating cell voltage E_{cell}, which is the electric potential difference between the positrode and negatrode terminal potentials ($E_{cell} = \Phi_p - \Phi_n$), can be evaluated as follows:

$$E_{cell} = E_p^{eq} - E_n^{eq} + \eta_{act,p} - \eta_{act,n} + \Phi_{el,p} - \Phi_{el,n}, \tag{12.29}$$

where $\eta_{act,p} = (\Phi_p - \Phi_{el,p}) - E_p^{eq}$ is the activation overpotential at the positrode–electrolyte interface, and $\eta_{act,n} = (\Phi_n - \Phi_{el,n}) - E_n^{eq}$ is the activation overpotential at the negatrode–electrolyte interface.

For the purposes of isolating and exploring membrane transport processes, we assume that the defect concentrations at the gas–electrolyte interfaces are equilibrated with the gas environment. We further assume that the concentration overpotentials within the composite porous electrodes can be neglected, along with any charge-transfer activation overpotentials at the electrode–electrolyte interfaces (i.e. $\eta_{act,n} = \eta_{act,p} = 0$). With these limiting-case assumptions, the cell voltage E_{cell} can be rewritten as

$$E_{cell} = \frac{RT}{F}\left(\ln\frac{\left[\mathrm{O_O^\bullet}\right]_{e,\,p}^{eq}}{\left[\mathrm{O_O^\times}\right]_{e,\,p}^{eq}} - \ln\frac{\left[\mathrm{O_O^\bullet}\right]_{e,\,n}^{eq}}{\left[\mathrm{O_O^\times}\right]_{e,\,n}^{eq}}\right) + \left(\Phi_{e,p} - \Phi_{e,n}\right) = E_{pn}^{eq} + \Delta\Phi_{el}, \tag{12.30}$$

where

$$E_{pn}^{eq} = \frac{RT}{F}\left(\ln\frac{\left[O_O^{\bullet}\right]_{e,\,p}^{eq}}{\left[O_O^{\times}\right]_{e,\,p}^{eq}} - \ln\frac{\left[O_O^{\bullet}\right]_{e,\,n}^{eq}}{\left[O_O^{\times}\right]_{e,\,n}^{eq}}\right), \tag{12.31}$$

and $[O_O^{\bullet}]_{e,\,p}^{eq}$ and $[O_O^{\times}]_{e,\,p}^{eq}$ are the equilibrated molar concentrations of $O_O^{\bullet}$ and $O_O^{\times}$ at the positrode–electrolyte interface, and $[O_O^{\bullet}]_{e,\,n}^{eq}$ and $[O_O^{\times}]_{e,\,n}^{eq}$ are the equilibrated molar concentrations of $O_O^{\bullet}$ and $O_O^{\times}$ at the negatrode–electrolyte interface.

12.4.4 BCZYYb membrane transport performance

To illustrate the behaviors of proton conduction and mixed conduction, consider a 20 μm thick BCZYYb membrane. As a limiting case, we assume that the defect concentrations at the gas–electrolyte interfaces are equilibrated with the gas environment and all polarization losses through the electrodes and at the electrode–electrolyte interfaces can be neglected. Under these limiting conditions, the faradaic efficiency is unaffected by membrane thickness. However, greater membrane thickness causes higher ohmic polarization within the membrane, leading to higher cell voltage and lower energy efficiency. For this illustration, the nominal gas mixture is 50.0% O_2 and 50.0% H_2O on the oxidizing (positrode) side and 97.0% H_2 and 3.0% H_2O on the reducing (negatrode) side. Both gas-phase pressures are assumed to be $p = 1.0$ atm and the operating temperature is uniform at $T = 650$ °C. Table 12.4 lists the equilibrium concentrations, conductivities, and transference numbers of the mobile defects under these reducing and oxidizing environments. The operating current density varies from $i_e = -2.0$ A cm^{-2} in electrolysis mode to $i_e = +1.0$ A cm^{-2} in the fuel-cell mode.

Figure 12.3 illustrates the spatial profiles of the electric potential and defect concentrations at six current densities. Figure 12.3(a) shows that the electrostatic potential Φ_{el} in electrolysis mode ($i_e < 0$) on the oxidizing (positrode) side is higher than it is on the reducing side. Thus, the electrostatic potential gradient $\nabla\Phi_{el}$ tends to drive the mobile charged defects from the oxidizing side toward the reducing side. Sustaining a high current density requires a large electrostatic potential gradient to drive the defect transport through the membrane, thus leading to large ohmic overpotential losses $\Delta\Phi_{el}$ (i.e. $\Delta\Phi_{el} = 0.078\,76$ V at $i_e = -1.0$ A cm^{-2}, and $\Delta\Phi_{el} = 0.1807$ V at $i_e = -2.0$ A cm^{-2}).

Table 12.4. Mobile defect concentrations, conductivities, and transference numbers under equilibrium conditions.

	97.0% H_2 + 3.0% H_2O at 650° C and 1.0 atm.			50.0% O_2 + 50.0% H_2O at 650° C and 1.0 atm.		
	$[X_k]_L$	σ_k (S m^{-1})	t_k	$[X_k]_L$	σ_k (S m^{-1})	t_k
$OH_O^{\bullet}$	1.2965×10^{-01}	$1.212\,7\times10^{-00}$	$8.343\,7\times10^{-01}$	$1.853\,7\times10^{-01}$	$1.733\,8\times10^{-00}$	$8.547\,3\times10^{-01}$
$V_O^{\bullet\bullet}$	$3.517\,3\times10^{-02}$	$2.407\,3\times10^{-01}$	$1.656\,3\times10^{-01}$	$4.361\,3\times10^{-03}$	$2.984\,9\times10^{-02}$	$1.471\,5\times10^{-02}$
$O_O^{\bullet}$	$2.248\,9\times10^{-12}$	$4.280\,2\times10^{-07}$	$2.944\,9\times10^{-07}$	$1.391\,4\times10^{-06}$	$2.648\,2\times10^{-01}$	$1.305\,5\times10^{-01}$

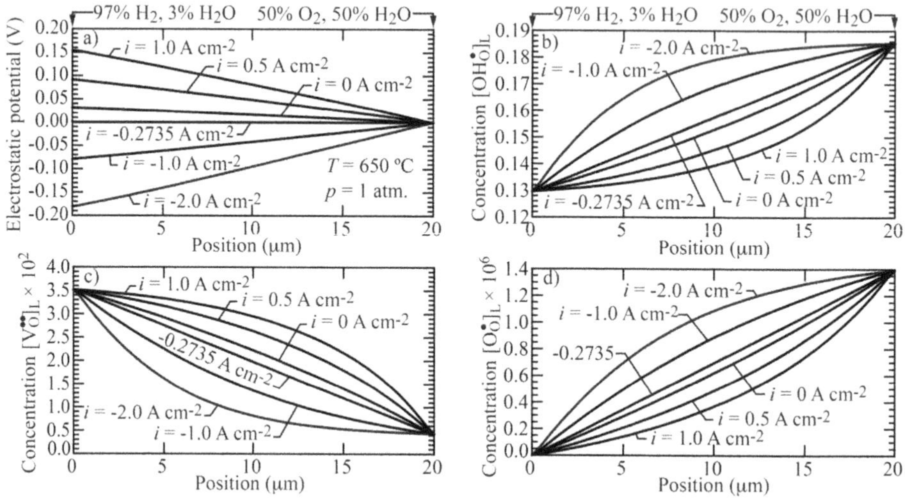

Figure 12.3. Predicted spatial profiles of the electrostatic potential and defect concentrations at current densities of −2.0, −1.0, −0.2735, 0.0, 0.5, and 1.0 A cm^{-2}. (a) Electrostatic potential profiles Φ_{el}. (b) Proton lattice concentration profiles $[OH_O^{\bullet}]_L$. (c) Oxygen vacancy lattice concentration profiles $[V_O^{\bullet\bullet}]_L$. (d) Polaron lattice concentration profiles $[O_O^{\bullet}]_L$. The operating temperature is 650 °C and the pressure is atmospheric. The gas-phase composition is fixed at 97.0% H_2 and 3.0% H_2O on the negatrode (H_2 collection) side and 50.0% O_2 and 50.0% H_2O on the positrode (H_2O feed and O_2 production) side.

Figure 12.3(a) shows that large electrostatic potential gradients $\nabla\Phi_e$ at high electrolysis current densities drive positively charged defects toward the reducing side of the membrane, forming significant defect-concentration gradients and leaving relatively shallow concentration gradients toward the oxidizing side. Because of the relatively small concentration gradients on the oxidizing side at high current density, the defect fluxes leaving the oxidizing side are mainly driven by $\nabla\Phi_{el}$. Thus, the defect fluxes $\mathbf{J}_k$ are approximately proportional to the defect conductivities σ_k on the oxidizing side (see equation (12.21)). The maximum achievable faradaic efficiency in electrolysis mode is largely determined by the summation of the proton transference number $t_{OH_O^{\bullet}}$ and the oxygen vacancy transference number $t_{V_O^{\bullet\bullet}}$ on the oxidizing side. At $i_e = -2.0$ A cm^{-2}, $\eta_{F,EC} = 86.00\%$, which is close to $t_{OH_O^{\bullet}} + t_{V_O^{\bullet\bullet}} \approx 0.87$ on the oxidizing side.

As illustrated in figure 12.3(a), $\nabla\Phi_{el}$ in the fuel-cell mode ($i_e > 0$) tends to drive the positively charged defects toward the oxidizing (positrode) side. At high fuel-cell current densities, the defect concentration gradients become shallower near the reducing (negatrode) side. Therefore, the fuel-cell faradaic efficiency is largely determined by the summation of the proton and oxygen vacancy transference numbers on the reducing (negatrode) side. At $i_e = 1.0$ A cm^{-2}, $\eta_{F,FC} = 96.61\%$, which is slightly less than $t_{OH_O^{\bullet}} + t_{V_O^{\bullet\bullet}} \approx 1.0$ on the reducing side. Figure 12.2 shows that $t_{O_O^{\bullet}} = 1.0 - (t_{OH_O^{\bullet}} + t_{V_O^{\bullet\bullet}})$ in oxidizing environments is much larger than it is in reducing environments. It is easier to achieve high faradaic efficiency with a BCZYYb membrane in the fuel-cell mode than in electrolysis mode. Figure 12.3 shows defect concentration profiles at a current density of $i_e \approx -0.2735$ A cm^{-2}, at which $\Delta\Phi_{el} \approx 0.0$ V ($E_{cell} \approx E_{pn}^{eq} = 1.0616$ V, $\eta_{F,EC} \approx 61.09\%$). Because the migration contribution to the defect fluxes vanishes under these conditions, the defect fluxes $\mathbf{J}_k$ are driven entirely by defect concentration gradients (see equation (12.21)), leading to nearly linear defect concentration profiles. For current densities $i_e > -0.2735$ A cm^{-2} and $\Delta\Phi_{el} < 0.0$ V, the electrostatic potential gradient $\nabla\Phi_{el}$ tends to drive positively charged defects from the reducing (negatrode) side toward the oxidizing (positrode) side, thus counteracting concentration-driven diffusion transport of $OH_O^{\bullet}$ and $O_O^{\bullet}$ from the oxidizing side toward the reducing side. At a current density of $i_e \approx -0.02515$ A cm^{-2} with $E_{cell} = 1.0299$ V and $\Delta\Phi_{el} \approx -0.02854$ V, the proton flux vanishes, $\mathbf{J}_{OH_O^{\bullet}} \approx 0.0$.

It is interesting to explore the relative effects of defect transport on the open-circuit cell voltage. Under open-circuit conditions ($i_e = 0.0$ A cm^{-2}), $\Delta\Phi_{el} \approx -0.03165$ V (figure 12.3(a)), the cell voltage is $E_{cell} = E_{pn}^{eq} + \Delta\Phi_{el} \approx 1.0301$ V, and $E_{pn}^{eq} = 1.0616$ V. If O-site polaron transport is neglected, $\Delta\Phi_{el} \approx -0.02065$ V and the open-circuit cell voltage becomes $E_{cell} \approx 1.0409$ V. If we consider only the proton transport, then $\Delta\Phi_{el} \approx -0.02849$ V and the open-circuit potential is $E_{cell} \approx 1.0331$ V, which is very close to the Nernst potential of 1.03268 V for a pure protonic conductor. If we consider only the oxygen ion transport, $\Delta\Phi_{el} \approx 0.08429$ V and the open-circuit potential is $E_{cell} = 1.1515$ V, which is also very close to the Nernst potential (1.1446 V) of the pure oxygen ion conductor.

12.4.4.1 Temperature effects

Figure 12.4 illustrates variations of cell voltage E_{cell}, potential difference across the membrane $\Delta\Phi_{el}$ (ohmic overpotential), faradaic efficiency η_F, and energy efficiency

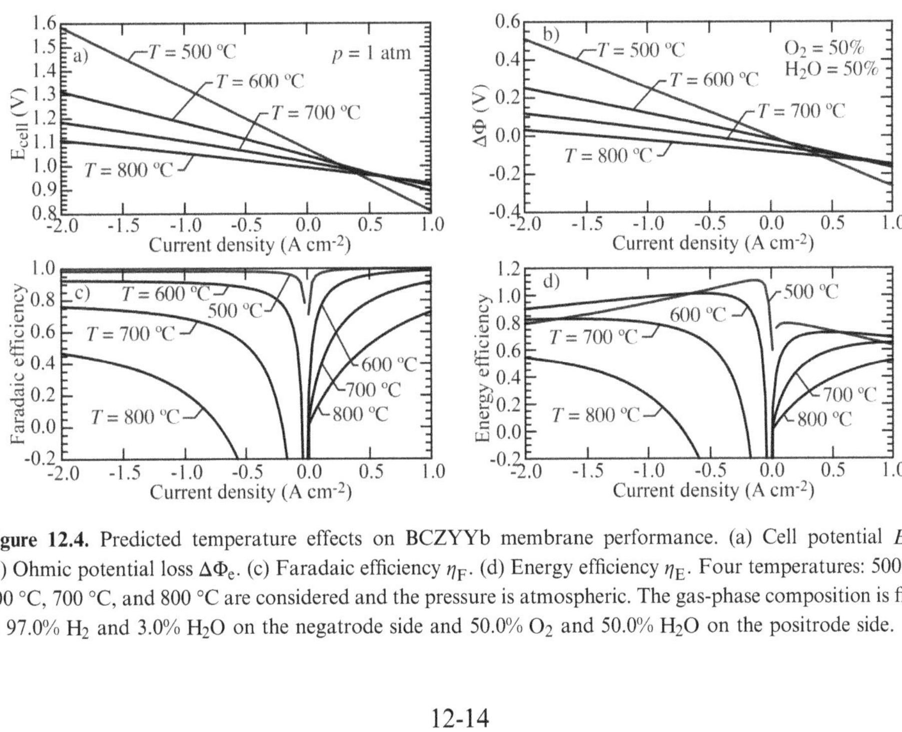

Figure 12.4. Predicted temperature effects on BCZYYb membrane performance. (a) Cell potential E_{cell}. (b) Ohmic potential loss $\Delta\Phi_e$. (c) Faradaic efficiency η_F. (d) Energy efficiency η_E. Four temperatures: 500 °C, 600 °C, 700 °C, and 800 °C are considered and the pressure is atmospheric. The gas-phase composition is fixed at 97.0% H_2 and 3.0% H_2O on the negatrode side and 50.0% O_2 and 50.0% H_2O on the positrode side.

η_E as functions of current density and temperature. We continue to assume that the defect concentrations at the gas–electrolyte interfaces are equilibrated with the gas environment and that all polarization losses through the electrodes and at the electrode–electrolyte interfaces can be neglected. At each operating temperature, figure 12.4 indicates that higher current densities deliver higher faradaic efficiencies in both electrolysis and fuel-cell modes. Higher current densities also produce higher ohmic losses, which have derivative consequences. In the electrolysis mode, higher cell potentials are needed to drive defect transport through the membrane. In fuel-cell mode, the higher ohmic overpotential reduces the operating cell voltage.

Faradaic efficiency at 500 °C is approximately 98.4% at $i_e = -2.0$ A cm^{-2} in electrolysis mode and about 99.9% at $i_e = 1.0$ A cm^{-2} in the fuel-cell mode. However, the faradaic efficiency of both modes drops greatly when the operating temperature increases from 500 °C to 800 °C, which is consistent with the significant decrease in the proton transference number and an increase in the polaron transference numbers at higher temperatures (figure 12.2). Figure 12.4(d) illustrates that the energy efficiency generally increases as the current density increases, but decreases as the temperature increases.

12.4.4.2 Pressure and oxygen effects

We continue to assume negligible polarization losses through the electrodes and at the electrode–electrolyte interfaces along with equilibrated defect concentrations at the gas–electrolyte interfaces. Under these conditions, figure 12.5 illustrates the effects of pressure and gas-phase composition on the oxidizing (positrode) side on the faradaic efficiency. The gas-phase composition on the fuel (negatrode) side is fixed at 97.0% H_2 and 3.0% H_2O and the operating temperature is fixed at 650 °C. Figure 12.5(a) shows that increasing the pressure can slightly improve the faradaic efficiency. The faradaic efficiency for $i_e = -2.0$ A cm^{-2} is about 85.9% at 1.0 atm, 89.9% at 5.0 atm, 91.3% at 10.0 atm, and 92.6% at 20.0 atm. In the fuel-cell mode, pressure has a weaker effect on the faradaic efficiency. Figure 12.5(b) shows a dramatic decrease in the faradaic efficiency as oxygen partial pressure p_{O_2} increases on the oxidizing (positrode) side. As illustrated in figure 12.2, the polaron transference number in the oxidizing environment increases significantly as p_{O_2} increases.

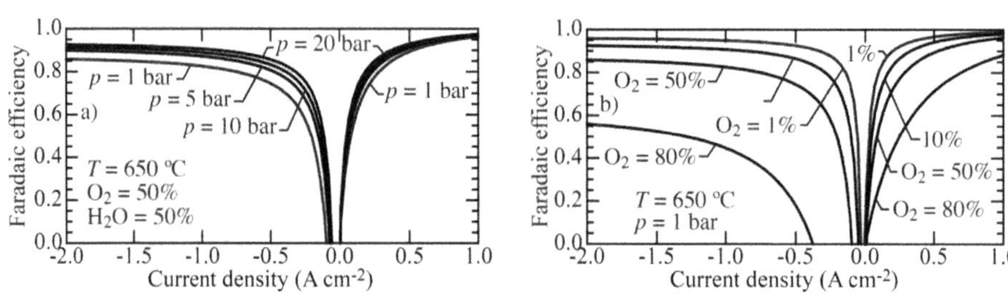

Figure 12.5. Predicted faradaic efficiencies as a function of current density. The operating temperature is 650 °C and the gas-phase composition on the fuel (negatrode) side is 97.0% H_2 and 3.0% H_2O. (a) Four pressures: 1.0, 5.0, 10.0, and 20.0 atm. (b) Four gas-phase compositions on the oxidizing (positrode) side: 1.0%, 10.0%, 50.0%, and 80.0% O_2 balanced with H_2O.

Thus, the increasing leakage current associated with polaron transport reduces the faradaic efficiency.

12.5 Electrochemical cells

In order to focus on defect transport within the electrolyte membrane, the foregoing discussion neglected the kinetics of the *defect-incorporation* and *charge-transfer* reactions. As discussed below, the rate limitations associated with reaction kinetics can significantly affect cell performance. It is important to note that the functional behaviors and rates of the defect-incorporation and charge-transfer reactions are fundamentally different. The defect-incorporation reactions (shown in table 12.3) involve the gas phase and the electrolyte phase, but there is no charge exchange between the phases. By contrast, the charge-transfer reactions (i.e. equations (12.3) and (12.4)) involve the gas phase, the electrode phase, and the BCZYYb electrolyte phase, and charges are transferred across the electrode–electrolyte interfaces.

Defect concentrations at the gas–electrode or gas–electrolyte interfaces may not be equilibrated. Three fluxes influence the mobile defect concentrations at the edges of the electrolyte membrane:

- Defect incorporation at the electrolyte surfaces that are exposed to gas-phase environments (see table 12.3),
- Charge-transfer reactions between the electrode and electrolyte phases (see equations (12.3) and (12.4)),
- Defect transport within the membrane (see equations (12.21), (12.22), and (12.23) and table 12.3).

Working with laboratory-scale cells assists our understanding of the chemistry at electrode–electrolyte interfaces and hence overall cell performance. The membrane–electrode assembly (MEA) structure essentially consists of a proton-conducting electrolyte membrane that separates the composite negatrode and positrode structures (figure 12.6). Both electrodes have percolating phases that enable gas transport to take place in addition to electronic and ionic conduction. As illustrated in figure 12.6, the negatrode is a porous composite of BCZYYb (the ion-conducting phase) and Ni (the electron-conducting phase). The porous positrode is composed of $BaCo_{0.4}Fe_{0.4}Zr_{0.1}Y_{0.1}O_{3-\delta}$ (BCFZY), which is a triple-conducting oxide, meaning that the single solid phase has significant conductivity for protons, oxygen vacancies, and electrons. The electrode and electrolyte models discussed here were previously developed by Zhu, Kee, and colleagues [19, 33, 35–37, 40].

12.5.1 Pore phase gas-phase transport

Gas-phase species transport within the electrode pore volumes (e.g. H_2, O_2, H_2O, N_2) can be represented as

$$\frac{\partial(\phi_g \rho)}{\partial t} + \sum_{k=1}^{K_g} \nabla \cdot \mathbf{j}_k = \sum_{k=1}^{K_g} \dot{r}_k W_k, \tag{12.32}$$

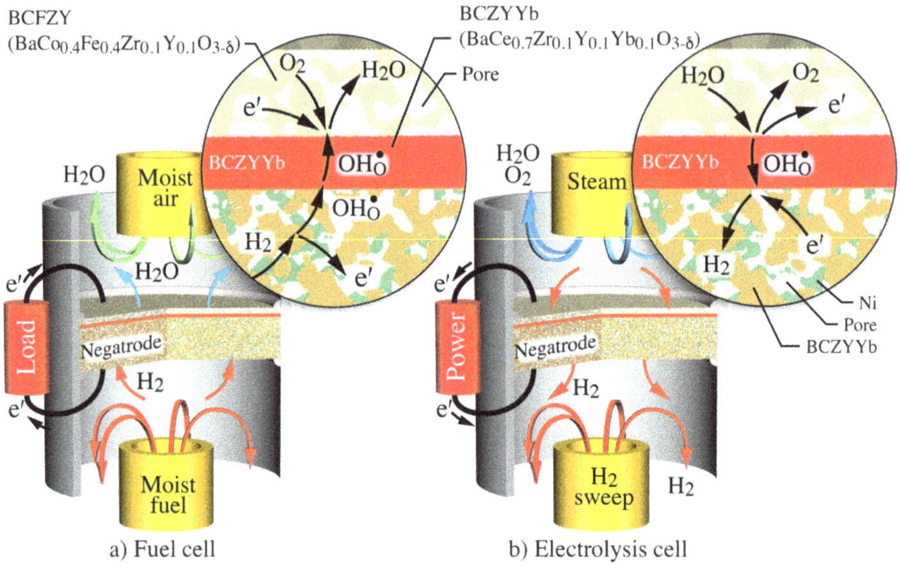

Figure 12.6. Illustration of the operational principles of fuel cells and electrolysis cells.

$$\frac{\partial}{\partial t}\left(\phi_{\mathrm{g}}\rho Y_k\right) + \nabla \cdot \mathbf{j}_k = \dot{r}_k W_k, \tag{12.33}$$

where ϕ_{g} is the gas-phase porosity, ρ is the gas-phase density, Y_k are mass fractions, and W_k are the gas-phase molecular weights. The factor $\dot{r}_k$ represents the rate at which gas-phase species are produced (consumed) by homogeneous and/or heterogeneous reactions within the electrode pore volume. The gas-phase relationships among pressure, temperature, density, and composition obey the ideal-gas equation of state ($p = \rho RT \sum_k Y_k / W_k$). In a fuel cell, heterogeneous catalytic chemistry may take place, such as hydrocarbon fuel reforming. This is rarely the case in steam electrolysis cells. If there is significant heterogeneous catalysis, the system of equations must be augmented to predict the surface-adsorbate coverages [41].

The gas-phase diffusive mass fluxes $\mathbf{j}_k$ within the pore volume can be evaluated using the dusty gas model (DGM) [42, 43], which represents bulk molecular diffusion, Knudsen diffusion, and pressure-driven Darcy flow. The DGM can be written as an implicit relationship between the gas-phase species' molar fluxes $\mathbf{J}_k$, molar concentrations $[X_k]$, concentration gradients, and the pressure p gradient, as follows:

$$\sum_{\ell \neq k} \frac{[X_\ell]\mathbf{J}_k - [X_k]\mathbf{J}_\ell}{[X_{\mathrm{T}}] D^{\mathrm{e}}_{k\ell}} + \frac{\mathbf{J}_k}{D^{\mathrm{e}}_{k,\,\mathrm{Kn}}} = -\nabla[X_k] - \frac{[X_k]}{D^{\mathrm{e}}_{k,\,\mathrm{Kn}}} \frac{B_{\mathrm{g}}}{\mu_{\mathrm{g}}} \nabla p, \tag{12.34}$$

where $[X_{\mathrm{T}}] = p/RT$ is the total molar concentration, R is the gas constant, and T is temperature. The gas-phase mixture viscosity is denoted by μ_{g}, and the pore permeability is denoted by B_g. The effective ordinary and Knudsen diffusion coefficients ($D_{k\ell}^{e}$ and $D_{k,\,Kn}^{e}$) may be evaluated as follows [43]:

$$D_{k\ell}^{e} = \frac{\phi_g}{\tau} D_{k\ell}, \qquad D_{k,\,Kn}^{e} = \frac{2}{3} r_p \frac{\phi_g}{\tau} \sqrt{\frac{8RT}{\pi W_k}}, \tag{12.35}$$

where τ is tortuosity, $D_{k\ell}$ are the ordinary multicomponent diffusion coefficients (note: not the binary diffusion coefficients), and r_p is the mean pore radius. The mass and molar fluxes are simply related as follows: $\mathbf{J}_k = \mathbf{j}_k / W_k$.

The production rates of gas-phase species $\dot{r}_k$ within the porous electrodes result from the catalytic chemistry, defect chemistry, and electrochemical charge-transfer reactions [37, 40]. In general, the production rates $\dot{r}_k$ are functions of temperature, gas composition, and possibly surface-adsorbate coverages. The electrochemical charge-transfer reactions are also functions of the local electrostatic potential differences between the electrode and electrolyte phases.

Boundary conditions are required in order to solve equations (12.32) and (12.33). At the interfaces between the composite electrodes and the dense electrolyte membrane, the gas-phase species' fluxes vanish. At the interfaces between the composite electrodes and the gas compartments, the gas-phase composition is usually assumed to be that within the compartments.

12.5.2 Charge conservation within the electron-conducting phase

Within the electron-conducting phase of the electrode, the charge balance equation for the local electrostatic potential Φ_{ed} can be written as

$$\nabla \cdot \mathbf{i}_{ed} = 0, \tag{12.36}$$

where the local current density $\mathbf{i}_{ed}$ is represented using Ohm's law by

$$\mathbf{i}_{ed} = -\sigma_{ed} \nabla \Phi_{ed}, \tag{12.37}$$

and σ_{ed} is the electronic conductivity.

The operating cell voltage E_{cell} can be defined as

$$E_{cell} = \Phi_{p,pc} - \Phi_{n,nc}, \tag{12.38}$$

where $\Phi_{p,pc}$ is the electrostatic potential at the positrode-side current collector and $\Phi_{n,nc}$ is the electrostatic potential at the negatrode-side current collector.

Boundary conditions are required in order to solve equations (12.26) and (12.36). At the electrode–electrolyte interfaces, $\mathbf{i}_{el}$ vanishes. The electrostatic potential at the positrode-side current collector $\Phi_{p,pc}$ is set to zero as the reference potential. At any operating cell voltage E_{cell}, the electrostatic potential at the negatrode-side current collector $\Phi_{n,nc}$ is set to $\Phi_{n,nc} = \Phi_{p,pc} - E_{cell} = -E_{cell}$.

12.5.3 Defect-incorporation chemistry

We now discuss a mixed conductor such as BCZYYb, which has three mobile defects. The defect production rates (mol m^{-2} s^{-1}) on the BCZYYb surfaces can be represented as

$$\dot{s}_{V_O^{\bullet\bullet}} = -\dot{q}_{O_2} - \dot{q}_{H_2O}, \tag{12.39}$$

$$\dot{s}_{OH_O^{\bullet}} = 2\dot{q}_{H_2O} + \dot{q}_{H_2}, \tag{12.40}$$

$$\dot{s}_{O_O^{\bullet}} = 2\dot{q}_{O_2} - \dot{q}_{H_2} - \dot{q}_{Trap}, \tag{12.41}$$

$$\dot{s}_{O_O^{\times}} = -\dot{q}_{O_2} - \dot{q}_{H_2O}, \tag{12.42}$$

$$\dot{s}_{\left(X_B'-O_O^{\bullet}\right)} = \dot{q}_{Trap}, \tag{12.43}$$

where the reaction rates of progress are defined as

$$\dot{q}_{H_2} = k_{f,H_2}[H_2]^{1/2}[O_O^{\bullet}] - k_{b,H_2}[OH_O^{\bullet}], \tag{12.44}$$

$$\dot{q}_{O_2} = k_{f,O_2}[O_2]^{1/2}[O_O^{\times}][V_O^{\bullet\bullet}] - k_{b,O_2}[O_O^{\bullet}]^2, \tag{12.45}$$

$$\dot{q}_{H_2O} = k_{f,H_2O}[H_2O][V_O^{\bullet\bullet}][O_O^{\times}] - k_{b,H_2O}[OH_O^{\bullet}]^2, \tag{12.46}$$

$$\dot{q}_{Trap} = k_{f,Trap}[X_B'][O_O^{\bullet}] - k_{b,Trap}[(X_B' - O_O^{\bullet})]. \tag{12.47}$$

The forward and backward rate constants are denoted by $k_{f,i}$ and $k_{b,i}$ and are related via the equilibrium constants as follows: $K_{c,i} = k_{f,i}/k_{b,i}$. Unfortunately, although these rates affect cell performance, the experimental data are insufficient to independently evaluate the defect-incorporation rates for the materials of interest here. Experiments such as conductivity-relaxation measurements could help to establish the needed rates [36, 44, 45].

The present model simply assumes that the same rate constant k_f can be used to express the forward rate for all the defect-incorporation reactions at the gas–BCZYYb interfaces. It should be noted that this *assumed*, *empirical* rate k_f incorporates both the actual defect reaction rates and the effective BCZYYb surface area. Each $k_{f,i}$ has different units for each defect reaction, depending on the reaction stoichiometry. The present model uses a nominal rate of $k_f = 10^3$, but also explores the effects of the assumed rate.

12.5.4 Charge-transfer chemistry

The present model assumes that electrochemical charge-transfer processes are limited to very narrow (typically tens of microns) regions near electrode–electrolyte interfaces and that the charge-transfer rates i_{BV} can be represented in terms of a Butler–Volmer formulation

$$i_{\mathrm{BV}} = i_0\left[\exp\left(\frac{\alpha_{\mathrm{a}}F\eta_{\mathrm{act}}}{RT}\right) - \exp\left(-\frac{\alpha_{\mathrm{c}}F\eta_{\mathrm{act}}}{RT}\right)\right], \tag{12.48}$$

where α_a and α_c are the apparent anodic and cathodic symmetry factors, respectively, and i_0 is the exchange current density. The Butler–Volmer expression is derived by assuming that the reversible charge-transfer reaction is written in the anodic direction, meaning it produces electrons. The first term on the right-hand side represents the anodic rate and the second term represents the cathodic rate. If a reaction is proceeding in the cathodic direction (i.e. consuming electrons) then the second term must dominate and the activation overpotential η_{act} must be negative.

As summarized below, Zhu and Kee [37] derived exchange-current–density i_0 expressions for charge-transfer processes occurring in PCECs. At the interfaces between MIEC electrodes and the gas phase, charged surface-adsorbed species may participate in the charge-transfer processes. In such cases, the Butler–Volmer representation of charge-transfer rates is further complicated [46].

The charge-transfer reaction for the oxidization of hydrogen to protons in the negatrode–electrolyte TPB regions can be globally written as

$$\mathrm{H_2(g)} + 2\mathrm{O_O^{\times}(el)} \rightleftharpoons 2\mathrm{OH_O^{\bullet}(el)} + 2\mathrm{e'(ed)}, \tag{12.49}$$

which involves several physical and chemical processes, including the adsorption and dissociation of hydrogen on the metallic surfaces, diffusive transport of adsorbed species on the electrode surfaces, bulk phase transport within the ionic and electronic phases, and a charge-transfer reaction step that incorporates protons into the electrolyte.

In terms of the activation overpotential, $\eta_{act,n} = E_n - E_n^{eq}$, the charge-transfer rate can be represented by a Butler–Volmer formulation of equation (12.48). The exchange current density i_0 can be expressed as [37]

$$i_0 = i_0^* \frac{\left(p_{\mathrm{H_2}}/p_{\mathrm{H_2}}^*\right)^{\beta_c/2}}{1 + \left(p_{\mathrm{H_2}}/p_{\mathrm{H_2}}^*\right)^{1/2}} [\mathrm{O_O^{\times}(el)}]^{\beta_c}[\mathrm{OH_O^{\bullet}(el)}]^{\beta_a}, \tag{12.50}$$

where β_a and β_c are the anodic and cathodic symmetric factors of the elementary charge-transfer step with $\beta_a + \beta_c = 1$. The apparent symmetric factors follow $\alpha_a = \beta_a$ and $\alpha_c = \beta_c$. The reference pressure of hydrogen $p_{H_2}^* = 1/K_{H_2}$, while K_{H_2} is the equilibrium constant of hydrogen adsorption on the metal surface. The temperature dependence of i_0^* can be expressed in Arrhenius form as $i_0^* = i_0^0 \exp(-E/RT)$ with i_0^0 and E as fitting parameters.

At the positrode–electrolyte interface, the overall charge-transfer reaction of protons with oxygen to form water can be globally expressed as

$$2\mathrm{H_2O(g)} + 4\mathrm{O_O^{\times}(el)} \rightleftharpoons \mathrm{O_2(g)} + 4\mathrm{OH_O^{\bullet}(el)} + 4\mathrm{e'(ed)}, \tag{12.51}$$

which requires gas-phase transport within the pore system, ion conduction within the electrolyte phase, electron conduction within the electrode phase, and that the reactants all meet at TPBs.

The overall charge-transfer reaction within the positrode (equations (12.3) and (12.51)) generally involves a sequence of steps, beginning with oxygen dissociative adsorption on the electrode surface, followed by oxygen surface diffusion, proton transfer from the electrolyte to the electrode surface, water formation on the electrode surface, and desorption of H_2O to the gas phase. In terms of the Butler–Volmer formulation for the charge-transfer rate (equation (12.48)), the exchange current density can be derived as [37]

$$i_0 = i_0^* \frac{\left(p_{O_2}/p_{O_2}^*\right)^{(1/2-\beta_c/4)} \left(p_{H_2O}/p_{H_2O}^*\right)^{\beta_c/2}}{1 + \left(p_{O_2}/p_{O_2}^*\right)^{1/2} + \left(p_{H_2O}/p_{H_2O}^*\right)} [O_O^{\times}]^{\beta_c} [OH_O^{\bullet}]^{\beta_a}, \tag{12.52}$$

where β_a and β_c are the anodic and cathodic symmetric factors of the rate-limiting elementary charge-transfer step with $\beta_a + \beta_c = 1$. The apparent symmetric factors are derived using $\alpha_a = 1 + \beta_a$ and $\alpha_c = \beta_c$. The parameters $p_{O_2}^* = 1/K_{O_2}$ and $p_{H_2O}^* = 1/K_{H_2O}$, where K_{O_2} and K_{H_2O} are the equilibrium constants of O_2 and H_2O adsorption on the electrode surface. The temperature dependence of i_0^* can also be expressed in Arrhenius form.

12.5.5 Parameter fitting

Liang *et al* [47] measured the cell voltages and power densities as functions of the current density for a Ni-BCZYYb|BCZYYb|BCFZY button cell. The Ni-BCZYYb negatrode of the button cell was taken to be 500 μm thick, the proton-conducting BCZYYb electrolyte membrane was 15 μm thick, and the BCFZY positrode was taken to be 20 μm thick. The fuel composition was moist H_2, and the air composition was moist air. The gas-phase pressure is atmospheric, and four operating temperatures of 500 °C, 550 °C, 600 °C, and 650 °C were considered. Figure 12.7(b) shows that the present model can very accurately represent the measured polarization behavior. Table 12.5 lists the best-fit button-cell physical and chemical parameters, which are used here. Figure 12.7(c) shows the predicted profiles of three overpotentials: the negatrode activation overpotential η_n, the positrode activation overpotential η_p, and the electrolyte ohmic overpotential η_{ohm} under the same operating conditions. As the temperature increases, the magnitudes of all three overpotentials decrease at the same current density. Thus, higher cell potentials are achieved at higher temperatures (figure 12.7(b)). At each operating temperature, the ohmic overpotential increases linearly as the current density increases. The negatrode overpotential increases significantly at low current density, but the rate of increase decreases at higher currents. The magnitude of the positrode overpotential increases faster at higher current densities. Note that the positrode activation overpotentials are negative, meaning that the charge-transfer reaction (equation (12.51)) is cathodic (i.e. consuming electrons). Based on the Butler–Volmer equation (equation (12.48)), the activation overpotential must be negative, thus making the second term dominant.

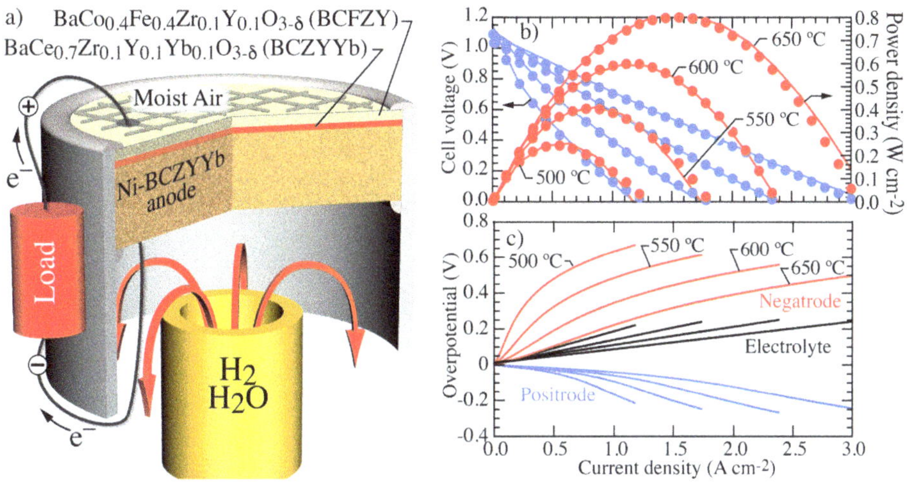

Figure 12.7. Fitting MEA parameters: (a) illustration of an MEA experimental setup; (b) comparison of the MEA voltage and power density as functions of current density predicted by the model (solid lines) versus experimentally measured values (solid dots) [47]; (c) predicted negatrode activation overpotential η_n, positrode activation overpotential η_p, and ohmic overpotential η_{ohm}. The fuel composition is moist H_2 presented to the negatrode side and moist air presented to the positrode side. The cell is operating at atmospheric pressure.

Table 12.5. Parameters used to model the MEA structure.

Negatrode		**Positrode**	
Thickness (L_a)	500 μm	Thickness (L_c)	40 μm
Porosity (ϕ_g)	0.35	Porosity (ϕ_g)	0.35
Ni volume fraction (ϕ_{Ni})	0.35	BCFZY volume fraction (ϕ_{BCFZY})	0.65
BCZYYb volume fraction (ϕ_{BCZYYb})	0.30	Tortuosity (τ_g)	4.50
Tortuosity (τ_g)	4.50	BCFZY particle radius (r_{BCFZY})	0.50 μm
Ni particle radius (r_{Ni})	0.50 μm	Exchange current factor (i_0^0)	2.48×10^5 A cm^{-1}
BCZYYb particle radius (r_{BCZYYb})	0.50 μm	Activation energy (E)	43.36 kJ mol^{-1} K^{-1}
Specific catalyst area (A_s)	4.0×10^4 cm^{-1}	Anodic symmetry factor (α_a)	0.80
Exchange current factor (i_0^0)	1.72×10^6 A cm^{-1}	Cathodic symmetry factor (α_c)	0.20
Activation energy (E)	82.60 kJ mol^{-1} K^{-1}		
Anodic symmetry factor (α_a)	0.30	**Electrolyte**	
Cathodic symmetry factor (α_c)	0.70	Thickness (L_{el})	20 μm

12.5.6 Defect-incorporation rates

The defect-incorporation rates (i.e. k_f) have a substantial effect on cell performance. Consider gas compositions of 97% H_2 and 3% H_2O on the reducing side and 50% O_2 and 50% H_2O on the oxidizing side at T = 650 °C and atmospheric pressure. Consider further that a fixed electrolysis current density is imposed, namely, $i_e = -0.5$ A cm^{-2}. It should be noted that the results in this section are based on the charge-transfer parameters listed in table 12.5. Figures 12.8(a) and 12.8(b) show the predicted concentrations of the three mobile defects ($OH_O^{\bullet}$, $V_O^{\bullet\bullet}$, and $O_O^{\bullet}$) within the BCZYYb electrolyte membrane at the positrode–electrolyte and negatrode–electrolyte interfaces for a wide range of defect-incorporation rates. Figure 12.8(c) illustrates the corresponding defect fluxes through the membrane, in which a negative flux denotes transport from the positrode toward the negatrode. As shown in figure 12.8(d), the faradaic efficiency depends significantly on the defect-incorporation rate.

At very low defect-incorporation rates (i.e. $k_f = 1.0$), the proton flux $J_{OH_O^{\bullet}}$ through the BCZYYb electrolyte membrane dominates and both the polaron flux $J_{O_O^{\bullet}}$ and oxygen vacancy flux $J_{V_O^{\bullet\bullet}}$ are negligibly small (figure 12.8(c)). Therefore, the faradaic efficiency is about 99.8% at $k_f = 1.0$, and the electrolyte membrane essentially behaves as a pure proton-conducting membrane. As indicated by equations (12.4) and (12.3), the charge-transfer reaction at the positrode–electrolyte interface in electrolysis mode splits H_2O to deliver $OH_O^{\bullet}$ into the electrolyte, while the charge-transfer reaction at the negatrode–electrolyte interface (i.e. equation (12.4)) consumes $OH_O^{\bullet}$ from the electrolyte to produce H_2 in the negatrode pore volume.

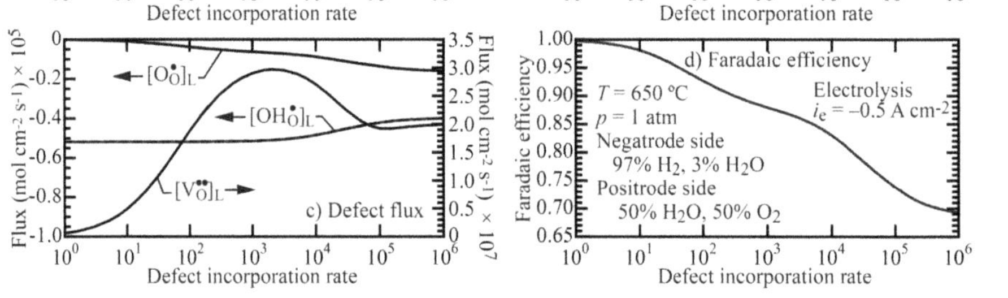

Figure 12.8. Effects of defect-incorporation reaction rates. (a) Defect concentrations in the BCZYYb electrolyte membrane at the negatrode–electrolyte interface. (b) Defect concentrations in the BCZYYb electrolyte membrane at the positrode–electrolyte interface. (c) Defect fluxes (mol cm^{-2} s^{-1}) through the electrolyte membrane. (d) Faradaic efficiency. The gas compositions are 97% H_2 and 3% H_2O on the reducing side and 50% O_2 and 50% H_2O on the oxidizing side at 650 °C and 1.0 atm. The imposed current density is fixed at $i = -0.5$ A cm^{-2}.

The present charge-transfer models only involve the $OH_O^{\bullet}$ entering (producing) or leaving (consuming) the electrolyte membrane (at the electrode–electrolyte interfaces). Thus, $V_O^{\bullet\bullet}$ and $O_O^{\bullet}$ can only be produced or consumed through the defect-incorporation reactions on the surface of the BCZYYb electrolyte membrane exposed to the gas environment. As long as the surface area of the dense electrolyte membrane is small or the defect reaction rate is small (i.e. low values of k_f), the transport of $V_O^{\bullet\bullet}$ and $O_O^{\bullet}$ through the electrolyte membrane is actually blocked at the electrode–electrolyte interfaces, leading to essentially pure proton transport through the electrolyte membrane and high faradaic efficiency.

On the BCZYYb electrolyte membrane surface exposed to the O_2–H_2O oxidizing environment (positrode side), the O_2 incorporation reaction ($\frac{1}{2}O_2 + O_O^{\times} + V_O^{\bullet\bullet} \rightleftharpoons 2O_O^{\bullet}$) delivers $O_O^{\bullet}$ into the BCZYYb electrolyte membrane. The reverse H_2 incorporation reaction ($OH_O^{\bullet} \rightleftharpoons \frac{1}{2}H_2 + O_O^{\bullet}$) and the reverse H_2O incorporation reaction ($2OH_O^{\bullet} \rightleftharpoons H_2O + V_O^{\bullet\bullet} + O_O^{\times}$) consume $OH_O^{\bullet}$ within the electrolyte to produce H_2 and H_2O in the gas environment within the positrode. It should be noted that the net rate of H_2 incorporation is small compared to the O_2 and H_2O incorporation rates as k_f increases. Figure 12.8(b) shows that the defect reactions lead to decreasing $[OH_O^{\bullet}]$ and increasing $[O_O^{\bullet}]$ and $[V_O^{\bullet\bullet}]$ on the positrode side.

On the BCZYYb electrolyte membrane surface exposed to the H_2–H_2O reducing environment (hydrogen collection, negatrode side), the O_2 incorporation reaction is reversed to consume $O_O^{\bullet}$ from the BCZYYb electrolyte membrane and release $O_2(g)$ into the negatrode pore volume. The H_2 reacts with $O_O^{\bullet}$ to produce $OH_O^{\bullet}$. The H_2O incorporation reaction is reversed to produce H_2O. The O_2 produced by reversing the O_2 incorporation reaction can react with H_2 to produce H_2O on the Ni surface within the negatrode microstructure. Figure 12.8(a) shows that $[O_O^{\bullet}]$ decreases as k_f increases, but $[OH_O^{\bullet}]$ and $[V_O^{\bullet\bullet}]$ do not vary monotonically because of competition between the defect-incorporation reactions.

As illustrated in figure 12.8, it is clear that $[O_O^{\bullet}]$ increases on the oxidizing side, but decreases on the reducing side as k_f increases. As a result, the $[O_O^{\bullet}]$ gradient across the BCZYYb membrane increases, leading to increasing polaron flux $J_{O_O^{\bullet}}$ but reducing proton flux $J_{OH_O^{\bullet}}$ (figure 12.8(c)) and therefore decreasing faradaic efficiency, as illustrated in figure 12.8(d). At the highest rate, i.e. $k_f = 10^6$, the faradaic efficiency drops to about 69.3%.

Figure 12.9 shows the effects of various current densities on the faradaic efficiency at fixed defect-incorporation reaction rates on BCZYYb surfaces. As the current density increases in both the fuel-cell and electrolysis modes, the proton fluxes from the charge-transfer processes increase while the defect reaction rates on the BCZYYb surface are fixed. Thus, the faradaic efficiency increases as the current density increases. Figure 12.9 also compares the faradaic efficiency at specified defect-incorporation forward rates of $k_f = 10^0$, 10^2, 10^4, and 10^6. The faradaic efficiency decreases as k_f increases at the same current density, which is consistent with figure 12.8.

Figure 12.9 shows that the cell voltage and power density drop slightly at the same current density as k_f increases. Since the small polaron $O_O^{\bullet}$ has higher mobility

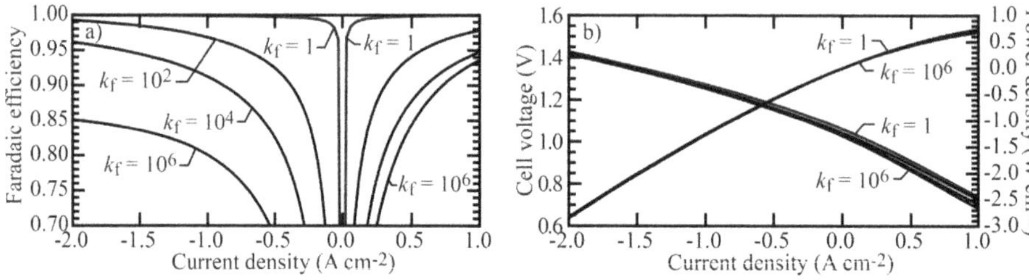

Figure 12.9. Effects of current density and defect-incorporation rates on cell performance. (a) Predicted faradaic efficiency. (b) Predicted cell voltage and power density. The defect-incorporation forward rate constants are set to $k_f = 10^0$, 10^2, 10^4, and 10^6. The gas-phase composition is 97% H_2 and 3% H_2O on the reducing side and 50% O_2 and 50% H_2O on the oxidizing side. The operating temperature is fixed at 650 °C and the pressure is atmospheric.

than that of the proton $OH_O^{\bullet}$, the higher $O_O^{\bullet}$ flux at higher k_f generally leads to higher electronic conductivity through the BCZYYb membrane, and hence lower cell voltage.

12.6 Concluding remarks

Although the technology is less mature than polymer-electrolyte membrane (PEM) electrolysis and SOECs, PCECs offer potentially significant advantages. Their benefits include higher efficiency than PEM electrolysis, especially if the steam can be raised using otherwise wasted heat. Compared to SOECs, PCECs can deliver nearly dry H_2 and thus eliminate a further separation process.

The proton-conducting materials (e.g. BCZYYb) are, in fact, MIECs. Charge-defect flux within the protonic ceramic electrolyte, including proton transport, may be represented in terms of Nernst–Planck theory. Electronic conduction via small polarons permits electronic leakage that tends to reduce faradaic efficiency. The mixed conduction increases modeling complexity, as exemplified by the prediction of the local electrostatic potential profiles needed to maintain charge neutrality.

Electrochemical cell performance depends on charge-transfer chemistry and defect-incorporation reactions. Charged defects (e.g. protons, oxygen vacancies, and small polarons) can be introduced into the protonic ceramic electrolyte via either charge transfer or defect incorporation. Charge-transfer rates are usually represented in the form of Butler–Volmer kinetics, with empirically derived exchange current densities and activation overpotentials that are fitted to measured cell performance. In this chapter, the charge-transfer processes are assumed to occur at the interfaces between the electrolyte membrane surfaces and the composite electrodes (i.e. at TPBs). However, the charge-transfer reactions for triple-conducting (i.e. protons, oxygen vacancies, and electrons) electrode materials can be extended by tens of microns from the membrane interface into the composite electrodes [37], significantly reducing activation overpotential losses and improving cell performance.

The kinetics of defect-incorporation reactions are less well known than the Butler–Volmer kinetics. The thermodynamics of the defect reactions (i.e. enthalpy and entropy) can be evaluated from measured conductivities as functions of the oxygen partial pressure and temperature. Although the reaction equilibrium constants can be evaluated from the thermodynamics, there is relatively little quantitative knowledge about the potentially rate-determining kinetics of specific proton-conducting electrolytes. Conductivity-relaxation experiments may be one way to determine the needed defect-incorporation rates. This chapter discusses the potential significance of kinetic competition between charge-transfer and defect-incorporation kinetics.

Although this chapter formulates and discusses electrochemical theory and modeling in a general setting, the illustrative examples are drawn from the one-dimensional laboratory-scale button cell. This chapter does not address the important issues associated with gas flows and thermal behaviors. Achieving practical technology requires the development of significantly upscaled reactors, usually in either planar or tubular stacks. In these cases, engineering design and development must be concerned with the coupled effects of fluid and thermal interactions. However, although design and development are affected by gas flows and thermal balances, the underpinning electrochemical concepts remain unchanged.

Acknowledgments

Our research into protonic ceramics and electrolysis was supported by the Colorado School of Mines Foundation via the Angel Research Fund and by ARPA-E via the REFUEL program. We gratefully acknowledge insightful discussions about protonic ceramic materials and electrolysis with our colleagues Dr Grover Coors (Hydrogène Hélix, SAS) and Profs. Robert Braun, Neal Sullivan, and Ryan O'Hayre (Colorado School of Mines).

References and additional reading

[1] Kee R, Ricote S, Zhu H, Braun R, Carins G and Persky J 2022 Perspectives on technical challenges and scaling considerations for tubular protonic-ceramic electrolysis cells and stacks *J. Electrochem. Soc.* **169** 054525

[2] Marrony M 2016 *Proton-Conducting Ceramics: From Fundamentals to Applied Research* (Boca Raton, FL: CRC Press)

[3] Katahira K, Kohchi Y, Shimura T and Iwahara H 2000 Protonic conduction in Zr-substituted $BaCeO_3$ *Solid State Ionics* **138** 90–8

[4] Fabbri E, D'Epifanio A, Bartolomeo E D, Licoccia S and Traversa E 2008 Tailoring the chemical stability of $Ba(Ce_{0.8-x}Zr_x)Y_{0.2}O_{3-\delta}$ protonic conductors for intermediate temperature solid oxide fuel cells (IT-SOFCs) *Solid State Ionics* **179** 558–64

[5] Medvedev D 2019 Trends in research and development of protonic ceramic electrolysis cells *Int. J. Hydrogen Energy* **44** 26711–40

[6] Lei L, Zhang J, Yuan Z, Liu J, Ni M and Chen F 2019 Progress report on proton conducting solid oxide electrolysis cell *Adv. Funct. Mater.* **29** 1903805

[7] Omata T, Okuda K, Tsugimoto S and Otsuka-Matsuo-Yao S 1997 Water and hydrogen evolution properties and protonic conducting behavior of Ca^{2+}-doped $La_2Zr_2O_7$ with a pyrochlore structure *Solid State Ionics* **104** 249–58

[8] Bjørheim T, Besikiotis V and Haugsrud R 2012 Hydration thermodynamics of pyrochlore structured oxides from TG and first principles calculations *Dalton Trans.* **41** 13343–51

[9] Haugsrud R and Norby T 2006 High-temperature proton conductivity in acceptor-doped $LaNbO_4$ *Solid State Ionics* **177** 1129–35

[10] Magraso A 2013 Transport number measurements and fuel cell testing of undoped and Mo-substituted lanthanum tungstate *J. Power Sources* **240** 583–8

[11] Malavasi L, Fisher C and Islam M 2010 Oxide-ion and proton conducting electrolyte materials for clean energy applications: structural and mechanistic features *Chem. Soc. Rev.* **39** 4370–87

[12] Kreuer K 2003 Proton-conducting oxides *Annu. Rev. Mater. Res.* **33** 333–59

[13] Norby T, Widerœ M, Glöckner R and Larring Y 2004 Hydrogen in oxides *Dalton Trans.* **19** 3012–8

[14] Han D, Liu X, Bjørheim T and Uda T 2021 Yttrium-doped barium zirconate-cerate solid solution as proton conducting electrolyte: why higher cerium concentration leads to better performance for fuel cells and electrolysis cells *Adv. Energy Mater* **11** 2003149

[15] Ryu K and Haile S 1999 Chemical stability and proton conductivity of doped $BaCeO_3$-$BaZrO_3$ solid solutions *Solid State Ionics* **125** 355–67

[16] Ricote S, Bonanos N and Caboche G 2009 Water vapour solubility and conductivity study of the proton conductor $BaCe_{(0.9-x)}Zr_xY_{0.1}O_{(3-\delta)}$ *Solid State Ionics* **180** 990–7

[17] Leonard K, Lee Y S, Okuyama Y, Miyazaki K and Matsumoto H 2017 Influence of dopant levels on the hydration properties of SZCY and BZCY proton conducting ceramics for hydrogen production *J. Hydrogen Energy* **42** 3926–37

[18] Zhu H, Ricote S, Duan C, O'Hayre R, Tsvetkov D and Kee R 2018 Defect incorporation and transport within dense $BaZr_{0.8}Y_{0.2}O_{3-\delta}$ (BZY20) proton-conducting membranes *J. Electrochem. Soc.* **165** F581–8

[19] Zhu H, Ricote S, Duan C, O'Hayre R and Kee R 2018 Defect chemistry and transport within dense $BaCe_{0.7}Zr_{0.1}Y_{0.1}Yb_{0.1}O_{3-\delta}$ (BCZYYb) proton-conducting membranes *J. Electrochem. Soc.* **165** F845–53

[20] Bonanos N 2001 Oxide-based protonic conductors: point defects and transport properties *Solid State Ionics* **145** 265–74

[21] Dippon M, Babiniec S, Ding H, Ricote S and Sullivan N 2016 Exploring electronic conduction through $BaCe_xZr_{0.9-x}Y_{0.1}O_{3-\delta}$ *Solid State Ionics* **286** 117–21

[22] Han D and Uda T 2018 The best composition of an Y-doped $BaZrO_3$ electrolyte: selection criteria from transport properties, microstructure, and phase behavior *J. Mater. Chem.* A **6** 18571–82

[23] Li W, Guan B, Ma L, Tian H and Liu X 2019 Synergistic coupling of proton conductors BZCYYb and $La_2Ce_2O_7$ to create chemical stable, interface active electrolyte for steam electrolysis cell *ACS Appl. Mater. Interfaces* **11** 18323–30

[24] He F, Song D, Peng R, Meng G and Yang S 2010 Electrode performance and analysis of reversible solid oxide fuel cells with proton conducting electrolyte of $BaCe_{0.5}Zr_{0.3}Y_{0.2}O_3$ *J. Power Sources* **195** 3359–64

[25] Duan C, Kee R, Zhu H, Sullivan N, Zhu L, Bian L, Jennings D and O'Hayre R 2019 Highly efficient reversible protonic ceramic electrochemical cells for power generation and fuel production *Nat. Energy* **4** 230–40

[26] Duffy J, Meng Y, Abernathy H and Brinkman K 2021 Surface and bulk oxygen kinetics of $BaCo_{0.4}Fe_{0.4}Zr_{0.2-x}Y_xO_{3-\delta}$ triple conducting electrode materials *Membranes* **11** 766

[27] Zohourian R, Merkle R, Raimondi G and Maier J 2018 Mixed-conducting perovskites as cathode materials for protonic ceramic fuel cells: understanding the trends in proton uptake *Adv. Funct. Mater.* **28** 1801241
[28] Strandbakke R, Cherepanov V, Zuev A, Tsvetkov D, Argirusis C, Sourkouni G, Prünte S and Norby T 2015 Gd- and Pr-based double perovskite cobaltites as oxygen electrodes for proton ceramic fuel cells and electrolyser cells *Solid State Ion* **278** 120–32
[29] Choi S, Davenport T and Haile S 2019 Protonic ceramic electrochemical cells for hydrogen production and electricity generation: exceptional reversibility, stability, and demonstrated faradaic efficiency *Energy Environ. Sci.* **12** 206–15
[30] Bi L, Boulfrad S and Traversa E 2014 Steam electrolysis by solid oxide electrolysis cells (SOECs) with proton-conducting oxides *Chem. Soc. Rev.* **43** 8255–70
[31] Zhu H, Ricote S and Kee R 2022 Faradaic efficiency in protonic-ceramic electrolysis cells *J. Phys. Energy* **4** 014002
[32] Winiarz P, Covarrubias M, Sriubas M, Bockute K, Miruszewski T, Skubida W, Jaworski D, Laukaitis G and Gazda M 2021 Properties of barium cerate-zirconate thin films *Crystals* **11** 1005
[33] Zhu H and Kee R 2016 Membrane polarization in mixed-conducting ceramic fuel cells and electrolyzers *Int. J. Hydrogen Energy* **41** 2931–43
[34] Kee R, Zhu H, Hildenbrand B, Vøllestad E, Sanders M and O'Hayre R 2013 Modeling the steady-state and transient response or polarized and non-polarized proton-conducting doped-perovskite membranes *J. Electrochem. Soc.* **160** F290–300
[35] Vøllestad E, Zhu H and Kee R 2014 Interpretation of defect and gas-phase fluxes through mixed-conducting ceramics using Nernst-Planck-Poisson and Integral formulations *J. Electrochem. Soc.* **161** F114–24
[36] Zhu H, Ricote S, Coors W and Kee R 2015 Interpreting equilibrium-conductivity and conductivity-relaxation measurements to establish thermodynamic and transport properties for multiple charged defect conducting ceramics *Faraday Discuss.* **182** 49–74
[37] Zhu H and Kee R 2017 Modeling protonic-ceramic fuel cells with porous composite electrodes in a button-cell configuration *J. Electrochem. Soc.* **164** F1400–11
[38] Liu M 1998 Equivalent circuit approximation to porous mixed-conducting oxygen electrodes in solid-state cells *J. Electrochem. Soc.* **145** 142–54
[39] Coffey G, Pederson L and Rieke P 2003 Competition between bulk and surface pathways in mixed ionic electronic conducting oxygen electrodes *J. Electrochem. Soc.* **150** A1139–51
[40] Zhu H and Kee R 2008 Modeling distributed charge-transfer processes in sofc membrane electrode assemblies *J. Electrochem. Soc.* **32** 1486–91
[41] Kee R, Coltrin M, Glarborg P and Zhu H 2018 *Chemically Reacting Flow: Theory, Modeling, and Simulation* 2nd edn (Hoboken, NJ: Wiley)
[42] Mason E and Malinauskas A 1983 Gas *Transport in Porous Media: The Dusty-Gas Model* (New York: American Elsevier)
[43] Zhu H, Kee R, Janardhanan V, Deutschmann O and Goodwin D 2005 Modeling elementary heterogeneous chemistry and electrochemistry in solid-oxide fuel cells *J. Electrochem. Soc.* **152** A2427–40
[44] Ricote S, Zhu H, Coors W, Chatzichristodoulou C and Kee R 2014 Equilibrium and transient conductivity for gadolinium-doped ceria under large perturbations: I Experiments *Solid State Ionics* **265** 22–8

[45] Zhu H, Ricote S, Coors W, Chatzichristodoulou C and Kee R 2014 Equilibrium and transient conductivity for gadolinium-doped ceria under large perturbations: II *Model. Solid State Ionics* **268** 198–207
[46] Fleig J, Merkle R and Maier J 2007 The $p(O_2)$ dependence of oxygen surface coverage and exchange current density of mixed conducting oxide electrodes: model considerations *Phys. Chem. Chem. Phys.* **9** 2713–23
[47] Liang M, He F, Zhou C, Chen Y, Ran R, Yang G, Zhou W and Shao Z 2021 Nickel-doped $BaCo_{0.4}Fe_{0.4}Zr_{0.1}Y_{0.1}O_{3-\delta}$ as a new high-performance cathode for both oxygen-ion and proton conducting fuel cells *Chem. Eng. J.* **420** 127717

IOP Publishing

High-Temperature Electrolysis
From fundamentals to applications
Werner Sitte and Rotraut Merkle

Chapter 13

Tubular protonic ceramic electrolysis cells and direct hydrogen compression

Einar Vøllestad

The pressurized operation of protonic ceramic electrolyzer cells (PCECs) is particularly advantageous due to a combination of favorable thermodynamics for proton transport in the PCEC electrolyte and the possibility of directly producing dry pressurized hydrogen without any downstream separation or compression. The smaller sealing area and superior tolerance to pressure and temperature gradients makes tubular PCECs preferable for pressurized operation, even though this configuration presents longer current paths and a lower power density than a planar configuration. While this technology is still in its infancy, the development and upscaling of tubular PCECs is ongoing and has a particular emphasis on scalable manufacturing, novel cell architectures, and current-collection solutions.

13.1 Introduction

The current trend in most electrolysis technologies is to enable pressurized operation and the direct delivery of pressurized hydrogen. This is primarily done to minimize the need for downstream mechanical compression—which is challenging and expensive in terms of components and maintenance—and to reduce the overall footprint of the stack. For high-temperature electrolysis—and protonic ceramic electrolyzers in particular—pressurized operation may also enable direct integration with synthesis reactions and chemical production processes currently operated at intermediate temperatures and elevated pressures (e.g. ammonia synthesis).

In this chapter we will discuss the operating principles of protonic ceramic electrolyzers—with a specific emphasis on tubular PCECs—and how these relate to pressurized electrolysis and the direct electrochemical compression of hydrogen.

doi:10.1088/978-0-7503-3951-3ch13

13.1.1 PCEC operating principles

The operating principle of a PCEC can be compared to those of high-temperature solid oxide electrolyzer cells (SOECs) and low-temperature proton-exchange membrane electrolyzer cells (PEMECs)—see figure 13.1. SOECs transport oxide ions from the cathode—where steam is reacted to form hydrogen and oxide ions—to the anode, where O_2 gas is formed, while proton-exchange membrane (PEM) electrolyzers transport protons—in the form of hydronium ions (H_3O^+)—from the anode to the cathode, where both H_2 and H_2O are released. In contrast, PCECs transport lattice protons (H^+)—generated from steam at the anode—to the cathode, where hydrogen is formed. For both technologies, the produced hydrogen is diluted by H_2O in the form of water drag (PEM) or unreacted steam (SOEC), which requires considerable water management and separation downstream, adding to the system complexity and cost. In contrast, the lattice transport of protons in PCECs enables the direct production of undiluted and pure hydrogen at the cathode, which presents a unique technological advantage. PCECs further allow for an intermediate operating temperature (400 °C–700 °C) due to a lower lattice proton migration barrier compared to that of oxide ions, while ceramic electrolytes can retain their protons at significantly higher temperatures (and lower relative humidities) than the technologically more mature polymeric electrolytes employed in PEMs. These specific operational features of PCECs are particularly important for pressurized operation, as they allow for the direct electrochemical compression of H_2 and potentially direct thermal and electrochemical integration with downstream chemical processes, such as hydrogenation reactions and ammonia synthesis, which require dry pressurized hydrogen as a reactant at intermediate temperatures.

13.1.2 Cell geometries for pressurized operation

Two main cell and stack configurations are employed for high-temperature electrolyzers based on planar and tubular geometries. A planar stack is composed

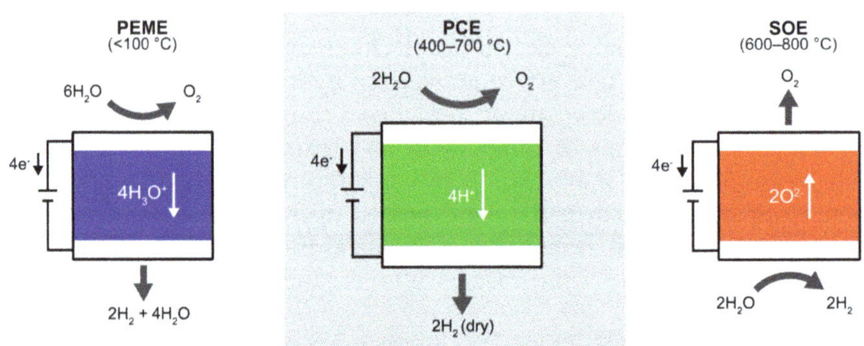

Figure 13.1. Schematic illustrations of three membrane-based electrolysis technologies and the corresponding gas compositions on both electrodes; PEM electrolysis (left), a protonic ceramic electrolyzer (middle), and a solid oxide electrolyzer (right). Reproduced from [1] with permission from Springer Nature.

of metal plate interconnects, glass-based sealing materials shaped as gaskets, and planar unit cells. It offers compact cell stacking, and the use of an external mechanical load ensures efficient mechanical contact between the interconnect plates and the cells (and thereby current flow). Furthermore, the short current paths result in low electrical resistance and high volumetric power. On the other hand, heat management in planar stacks is challenging and often contributes to temperature gradients within the stacks. This, combined with the use of sealing gaskets that create large sealing areas, induce thermomechanical stresses that challenge cycling stability in pressurized operation as well as in pressure gradients. In the tubular configuration, the sealants are applied to a reduced surface area, with the possibility of having the manifolds connected to one side of the cell only. This configuration facilitates heat management and is significantly more robust to thermomechanical stresses. It can more easily tolerate rapid thermal cycling events and has a higher tolerance for pressure gradients across the cell. This is particularly helpful if electrochemical pumping and the direct compression of hydrogen in the cell are used. However, tubular cells feature longer current paths and lower packing densities, resulting in larger electrical resistance and lower volumetric power density at the stack/assembly level compared to planar stacks.

To date, the planar configuration is the most predominant in solid oxide electrolyzer (SOE) stacks operating at ambient pressure, and substantial upscaling of the technology has been demonstrated at the 45–55 kW kg^{-1} H_2 scale [2]. Several recent studies have also reported the performances of short planar SOE stacks operated at 8–10 bar, which showed a significant decrease in the internal resistance ascribed to an increased frequency of reactants hitting the triple-phase boundary and decreased diffusion resistance [3, 4].

Protonic ceramic electrolyzer (PCE) applications have been reported for both planar and tubular button cells, whereas the upscaled production of the technology is dominated by tubular cells that have active surface areas of up to 60 cm^2 operating at high pressure (up to 10 bar) as well as in pressure gradients. For instance, pressure gradients of >50 bar (overpressure on the outside of the tubular cells) have been applied to a protonic ceramic tubular cell without any mechanical degradation or increased leakage rate across the cell [5]. In comparison to SOEs, the PCE technology is at its infancy with regards to stack and system development. There are currently limited public reports of PCE stack technology, and pioneering development in current R&D projects is focused on tubular PCE stacks of less than 10 kW with a targeted operating pressure of 30 bar [6, 7]. In these projects, the repeating unit is designed as a 'tube-in-a-shell,' in which the tubular electrochemical cell is contained in a steel vessel with the necessary seal and electrical and gas manifolds, as shown in figure 13.2. These units can then be assembled in series to build up power. More R&D efforts dedicated to stacks, system designs, and operational optimization are required to leverage the advantages of PCE materials and technology.

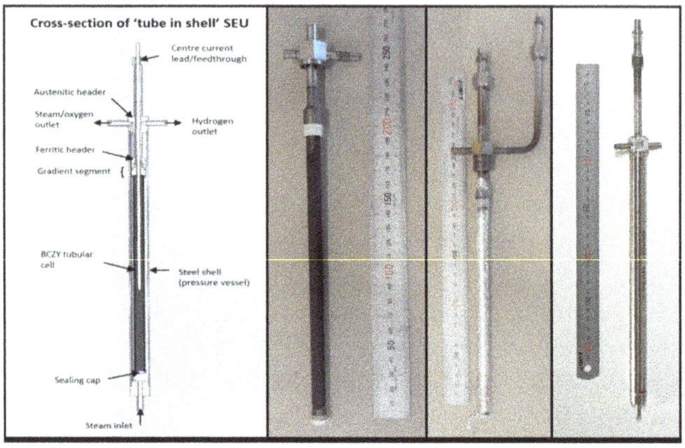

Figure 13.2. Schematic illustration a single engineering unit (left) and images of upscaled tubular PCEs with an active area of 60 cm^2 designed for pressurized operation as part of the ongoing FCH-JU2 project GAMER.

13.2 The thermodynamics and kinetics of pressurized PCECs

Operating the electrolyzer under pressure is beneficial from a system-level perspective, as the hydrogen product is normally used at pressure in downstream processes. The mechanical compression of gases is carried out in multi-step polytropic compressors with intercooling and a typical efficiency range of 70%–80%, depending on the application, whereas the pressurization of liquid water is much simpler and requires little energy. While the Nernst potential of the electrolysis cell (and thus the energy cost per hydrogen produced) generally increases with increased pressure, electrochemical compression is still more energy efficient than conventional mechanical compression. Another important argument in favor of pressurized operation is the improvement in heat transfer in the heat exchangers of the balance of plant. For large-scale fuel production plants, heat exchanger sizes become unmanageable at atmospheric pressure. Furthermore, at higher operating voltages, the increase in pressure lowers the overpotential due to improved electrode kinetics and reduced diffusion restrictions, as experimentally shown for SOEs operated at elevated pressures [3]. For PCECs in particular, higher pressures further result in a higher degree of hydration—and thus enhanced proton conductivity—of the cell component materials, which should enhance the overall performance. The impacts of pressurized operation on the thermodynamics and kinetics at the cell level and the component material level will be discussed in more detail in the following sections.

13.2.1 Cell-level thermodynamics of pressurized operation

The temperature-dependent energy demands of electrolyzers operating at 1 bar and 30 bar are shown in figure 13.3. ΔH is the total energy demand, ΔG is the energy demand required for electrical work, and $T\Delta S$ is the heat demand. Increased operating temperatures reduce the electricity demand and increase the heat demand in all cases, resulting in an overall increase in electrical efficiency—assuming there is

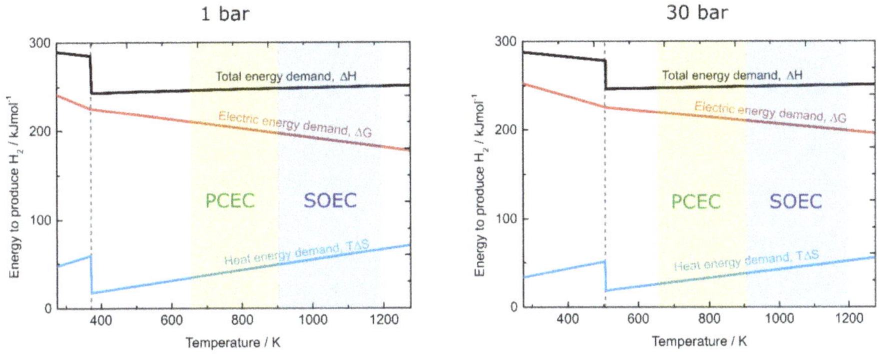

Figure 13.3. Energy demand versus temperature for high-temperature electrolysis operated at 1 bar (left) and 30 bar (right).

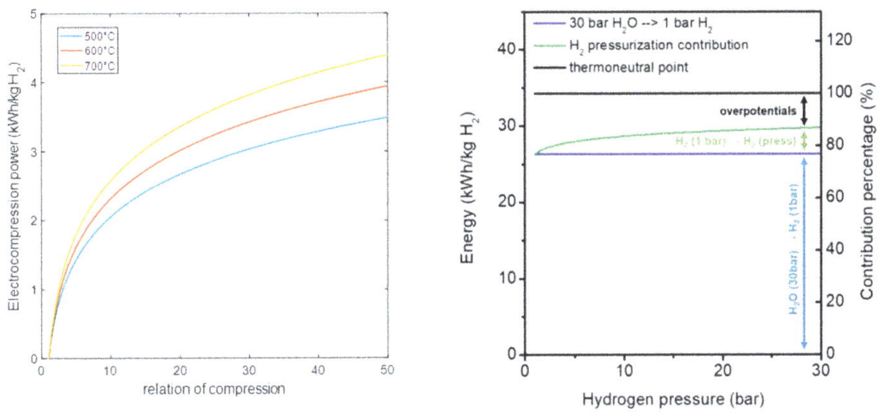

Figure 13.4. (Left) Electrochemical compression work as a function of the compression ratio and (right) different contributions to the total energy demand of the electrolyzer. Reproduced from [1] with permission from Springer Nature.

sufficient heat available to meet the required heat energy demand. As the figure also illustrates, increasing the operating pressure of high-temperature electrolyzers results in an increased electrical energy demand, due to the additional electrical work needed for the electrochemical compression of hydrogen.

However, the compression work is only a fraction of the total electrical energy consumed in the electrolyzer. As can be seen from figure 13.4, a compression ratio of 40:1 only consumes ~4 kWh kg^{-1} H_2 when operated at 600 °C—which accounts for approximately 10% of the overall energy demand during thermoneutral operation. In a simplified picture, the three sources of energy demand in pressurized high-temperature electrolysis are: (i) the electrolysis of $H_2O(g)$, (ii) Joule heating (overpotential related), and (iii) the formation of H_2 at different pressures. In thermoneutral operation, the two latter contributions provide the heat needed for the enthalpy of the electrolysis reaction. At higher hydrogen delivery pressures, additional compression heat is provided to the system which necessitates operation at lower overpotentials—and thus lower current density—to maintain thermoneutral operation.

As the above discussion illustrates, operation at elevated pressures results in a higher electrical energy demand per kg H_2 than ambient-pressure electrolysis (H_2 supplied at ambient pressure). However, this energy penalty is mostly offset by the need for a downstream compression step in most ambient-pressure electrolysis systems, which also adds considerable cost, complexity, and maintenance at the system level. Generally, the *in situ* utilization of compression heat enables increased overall energy efficiency of the electrolysis plant when compared with external compression. In addition, there are specific thermodynamic and kinetic benefits of operating PCECs at pressure, which will be discussed in the following sections.

13.2.2 Thermodynamics and kinetics of cell components

Operating PCECs at high pressure also impacts the kinetics and thermodynamics at the component and material levels. It is well established that electrode kinetics are improved at higher pressures due to the enhanced activity of reactants in the electrochemical reactions, which consequently results in lower electrode polarization resistance and improved cell performance (higher current density at the thermoneutral potential).

In addition to a positive effect of the electrode kinetics, operation at elevated temperatures also directly influences the defect chemistry and transport properties of the component materials in PCECs. In contrast to most oxide-ion-conducting electrolytes, in which the concentration of mobile oxygen vacancies ($V_O^{\bullet\bullet}$) is fixed by the acceptor concentration (Y'_{Zr}), the equilibrium concentration of protons ($OH_O^{\bullet}$) in protonic ceramic electrolytes are also dependent on the partial pressure of steam (see chapter 11):

$$H_2O(g) + O_O^x + V_O^{\bullet\bullet} \rightleftharpoons 2OH_O^{\bullet} \tag{13.1}$$

$$[OH_O^{\bullet}] = K_{hyd}^{-1/2}\,[O_O^x]^{1/2}\,[V_O^{\bullet\bullet}]^{1/2} pH_2O^{1/2} \tag{13.2}$$

$$[OH_O^{\bullet}] = [Y'_{Zr}] + 2[V_O^{\bullet\bullet}]. \tag{13.3}$$

To illustrate the impact of steam pressure on the transport properties of protonic ceramic electrolytes, figure 13.5 depicts calculated proton concentrations at 600 °C for $BaZr_{1-y}Y_yO_{3-y/2}$ (BZY) and $BaCe_{1-y}Y_yO_{3-y/2}$ (BCY) based on hydration thermodynamics described in the literature [8]. As can be seen, there is a substantial increase in the proton concentration of a BZY-based electrolyte upon increasing the steam pressure from 0.2 to 0.5 bar (commonly used in ambient-pressure PCECs with 20%–50% H_2O) to >5 bar, at which the electrolyte is essentially fully hydrated. From figure 13.5 it is also evident that increased steam pressure has a larger impact on BZY-type electrolytes with large Zr contents due to the less favorable hydration thermodynamics at these temperatures. Most electrolytes employed today are solid solutions of BCY–BZY ($BaZr_{1-x-y}Ce_xY_yO_{3-y/2}$, BZCY), for which no consistent

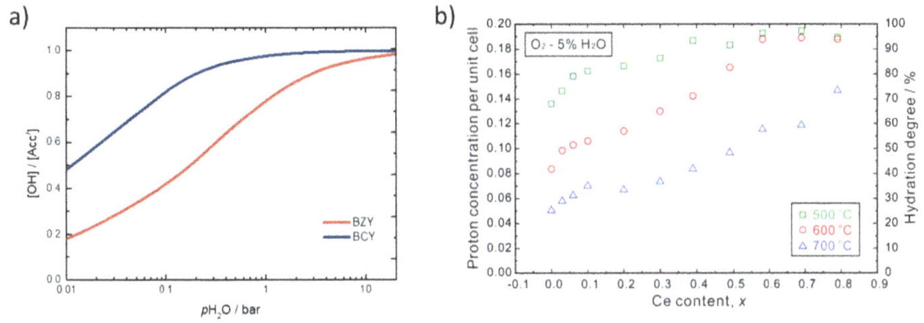

Figure 13.5. (a) Calculated proton concentrations for BZY and BCY at 600 °C as a function of steam pressure based on hydration thermodynamics from [8]. (b) Measured proton concentrations as a function of Ce content in $BaZr_{0.8-y}Ce_yY_{0.2}O_{2.9}$ measured in oxygen with a pH_2O of 0.05 bar. Reproduced from [9] with permission from Wiley Materials.

thermodynamic data for hydration exist. However, recent reports on the effect of the Zr/Ce ratio on proton conductivity indicate that the proton concentrations and hydration thermodynamics of BZCY-based electrolytes change linearly in between the BZY and BCY extremes of solid solutions [9]. Accordingly, the pressurized operation of PCECs is expected to result in higher proton concentrations and higher proton conductivity in the electrolyte, while ambient-pressure electrolysis operated at 600 °C will—for most BZCY-based electrolytes with Ce concentrations below 60%—result in a co-ionic electrolyte with both protonic and oxide ionic conduction through the electrolyte.

Another important factor in determining the efficiency and performance of PCECs is the faradaic efficiency of the cells—which to a large extent is determined by the electronic transference number of the electrolyte under operating conditions. For BZCY-based electrolytes, it is primarily the p-type electronic conduction under oxidizing conditions which is the determining parameter for electronic leakage (and thus loss in faradaic efficiency) in electrolytic operation (see chapter 12). Assuming an essentially fully hydrated electrolyte ($[OH_O^\bullet] = [Y'_{Zr}]$), the concentration of electron holes ($h^\bullet$) in BZCY-based electrolytes can be expressed through its equilibrium reaction with O_2 and H_2O as follows:

$$H_2O(g) + 2O_O^x + 2h^\bullet \rightleftharpoons 2OH_O^\bullet + \frac{1}{2}O_2(g) \tag{13.4}$$

$$[h^\bullet] = K_{ox}[OH_O^\bullet]\, pH_2O^{-1/2} pO_2^{1/4}. \tag{13.5}$$

As can be seen from equation (13.5), the concentration of electron holes scales positively with $pO_2^{1/4}$ and negatively with $pH_2O^{-1/2}$. The different exponents in the two terms result in a negative dependency on the total pressure, assuming a constant O_2/H_2O ratio. In a PCEC with constant steam utilization (and thus a constant O_2/H_2O ratio), the concentration of electron holes in the electrolyte on the steam side decreases with increasing total pressure, as illustrated in figure 13.6(a). Consequently, we would expect decreased electronic leakage and increased faradaic

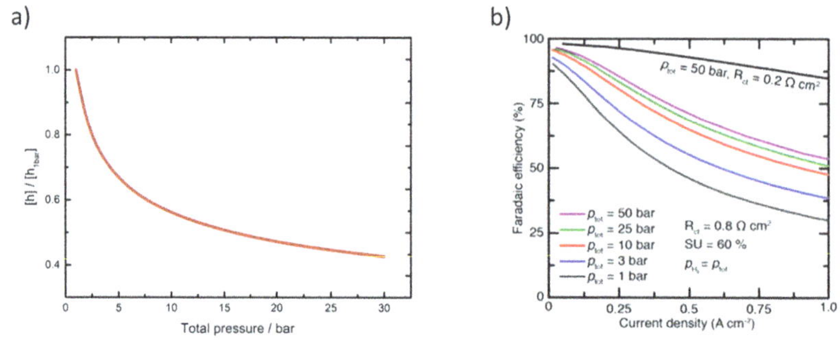

Figure 13.6. (a) Normalized concentration of electron holes in the electrolyte at the steam electrode interface as a function of total pressure calculated using equation (13.5). (b) Calculated impact of increased total pressure on faradaic efficiency at a constant steam utilization rate. Reproduced from [1] with permission from Springer Nature.

efficiency upon operating PCECs at higher total pressures, as indicated in figure 13.6(b).

Recent developments in protonic ceramic cells have seen the emergence of mixed protonic electronic oxides as highly active steam/air electrodes. While the actual proton concentrations (and also proton conductivities) and hydration thermodynamics are not yet established for most candidate materials, it is clear that increased steam pressure should enhance the partial proton conductivity in these materials and facilitate an enhanced electrochemically active surface area of the electrodes. However, increased pH_2O may also facilitate a larger surface coverage of OH^- species on the surface, which may occupy active sites for the oxygen evolution reaction (OER) and thus reduce the catalytic activity per surface area. At this point, there are no consistent reports in the literature on the impact of steam pressure on electrode polarization resistance in protonic ceramic cells, and this thus remains an open question for further fundamental research as part of the development of pressurized PCECs.

13.3 Materials, cell architectures, and assembly

13.3.1 Materials for pressurized operation

Operating PCECs at pressure imposes an additional and important constraint on the selection of component materials in a final cell design—namely, their chemical and mechanical stability under high steam pressures. Steam is an acidic gas which can readily react with basic oxides to form hydroxides. Perovskite-based proton-conducting oxides typically have a basic A-site cation to facilitate the hydration of oxygen vacancies and the incorporation of protons [10]. However this may then come at the expense of resistance to decomposition at high steam pressures. The decomposition reaction in the proton-conducting perovskite $BaCeO_3$ can be written as:

$$H_2O(g) + BaCeO_3(s) \rightleftharpoons Ba(OH)_2(s) + CeO_2(s) \tag{13.6}$$

For perovskites, the thermodynamic stability is generally related to the Goldschmidt tolerance factor, τ:

$$\tau = \frac{r_A + r_O}{\sqrt{2}(r_B + r_O)}. \tag{13.7}$$

A tolerance factor closer to unity results in a more cubic and typically more stable perovskite structure. This is also the case for the $BaZr_{1-x-y}Ce_xY_yO_3$ solid solution, in which the $BaZr_{1-y}Y_yO_3$ end member is a fully cubic perovskite with excellent resistance to acidic gases such as CO_2 and H_2O, whereas the thermodynamic stability decreases with increasing Ce contents, as the crystal structure becomes less cubic [11]. Unfortunately, the chemical stability in acidic gases is negatively correlated with the measured proton conductivity in the BZCY solid solution, and the highest proton conductivities are typically reported for Ce concentrations above 60%. Accordingly, there is necessarily a compromise between chemical stability and performance when selecting the electrolyte composition for PCECs—in which higher steam pressures and lower operating temperatures require a higher Zr content in the electrolyte to achieve sufficient stability. The chemical stability may also be influenced by replacing the acceptor dopant (Y) in BZCY with another trivalent cation with a different ionic radius, thereby altering the tolerance factor of the perovskite while still maintaining a constant Zr/Ce ratio—as has been suggested in the case of co-doping with Yb in [12, 13]

A similar challenge is also present for the steam electrode and the current-collection system, which are also exposed to high steam pressures. The most promising steam electrodes reported to date can generally be described as ABO_3-type perovskites (or double perovskites) with a large basic A site and a smaller transition-metal B-site with a more acidic character. At present, there are only a few reports regarding the chemical stability of steam electrode materials at high steam pressures—but there are several indications that stability at high steam concentrations is a challenge for many electrode materials with predominantly large and basic A sites comprising, for instance Ba^{2+} and Sr^{2+} [14, 15]. One of the few available reports showed that the partial substitution of Ba^{2+} with La^{3+} on the A site of the double perovskite $Ba_{1-x}La_xGd_{0.8}La_{0.2}Co_2O_{6-\delta}$ (BGLC) resulted in significantly increased stability at high steam pressures. At 50% substitution ($x = 0.5$), the double perovskite was shown to be stable in steam pressures of up to 21 bar at 600 °C for 100 h (figure 13.7) [1, 16].

13.3.2 Tubular cell fabrication and assemblies

This section reports the development of tubular cells that integrate BZCY-based electrolyte materials. The tubular cells are mechanically supported by the comparably thick Ni–BZCY cermet electrode. Their assembly starts with the production of the tubular BZCY–NiO electrode (typically containing 60 wt% NiO) by plastic extrusion or slip casting; both techniques enable volume production. After drying and cutting the electrode material to the desired length, the resulting green (unsintered) tubular NiO–BZCY electrode is coated with the electrolyte layer by

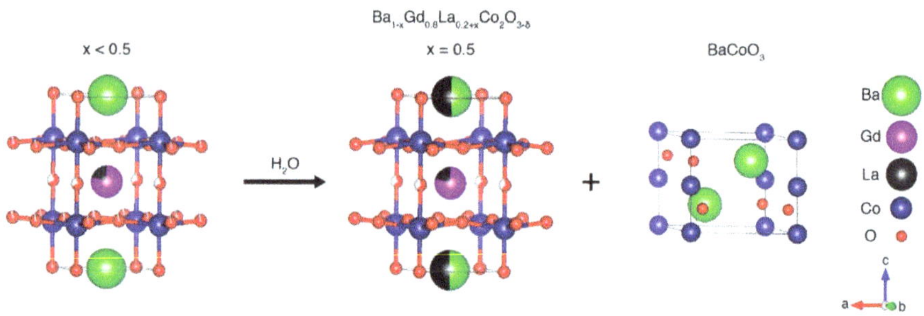

Figure 13.7. Decomposition reaction of the double perovskite $Ba_{1-x}La_xGd_{0.8}La_{0.2}Co_2O_{6-\delta}$ (BGLC) at high steam pressures. The end-member composition, which has 50% La substitution at the A site, remains stable and has a single phase after exposure to 75% steam and 25% air at a total pressure of 28 bar at 600 °C for 100 h. Reproduced from [1] with permission from Springer Nature.

either spray coating or dip coating. The two functional layers are then co-sintered to obtain a half-cell architecture comprising NiO–BZCY/BZCY. An important feature of these production steps relies on the use of BZCY precursor materials in the form of oxides (ZrO_2, CeO_2, Y_2O_3, etc.) and sulfates ($BaSO_4$), which yield the desired stoichiometric oxides upon solid state reactive sintering (SSRS). This contributes to reduced production costs and lower CO_2 emissions due to the reduction in the number of processing steps and the lower sintering temperature of the electrolyte. The steam electrode can then be applied to the tubular half-cells by dip coating, spray coating, or screen printing. Several pathways are employed to consolidate the steam electrode.

In several published reports, the steam electrode was directly coated on the sintered (unreduced) half-cell prior to the sealing and reduction of the NiO–BZCY electrode. In these reports, a commercial product, Ceramabond 552-VFG, was used as the sealant to join the cell to a ceramic sample holder. The complete cell was then sealed and reduced. This pathway works sufficiently for laboratory-scale experiments at ambient pressures on short cells. However, it is not robust and would not be suitable for operating at pressure, as this type of seal is not gastight. In the field of solid oxide cell technologies, a large amount of effort is dedicated to the development of gastight seals using glass and glass ceramic materials.

The benefit of this alternative approach is to utilize such seals for high-pressure applications. CTMS has developed a sealing technology that enables the tubular cells to be sealed to various ceramic and metallic parts [5]. In this process, the tubular half-cell is first reduced to Ni–BZCY/BZCY to reach the desired thermal expansion. The reduced cell is then sealed to a dense cap on one side, and a ceramic riser on the other side using thermally matched glass ceramic sealants with high pressure tolerance and excellent thermal cycling capability. This step is carried out in a reducing atmosphere to avoid oxidation of the electrode. This overall process necessitates application of the steam electrode after the tubular half-cells are fully reduced and sealed to avoid any reduction and decomposition of the redox-active steam electrode materials (which typically contain transition-metal cations such as

Co and Fe). After the electrode is coated and dried, the complete cell is fired in a dual atmosphere to achieve the desired microstructure and adhesion of the steam electrode. While this route enables superior pressure tolerance and thermal cycling capability, it also imposes some challenges associated with the annealing and consolidation of the steam electrode under dual atmosphere conditions (to maintain reduced Ni–BZCY).

The final step consists of applying the current-collector system. On the steam electrode side, Ag or Au inks are typically applied to the tubular cells by brush painting, followed by wrapping the cell with Ag or Au wires. The complete assemblies are *in situ* annealed during the conditioning phase of the cells. The inner electrode is then contacted using a metal wool (typically Ni or Cu), which directly contacts the Ni–BZCY cermet. Here, it is emphasized that the outer current-collection system, while operational, is not optimal for scaled-up production, as it is too costly. A significant challenge here is to ensure that there are sufficient mechanical contact points between the current collector and the electrode layer it contacts—and that these still remain in good contact after thermal cycling. This is particularly challenging for tubular cells that are not subjected to an external mechanical force.

13.4 Status of tubular PCEC technology

Until recently, the literature on tubular protonic ceramic cells was very scarce. However, recent years have seen an emergence of several groups working on developing—and to some extent scaling—tubular protonic ceramic electrochemical cells, although most have focused primarily on the fuel-cell application [1, 17–21]. Among the published reports, only three reports available in the recent literature present electrochemical data on tubular protonic ceramic electrolysis. Here, we will present and discuss these recent results to highlight what has been achieved thus far and the questions and challenges that remain to be solved for this technology.

13.4.1 Ambient-pressure cell testing

While the main promise of tubular PCECs is the feasibility of pressurized operation, most of the reported electrochemical data from the literature were measured at ambient pressure. Below, we will summarize the key findings from the most relevant studies of ambient-pressure operation.

Tarutin and colleagues [20] developed a tubular electrolysis with all-Ni-based functional electrodes using a combination of tape calendering and spray coating. The cell employed a 25 μm thick spray-coated $BaCe_{0.5}Zr_{0.3}Dy_{0.2}O_{3-\delta}$ (BCZD) electrolyte layer supported on a Ni–$BaCe_{0.5}Zr_{0.3}Dy_{0.2}O_{3-\delta}$ support made by tape calendering. The steam electrode employed in this work was a Ruddlesden–Popper oxide with the formula $Pr_{1.95}Ba_{0.05}NiO_{4+\delta}$ covered by a $LaNi_{0.6}Fe_{0.4}O_{3-\delta}$ (LNF) current-collector layer. The tube was sealed towards a BCZD cap on top and a zirconia support tube on the bottom using a borosilicate glass. The same glass was used across the film junctions (rolled layers) to avoid any gas crossover (see figure 13.8). This cell achieved a current density of 615 mA cm^{-2} under thermoneutral

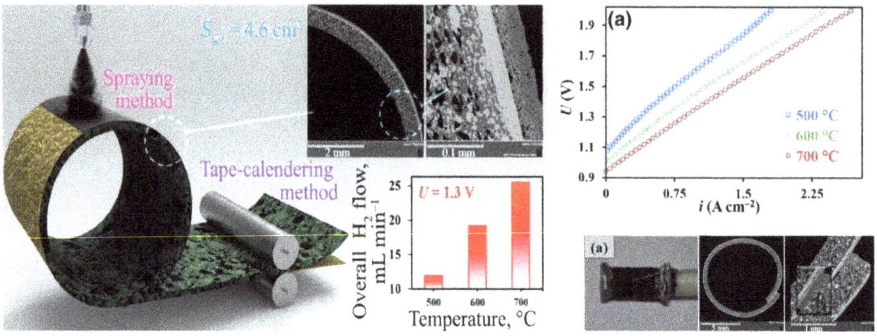

Figure 13.8. Schematic illustration of the combined tape-calendering and spray-coating method used by Tarutin *et al* and corresponding images and electrochemical results of the tubular cells tested at ambient pressure. Reproduced from [20] with permission from Elsevier.

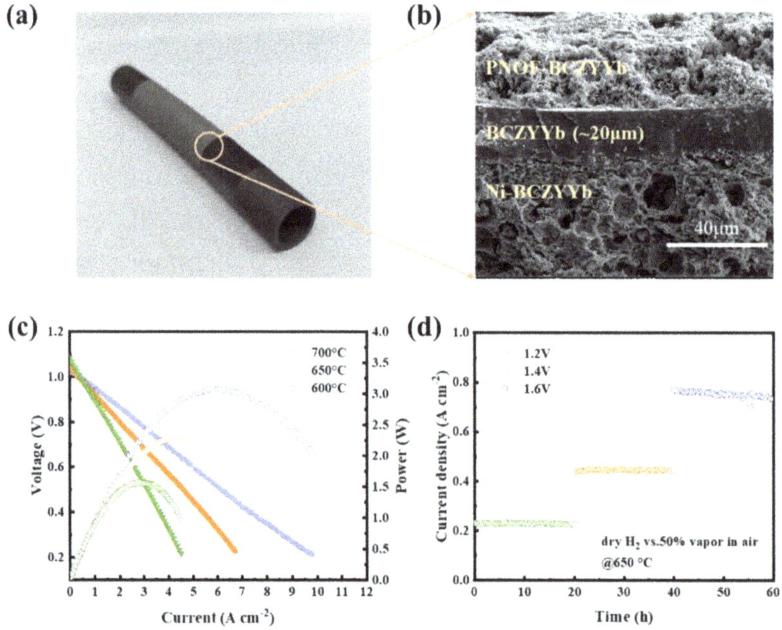

Figure 13.9. (a) Picture and (b) scanning electron microscopy (SEM) cross-section of the tubular cell made by Li *et al.* Electrochemical results of ambient-pressure testing in fuel-cell mode (c) and electrolysis mode (d). Reproduced from [19] with permission from Elsevier.

operation (1.3 V) at a temperature of 600 °C. The feed gases were humidified H_2 on the hydrogen electrode (0.03 atm H_2O) and wet air (0.3 atm H_2O) on the steam electrode side. Unfortunately, there was no direct measurement of hydrogen production in this work—and the reported H_2 flow was calculated based only on the current density and an expected ionic transference number extrapolated from the open-circuit voltage. The stability of the complete cell is also challenging to assess, as it was only tested continuously for 60 h with an extrapolated degradation rate of 8%/khr.

Li *et al* [19] recently reported a reversible tubular PCEC made by extrusion based on a $BaCe_{0.7}Zr_{0.1}Y_{0.1}Yb_{0.1}O_{3-\delta}$ (BCZYYb) electrolyte and a fluorinated Ruddlesden–Popper type steam electrode, $Pr_2NiO_{3.9+\delta}F_{0.1}$ (PNOF), depicted in figure 13.9. In electrolysis mode at 650 °C, this tubular cell achieved a current density of 410 mA cm^{-2} at a cell potential of 1.4 V, using dry hydrogen on the H_2 electrode and 50% H_2O in air on the steam electrode at atmospheric pressure. However, this work did not measure faradaic efficiency, so it is challenging to assess the actual hydrogen production rate or the electronic leakage of the cell under these conditions. The cell was tested over 60 h of operation and showed a decline in current density during the last 20 h of operation at 1.6 V (figure 13.9).

13.4.2 Pressurized tubular PCEs

A paper by Vøllestad *et al* from 2019 is currently the only available report in the literature on the pressurized operation of PCECs using tubular cell segments with active areas in the range of 10–16 cm^2 [1]. These cells comprise an extruded Ni–BZCY72 support with a spray-coated $BaZr_{0.8}Ce_{0.1}Y_{0.1}O_{3-\delta}$ (BZCY81) electrolyte layer ~25 μm thick. The high Zr content and low Y content of the electrolyte are selected to provide chemical and mechanical robustness and stability, even though the resulting proton conductivity is lower than those of many of the Ce-rich compositions used for ambient-pressure electrolysis. The steam electrode is based on the double perovskite $Ba_{0.5}La_{0.5}Gd_{0.8}La_{0.2}Co_2O_{6-\delta}$ (BGLC587) due to its demonstrated stability under high steam pressures (see figure 13.7). In this work, cells approximately 4–6 cm long were tested at total pressures ranging from 2 to 6 bar, with steam pressures ranging from 1 to 4 bar. BGLC was employed as both a single-phase electrode and as a composite electrode with 50 vol% BZCY. As can be seen from figure 13.10, the composite electrode displayed significantly better performance compared to the single-phase electrode. This was primarily attributed to insufficient adhesion (and partial delamination) of the single-phase electrode due to the severe thermal expansion mismatch between the BGLC electrode and the

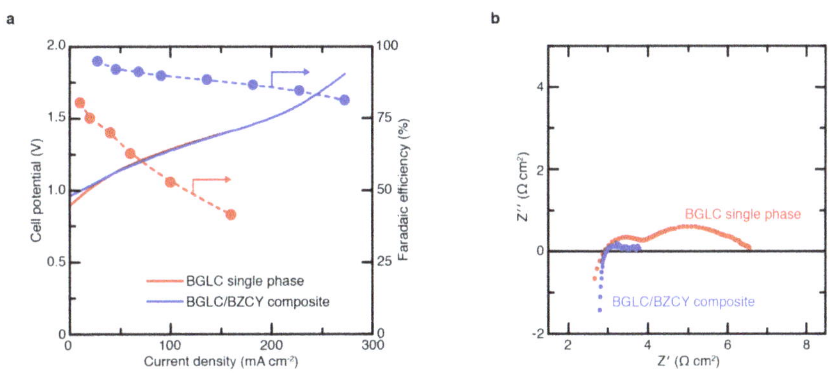

Figure 13.10. (a) *IV*-characteristics and faradaic efficiency of a tubular PCEC operated at a total pressure of 3 bar and a steam pressure of 1.5 bar with either a single-phase or a composite BGLC steam electrode, and (b) the associated electrochemical impedance spectra, showing a significantly larger area specific resistance for the single-phase electrode. Reproduced from [1] with permission from Springer Nature.

tubular support—which is particularly challenging for a tubular configuration that does not allow for an external mechanical load to maintain mechanical contact. The thermomechanical mismatch was sufficiently alleviated by the introduction of a composite electrode which effectively reduced the thermal expansion of the electrode layer and thus maintained good mechanical contact with the electrolyte layer after thermal cycling. Figure 13.10 also highlights the impact of the electrode performance on the faradaic efficiency of PCECs, showing a significantly enhanced faradaic efficiency for the cell with the composite electrode. Stable operation was further demonstrated over 700 h at a steam pressure of 1.5 bar at 600 °C, highlighting the robustness of the tubular cell in pressurized operation.

This work also highlighted some of the challenges associated with current collection in larger tubular cells due to the lack of an external mechanical load applied on either electrode, where the ohmic resistance was significantly affected by the choice of current-collection system used on the outer electrode. Recent improvements in the overall current-collection system using a combination of silver paste and thin silver wires on the steam electrode have resulted in a significantly reduced ohmic resistance—below 1 Ω cm^2—corresponding to the expected ohmic resistance of the 25 μm thick BZCY81 electrolyte layer (figure 13.11). This configuration has also been successfully scaled to larger tubular cells with an active area of 60 cm^2 as part of the ongoing FCH-JU project, GAMER (see figure 13.2).

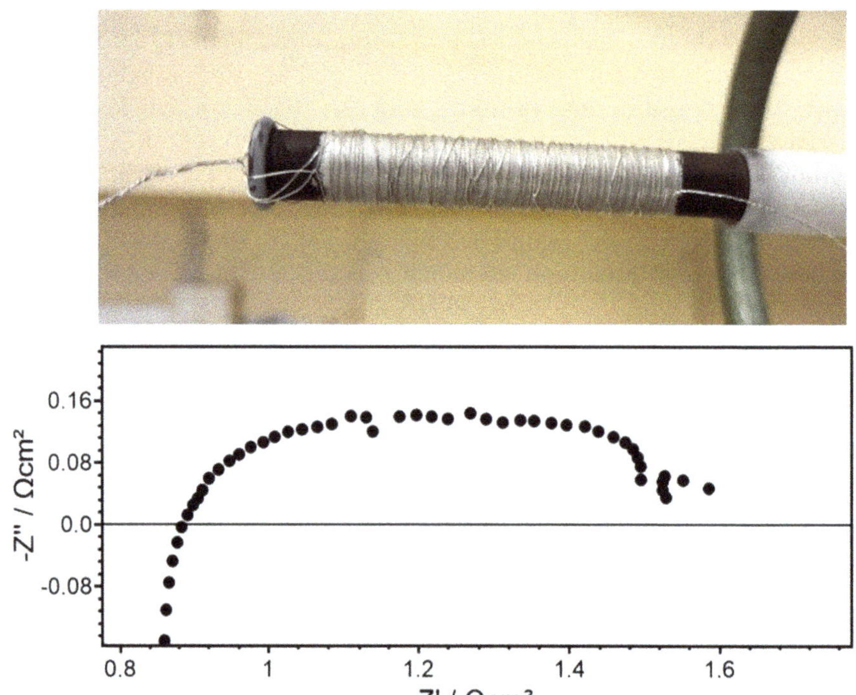

Figure 13.11. Image (top) and impedance spectrum (bottom) of a tubular cell with an optimized Ag current-collection system showing an ohmic resistance of less than 1 Ω cm^2 at 600 °C measured at a total pressure of 3 bar and 1.5 bar of steam on the steam electrode.

13.4.3 Future prospects for pressurized tubular PCECs

While the above results have successfully demonstrated the feasibility and robustness of pressurized steam electrolysis using tubular PCECs, there are still several challenges to overcome. One important focus will be to further improve the ohmic resistance of the tubular cells. The electrolyte composition (BZCY81) employed by Vøllestad *et al* [1] is limited in terms of its proton conductivity—particularly if a lower operating temperature is desired. Increasing the concentration of acceptor dopants is a natural step towards higher proton conductivity, in addition to the expected positive effect of increased Ce content in the electrolyte. Considering the latter, it is then important to verify that this does not come at the cost of chemical stability for the specific steam pressure and temperature envisioned for the electrolyzer operation. An increased Ce content in the electrolyte has also been shown to positively impact the faradaic efficiency by increasing the ionic transference number under oxidizing conditions [9, 22]. A change in the electrolyte composition used in the Ni–BZCY tubular support would also impact the thermomechanical properties of the tubular cell itself—and would thus necessitate tailoring of the glass ceramic sealant material to properly match the thermal expansion behavior of the tubular cell with a new electrolyte composition.

The lack of mechanical force applied to the tubular cells during operation (in contrast to the case of planar cells) imposes significant requirements on both the mechanical and electrical contacts between the individual layers to maintain low contact resistance across each interface of the cell assembly. This is particularly challenging at the electrode–current-collector interface on the steam side of the cell. At the electrode–electrolyte interface, there is also a need to increase the number of mechanical contact points. This can be solved by introducing a thin interfacial layer which bonds well to both the electrode and electrolyte—as demonstrated for planar cells by the introduction of a thin $PrBa_{0.5}Sr_{0.5}Co_{1.5}Fe_{0.5}O_{5+\delta}$ (PBSCF) layer deposited by pulsed laser deposition (PLD) [12]. In the current-collection system on the steam side, it will be necessary to move away from the use of precious metals such as silver or gold to reduce costs. While several materials (e.g., high-temperature steels) are available that have sufficient stability and conductivity under PCEC operating conditions, it remains a challenge to secure adequate attachment between the current-collection material and the underlying electrode. The development of a robust and cost-efficient current-collection system for tubular PCECs is critical for any upscaling of the technology—and this is currently being investigated and developed as part of the Horizon 2020 project WINNER [7].

13.5 Concluding remarks

The specific thermodynamics and transport properties of proton-conducting ceramics offer specific advantages for pressurized steam electrolysis using PCECs, such as the direct production of dry electrochemically compressed hydrogen. Higher-pressure operation will further alleviate some of the challenges associated with electronic leakage observed in PCECs—but may also impose challenges associated with material integrity and chemical stability at high steam pressures.

PCEC technology is currently relatively immature compared to the more established electrolysis technologies, but recent years have shown rapid development in both cell performance and stability at the single-cell level, underscoring the significant potential of the technology. At this juncture, it is now critical to consolidate these developments and place a larger emphasis on moving the technology from laboratory-scale testing towards larger-scale demonstrators based on robust and reproducible production routes, as well as on dedicated studies to evaluate and understand the long-term durability of PCECs operated in pressurized environments.

Acknowledgments

This contribution was made possible through funding from the Fuel Cell and Hydrogen 2 Joint Undertaking under Grant Agreements 101007165 ('WINNER') and 779486 ('GAMER'). This Joint Undertaking receives support from the European Union's Horizon 2020 Research and Innovation programme, Hydrogen Europe and Hydrogen Europe Research.

References

[1] Vøllestad E, Strandbakke R, Tarach M, Catalán-Martínez D, Fontaine M-L, Beeaff D, Clark D R, Serra J M and Norby T 2019 Mixed proton and electron conducting double perovskite anodes for stable and efficient tubular proton ceramic electrolysers *Nat. Mater.* **18** 752–9

[2] IRENA 2021 *Making the Breakthrough: Green hydrogen policies and technology costs* International Renewable Energy Agency (IRENA) 1–68 https://www.irena.org/-/media/Files/IRENA/Agency/Publication/2020/Nov/IRENA_Green_Hydrogen_breakthrough_2021.pdf

[3] Jensen S H, Sun X, Ebbesen S D and Chen M 2016 Pressurized operation of a planar solid oxide cell stack *Fuel Cells* **16** 205–18

[4] Riedel M, Heddrich M P, Ansar A, Fang Q, Blum L and Friedrich K A 2020 Pressurized operation of solid oxide electrolysis stacks: an experimental comparison of the performance of 10-layer stacks with fuel electrode and electrolyte supported cell concepts *J. Power Sources* **475** 228682

[5] Malerød-Fjeld H *et al* 2017 Thermo-electrochemical production of compressed hydrogen from methane with near-zero energy loss *Nat. Energy* **2** 923–31

[6] FCH-JU2 GAMER project – co-financed by Horizon 2020 https://www.sintef.no/projectweb/gamer/

[7] FCH-JU2 WINNER https://www.sintef.no/projectweb/winner/

[8] Bjørheim T S, Løken A and Haugsrud R 2016 On the relationship between chemical expansion and hydration thermodynamics of proton conducting perovskites *J. Mater. Chem.* A **4** 5917–24

[9] Han D, Liu X, Bjørheim T S and Uda T 2021 Yttrium-doped barium zirconate-cerate solid solution as proton conducting electrolyte: why higher cerium concentration leads to better performance for fuel cells and electrolysis cells *Adv. Energy Mater.* **11** 2003149

[10] Zohourian R, Merkle R, Raimondi G and Maier J 2018 Mixed-conducting perovskites as cathode materials for protonic ceramic fuel cells: understanding the trends in proton uptake *Adv. Funct. Mater.* **28** 1801241

[11] Zhong Z 2007 Stability and conductivity study of the $BaCe_{0.9-x}Zr_xY_{0.1}O_{2.95}$ systems *Solid State Ionics* **178** 213–20

[12] Choi S, Kucharczyk C J, Liang Y, Zhang X, Takeuchi I, Ji H-I and Haile S M 2018 Exceptional power density and stability at intermediate temperatures in protonic ceramic fuel cells *Nat. Energy* **3** 202–10

[13] Nguyen N T Q and Yoon H H 2013 Preparation and evaluation of $BaZr_{0.1}Ce_{0.7}Y_{0.1}Yb_{0.1}O_{3-\delta}$ (BZCYYb) electrolyte and BZCYYb-based solid oxide fuel cells *J. Power Sources* **231** 213–8

[14] Duan C, Huang J, Sullivan N and O'Hayre R 2020 Proton-conducting oxides for energy conversion and storage *Appl. Phys. Rev.* **7** 011314

[15] Papac M, Stevanović V, Zakutayev A and O'Hayre R 2021 Triple ionic–electronic conducting oxides for next-generation electrochemical devices *Nat. Mater.* **20** 301–13

[16] Vøllestad E, Schrade M, Segalini J, Strandbakke R and Norby T 2017 Relating defect chemistry and electronic transport in the double perovskite $Ba_{1-x}Gd_{0.8}La_{0.2+x}Co_2O_{6-\delta}$ (BGLC) *J. Mater. Chem.* A **5** 15743–51

[17] Cao D, Zhou M, Yan X, Liu Z and Liu J 2021 High performance low-temperature tubular protonic ceramic fuel cells based on barium cerate-zirconate electrolyte *Electrochem. Commun.* **125** 106986

[18] Hanifi A R, Sandhu N K, Etsell T H, Luo J-L and Sarkar P 2017 Fabrication and characterization of a tubular ceramic fuel cell based on $BaZr_{0.1}Ce_{0.7}Y_{0.1}Yb_{0.1}O_{3-\delta}$ proton conducting electrolyte *J. Power Sources* **341** 264–9

[19] Li G, Gou Y, Ren R, Xu C, Qiao J, Sun W, Wang Z and Sun K 2021 Fluorinated $Pr_2NiO_{4+\delta}$ as high-performance air electrode for tubular reversible protonic ceramic cells *J. Power Sources* **508** 230343

[20] Tarutin A, Kasyanova A, Lyagaeva J, Vdovin G and Medvedev D 2020 Towards high-performance tubular-type protonic ceramic electrolysis cells with all-Ni-based functional electrodes *J. Energy Chem.* **40** 65–74

[21] Zhu L, O'Hayre R and Sullivan N P 2021 High performance tubular protonic ceramic fuel cells via highly-scalable extrusion process *Int. J. Hydrogen Energy* **46** 27784–92

[22] Duan C, Kee R, Zhu H, Sullivan N, Zhu L, Bian L, Jennings D and O'Hayre R 2019 Highly efficient reversible protonic ceramic electrochemical cells for power generation and fuel production *Nat. Energy* **4** 230

IOP Publishing

High-Temperature Electrolysis
From fundamentals to applications
Werner Sitte and Rotraut Merkle

Chapter 14

Planar protonic ceramic electrolysis cells for H_2 production and CO_2 conversion

Fan Liu and Chuancheng Duan

Protonic ceramic electrolysis cells (PCECs) for H_2 production and CO_2 conversion are truly remarkable technologies. Prior attempts to achieve high-performance H_2O electrolysis and CO_2 conversion have been plagued by the low conductivity of proton-conducting oxides and the lack of highly active electrode materials. Extensive efforts have been devoted to the development of advanced electrolyte and electrode materials, allowing the efficient production of H_2 via H_2O electrolysis and the conversion of CO_2 into value-added chemicals (e.g. CO and CH_4). This chapter introduces planar PCECs for H_2 production and CO_2 conversion and has a particular focus on highlighting the advantages and challenges of employing PCECs for H_2 production and CO_2 conversion. Recent progress in the field of planar PCECs is comprehensively reviewed and critically evaluated. Additionally, the challenges and opportunities of PCECs are analyzed, with the aim of guiding future PCEC development.

14.1 H_2 production and CO_2 conversion in PCECs

14.1.1 PCECs for H_2 production

There is an increasing demand for H_2, while the cost of renewable electricity (e.g. solar and wind) has drastically reduced, advancing the development of electrolyzers for green H_2 production. In particular, PCECs, as shown in figure 14.1 [1], have received growing attention due to their ability to efficiently electrolyze H_2O with an energy efficiency comparable to those of oxygen ion solid oxide electrolysis cells (O-SOECs), while operating at lower temperatures. Figure 14.1 displays a planar PCEC which consists of three layers, including the fuel electrode (i.e. the cathode or negative electrode) where the hydrogen evolution reaction occurs, the dense proton-conducting ceramic electrolyte membrane, and the air electrode (i.e. the anode or positive electrode) that splits water and produces oxygen. Therefore, unlike

doi:10.1088/978-0-7503-3951-3ch14

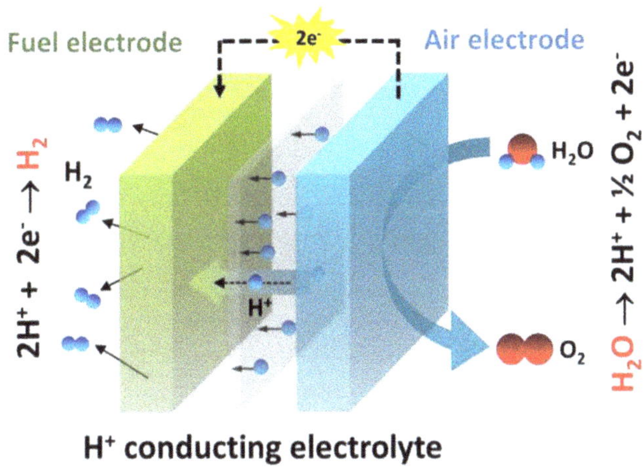

Figure 14.1. Schematic illustration of protonic ceramic electrolysis cells (PCECs) used for H_2 production. Reproduced from [1] with permission from the Royal Society of Chemistry.

O-SOECs, in which H_2O is fed to the cathode while H_2 is also produced at the cathode, H_2O is delivered to the anode of PCECs and H_2 is generated at the cathode. This unique characteristic offers the opportunity to directly produce high-purity, dry H_2, simplifying the PCEC system design. However, it also imposes additional requirements on the design and development of anode materials for PCECs. While many research articles have been published on this topic (table 14.1), here, we aim to comprehensively summarize and analyze the advancements in the field of PCECs for H_2 production.

14.1.2 CO_2 conversion in PCECs

The second important application of PCECs is converting CO_2 to carbon-containing compounds, which gives rise to substantial environmental and economic benefits. In the context of growing concern about the high global atmospheric concentration of CO_2, economically viable and efficient technologies for large-scale and reliable CO_2 conversion and utilization must be widely deployed to minimize the environmental risks imposed by human activities that have high CO_2 emissions (e.g. coal-fired power plants). Fortunately, PCECs driven by renewable power can provide opportunities to utilize CO_2 for energy storage and the synthesis of sustainable chemicals [2–8]. This modular and electrochemical approach is particularly promising for the reduction of point-source CO_2 emissions due to its versatility for the manufacture of chemical feedstocks on site [9]. PCECs can operate at intermediate temperatures (e.g. 400 °C–500 °C), which favors the production of chemicals beyond carbon monoxide, such as CH_4, which can be used as sustainable feedstocks for various industries [10]. Moreover, reversible methane fuel cells based on PCECs can be used for energy storage [10, 11], yielding a theoretical round-trip efficiency that approaches 100% and allowing the creation of a closed cycle with a minimal carbon footprint. Thus, creating a closed cycle for methane has significant economic and

Table 14.1. Summary of the PCECs developed for H_2O electrolysis

Year	Configuration of PCEC: Anode|Electrolyte|Cathode	Temper-ature (°C)	Current density at 1.4V (mA/cm^2)	Faradaic efficiency	Positive electrode (anode) steam concentration	Stability	Ref.
2021	PBCC-BCO|BZCYYb1711 (10 μm) |Ni-BZCYYb1711	600	2947	62.5%	30% H_2O	3.3%/1000 h for 1833 h, 1 A cm^{-2} at 650 °C	[33]
2021	SCFN|BZCYYb1711 (26 μm) |Ni-BZCYYb1711	600	1331	No data	3% H_2O	No degradation for 120 h in fuel cell/electrolysis cycles at 550 °C	[34]
2021	YEBCG-BZCYYb1711|BZCYYb1711 (14 μm)|Ni-BZCYYb1711	700 600	4517 1685	No data	Air (20% H_2O)	No degradation for 1000 h reversible operation, 0.2 A cm^{-2} at 600 °C	[35]
2020	PB5N-BZCD35|BZCD35 (25 μm) |Ni-BZCD35	700 600	1228 938	61.8% 60.5%	Air (30% H_2O)	0.5% degradation for 60 h, 1.3 V at 600 °C	[36]
2020	SFM-BZY20|BZY20 (18 μm)|Ni-BZCY17	650 600	821 573	27.4% 32.8%	Air (3% H_2O)	No degradation for 10 h, 1.3 V at 600 °C	[37]
2020	Hollow PNC fibers|BZCYYb4411 |Ni-BZCYYb4411	600	1287	68.4%	Air (15% H_2O)	No degradation for 220 h, 1.4 V at 500 °C	[32]
2020	LSCN8273|BZCY44 (24 μm)|Ni-BZCY44	700 600	3161 1698	No data	Air (3% H_2O)	No degradation for 60 h, 1.3 V at 600 °C	[38]
2020	PNC|BCZYYbGd (25 μm) |Ni-BZCYYb1711	600	1036	No data	20% H_2O	No data	[39]
2020	PBCC95|BZCYYb4411 (20 μm) |Ni-BZCYYb4411	600	1339	No data	O_2 (20% H_2O)	No degradation for 160 h, 1.3 V at 500 °C	[40]
2020	NBN-BZCD35|BZCD35 (15 μm) |Ni-BZCD35	750 600	1372 944	No data	Air (30% H_2O)	No data	[41]
2020	NBNF|BZCYYb35 (25 μm) |Ni-BZCYYb35	750 600	2341 552	No data	Air (50% H_2O)	No data	[42]
2020	LSN infiltrated BCZYYC2 |BCZYYC2 (13 μm) |Ni- BCZYYC2 FL|Ni- BCZYYC2	700 600	4792 1913	No data	Air (20% H_2O)	No degradation for 100 h, 1.3 V at 700 °C	[43]
2020	LSFCu-BZCY17|BZCY17 (211 μm) | LSFCu-BZCY17	800 700	2202 843	No data	N_2 (53.2% H_2O)	No degradation for 10 h, OCV at 800 °C	[44]

(Continued)

Table 14.1. (*Continued*)

Year	Configuration of PCEC: Anode|Electrolyte|Cathode	Temper-ature (°C)	Current density at 1.4V (mA/cm^2)	Faradaic efficiency	Positive electrode (anode) steam concentration	Stability	Ref.
2019	BGLC-BZCY72|BZCY72 (30 μm) |Ni-BZCY72	700 600	299 172	53.9% 88.9%	42.8%H_2O (1.5 bar H_2O, 80 mbar O_2, 2 bar Ar)	No degradation for 700 h, 62.5 mA cm^{-2} at 600 °C	[45]
2019	$Ba_3(MnO_4)_2$-$Sb_{0.05}Sn_{0.95}O_{2-\delta}$ |BZCY44 (1.2 mm)|Pt	700 600	48 12	80.2% 84.2%	Air (20% H_2O)	No data	[38]
2019	PBSCF|PBSCF PLD|BZCYYb4411 (15 μm)|Ni-BZCYYb4411	650 500	4813 1220	74.0%@1.3 V 82.0%@ 1.3V	Air (3% H_2O)	No degradation for 300 h, 1.3 V at 550 °C	[1]
2019	BCFZY+BZCYYb|BZCYYb7111| BCZYYb7111+Ni	600 600	~1300 ~1300	~86.0% ~94.0%	Air (10% H_2O) Air (20% H_2O)	No severe degradation for >1000 hours at 550 °C	[11]
2019	SDC infiltrated BCFZY-BZCYYb1711| BZCYYb1711 (17 μm) |Ni-BZCYYb1711	700 600	1096 368	No data	Air (20% H_2O)	0.2% degradation for 250 h, 0.45 A cm^{-2} at 650 °C	[46]
2019	BCFZY-BZCY36|BCZYSm10 (25 μm) |Ni-BCZYSm10	600	514	No data	Air (12% H_2O)	No data	[47]
2018	SSC|BZCY44 (1.5 mm)|Pt	700 600	45 15	72.2% 78.5%@11 mA/cm^2 76.5%@23.6 mA/cm^2	Air (20% H_2O)	No data	[48]
2018	BLC|BZCY(54)$_{8/9}$2 (12 μm)| Ni–SZCY	600	377	86.3%	1% O_2 / 80% H_2O	No data	[31]
2018	BLC|SZCY (12 μm)|Ni–SZCY	600	278	92.0%	1% O_2 / 80% H_2O	No data	[31]
2018	BGLC|SZCY (20 μm)|Ni–SZCY	600	179	69.9%	80% H_2O	No data	[30]
2018	3D PBSCF|BZCYYb1711 (20 μm) |Ni-BZCYYb1711	600	1036.2	No data	O_2 (12% H_2O)	No degradation for 78 h, 1.6 V at 500 °C	[49]
2018	NBSCF-BZCYYb1711|BZCYYb1711 (20 μm)|Ni-BZCYYb1711	750 600	4213 1195	No data	Air (10% H_2O)	No degradation for 60 h, 0.45 A cm^{-2} at 550 °C	[50]
2018	NBN-BZCD53|BZCD53 (50 μm) |Ni-BZCD53	700	707	No data	Air (30% H_2O)	No data	[51]
2018	LSN|BCZY17 (16 μm)|Ni-BCZY17 FL|Ni-BCZY17	700 600	2323 868	No data	Air (20% H_2O)	No data	[52]

2018	LSN-BCZYYC2|BCZYYC2 (13 μm) |Ni-BCZYYC2 FL|Ni- BCZYYC2	700	2566	No data	Air (20% H_2O)	No degradation for 60 h, 1.3 V at 700 °C	[53]
2018	PNO-BCZY26|BCZY26 (20 μm) |Ni-BCZY26	700	1250	No data	Air (40% H_2O)	No data	[54]
		600	468				
2018	SLF|BCZY35 (20 μm)|Ni-BCZY35	700	1526	No data	Air (20% H_2O)	No degradation for 30 h, 0.21 A cm^{-2} at 650 °C	[55]
		600	741				
2018	SEFC-BCZY17|BCZY17 (15 μm) |Ni-BCZY17	700	1567	No data	Air (10% H_2O)	No degradation for 230 h, 1.3 V at 600 °C	[56]
		600	675				
2017	SFM-BZY20|BZY20 (16 μm)|Ni-BZY20	650	412	63.6%@1.3V	Air (3% H_2O)	No degradation for 100 h, 1.3 V at 600 °C	[57]
		600	291	No data			
2017	LSM-BCZI3|BCZI3 (15 μm)|Ni-BCZI3	700	223	No data	Air (20% H_2O)	No degradation for 60 h, 1.1 V at 700 °C	[58]
		600	597				

Abbreviations: FL = functional layer, PLD = pulsed laser deposition.

Electrolytes BZCYYb1711 = $BaZr_{0.1}Ce_{0.7}Y_{0.1}Yb_{0.1}O_{2.9}$, YEBCG = $Y_{0.8}Er_{0.2}BaCo_{3.2}Ga_{0.8}O_{7+\delta}$, BZCD35= $BaCe_{0.5}Zr_{0.3}Dy_{0.2}O_{2.9}$, BZY20 = $BaZr_{0.8}Y_{0.2}O_{2.9}$, BZCYYb4411 = $BaCe_{0.4}Zr_{0.4}Y_{0.1}Yb_{0.1}O_{2.9}$, BZCY44 = $BaZr_{0.4}Ce_{0.4}Y_{0.2}O_{2.9}$, BCZYYbGd = $BaCe_{0.5}Zr_{0.2}Y_{0.1}Yb_{0.1}Gd_{0.1}O_{2.85}$, BZCYYb35 = $BaZr_{0.3}Ce_{0.5}Y_{0.1}Yb_{0.1}O_{2.9}$, BCZYYC2 = $BaCe_{0.68}Zr_{0.1}Y_{0.1}Yb_{0.1}Cu_{0.02}O_{2.9}$, BZCY72 = $BaZr_{0.7}Ce_{0.2}Y_{0.1}O_{2.95}$, SDC = $Sm_{0.2}Ce_{0.8}O_{1.9}$, BZCY36 = $BaZr_{0.3}Ce_{0.6}Y_{0.1}O_{2.95}$, BZCYSm10 = $BaCe_{0.7}Zr_{0.1}Y_{0.2-x}Sm_xO_{2.9}$, BZCY(54)$_{8/9}$2 = $Ba(Zr_{0.5}Ce_{0.4})_{8/9}Y_{0.2}O_{2.9}$, SZCY = $SrZr_{0.5}Ce_{0.4}Y_{0.1}O_{2.95}$, BZCD53 = $BaCe_{0.3}Zr_{0.5}Dy_{0.2}O_{2.9}$, BZCY26 = $BaZr_{0.2}Ce_{0.6}Y_{0.2}O_{2.9}$, BZCY35 = $BaZr_{0.3}Ce_{0.5}Y_{0.2}O_{2.9}$, BCZI3 = $BaCe_{0.8-x}Zr_{0.2}In_xO_{3-\delta}$.
Electrodes: PBCC = $PrBa_{0.8}Ca_{0.2}Co_2O_{5+\delta}$, BCO = $BaCoO_{3-\delta}$, SCFN = $Sr_{0.9}Ce_{0.1}Fe_{0.8}Ni_{0.2}O_{3-\delta}$, PB5N = $Pr_{1.95}Ba_{0.05}NiO_{4+\delta}$, SFM = $Sr_2Fe_{1.5}Mo_{0.5}O_{6-\delta}$, PNC = $PrNi_{0.5}Co_{0.5}O_{3-\delta}$, LSCN8273 = $La_{0.8}Sr_{0.2}Co_{0.7}Ni_{0.3}O_{3-\delta}$, PBCC95 = $(PrBa_{0.8}Ca_{0.2})_{0.95}Co_2O_{6-\delta}$, NBN = $Nd_{1.95}Ba_{0.05}NiO_{4+\delta}$, NBNF = $Nd_{1.9}Ba_{0.1}NiO_{4+\delta}F_\gamma$, LSN = $La_{1.2}Sr_{0.8}NiO_{4-\delta}$, LSFCu = $(LaSr)_{0.9}Fe_{0.9}Cu_{0.1}O_4$, BGLC = $Ba_{1-x}Gd_{0.8}La_{0.2+x}Co_2O_{6-\delta}$, PBSCF = $PrBa_{0.5}Sr_{0.5}Co_{1.5}Fe_{0.5}O_{5+\delta}$, BCFZY = $BaCo_{0.4}Zr_{0.1}Fe_{0.4}Y_{0.1}O_{3-\delta}$, SSC = $Sm_{0.5}Sr_{0.5}CoO_{3-\delta}$, BLC = $Ba_{0.5}La_{0.5}CoO_{3-\delta}$, NBSCF = $NdBa_{0.5}Sr_{0.5}Co_{1.5}Fe_{0.5}O_{5+\delta}$, PNO = $Pr_2NiO_{4+\delta}$, SLF = $Sr_{2.8}La_{0.2}Fe_2O_{7-\delta}$, SEFC = $SrEu_2Fe_{1.8}Co_{0.2}O_{7-\delta}$, LSM = $(La_{0.8}Sr_{0.2})_{0.98}MnO_3$.

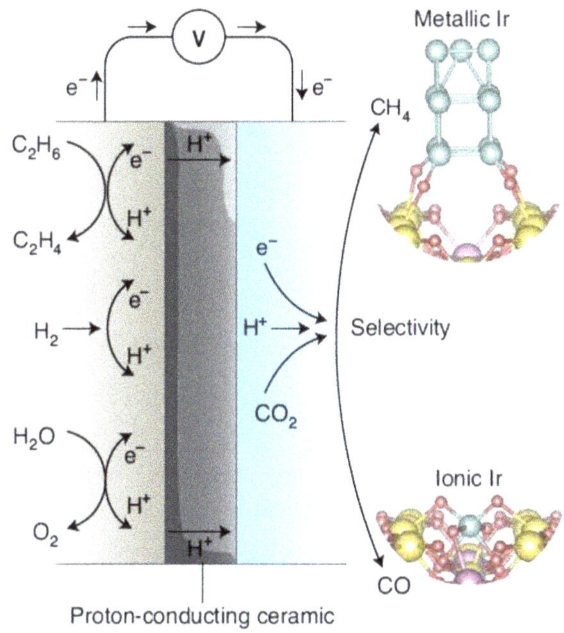

Figure 14.2. Schematic illustration of protonic ceramic electrolysis cells for CO_2 conversion. Reprinted and adapted by permission from Springer Nature [12].

environmental benefits [10]. Therefore, integrating PCECs with carefully designed CO_2 reduction electrodes can achieve sustainable chemical manufacture, enhance CO_2 conversion and desired product yield, enable long-term energy storage, and reduce greenhouse gas emissions.

Figure 14.2 shows a typical PCEC configured to convert CO_2 into value-added chemicals. The middle layer of the PCEC is a dense proton-conducting ceramic electrolyte membrane which exhibits high proton conductivity at 400 °C–500 °C (10^{-5}–10^{-3} S cm) and separates the gas streams fed to the two electrodes [13–17]. The porous positive electrode releases electrons to the outer circuit, oxidizes water, and produces oxygen and protons. The porous negative electrode functions as the CO_2 reduction electrode, activating and converting CO_2. The core reaction is that CO_2 molecules are reduced by protons and electrons. Depending on the chemicals fed to the positive electrode, the electrochemical cell configurations fall into three categories: alkanes, renewable H_2, and water. These feedstocks do not impact the CO_2 reduction mechanisms in the negative electrode; thus, we focus on discussing the recent progress in the field of negative electrode materials.

14.1.3 Thermodynamics of H_2O electrolysis and CO_2 conversion in PCECs

Figure 14.3 presents the thermodynamics of H_2O electrolysis. ΔH_R denotes the enthalpy change and the total energy demand of the corresponding reaction. ΔG_R is the Gibbs free energy change of the chemical reaction and the minimum electrical energy required to operate the electrolyzers. $T\Delta S$ represents heat production.

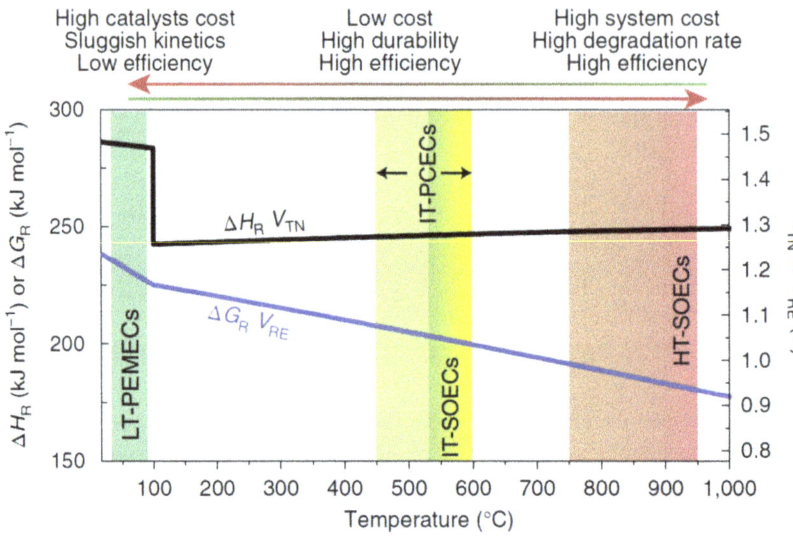

Figure 14.3. Thermodynamics of H_2O electrolysis as a function of temperature. ΔG_R is correlated with the reversible voltage (V_{RE}), while ΔH_R is correlated with the thermal neutral voltage (V_{TN}) in electrolysis mode. ΔG_R and V_{RE} share the same plot and different y-axes. ΔH_R and V_{TN} share the same plot and different y-axes. PEMECs: proton exchange membrane electrolysis cells; IT-SOECs: intermediate-temperature solid oxide electrolysis cells; HT-SOECs, high-temperature solid oxide electrolysis cells. Reprinted and adapted by permission from Springer Nature [18].

The reversible voltage (V_{RE}) and thermoneutral voltage (V_{TN}) are also displayed in figure 14.3 (equations (14.1) and (14.2)).

$$V_{RE} = \Delta G / nF. \tag{14.1}$$

$$V_{TN} = \Delta H / nF \tag{14.2}$$

Both electrical energy (ΔG_R) and thermal energy ($T\Delta S$) are needed to perform H_2O electrolysis. Therefore, the total energy (ΔH_R) required for H_2O electrolysis consists of ΔG_R and $T\Delta S$. As shown in figure 14.3, increasing the operating temperature allows more thermal energy to be used for H_2O electrolysis and reduces the electrical energy consumption, enhancing the efficiency of electrical-to-chemical energy conversion and offering an opportunity to use waste heat. However, high operating temperatures can lead to high system costs and degradation. Therefore, it is expected that intermediate temperatures are more favorable for low-cost, highly efficient, and durable H_2O electrolysis.

Figure 14.4 presents the thermodynamics of CO_2 conversion in PCECs using hydrogen/protons sourced from three different reactants fed to the positive electrode, which include H_2O, alkanes (e.g. C_2H_6), and H_2. The overall electrical energy consumption depends on the feedstock delivered to the positive electrode. The co-conversion of CO_2 and H_2O requires the highest amount of electrical energy (figures 14.4(a) and (c)). The electrical energy consumption of CO_2-to-CO conversion is lower than that of CO_2-to-CH_4 conversion, as the total energy demand

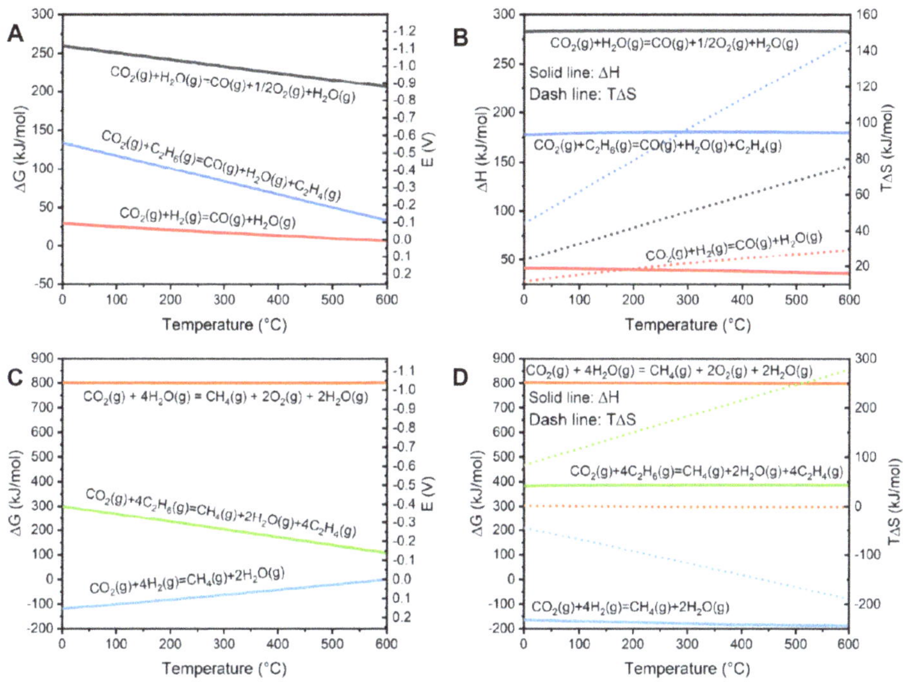

Figure 14.4. Thermodynamics of CO_2 conversion as a function of temperature using hydrogen/protons sourced from three reactants fed to the positive electrode, namely, H_2O, C_2H_6, and H_2. ΔG is correlated with the reversible voltage (V). ΔG and V share the same plot and different y-axes. The results are calculated using HSC software. (a) ΔG and V of CO_2 reduction to CO as a function of temperature. (b) ΔH and $T\Delta S$ of CO_2 reduction to CO as a function of temperature. (c) ΔG and V of CO_2 reduction to CH_4 as a function of temperature. (d) ΔH and $T\Delta S$ of CO_2 reduction to CH_4 as a function of temperature.

(ΔH) of CO production is lower than for CH_4 (figures 14.4(b) and (d)). Additionally, the heat demand ($T\Delta S$) of CO production increases with increasing operating temperatures (figure 14.4(b)), enabling heat to be used for CO production. As the total energy demand of CO production is relatively independent of the operating temperature, the electrical energy consumption and reversible voltage used for CO production decrease with increasing temperature. However, the heat demand of CH_4 production strongly depends on the reactants delivered to the positive electrode (figure 14.4(d)).

14.1.4 Advantages of employing PCECs for H_2 production and CO_2 conversion

Although PCECs are a nascent technology, they already show significant promise for efficient and high-performance H_2O electrolysis and CO_2 conversion [10, 11, 19]. PCECs offer several important benefits compared to other technologies for H_2 production and sustainable chemical manufacture:

(1) Unlike O-SOECs or polymer membrane-based electrolysis cells, PCECs can produce pure, dry H_2 that is amenable to direct electrochemical compression (without dehumidification) inside the cell [10, 18, 20], potentially

reducing system complexity and cost (i.e. no external condenser is required, and the H_2 compressor could be greatly simplified), as well as enhancing overall energy efficiency (electrochemical compression could be more efficient than mechanical compression) [20].

(2) Operating intermediate-temperature PCECs (600 °C) is more efficient than operating low-temperature (LT, 50 °C–100 °C) polymer membrane-based electrolysis cells; the efficiency of intermediate-temperature PCECs can approach those of high-temperature (HT, 700 °C–900 °C) [4–8] and intermediate-temperature (IT, 600 °C) [9–13] SOEC-based cells. Reduced operating temperatures (versus HT-SOEC-based cells) enable hybridization with a broader range of waste heat sources and relax stack and balance-of-plant constraints, potentially lowering costs while improving the reliability, thermal cycling tolerance, and dynamic response. For example, PCECs could be integrated with fossil power plants and nuclear power plants and thus utilize waste heat for H_2 production.

(3) Under an external potential, as illustrated in figure 14.2, PCECs reduce CO_2 at the negative electrode using renewable electrons and protons generated by converting various feedstocks at the positive electrode. PCECs uniquely achieve a high proton flux at moderate operating temperatures (400 °C–500 °C), placing them in a thermodynamic and kinetic 'sweet spot' for CO_2 reduction that overlaps with temperatures relevant to CO_2 hydrogenation. Thus, the co-conversion of CO_2 and H_2O into CH_4 or other hydrocarbons is feasible.

(4) The electrochemical conversion of CO_2 in ceramic electrochemical reactors can lead to coking (ascribed to the Boudouard reaction ($2CO = C + CO_2$)) and the pyrolysis of hydrocarbons, while the unique characteristics of protonic ceramics (e.g. relatively basic surfaces) can mitigate carbon formation; this, in turn, enables excellent coking tolerance and may even eliminate the coking entirely [21, 22].

14.2 Current progress in the field of PCECs for H_2 production and CO_2 conversion

14.2.1 PCECs for H_2 production

As shown in figure 14.5, PCECs were pioneered for H_2O electrolysis by Iwahara *et al* from the early 1980s to the 1990s [23–26]. However, there has been little subsequent work on this topic, which has been due to the lack of highly proton-conductive ceramic membranes. Additionally, their poor sinterability and high sintering temperatures have hindered their further development. Therefore, subsequent research has focused on developing new compositions and novel fabrication methods to enhance conductivity and improve densification. It has been widely recognized that adding a sintering aid (e.g. NiO) can significantly benefit densification and conductivity [16, 27, 28]. With the most recent advancements in the field of proton-conducting ceramics, which have enhanced conductivity and sinterability [11, 13, 19, 29], interest in the use of PCECs for H_2 production has been reignited in

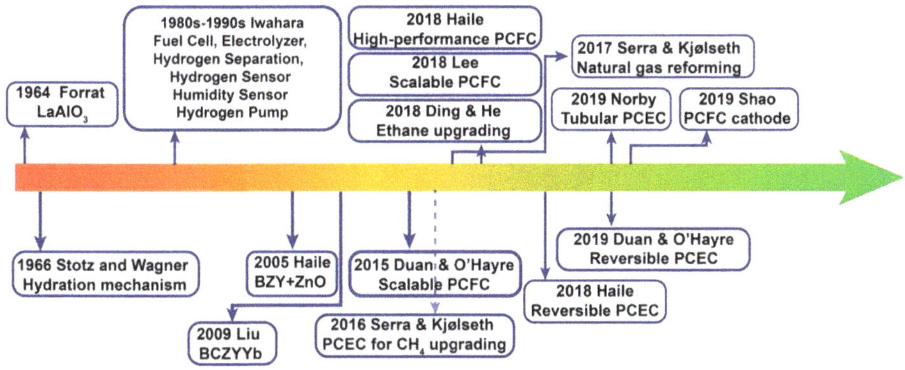

Figure 14.5. Historic overview of proton-conducting oxides, applications, and remarkable breakthroughs. Reprinted and adapted from [10] by permission from AIP Publishing.

2018 and 2019 [1, 11, 30, 31]. Since then, a large body of great work has demonstrated the use of PCECs for H_2 production and improved their performances so that they are now comparable with those of O-SOECs (table 14.1).

In 2019, as shown in figure 14.5, Haile *et al* first demonstrated intermediate-temperature protonic ceramic electrolyzers that delivered comparable performances to those of O-SOECs [1]. In 2018, they developed a new electrolyte composition, $BaCe_{0.4}Zr_{0.4}Y_{0.1}Yb_{0.1}O_3$ (BCZYYb4411), which reduced ohmic loss and enhanced long-term stability. A state-of-the-art triple-conducting oxide, $PrBa_{0.5}Sr_{0.5}Co_{1.5}Fe_{0.5}O_{5+\delta}$ (PBSCF), was used as the positive electrode [29]. At an operating temperature of 600 °C, an exceptional current density of >2.0 A cm^{-2} was achieved at 1.4 V. Even at an operating temperature of 500 °C, this PCEC was able to deliver a current density of >0.5 A cm^{-2} at 1.4 V. A concurrent work conducted by Duan and O'Hayre *et al* in 2019 also presented similar PCEC performances using $BaCe_{0.7}Zr_{0.1}Y_{0.1}Yb_{0.1}O_3$ (BCZYYb7111) as the electrolyte membrane and BCFZY4411 as the positive electrode (figure 14.6(b)) [18]. To further reduce the overpotential ascribed to the positive electrode, Ding *et al* pioneered a novel positive electrode, $PrNi_{0.5}Co_{0.5}O_3$, which had enhanced proton conduction that improved PCEC performance [32]. Furthermore, Ding *et al* optimized the morphology of the positive electrode using a template method. As shown in figure 14.6(c), the PCEC performance was significantly enhanced and a reasonable current density was achieved at 400 °C.

Significant progress has been made in optimizing PCECs and developing positive electrode materials to improve PCEC performances [1, 10, 11, 32, 38, 45, 47, 49, 59]. Although compelling current densities have been obtained which are comparable to or even higher than those of O-SOECs, the electronic leakage and low faradaic efficiency of PCECs have been noted [1, 11, 32]. Figure 14.6(d) shows the voltages of a PCEC as a function of current density under various steam concentrations at 600 °C, which do not change with variations in the steam concentration. However, the steam concentration significantly affects the faradaic efficiency, especially a relatively low current density (figure 14.7(a)). The faradaic efficiency tends to decrease as the steam concentration is reduced at the positive electrode. This impact

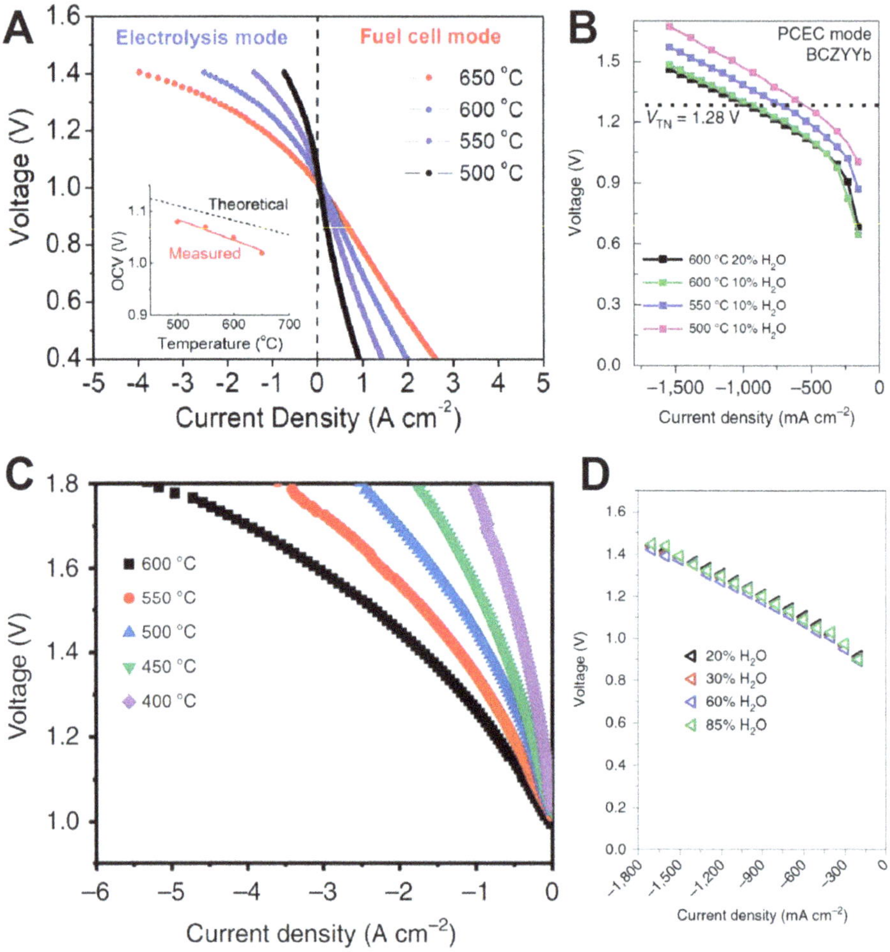

Figure 14.6. Performances of protonic ceramic electrolysis cells for H_2 production: (a) BCZYYb4411-based PCECs with PBSCF as the positive electrode. Reprinted and adapted from [1] with permission from the Royal Society of Chemistry. (b) BCZYYb7111-based PCECs with BCFZY4411 as the positive electrode. Reprinted and adapted by permission from Springer Nature [18]. (c) BCZYYb4411-based PCECs with $PrNi_{0.5}Co_{0.5}O_{3-\delta}$ (PNC) as the positive electrode. Reprinted under the terms of a CC-BY licence. (d) Impact of steam concentration on the performances of BCZYYb7111-based PCECs with BCFZY4411 as the positive electrode. Reprinted and adapted by permission from Springer Nature [18].

of steam concentration on faradaic efficiency has been widely recognized [32]. Haile *et al* also noted that reducing the air flow rate from 150 ml min^{-1} to 100 ml min^{-1} can slightly increase the faradaic efficiency (figure 14.7(b)) [1]. As the steam concentration is controlled using a gas bubbler and air is the carrier gas, a high carrier gas flow rate can slightly reduce the steam concentration; thus, reducing the carrier gas flow rate can slightly enhance the faradaic efficiency, which is consistent with the effect observed by Duan and O'Hayre *et al* [11].

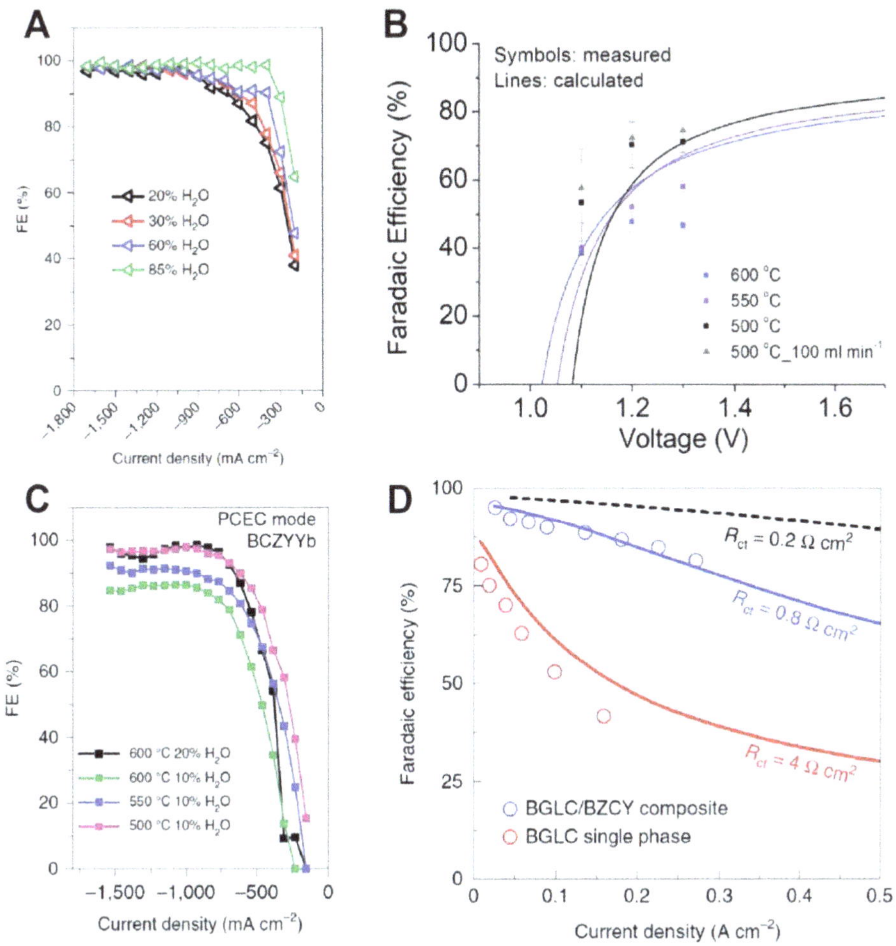

Figure 14.7. Faradaic efficiency of PCECs as a function of current density or applied voltage. (a) Impacts of steam concentration on the performances of BCZYYb7111-based PCECs with BCFZY4411 as the positive electrode. Reprinted and adapted by permission from Springer Nature [18]. (b) Impacts of operating temperature on the faradaic efficiency of BCZYYb4411-based PCECs with PBSCF as the positive electrode. Reprinted and adapted with permission from the Royal Society of Chemistry [1]. (c) Impacts of operating temperature on the faradaic efficiency of BCZYYb7111-based PCECs with BCFZY4411 as the positive electrode. Reprinted and adapted by permission from Springer Nature [18]. (d) Faradaic efficiency of BZCY-based PCECs with $Ba_{1-x}Gd_{0.8}La_{0.2+x}Co_2O_{6-\delta}$ (BGLC) as the positive electrode. Reprinted and adapted by permission from Springer Nature [45].

It should be noted that the operating temperature can significantly affect the faradaic efficiency. As shown in figures 14.7(b) and (c), increasing the operating temperature tends to decrease the faradaic efficiency. While high operating temperatures can improve the current density, the overall H_2 production rate might not increase, due to more severe electronic leakage, leading to reduced energy efficiency.

Moreover, it is important to recognize that the faradaic efficiency of PCECs shows a high dependency on the current density or applied voltage. Figures 14.7(b) and (c) show that the faradic efficiency is extremely low at relatively low current densities or voltages, while increasing the current density can quickly increase the faradaic efficiency until it shows a propensity to be constant. At a current density approving zero, the faradaic efficiency is undetectable, and reaches zero at a low current density (e.g., <100 mA cm^{-2}). The impacts of current density or potential on faradaic efficiency have been illuminated by determining the H_2 production rate and correlating it with the current density, which has been recognized in a great body of work in this field (table 14.1). However, some results show an opposite trend. For example, as shown in figure 14.7(d), the faradaic efficiency tends to decrease with increasing current density. This kind of dependency was also noted in [32] and detailed in chapter 12.

It has been recognized that proton-conducting ceramics exhibit mixed electronic and ionic conduction, especially in an oxidizing atmosphere [60, 61]. As shown in figure 14.8(a), p-type electronic conduction can lead to a net flux of protons from the negative electrode to the positive electrode. Thus, the protonic ceramic membrane functions as a hydrogen permeation membrane, suggesting that the faradaic efficiency is negative at a current density close to zero. Increasing the applied current density in electrolysis mode produces additional hydrogen via H_2O electrolysis (figure 14.8(b)). However, the direction of proton flux ascribed to H_2O electrolysis is opposite to the direction of the proton flux due to permeation. Thus, these two proton fluxes act to offset each other. If the permeation flux is higher than the H_2O electrolysis proton flux, this leads to negative faradaic efficiency. The faradaic efficiency tends to sharply increase once the H_2O electrolysis proton flux dominates the proton flux (figures 14.7(a) and 14.7(b)).

Therefore, the transference number of the electron holes is essential for the faradaic efficiency of PCECs. The electron holes are attributed to the oxidation of the electrolyte membrane, indicating that the electrolyte material and operating conditions can be modulated to reduce electronic leakage and enhance faradaic efficiency. The strategies include: (1) developing proton-conducting ceramics that favor hydration and suppress oxidation; (2) supplying a gas stream to the positive

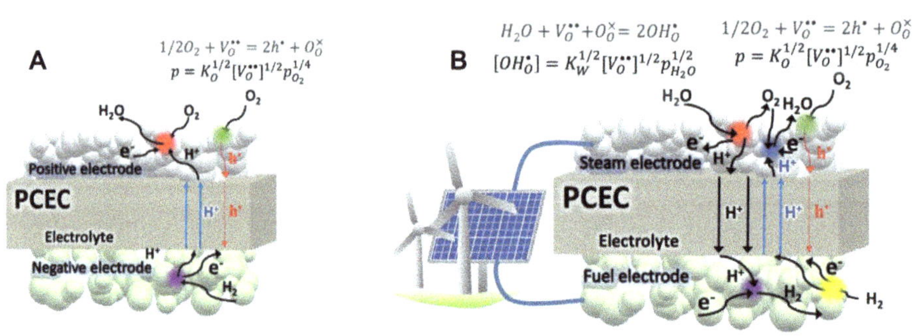

Figure 14.8. Schematic illustration showing why the faradaic efficiency drops sharply when the current density is close to zero. (a). Under OCV conditions. (b) Under charging conditions at a relatively low current density. Reprinted and adapted by permission from Springer Nature [18].

electrode that has a low oxygen partial pressure and high steam concentration to further enhance hydration and suppress oxidation; and (3) operating PCECs at appropriately low operating temperatures and high current densities to achieve enhanced faradaic efficiency [10, 18].

The durability of PCECs is key to their implementation for H_2 production. The long-term durability of PCECs has been studied in various works (table 14.1). Figure 14.9 presents the most recent work on the long-term stability testing of PCECs, which shows that PCECs are stable at high current densities and intermediate operating temperatures. For example, Haile *et al* tested BCZYYb4411-based PCECs at a constant voltage of 1.3 V at 550 °C for >500 h. The corresponding current density was stable over 500 h [1]. Duan and O'Hayre *et al* also tested the stability of BCZYYb7111-based PCECs at a constant current density of ~1.4 A cm^{-2} at 600 °C and 550 °C. The recorded terminal voltage did not significantly degrade over >1000 h [18]. A similar stability test was conducted by Ding *et al* using the constant-voltage approach [32].

The approach that is used to evaluate the PCEC stability is to operate the PCEC in either constant-current or constant-voltage mode while recording the cell voltage or current as a function of time. As PCECs are a relatively nascent technology, it has been suggested that it would be beneficial to simultaneously conduct additional measurements to better analyze the stability of PCECs and probe their potential

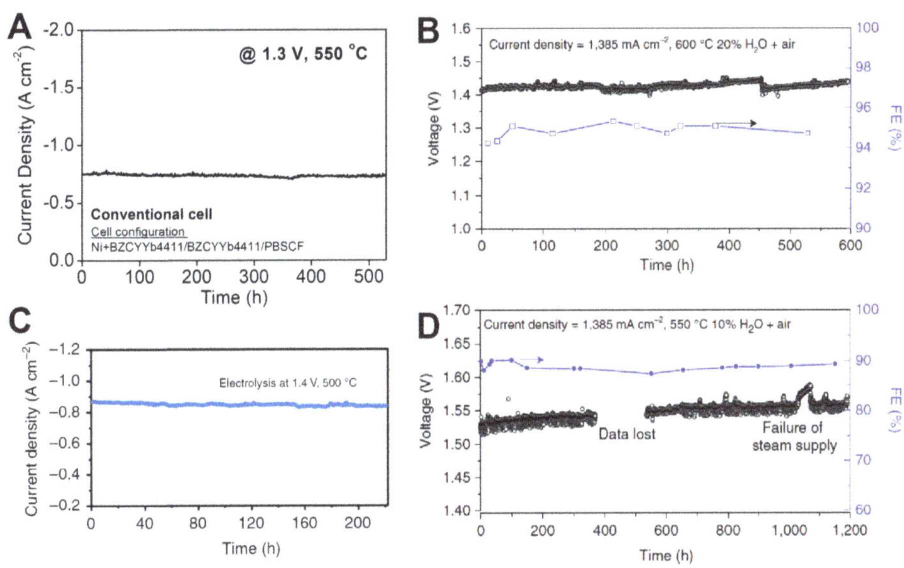

Figure 14.9. Stability of PCECs for H_2 production: (a) BCZYYb4411-based PCECs with PBSCF as the positive electrode. Reprinted and adapted with permission from the Royal Society of Chemistry [1]. (b) BCZYYb7111-based PCECs with BCFZY4411 as the positive electrode. Reprinted and adapted by permission from Springer Nature [18]. (c) BCZYYb4411-based PCECs with PNC as the positive electrode. Reprinted under the terms of a CC-BY licence. (d) Impacts of steam concentration on the performances of BCZYYb7111-based PCECs with BCFZY4411 as the positive electrode. Reprinted and adapted by permission from Springer Nature [18].

degradation mechanisms. These measurements include time-resolved electrochemical impedance spectroscopy, which could identify the degradation mechanism, and measuring the H_2 production rate and faradaic efficiency as a function of time. As mentioned above, unlike O-SOECs that typically have negligible electronic leakage, PCECs can potentially have severe electronic leakage. Therefore, stability measurements made at a constant current or voltage might not be able to comprehensively represent PCEC stability.

As electronic leakage is widely recognized as the primary issue that affects PCECs, Duan and O'Hayre *et al* have concurrently measured the faradaic efficiency as a function of operational time. Although the terminal voltage increases slightly at a constant current, the faradaic efficiency is relatively constant, suggesting that the transport properties of the proton-conducting electrolyte membrane are stable. Long-term operation did not give rise to increased electronic leakage. However, only a handful of published articles discuss measuring the faradaic efficiency as a function of operating time (table 14.1). Furthermore, operating the PCECs in the constant-voltage mode complicates the measurement of faradaic efficiency, suggesting that the stability of PCECs should be investigated using the constant-current mode to better determine the faradaic efficiency as a function of time.

To probe the potential degradation mechanisms, Duan and O'Hayre *et al* used electrochemical impedance spectroscopy before and after >500 h of operation at a current density of 1385 mA cm^{-2} at 600 °C and 550 °C [18]. Figures 14.10(a) and (b) clearly show that the ohmic resistance did not change after >500 h of operation, but the electrode polarization resistance increased, indicating that the degradation of PCECs can be ascribed to the electrode. They also recognized that the degradation rate is higher at higher steam concentrations, implying that the degradation could be due to the positive electrode. Referring to figure 14.1, in contrast to O-SOECs, in which steam is fed to the negative electrode, steam is fed to the positive electrode of PCECs. It should be noted that the positive electrodes of PCECs are typically made of an oxide with a perovskite or double perovskite crystal structure [1, 19, 32, 45, 62]. The positive electrode might be prone to decomposition at high steam concentrations, particularly at reduced operating temperatures. Therefore, the degradation shown in figures 14.9(b) and (d) may be attributed to the positive electrode. Ding *et al* studied the chemical stability of PNC, which is stable at high steam concentrations, suggesting that the chemical stability of positive electrode materials can be improved to support long-term durable PCEC operation [32]. Therefore, it is necessary to conduct additional measurements beyond constant-current/voltage-stability testing, i.e. measurements which can comprehensively evaluate the stability of PCECs and probe their degradation mechanisms.

14.2.2 PCECs for CO_2 conversion

With the advancements in the field of PCECs for H_2 production, CO_2 reduction in PCECs for high-value chemical production has tended to gain more attention, as converting CO_2 in PCECs does yield various benefits. Duan and O'Hayre *et al* directly used PCECs that were developed for H_2O electrolysis for CO_2 conversion [11].

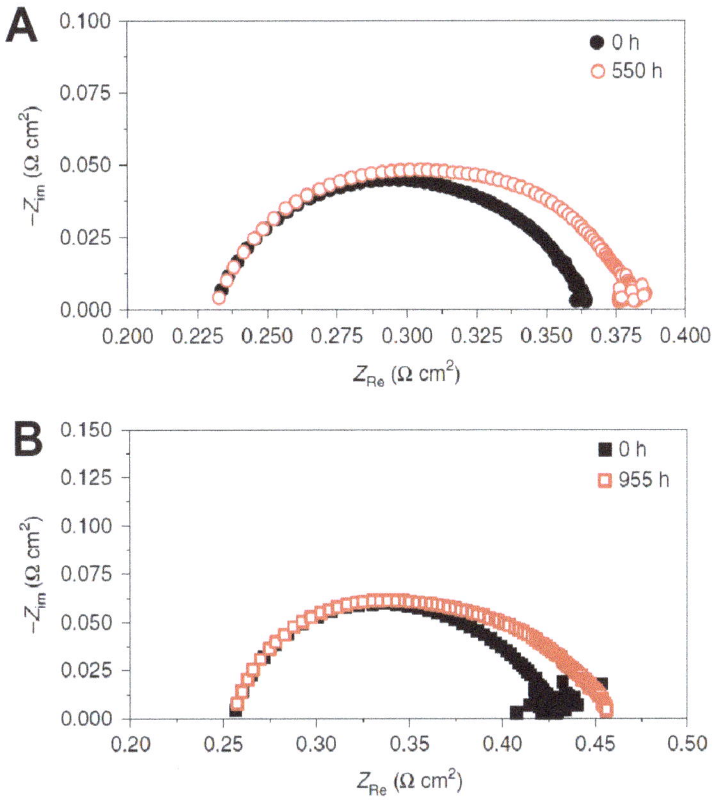

Figure 14.10. Electrochemical impedance spectroscopy of PCECs: (A) electrochemical impedance spectroscopy (EIS) of BCZYYb7111-based PCECs with BCFZY4411 as the positive electrode before and after 550 h of operation at a current density of 1385 mA cm^{-2} at 600 °C. (B) Electrochemical impedance spectroscopy of cell no. 8 before and after 955 h of operation at a current density of 1385 mA cm^{-2} at 550 °C. Reprinted and adapted by permission from Springer Nature [18].

As shown in figure 14.11(a), CO_2 could be reduced to both CO and CH_4 with a high rate of CO_2 conversion at an intermediate operating temperature. The intermediate operating temperature allows CH_4 to be produced, which is one of the main benefits of converting CO_2 in PCECs. However, it also leads to poor selectivity, producing a mixture of CO and CH_4. Furthermore, Duan and O'Hayre *et al* recognized that the negative electrode favors the hydrogen evolution reaction (HER), leading to high faradaic selectivity toward H_2. As shown in figure 14.11(a), the current density does not proportionally enhance CO_2 conversion, implying that the HER is more active at a higher current density or potential. This study suggests that PCECs designed for H_2O electrolysis might not lead to the most favorable CO_2 conversion, as they could give rise to a substantial HER and poor selectivity. Thus, the negative electrode should be rationally designed to suppress the HER and selectively reduce CO_2. For example, Ding *et al* pioneered an approach to modulating the CO_2 reduction pathway to selectively produce either CO or CH_4 [63]. Using a metallic Ir-based negative

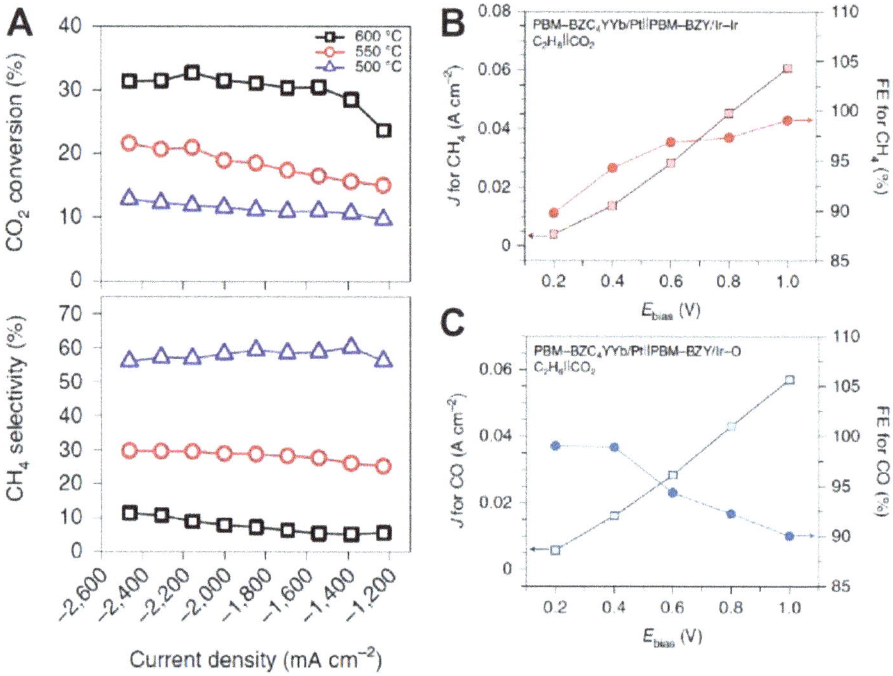

Figure 14.11. Performances of CO_2 conversion in PCECs: (a) BCZYYb7111-based PCEC with a Ni-based negative electrode. Reprinted and adapted by permission from Springer Nature [18]. (b) BCZYYb4411-based PCEC with a metallic Ir-based negative electrode. Reprinted and adapted by permission from Springer Nature. (c) BCZYYb4411-based PCEC with an Ir–O-based negative electrode. Reprinted and adapted by permission from Springer Nature.

electrode, CO_2 can be selectively reduced to CH_4 with an outstanding faradaic selectivity of >95%. The negative electrode can favor CO production if it is integrated with Ir–O-based catalysts. Furthermore, both catalysts drastically inhibited the HER and improved CO_2 conversion. The work conducted by Ding *et al* validates the concept that modulating the negative electrode can tune the CO_2 reduction pathway and selectively produce the desired product. Other teams have also demonstrated the reduction of CO_2 to value-added chemicals in PCECs, which has confirmed the feasibility of CO_2 conversion in PCECs [51, 64].

According to the thermodynamics of CO_2 conversion and H_2O electrolysis, reducing CO_2 to CO enables more heat to be used than H_2O electrolysis, leading to the provision of a lower potential to the PCEC and lower electrical energy consumption. Therefore, the conversion of CO_2 in PCECs allows the additional waste heat to be used for chemical manufacture, which has been demonstrated by Medvedev *et al* [51]. However, the reduction of CO_2 by hydrogen to form CH_4 is highly exothermic; thus, it cannot utilize heat. Integrating CO_2 reduction to CH_4 with an endothermic reaction, for instance H_2O electrolysis in endothermic mode, can directly use the heat produced by CO_2 reduction for H_2O electrolysis, potentially creating a thermoneutral zone and maximizing the overall energy efficiency.

The integration of CO_2 conversion with H_2O electrolysis, which can reduce the applied overpotential, has been experimentally demonstrated [18, 51].

CO_2 conversion in PCECs can also be realized by integrating it with dry reforming. Luo *et al* have intimately integrated CO_2 dry reforming with PCECs; this allows power to be generated while concurrently producing syngas [65]. Oxygen ion solid oxide fuel cells (O-SOFCs) can fully oxidize the fuel delivered to the anode in fuel-cell mode; thus, the dry reforming of methane in O-SOFCs cannot convert and utilize CO_2, as the CO could ultimately be oxidized to CO_2. As the electrolyte membranes of PCECs exhibit relatively low oxygen ion conduction at <600 °C, PCECs cannot further oxidize CO and can only utilize H_2 for power generation. Therefore, PCECs can enable the co-generation of power and syngas to take place via dry reforming. Luo *et al* rationally designed a $Ni_{0.8}Co_{0.2}$–$La_{0.2}Ce_{0.8}O_{1.9}$ (NiCo–LDC) composite to be used as an anode, which significantly improved the dry reforming activity and achieved a CO_2 conversion of 91.5% at 700 °C. Moreover, its long-term durability was also tested, and no degradation was observed, indicating that the PCEC constructed using this anode exhibits an excellent coking tolerance at a high CO/CO_2 ratio.

14.3 Challenges and opportunities of H_2 production in protonic ceramic electrochemical cells

14.3.1 Faradaic efficiency of PCECs for H_2 production

Despite the outstanding performance which PCECs have demonstrated for H_2 production, challenges related to efficient H_2 production remain. The primary challenge is their relatively low faradaic efficiency, which is attributed to electronic leakage. Referring to figure 14.8, the electronic leakage is due to the oxidation of the proton-conducting electrolyte membrane. However, the electrolyte is always exposed to an oxidizing atmosphere on the positive electrode side. Therefore, the oxidation of the electrolyte membrane is inevitable, but it can be suppressed. The oxidation of the electrolyte membrane is thermodynamically controlled, which in turn is intrinsically determined by its composition. Additionally, the operating temperature and oxygen partial pressure also affect oxidation. To minimize oxidation and electronic leakage, the electrolyte membrane's composition should be deliberately designed to hinder oxidation and favor hydration, reducing the transference number of p-type electronic charge carriers and enhancing faradaic efficiency. Extensive efforts have been devoted to studying and modulating the defect chemistry and transport properties of proton-conducting electrolyte membranes. For example, it has been noted that Sc-doped BCZY leads to lower electronic leakage than Yb-doped BCZY, implying that BCZYSc-based PCECs could potentially give rise to higher faradaic efficiencies than BCZYYb-based PCECs [66]. Furthermore, the dehydration temperature of BCY is much higher than that of BZY, indicating that BZY is more prone to being oxidized at 500 °C–600 °C [67]. Therefore, it is expected that PCECs with BCY electrolytes can achieve higher faradaic efficiencies than those of BZY-based PCECs. However, it has been noted that high Ce contents could be detrimental to their long-term durability.

Moreover, it should be recognized that a high sintering temperature could result in Ba vaporization and create an A-site deficiency, which could alter the defect chemistry of the electrolyte and its transport properties. The sintering aid (e.g. NiO) that is typically used to densify the electrolyte membrane could also modify the transport properties and affect the electronic leakage. Therefore, the electronic leakage of the electrolyte membrane is a complicated issue, which cannot be straightforwardly addressed by developing a new electrolyte. Both the electrolyte composition and the corresponding fabrication procedures can impact the faradaic efficiency of PCECs.

As the oxidation of protonic ceramics is endothermic, appropriately reducing the operating temperature can help to enhance efficiency. However, the reduced operating temperature increases the overpotential and decreases the energy efficiency. Therefore, it is suggested that the operating temperature should be carefully adjusted to achieve the highest energy efficiency. Additionally, the atmosphere at the positive electrode also impacts electrolyte oxidation. A high $p(O_2)$ and a low $p(H_2O)$ facilitate oxidation and decrease faradaic efficiency. It should be noted that the current density can also change the atmosphere and affect the faradaic efficiency. For example, operating the PCEC at a high current density can increase H_2O utilization, which reduces the $p(H_2O)$ and simultaneously produces O_2, which can exacerbate electronic leakage and reduce faradaic efficiency. Therefore, the operating current density should also be rigorously selected to obtain high energy efficiency.

In summary, the challenges related to low faradaic efficiency could be addressed by considering both the materials and operating conditions required to achieve high energy efficiencies.

14.3.2 Long-term durability

As mentioned above, the degradation of PCECs for H_2 production is primarily ascribed to the positive electrode. Therefore, enhancing the chemical stability of the positive electrode should be at the heart of optimizing PCECs. Since PCECs for H_2 production have been revigorated for less than ten years, only a handful of papers have been published on this topic (table 14.1). Other potential degradation mechanisms have not been identified, suggesting that additional work could be conducted to probe the degradation. Although some studies have performed long-term durability testing, there has been a lack of comprehensive postmortem characterization to better understand PCECs after long-term operation, which could assist in probing the degradation mechanisms. It is therefore suggested that it would be beneficial to conduct additional characterizations such as transmission electron microscopy, energy-dispersive x-ray spectroscopy, and x-ray photoelectron spectroscopy to study the electrodes, electrolyte, and interfaces.

14.4 Concluding remarks

In this chapter, we briefly introduced PCECs for both H_2 production and CO_2 conversion. We then comprehensively and critically analyzed the most up-to-date progress in the use of PCECs for both H_2 production and CO_2 conversion. We particularly highlighted the associated challenges and provided our perspectives on future development. The key conclusions of this chapter are outlined below:

(1) Employing PCECs for H_2 production via H_2O electrolysis can achieve an energy efficiency of >90%, which is higher than those of low-temperature polymer membrane-based electrolyzers and comparable to those of O-SOECs.
(2) PCECs may have a relatively low faradaic efficiency, which is attributed to the oxidation of the electrolyte membrane under an oxidizing atmosphere.
(3) Oxidation and electronic leakage could be inhibited by modulating the composition and carefully altering the operating conditions. A low oxygen partial pressure, a high steam concentration, an appropriately low operating temperature, and a slightly high current density could reduce electronic leakage and achieve high faradaic efficiency.
(4) The degradation of PCECs for H_2 production is primarily ascribed to the positive electrode, which can react with H_2O, especially at high steam concentrations.
(5) Using PCECs with conventional Ni-based negative electrodes for CO_2 conversion leads to a strong HER and poor selectivity.
(6) The HER and CO_2 reduction selectivity can be modulated by rationally designing negative electrodes for PCECs.

Acknowledgments

This work was financially supported by faculty research funding from Kansas State University.

References

[1] Choi S, Davenport T C and Haile S M 2019 Protonic ceramic electrochemical cells for hydrogen production and electricity generation: exceptional reversibility, stability, and demonstrated faradaic efficiency *Energy Environ. Sci.* **12** 206–15

[2] Sapountzi F M, Gracia J M, Kee-J Weststrate C J, Fredriksson H O A and (Hans) Niemantsverdriet J W W 2017 Electrocatalysts for the generation of hydrogen, oxygen and synthesis gas *Prog. Energy Combust. Sci.* **58** 1–35

[3] Hu B, Guild C and Suib S L 2013 Thermal, electrochemical, and photochemical conversion of CO_2 to fuels and value-added products *J. CO_2 Util.* **1** 18–27

[4] Huang Y, Handoko A D, Hirunsit P and Yeo B S 2017 Electrochemical reduction of CO_2 using copper single-crystal surfaces: effects of CO^* coverage on the selective formation of ethylene *ACS Catal.* **7** 1749–56

[5] Wang K *et al* 2019 Band alignment in Zn_2SnO_4/SnO_2 heterostructure enabling efficient CO_2 electrochemical reduction *Nano Energy* **64** 103954

[6] Duan X, Xu J, Wei Z, Ma J, Guo S, Wang S, Liu H and Dou S 2017 Metal-free carbon materials for CO_2 electrochemical reduction *Adv. Mater.* **29** 1701784

[7] Geng Z, Kong X, Chen W, Su H, Liu Y, Cai F, Wang G and Zeng J 2018 Oxygen vacancies in ZnO nanosheets Enhance CO_2 electrochemical reduction to CO *Angew. Chem.* **130** 6162–7

[8] Back S and Jung Y 2017 TiC- and TiN-supported single-atom catalysts for dramatic improvements in CO_2 electrochemical reduction to CH_4 *ACS Energy Lett.* **2** 969–75
[9] Bassano C, Deiana P, Lietti L and Visconti C G 2019 P2G movable modular plant operation on synthetic methane production from CO_2 and hydrogen from renewables sources *Fuel* **253** 1071–9
[10] Duan C, Huang J, Sullivan N and O'Hayre R 2020 Proton-conducting oxides for energy conversion and storage *Appl. Phys. Rev.* **7** 011314
[11] Duan C, Kee R, Zhu H, Sullivan N, Zhu L, Bian L, Jennings D and O'Hayre R 2019 Highly efficient reversible protonic ceramic electrochemical cells for power generation and fuel production *Nat. Energy* **4** 230–40
[12] Duan C 2021 Selective CO_2 electrohydrogenation *Nat. Catal.* **4** 264–5
[13] Liu M, Yang L, Wang S, Blinn K, Liu M, Liu Z and Cheng Z 2009 Enhanced sulfur and coking tolerance of a mixed ion conductor for SOFCs: $BaZr_{0.1}Ce_{0.7}Y_{0.2-x}Yb_xO_{3-\delta}$ *Science* **326** 126–9
[14] Babilo P and Haile S M 2005 Enhanced sintering of yttrium-doped barium zirconate by addition of ZnO *J. Am. Ceram. Soc.* **88** 2362–8
[15] Nikodemski S, Tong J, Duan C and O'Hayre R 2016 Ionic transport modification in proton conducting $BaCe_{0.6}Zr_{0.3}Y_{0.1}O_{3-\delta}$ with transition metal oxide dopants *Solid State Ionics* **294** 37–42
[16] Nikodemski S, Tong J and O'Hayre R 2013 Solid-state reactive sintering mechanism for proton conducting ceramics *Solid State Ionics* **253** 201–10
[17] Tong J, Clark D, Bernau L, Sanders M, O'Hayre R, Inui H, Awakura Y, Malavasi L and Gerbasi R 2010 Solid-state reactive sintering mechanism for large-grained yttrium-doped barium zirconate proton conducting ceramics *J. Mater. Chem.* **20** 6333
[18] Duan C, Kee R, Zhu H, Sullivan N, Zhu L, Bian L, Jennings D and O'Hayre R 2019 Highly efficient reversible protonic ceramic electrochemical cells for power generation and fuel production *Nat. Energy* **4** 230–30
[19] Duan C *et al* 2018 Highly durable, coking and sulfur tolerant, fuel-flexible protonic ceramic fuel cells *Nature* **557** 217–22
[20] Kee B L, Curran D, Zhu H, Braun R J, DeCaluwe S C, Kee R J and Ricote S 2019 Thermodynamic insights for electrochemical hydrogen compression with proton-conducting membranes *Membranes* **9** 77
[21] Skafte T L, Blennow P, Hjelm J and Graves C 2018 Carbon deposition and sulfur poisoning during CO_2 electrolysis in nickel-based solid oxide cell electrodes *J. Power Sources* **373** 54–60
[22] Li Y, Chen X, Yang Y, Jiang Y and Xia C 2017 Mixed-conductor $Sr_2Fe_{1.5}Mo_{0.5}O_{6-\delta}$ as robust fuel electrode for pure CO_2 reduction in solid oxide electrolysis cell *ACS Sustain. Chem. Eng.* **5** 11403–12
[23] Iwahara H, Esaka T, Uchida H and Maeda N 1981 Proton conduction in sintered oxides and its application to steam electrolysis for hydrogen production *Solid State Ionics* **3–4** 359–63
[24] Iwahara H 1988 High temperature proton conducting oxides and their applications to solid electrolyte fuel cells and steam electrolyzer for hydrogen production *Solid State Ionics* **28–30** 573–8
[25] Iwahara H, Uchida H and Yamasaki I 1987 High-temperature steam electrolysis using $SrCeO_3$-based proton conductive solid electrolyte *Int. J. Hydrogen Energy* **12** 73–7
[26] Iwahara H, Uchida H and Maeda N 1982 High temperature fuel and steam electrolysis cells using proton conductive solid electrolytes *J. Power Sources* **7** 293–301

[27] Tong J, Clark D, Bernau L, Sanders M and O'Hayre R 2010 Solid-state reactive sintering mechanism for large-grained yttrium-doped barium zirconate proton conducting ceramics *J. Mater. Chem.* **20** 6333
[28] Tong J, Clark D, Bernau L, Subramaniyan A and O'Hayre R 2010 Proton-conducting yttrium-doped barium cerate ceramics synthesized by a cost-effective solid-state reactive sintering method *Solid State Ionics* **181** 1486–98
[29] Choi S, Kucharczyk C J, Liang Y, Zhang X, Takeuchi I, Ji H-I and Haile S M 2018 Exceptional power density and stability at intermediate temperatures in protonic ceramic fuel cells *Nat. Energy* **3** 202–10
[30] Leonard K, Druce J, Thoreton V, Kilner J A and Matsumoto H 2018 Exploring mixed proton/electron conducting air electrode materials in protonic electrolysis cell *Solid State Ionics* **319** 218–22
[31] Leonard K, Okuyama Y, Takamura Y, Lee Y S, Miyazaki K, Ivanova M E, Meulenberg W A and Matsumoto H 2018 Efficient intermediate-temperature steam electrolysis with Y:$SrZrO_3$–$SrCeO_3$ and Y:$BaZrO_3$–$BaCeO_3$ proton conducting perovskites *J. Mater. Chem.* A **6** 19113–24
[32] Ding H *et al* 2020 Self-sustainable protonic ceramic electrochemical cells using a triple conducting electrode for hydrogen and power production *Nat. Commun.* **11** 1–11
[33] Zhou Y *et al* 2021 An active and robust air electrode for reversible protonic ceramic electrochemical cells *ACS Energy Lett.* **6** 1511–20
[34] Song Y *et al* 2021 Nanocomposites: a new opportunity for developing highly active and durable bifunctional air electrodes for reversible protonic ceramic cells *Adv. Energy Mater.* **11** 2101899
[35] Shin J S, Park H, Park K, Saqib M, Jo M, Kim J H, Lim H T, Kim M, Kim J and Park J Y 2021 Activity of layered swedenborgite structured $Y_{0.8}Er_{0.2}BaCo_{3.2}Ga_{0.8}O_{7+\delta}$ for oxygen electrode reactions in at intermediate temperature reversible ceramic cells *J. Mater. Chem.* A **9** 607–21
[36] Tarutin A, Kasyanova A, Lyagaeva J, Vdovin G and Medvedev D 2020 Towards high-performance tubular-type protonic ceramic electrolysis cells with all-Ni-based functional electrodes *J. Energy Chem.* **40** 65–74
[37] Lei L, Zhang J, Guan R, Liu J, Chen F and Tao Z 2020 Energy storage and hydrogen production by proton conducting solid oxide electrolysis cells with a novel heterogeneous design *Energy Convers. Manage.* **218** 113044
[38] Su H and Hu Y H 2021 Degradation issues and stabilization strategies of protonic ceramic electrolysis cells for steam electrolysis *Energy Sci. Eng.* **10** 1706–25
[39] Rajendran S, Thangavel N K, Ding H, Ding Y, Ding D and Reddy Arava L M 2020 Tri-doped $BaCeO_3$–$BaZrO_3$as a chemically stable electrolyte with high proton-conductivity for intermediate temperature solid oxide electrolysis cells (SOECs) *ACS Appl. Mater. Interfaces* **12** 38275–84
[40] Tang W, Ding H, Bian W, Wu W, Li W, Liu X, Gomez J Y, Regalado Vera C Y, Zhou M and Ding D 2020 Understanding of A-site deficiency in layered perovskites: promotion of dual reaction kinetics for water oxidation and oxygen reduction in protonic ceramic electrochemical cells *J. Mater. Chem.* A **8** 14600–8
[41] Danilov N, Lyagaeva J, Vdovin G, Pikalova E and Medvedev D 2018 Electricity/hydrogen conversion by the means of a protonic ceramic electrolysis cell with $Nd_2NiO_{4+\delta}$-based oxygen electrode *Energy Convers. Manage.* **172** 129–37

[42] Tarutin A P, Vdovin G K, Medvedev D A and Yaremchenko A A 2020 Fluorine-containing oxygen electrodes of the nickelate family for proton-conducting electrochemical cells *Electrochim. Acta* **337** 135808

[43] Sun C, Yang S, Lu Y, Wen J, Ye X and Wen Z 2020 Tailoring a micro-nanostructured electrolyte-oxygen electrode interface for proton-conducting reversible solid oxide cells *J. Power Sources* **449** 227498

[44] Fu L, Zhou J, Yang J, Lian Z, Wang J, Cheng Y and Wu K 2020 Exsolution of Cu nanoparticles in $(LaSr)_{0.9}Fe_{0.9}Cu_{0.1}O_4$ ruddlesden-popper oxide as symmetrical electrode for solid oxide cells *Appl. Surf. Sci.* **511** 145525

[45] Vøllestad E, Strandbakke R, Tarach M, Catalán-Martínez D, Fontaine M-L, Beeaff D, Clark D R, Serra J M and Norby T 2019 Mixed proton and electron conducting double perovskite anodes for stable and efficient tubular proton ceramic electrolysers *Nat. Mater.* **18** 752–9

[46] Saqib M, Lee J-I, Shin J-S, Park K, Kim Y-D, Kim K B, Kim J H, Lim H-T and Park J-Y 2019 Modification of oxygen-ionic transport barrier of $BaCo_{0.4}Zr_{0.1}Fe_{0.4}Y_{0.1}O_3$ steam (Air) electrode by impregnating samarium-doped ceria nanoparticles for proton-conducting reversible solid oxide cells *J. Electrochem. Soc.* **166** F746–54

[47] Meng Y, Gao J, Huang H, Zou M, Duffy J, Tong J and Brinkman K S 2019 A high-performance reversible protonic ceramic electrochemical cell based on a novel Sm-doped $BaCe_{0.7}Zr_{0.1}Y_{0.2}O_{3-\Delta}$ electrolyte *J. Power Sources* **439** 227093

[48] Kobayashi T, Kuroda K, Jeong S, Kwon H, Zhu C, Habazaki H and Aoki Y 2018 Analysis of the anode reaction of solid oxide electrolyzer cells with $BaZr_{0.4}Ce_{0.4}Y_{0.2}O_{3-\delta}$ electrolytes and $Sm_{0.5}Sr_{0.5}CoO_{3-\delta}$ anodes *J. Electrochem. Soc.* **165** F342–9

[49] Wu W, Ding H, Zhang Y, Ding Y, Katiyar P, Majumdar P K, He T and Ding D 2018 3D self-architectured steam electrode enabled efficient and durable hydrogen production in a proton-conducting solid oxide electrolysis cell at temperatures lower than 600 °C *Adv. Sci.* **5** 1800360

[50] Kim J, Jun A, Gwon O, Yoo S, Liu M, Shin J, Lim T H and Kim G 2018 Hybrid-solid oxide electrolysis cell: a new strategy for efficient hydrogen production *Nano Energy* **44** 121–6

[51] Danilov N, Tarutin A, Lyagaeva J, Vdovin G and Medvedev D 2018 CO_2-promoted hydrogen production in a protonic ceramic electrolysis cell *J. Mater. Chem.* A **6** 16341–6

[52] Yang S, Lu Y, Wang Q, Sun C, Ye X and Wen Z 2018 Effects of porous support microstructure enabled by the carbon microsphere pore former on the performance of proton-conducting reversible solid oxide cells *Int. J. Hydrogen Energy* **43** 20050–8

[53] Yang S, Zhang S, Sun C, Ye X and Wen Z 2018 Lattice incorporation of Cu^{2+} into the $BaCe_{0.7}Zr_{0.1}Y_{0.1}Yb_{0.1}O_3$ electrolyte on boosting its sintering and proton-conducting abilities for reversible solid oxide cells *ACS Appl. Mater. Interfaces* **10** 42387–96

[54] Li W, Guan B, Ma L, Hu S, Zhang N and Liu X 2018 High performing triple-conductive $Pr_2NiO_{4+\delta}$ anode for proton-conducting steam solid oxide electrolysis cell *J. Mater. Chem.* A **6** 18057–66

[55] Huan D, Wang W, Xie Y, Shi N, Wan Y, Xia C, Peng R and Lu Y 2018 Investigation of real polarization resistance for electrode performance in proton-conducting electrolysis cells *J. Mater. Chem.* A **6** 18508–17

[56] Huan D, Shi N, Zhang L, Tan W, Xie Y, Wang W, Xia C, Peng R and Lu Y 2018 New, efficient, and reliable air electrode material for proton-conducting reversible solid oxide cells *ACS Appl. Mater. Interfaces* **10** 1761–70

[57] Lei L, Tao Z, Wang X, Lemmon J P and Chen F 2017 Intermediate-temperature solid oxide electrolysis cells with thin proton-conducting electrolyte and a robust air electrode *J. Mater. Chem.* A **5** 22945–51
[58] Yang S, Wen Y, Zhang S, Gu S, Wen Z and Ye X 2017 Performance and stability of $BaCe_{0.8-x}Zr_{0.2}In_xO_{3-\delta}$-based materials and reversible solid oxide cells working at intermediate temperature *Int. J. Hydrogen Energy* **42** 28549–58
[59] Wu W, Ding H, Zhang Y, Ding Y, Katiyar P, Majumdar P K, He T and Ding D 2018 Hydrogen production: 3D self-architectured steam electrode enabled efficient and durable hydrogen production in a proton-conducting solid oxide electrolysis cell at temperatures lower than 600 °C *Adv. Sci.* **5** 1870070
[60] Zhu H, Ricote S, Duan C, O'Hayre R P, Tsvetkov D S and Kee R J 2018 Defect incorporation and transport within dense $BaZr_{0.8}Y_{0.2}O_{3-\delta}$ (BZY20) proton-conducting membranes *J. Electrochem. Soc.* **165** F581–8
[61] Zhu H, Ricote S, Duan C, O'Hayre R P and Kee R J 2018 Defect chemistry and transport within dense $BaCe_{0.7}Zr_{0.1}Y_{0.1}Yb_{0.1}O_{3-\delta}$ (BCZYYb) proton-conducting membranes *J. Electrochem. Soc.* **165** F845–53
[62] Zhou Y, Zhang W, Kane N, Luo Z, Pei K, Sasaki K, Choi Y M, Chen Y, Ding D and Liu M 2021 An efficient bifunctional air electrode for reversible protonic ceramic electrochemical cells *Adv. Funct. Mater.* **31** 1511–20
[63] Li M, Hua B, Wang L-C, Sugar J D, Wu W, Ding Y, Li J and Ding D 2021 Switching of metal–oxygen hybridization for selective CO_2 electrohydrogenation under mild temperature and pressure *Nat. Catal.* **4** 274–83
[64] Bausá N, Escolástico S and Serra J M 2019 Direct CO_2 conversion to syngas in a $BaCe_{0.2}Zr_{0.7}Y_{0.1}O_{3-\delta}$-based proton-conducting electrolysis cell *J. CO_2 Util.* **34** 231–8
[65] Hua B, Yan N, Li M, Zhang Y, Sun Y, Li J, Etsell T, Sarkar P, Chuang K and Luo J-L 2016 Novel layered solid oxide fuel cells with multiple-twinned $Ni_{0.8}Co_{0.2}$ nanoparticles: the key to thermally independent CO_2 utilization and power-chemical cogeneration *Energy Environ. Sci.* **9** 207–15
[66] Somekawa T, Matsuzaki Y, Sugahara M, Tachikawa Y, Matsumoto H, Taniguchi S and Sasaki K 2017 Physicochemical properties of $Ba(Zr,Ce)O_{3-\delta}$-based proton-conducting electrolytes for solid oxide fuel cells in terms of chemical stability and electrochemical performance *Int. J. Hydrogen Energy* **42** 16722–30
[67] Kreuer K D 2003 Proton-conducting Oxides *Annu. Rev. Mater. Res.* **33** 333–59

IOP Publishing

High-Temperature Electrolysis
From fundamentals to applications
Werner Sitte and Rotraut Merkle

Chapter 15

Co-solid oxide electrolysis and methanation

Andreas Krammer and Markus Lehner

A reduction in industrial CO_2 emissions can be achieved by recycling it back into the energy system using renewable energy sources. Using a combined co-solid oxide electrolysis cell (SOEC) and methanation power-to-gas (PtG) plant, it is possible to transform CO_2 into a valuable substitute natural gas for long-term energy storage. This chapter presents the requirements for the successful methanation of co-SOEC syngas at the catalyst, reactor, and plant levels. Reaction kinetics and thermodynamics define the baseline for well-balanced reaction conditions. The catalytically active materials, carrier materials, and catalyst forms used for chemical methanation need to be considered to maximize performance. Several reactor designs for co-SOEC syngas methanation are available, which differ in their combinations of phases, cooling characteristics, and complexity. Heat integration strategies and reactor arrangements substantially influence the methanation performance and the overall system efficiency. The economic viability of this combined co-SOEC and methanation system depends on an optimally tuned design at all system levels.

15.1 Power-to-Gas as an option for chemical storage of renewable energy

The decarbonization of our existing energy systems will be based on an enormous increase in sustainable but also fluctuating power sources, such as wind and solar energy. PtG is an efficient strategy for transforming surplus electric energy into valuable gaseous energy carriers suitable for long-term storage [1–3]. The key element of a PtG plant is the electrolysis unit. The electrolyzer uses electric energy to reduce water to hydrogen. As an option, when co-electrolysis is used, carbon dioxide can also be fed together with water to produce a syngas product. Either way, a subsequent methanation reactor can transform both hydrogen and carbon oxides (CO, CO_2) into methane. Figure 15.1 shows a schematic process diagram of a PtG system consisting of co-electrolysis and methanation.

doi:10.1088/978-0-7503-3951-3ch15

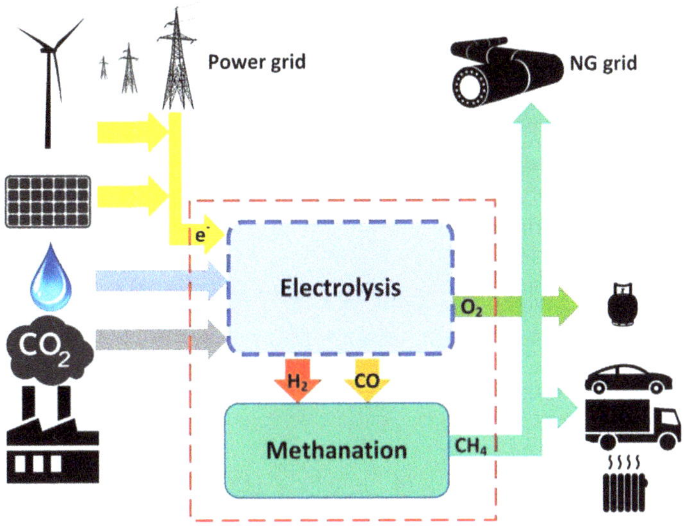

Figure 15.1. Schematic illustration of a PtG system that includes co-electrolysis and methanation.

A PtG unit allows electric energy to be stored in the form of highly calorific gases (at the terawatt-hour scale) for an almost unlimited time period in former natural gas reservoirs, for example. Hydrogen, as well as methane, can be transformed back to electric energy when needed, used directly in industrial processes, or used for heating or transport. While the production of methane requires an additional production step that leads to higher transformation losses, it benefits from important technical advantages. Existing infrastructure for the storage, distribution, and utilization of methane is already available in the form of the existing natural gas grid. Furthermore, CO_2 from industrial sources can be reused by methanation, thus recycling carbon back to the energy system. While some industrial CO_2 sources will certainly be replaced, e.g. by direct electrification, certain processes cannot be fully decarbonized. The carbonates in mineral raw materials, such as the limestone used for cement production, some sorts of iron ore used for steel production, or those used in the course of refractory material production, release CO_2-based materials due to calcination. Furthermore, waste incineration and biogas production are, if not indispensable, very valuable technologies which cause CO_2 emissions. Therefore, it is necessary to establish solutions that can handle unavoidable CO_2 emissions efficiently. CO_2 can be captured post-combustion (e.g. by amine scrubbing), transformed into CH_4, e.g. by a combined electrolysis and methanation system, and then injected into the natural gas grid as a substitute for primary natural gas. As a result, methanation will play a substantial role in the energy transformation from fossil fuels to renewables. Several different methanation process designs are possible in terms of reactor type, catalyst form, and thermal management. However, the focus of this chapter is the field of catalytic methanation, whereas biological methanation in stirred tank reactors plays a minor role in direct combination with high-temperature co-electrolysis.

15.2 The fundamentals of catalytic methanation

The methanation process is based on the Sabatier reactions first discovered by Sabatier and Senderens in 1902 [4]. These two Sabatier reactions, the CO_2 methanation reaction (equation (15.1)) and the CO methanation reaction (equation (15.2)), are always accompanied by a third, the reverse water gas shift reaction (equation (15.3)). The reverse water gas shift reaction links the two methanation reactions by linear combination.

$$\text{CO}_2\,\text{methanation: } CO_2 + 4H_2 \rightleftharpoons CH_4 + 2H_2O \quad \Delta H_R^{553\,K} = -176.4\frac{\text{kJ}}{\text{mol}} \tag{15.1}$$

$$\text{CO methanation: } CO + 3H_2 \rightleftharpoons CH_4 + H_2O \quad \Delta H_R^{553\,K} = -215.9\frac{\text{kJ}}{\text{mol}} \tag{15.2}$$

$$\text{Reverse water gas shift: } CO_2 + H_2 \rightleftharpoons CO + H_2O \quad \Delta H_R^{553\,K} = 39.5\frac{\text{kJ}}{\text{mol}} \tag{15.3}$$

The Boudouard reaction is an adverse side reaction that, among other side reactions, leads to carbon deposition [5]. It is favored at low pressures and low H_2/CO or H_2/CO_2 ratios in the methanation feed, which are addressed in greater detail in section 15.3.

$$\text{Boudouard reaction: } CO_2 + C \rightleftharpoons 2CO \quad \Delta H_R^{553\,K} = 173.8\frac{\text{kJ}}{\text{mol}} \tag{15.4}$$

The methanation process can be performed in several reactor types that differ in their operational methods and reactor cooling. Reactor performances can be compared using the resulting gas product quality, the conversion rate of carbon oxides, and temperature conditions. For comparability between reactor systems, the gas hourly space velocity (GHSV) must be considered, which represents the catalytic load and is always expressed under standard conditions as shown in equation (15.5).

$$\text{GHSV}_{\text{STP}} = \frac{\dot{V}_{\text{in}}}{V_{\text{catalyst}}}\frac{T_{\text{in}}}{T_{\text{STP}}}\frac{p_{\text{STP}}}{p_{\text{in}}} \tag{15.5}$$

15.2.1 Methanation reactors

In the most prevalent and mature packed-bed reactors, the catalyst is mostly applied in the form of a spherical or cylindrical bulk material that fills a tubular system. For adiabatic packed-bed reactors without direct cooling, intermediate cooling of the process gas can be used between several serial reactors. Further temperature control strategies for adiabatic reactor systems are product gas recycling, staged feed injection, or steam injection, which have the disadvantages of higher operational costs and lower CO_2/CO conversion. However, adiabatic systems are easily exposed to hot spots, lower space-time yields, and generally lower CO/CO_2 conversions. Due to highly active catalysts, the limiting factor of packed-bed methanation is heat removal from the catalyst bed (and to a lesser extent reaction kinetics), especially for high GHSV and CO-containing feed gases, such as co-SOEC syngas. Therefore, direct temperature control of packed-bed reactors is essential [2, 6]. Packed-bed systems with direct cooling can be realized in the form of tube-bundle reactors. Low-diameter reactor pipes are important in order to reduce the high radial heat gradients within the catalyst bed [7] and enhance radial heat transfer [6]. As an alternative to bulk catalysts, catalytically active materials can also be applied on ceramic or metallic honeycombs. These structured monolithic catalysts promise even higher radial heat transfer (by two or three orders of magnitude) and less pressure loss [2]. Microstructured reactor systems aim to achieve highly intensified heat transfer for improved temperature control. Although the maturity level of this reactor design is lower, plate reactors for intensified heat transfer are already commercially available for small-scale applications [8]. Furthermore, fluidized catalytic beds have been developed for enhanced heat and mass transfer. However, those reactors place a high demand on the attrition resistance of the catalyst. An additional reactor type is the three-phase bubble column reactor, which benefits from enhanced heat transfer but also results in additional mass transfer resistance between the reactants and the catalyst due to the required transfer from the gas to the liquid phase [9].

Trickle-bed or continuously stirred tank reactors are used for biological methanation, which makes use of microorganisms to catalyze the methanation reaction within the liquid phase. Biological methanation is conducted at moderate temperatures (30 °C–70 °C) and benefits from high resistance to catalyst poisons (such as sulfur components) compared to catalytic methanation. Figure 15.2 gives an overview of reactor and plant types, including technical readiness levels.

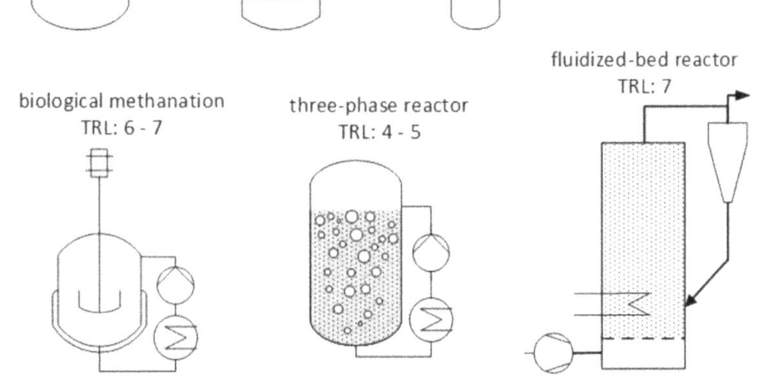

Figure 15.2. Overview of methanation reactor types including technology readiness levels (TRLs) based on [2, 10, 11].

15.2.2 Methanation catalysts

Active materials for catalytic methanation can be found in the VIII to X groups of metals. Although several metals have been investigated in terms of activity (Ru >Fe >Ni >Co >Rh >Pd >Pt >Ir), nickel remains the most important active material for methanation due to its high activity and high selectivity at low material cost [2, 12]. Mills and Steffgen [13] published a list of most relevant methanation catalysts:

Activity: Ru >Fe> Ni >Co >Mo

Selectivity: Ni >Co >Fe >Ru.

The most active catalyst material is ruthenium, but as a result of its immense costs, its use is infeasible for commercial applications. Although iron is also very active, its tendency to form higher hydrocarbons is highly adverse. Since cobalt is more expensive, less active, and less selective towards methane, it also fails to compete with nickel. Molybdenum has the highest reported sulfur stability in comparison to all the substances mentioned above. On the other hand, molybdenum is less active and less selective than nickel. As a result, nickel remains by far the most important catalyst for industrial methanation applications, providing great performance at low cost. Most commonly, the active substance is applied to metal oxide supports that usually consist of Al_2O_3, SiO_2, or TiO_2. As the main goal of the catalyst support material is to provide a high surface area for the application of active sites, the support material significantly influences the catalytic activity. Certain catalyst properties can be tuned by substances that provide enhancements. Ni/Al_2O_3 catalysts can be enhanced by MgO, which improves their carbon resistance [14] and thermal stability [15], whereas La_2O_3 enhances catalytic activity [16], and CeO_2 leads to higher reducibility and long-term stability [17]. The thermal stability, activity, and coke resistance of nickel catalysts can be improved by V_2O_3 [2, 18].

Industrial off-gases that serve as CO_2 sources for PtG plants often contain catalyst poisons such as sulfur compounds, heavy metals, tars, dust, chlorine compounds, ammonia, or alkalis. While dust or tar can deactivate the catalyst by mechanically blocking its active sites, other compounds such as sulfur can react with nickel to form irreversible inactivating bonds. In addition, high temperatures (especially in adiabatic reactors) can lead to thermal degradation by causing sintering effects. Furthermore, the carbon deposition caused by undesired side reactions that depend on the C–H–O ratio, temperature, and pressure can result in the loss of active sites. Countermeasures against the formation of carbon can be taken at the reactor and process design levels, as discussed in sections 15.3 and 15.4.

The exact methanation reaction mechanism on catalytic nickel or noble metal sites is still object of research. There appears to be a consensus among authors that a Langmuir–Hinshelwood–Hougen–Watson (LHHW) approach is the basis of the methanation mechanism. The LHHW mechanism assumes that both reaction partners first need to be adsorbed at the active sites to reach an activated state before they can form covalent bonds with each other. Nevertheless, the exact reaction mechanism and whether or not differences between CO_2 and CO methanation occur are not yet clear. However, two main reaction mechanisms are

Table 15.1. Two methanation mechanisms on nickel or noble metal catalysts discussed in the literature; '*' indicates adsorption sites [7, 19–23].

Intermediate surface carbon mechanism	Hydrogen-assisted mechanism
1: $CO_2 + 2^* \leftrightarrow CO^* + O^*$	1: $CO_2 + 2^* \leftrightarrow CO^* + O^*$
2: $H_2 + 2^* \leftrightarrow 2\ H^* +$	2: $H_2 + 2^* \leftrightarrow 2\ H^* +$
3: $CO^* + {}^* \leftrightarrow C^* + O^*$	3: $CO^* + H^* \leftrightarrow CHO^* + {}^*$
4: $C^* + H^* \leftrightarrow CH^* + {}^*$	4: $CHO^* + {}^* \leftrightarrow CH^* + O^*$
5: $O^* + H^* \leftrightarrow OH^* + {}^*$	5: $CH^* + 3H^* \leftrightarrow CH_4{}^* + 3^*$
6: $OH^* + H^* \leftrightarrow H_2O^* + {}^*$	6: $CH_4{}^* \leftrightarrow CH_4 + {}^*$
7: $H_2O^* \leftrightarrow H_2O + {}^*$	7: $O^* + H^* \leftrightarrow OH^* + {}^*$
8: $CH^* + 3H^* \leftrightarrow CH_4{}^* + 3^*$	8: $OH^* + H^* \leftrightarrow H_2O^* + {}^*$
9: $CH_4{}^* \leftrightarrow CH_4 + {}^*$	9: $H_2O^* \leftrightarrow H_2O + {}^*$

currently discussed among researchers: the 'intermediate surface carbon mechanism' and the 'hydrogen-assisted mechanism.' The rate-determining step (RDS) of the reaction mechanism, namely the formation of COH* or CO dissociation, is under ongoing discussion as well. The derivation of a formal kinetic expression is based on the RDS of the reaction mechanism [7] (table 15.1).

15.2.3 Methanation kinetics

The kinetic expressions represent the reaction rates of the underlying reactions. In general, in accordance with Arrhenius' law, the reaction rates increase at higher temperatures. However, the methanation reactions are characterized by an ignition temperature (225 °C–285 °C according to [24–26]) that depends on gas concentration and pressure. Below the ignition temperature, the reaction rate tends to zero. Therefore, the reactor temperature needs to be kept above this temperature level to keep the chemical reaction going. At the upper boundary of the temperature window, the reaction kinetics decelerates when thermodynamic equilibrium is approached at the particular pressure and temperature. The reaction rate is expressed by the formal kinetic equations as a function of temperature, pressure, and gas concentration. As the exact elementary reactions of the methanation mechanism are not yet fully understood, formal kinetic expressions can only be experimentally measured and derived. Formal kinetic formulations represent the reactions kinetics using a mathematical correlation to the observed experimental results. Although the mathematical structure of the kinetic law (LHHW or power law) is maintained, its physical parameters, such as the activation energy or the reaction order, serve only as adjustable values to limit the deviation between the experimental findings and the mathematical representation. Nevertheless, the form of the formal kinetic expression is influenced by the RDS of the reaction mechanism. Since the RDS is still part of the scientific discussion, as mentioned above, different forms of expression have been proposed. A number of power-law and LHHW kinetic approaches have been suggested for methanation in the literature, which differed in their measuring procedures (integral reactor or differential reactor), catalyst forms (powder, pellets), catalyst materials (nickel, ruthenium) as well as pressure and temperature ranges. Furthermore, the kinetics can be formulated as single-step kinetics representing only one reaction (generally CO_2 methanation) (equation (15.1)), or as multistep kinetics that includes two or three of the methanation reactions (equations (15.1)–(15.3)).

In principle, in a power-law approach, the reaction rate depends on the rate coefficient, the partial pressures of the contributing species, and the 'driving force' of a reaction. The 'driving force' represents the deviation from thermodynamic equilibrium and takes on a value between zero and one. The driving force and therefore the reaction kinetics tend to zero if the species concentrations reach the thermodynamic equilibrium. The principal power-law expression for the CO_2 methanation reaction rate is shown in equation (15.6) [26].

$$r_j = k_j \cdot p_{\mathrm{CO_2}}^{a} \cdot p_{\mathrm{H_2}}^{b} \cdot \left(1 - \frac{p_{\mathrm{CH_4}} \cdot p_{\mathrm{H_2O}}^{2}}{p_{\mathrm{CO_2}} \cdot p_{\mathrm{H_2}}^{4} \cdot K_{p,\,\mathrm{CO_2-Meth}}}\right) [\mathrm{mol\ kg_{cat}^{-1}\ s^{-1}}] \tag{15.6}$$

For the LHHW kinetic expression, the power-law expression is complemented by the 'adsorption term,' which reflects the suppressing effects of the molecular adsorption of reactants or the desorption of products such as methane or water. The exponent of the adsorption expression corresponds to the number of active sites taking part in the reaction [26, 27]. Equation (15.7) represents the principal LHHW reaction rate of the CO_2 methanation reaction.

$$r_j = \frac{k_j \cdot p_{\mathrm{CO_2}}^{a} \cdot p_{\mathrm{H_2}}^{b} \cdot \left(1 - \frac{p_{\mathrm{CH_4}} \cdot p_{\mathrm{H_2O}}^{2}}{p_{\mathrm{CO_2}} \cdot p_{\mathrm{H_2}}^{4} \cdot K_{p,\,\mathrm{CO_2-Meth}}}\right)}{\left(1 - K_{\mathrm{CO_2}} \cdot p_{\mathrm{CO_2}}^{c} + K_{\mathrm{H_2}} \cdot p_{\mathrm{H_2}}^{d} + K_{\mathrm{CH_4}} \cdot p_{\mathrm{CH_4}}^{e} + \ldots\right)^2} [\mathrm{mol\ kg_{cat}^{-1}\ s^{-1}}] \tag{15.7}$$

The kinetic factors are calculated based on the Arrhenius-type relation (equation (15.8)) and the adsorption factors by the van 't Hoff-type (equation (15.9)) relation. As already mentioned, the parameters of these equations ($E_{A,\,j}$, k_j^0, $\Delta H_{\mathrm{ads},\,i}$, K_i^0) lose their physical purposes and serve as fitting parameters to achieve best correlation between the formal kinetic function and the experimental results.

$$k_j = k_j^0 \cdot \exp\left(-\frac{E_{\mathrm{A},\,j}}{RT}\right) \tag{15.8}$$

$$K_i = K_i^0 \cdot \exp\left(-\frac{\Delta H_{\mathrm{ads},\,i}}{RT}\right) \tag{15.9}$$

As early as 1950, the first LHHW kinetics for methanation was published by Binder and White [28]. Numerous kinetic formulations have been published since then, most often using the LHHW approach in recent publications [7, 20]. Gruber [26] presented a list of detailed kinetic approaches for nickel catalysts with varying nickel loads (5 w% to >55 w%), temperature ranges, and pressure ranges, which can be regarded as the most important recent kinetic formulations in the literature (table 15.2).

Table 15.2. Overview of detailed methanation kinetic expressions taken from Gruber [26].

No.	Author	Ni content	*T* in °C	p in bar	Type
1	Martinez Molina [29]	54 w%	150–260	atmos.	Single-step power law
2	Schollenberger [30]	unknown	200–300	2–17	Single-step power law
3	Weatherbee [31]	3 w%	227–327	1.4–1.75	Single-step LHHW
4	Koschany [20]	58 w%	180–340	1–15	Single-step LHHW
5	Yang Lim [32]	12 w%	180–210	up to 20	Single-step LHHW
6	HELMETH kinetic [26, 33]	>55 w%	250–350	15–25	Single-step LHHW
7	Xu & Froment [34]	15.2 w%	300–400	3–10	Multistep LHHW
8	Kopyscinski [21]	50 w%	280–360	up to 2	Multistep LHHW
9	Ducamp [35]	14–17 w%	280–400	1–10	Multistep LHHW
10	Rönsch [19]	18/50 w%	275–360	1–5	Multistep LHHW

For packed-bed methanation reactor modeling, the kinetic approach described by Xu and Froment [34] was often used directly or in an adapted version [2]. The three-step kinetics used by Xu and Froment was later adapted by Parlikkad *et al* [36], Kang and Lee [37], Klose and Baerns [38], and Zhang *et al* [39] [2]. Rönsch *et al* [19] again recommended and adapted the latter two Xu and Froment adaptions. Kopyscinski [21] modified findings by Weatherbee and Bartholomew [31] for CO methanation. The authors' kinetic expression was established for highly isothermal conditions and is therefore especially suitable for fluidized-bed methanation. Furthermore, in the recent work by Koschany *et al* [20], a single-step LHHW kinetics for CO_2 methanation was elaborated. Based on 200 data points within a temperature range between 180 °C and 340 °C and pressures of up to 15 bar, the authors found the best experiment-model fit for a hydrogen-assisted reaction mechanism assumption. Comprehensive overviews and comparisons of methanation reaction kinetic expressions are given by Kopyscinski *et al* [21], Rönsch [19], Younas *et al* [12], Gruber [26], and Neubert [7].

The kinetic expressions discussed above represent intrinsic kinetics without any form of mass transfer limitation. For packed-bed reactors filled with industrially applicable catalyst particles, mass transfer limitations must often be taken into account [6, 24, 25, 35, 40–42]. Therefore, not only can the intrinsic kinetics be applied, but also external factors limiting the chemical reaction rate. It is necessary to consider mass transfer limitation due to diffusion from the gas bulk to the catalyst surface, as well as the intraparticle diffusion limitation. Gruber [40], Sun and Simakov [41], Kiewidt [6], and Kreitz [42] used catalytic effectiveness factors calculated based on a Thiele-modulus approach within their packed-bed methanation models to consider mass transfer limitations. Ducamp *et al* [35], Try *et al* [25], and Schlereth and Hinrichsen [24] included a particle model to take mass transfer limitations into account.

15.3 Thermodynamics of catalytic methanation

Figure 15.3 gives the thermodynamic equilibrium compositions at varying temperatures and pressures, based on assumed co-SOEC product concentrations of

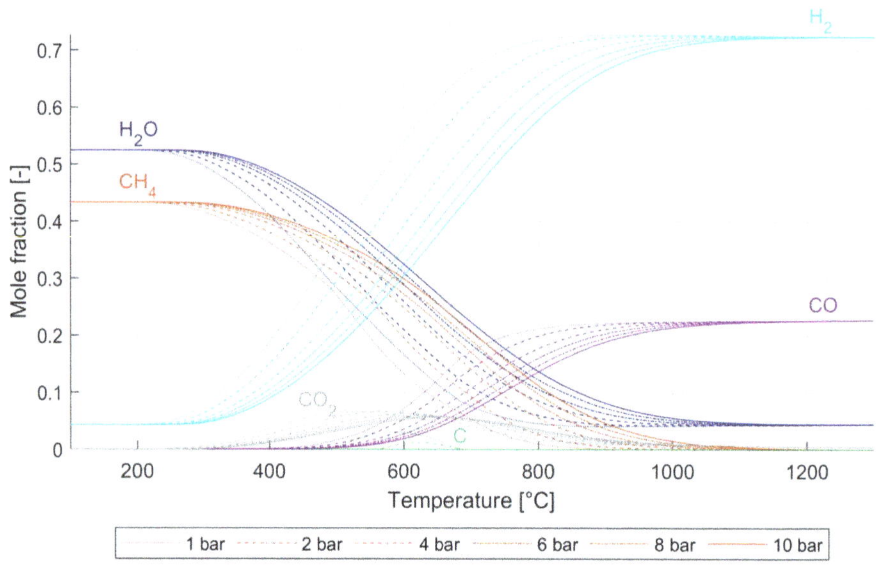

Figure 15.3. Dependence of the equilibrium composition of co-SOEC syngas on temperature and pressure based on a co-SOEC syngas composition with a 3% hydrogen excess (76.8 vol% H_2, 18.3 vol% CO, and 4.9 vol% CO_2) [43].

76.8 vol% H_2, 18.3 vol% CO, and 4.9 vol% CO_2. This composition contains hydrogen at more than 3% above the stoichiometric concentration according to equations (15.1) and (15.2) and is highly suitable for methanation purposes.

According to figure 15.3, moderate temperatures below 320 °C are necessary to achieve full conversion at maximum CH_4 concentration. In any event, temperatures below 400 °C are necessary to achieve sufficiently high methane concentrations, even at increased pressures above 4 bar. The beneficial effect of increased pressures between 1 and up to 10 bar is illustrated in figure 15.3 through different line styles. Increased pressures of 4 bar and higher not only lead to higher equilibrium methane concentrations but also reduce the tendency toward carbon formation between 500 °C and 700 °C.

In terms of the practical implications of the reactor temperature for reactor performance, contrary effects must be expected based on thermodynamics and kinetics. Starting from the ignition temperature (225 °C–285 °C [24–26]), the reaction rate increases with rising temperature until the temperature approaches the thermodynamic equilibrium value. On the other hand, based on thermodynamic equilibrium, the highest methane concentration can be expected at moderate temperatures, as already discussed according to figure 15.3. This antagonistic interplay between thermodynamics and kinetics makes temperature control in methanation systems highly important. A trade-off between high reaction rates at higher temperatures and high equilibrium product concentrations at lower temperatures is necessary to achieve optimum heat management and therefore reactor performance. Neubert [7] aptly called this crucial interaction between plant complexity, thermodynamics, and kinetics the 'trilemma of methanation.' The issue of optimum reactor temperature profiles is further addressed in section 15.4.

The Boudouard reaction (equation (15.4)) or the decomposition reactions of methane or higher hydrocarbons lead to the formation of solid carbon, which is deposited on the active catalyst sites. Different forms of carbon can be formed, such as graphitic carbon, amorphous carbon, vermicular carbon, bulk Ni_3C, or adsorbed carbon species. The tendency toward carbon formation is influenced by the C–H–O ratio and the temperature and the pressure of the process gas. High proportions of H and O reduce the risk of carbon formation. In consequence, the higher the over-stoichiometric excess of hydrogen, the lower the carbon formation risk. In addition, the injection of water into the methanation feed is an option that reduces the carbon formation tendency at the cost of hindering kinetic effects [26] and reducing equilibrium concentrations. [5]

In figure 15.4, the ternary C–H–O diagram is shown, which describes the risk of graphitic carbon formation. For each temperature-pressure pair, this diagram can be divided by an equilibrium curve into an upper section, in which carbon formation is thermodynamically possible, and a lower one, where it is not. Depending on the process gas's C–O–H ratio, the alignment within this diagram gives a good indication of whether or not carbon formation is likely to happen. With an increasing excess of hydrogen based on the same CO/CO_2 ratio, the tendency

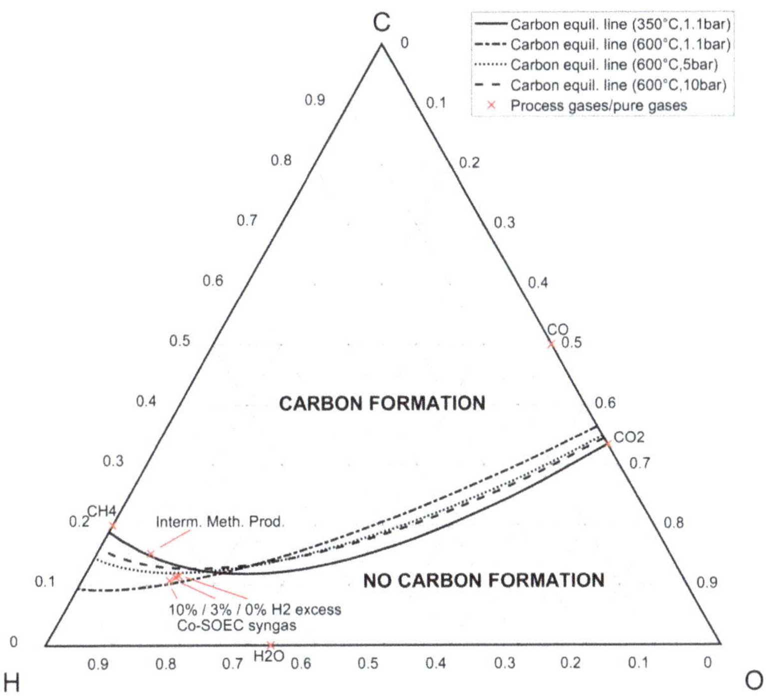

Figure 15.4. Ternary C–O–H diagram with pure gases and process gases (red 'x'-markings) including co-SOEC syngas with 0%, 3%, and 10% of excess H_2 and an intermediate methanation product after water removal. Graphitic carbon formation equilibrium lines for four temperature-pressure pairs (350 °C/1.1 bar; 600 °C/1.1 bar; 600 °C/5 bar; 600 °C/10 bar) divide the areas of thermodynamically possible carbon deposition. Figure adapted from Krammer *et al* [43].

toward carbon formation can be reduced, although even high proportions of hydrogen do not completely eliminate the risk. In fact, the pressure has to be kept elevated for typical co-SOEC compositions to prevent carbon deposition at any temperature. The increased potential for the formation of carbon at around 600 °C and 1.1 bar pressure in figure 15.4 is in alignment with stable equilibrium shares at low pressures and similar temperatures in figure 15.3.

15.4 Requirements for the successful methanation of co-SOEC syngas

The coordinated interplay between co-SOECs and methanation determines the success of a combined PtG plant. The syngas concentration influences not only the preferred reactor design, including the catalyst and the cooling method, but also the necessary pressure and resulting temperature profile that avoid carbon deposition and produce high-quality synthetic natural gas (SNG). This section discusses the requirements for co-SOEC syngas, based on the fundamentals of thermodynamics and kinetics elaborated in sections 15.2 and 15.3. Furthermore, a feasible reactor concept and promising operating conditions are derived.

The syngas generated by a co-SOEC without further process gas treatment other than water removal mainly consists of hydrogen and carbon monoxide, depending on the electrolyzer's feed concentration and recycling ratio. Small amounts of carbon dioxide downstream from the electrolyzer are the unproblematic result of incomplete CO_2 reduction [44]. A feed gas mixture for methanation purposes needs to follow certain criteria regarding minimum hydrogen content according to the methanation reactions. The H_2–CO ratio needs to be greater than three and the remaining hydrogen should result in a H_2–CO_2 ratio of more than four to guarantee an over-stoichiometric hydrogen concentration. As discussed in section 15.3 and illustrated in figure 15.4, the tendency for the reaction to form solid carbon can be reduced by increasing the hydrogen in the methanation feed to a level greater than the stoichiometric level. In addition, higher hydrogen contents lead to higher conversions of CO/CO_2. On the other hand, the higher the excess of hydrogen in the syngas, the more unconverted hydrogen remains in the SNG product downstream from the methanation, even at full CO/CO_2 conversion. Due to a reduction in the total process gas volume flow within the methanation reactor (a consequence of the mole-decreasing reactions and the subsequent removal of formed water), the relative volumetric proportion of hydrogen in the product is even higher than in the syngas. For example, a share of 10 vol% excess hydrogen in the co-SOEC syngas serving as the methanation feed would result in about 23 vol% of H_2 in the dry SNG, assuming complete methanation and full water removal. The option of hydrogen recycling from the SNG product back to the feed is suboptimal, since it requires gas separation equipment such as polymer membranes [45], which increase plant complexity and decrease efficiency. Therefore, it is most feasible to omit recycle lines and employ direct injection into the natural gas grid in one run, while allowing the excess hydrogen to remain in the SNG. According to a joint report by the European Network of Transmission System Operators for Gas (ENTSOG),

Hydrogen Europe, and Gas Infrastructure Europe, moderate proportions of 10 vol % of hydrogen are technically already feasible for natural gas grids in most areas of Europe with no or little adaption [46, 47]. In Austria, 10 vol% of hydrogen is tolerated since 2021, according to recent legislation [48]. In Germany, up to 10 vol% of hydrogen, and, based on a recently released technical code starting in fall 2021, even higher concentrations are tolerated, depending on local grid requirements [49]. In this regard, the aim should be a trade-off between reducing the carbon deposition tendency and high conversion versus acceptable hydrogen proportions in the SNG product. To give a concrete example, a typical co-SOEC concentration of 76.8 vol% H_2, 18.3 vol% CO, and 4.9 vol% CO_2 was tested in experiments and by modeling [43]. Based on this input concentration at full methanation, almost 90 vol% of methane and 10 vol% of hydrogen could be generated after a final drying step, which represents an attractive SNG product for direct grid injection. As a consequence, for this suggested co-SOEC composition, a hydrogen excess of 3% in the methanation feed results in a moderate hydrogen content of 10 vol% in the SNG product. However, an elevated pressure of at least 4 bar should be considered to prevent carbon deposition (figure 15.3).

The temperature profile within a catalytic two-phase reactor is an essential performance parameter that determines the reaction kinetics, catalyst degradation, and thermodynamic outlet concentration. The CO methanation reaction is accompanied by a higher exothermal reaction enthalpy compared to that of CO_2 methanation (compare equations (15.1) and (15.2)), which makes heat removal in co-SOEC syngas methanation even more relevant. As already discussed in section 15.2, the proven temperature-control concepts of adiabatic fixed beds are based on intermediate heat removal, staged heat injection, product recycling, or steam injection [2, 7]. However, the overall plant efficiency is higher and the required catalyst mass and reactor volume are lower if the reaction heat is directly removed from the catalyst in actively cooled polytropic reactors. Therefore, direct reactor cooling in fixed-bed, structured, or microstructured systems is a promising approach for cost-effective and high-capacity methanation. Recent research progress into temperature control strategies for catalytic methanation is addressed in the following section.

Reaction temperature control is an important area of recent research into catalytic methanation. For two-phase catalytic methanation, two main tasks can be accomplished by reactor cooling, as shown in figure 15.5. The first is to prevent the maximum temperature of the catalyst from being exceeded, thereby preventing sintering. The second is to guarantee a high thermodynamic equilibrium concentration of methane at the reactor outlet [7].

Kiewidt and Thöming [6] determined an optimal temperature profile based on a Semenov number-optimization method for single-stage methanation systems. The space–time yield could be increased twofold by temperature profile optimization based on the Semenov number, which can be tuned via the reactor diameter and gas load. The authors emphasized the importance of heat transfer to a cooling medium. Enhanced heat removal can also be realized with structured and microstructured reactors at the expense of higher investment costs. Using 1D modeling, El-Sibai *et al* [50]

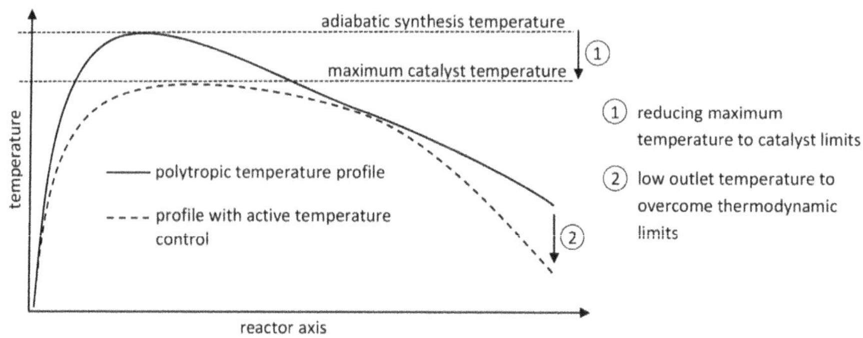

Figure 15.5. Schematic illustration of the two main purposes of active reactor cooling for methanation systems. Reproduced based on a figure by Neubert [7].

found that the most cost efficient number of cooled fixed-bed reactors in a cascade with intermediate water removal is two, whereas three reactors leads to a maximum space–time yield. In addition, the authors found that a minimization of reactor diameter is beneficial, since it enhances heat transfer to the cooling medium and increases the area-to-volume ratio. Although research activity for microstructured reactors is still high, plate reactors for small-scale applications are already commercially available with a GHSV of 31 500 h^{-1} for CO_2 methanation. The catalyst is placed in microbeds stacked sequentially between cooling foils, ensuring that the surfaces have very high heat transfer capabilities [8]. In contrast to two-phase reactors, three-phase methanation allows for almost isothermal conditions at the expense of lower gas/liquid mass transfer rates. As a result, three-phase systems are not limited by heat transfer, but by mass transfer—to an extent that makes them inferior in terms of space-time yield. A comparison of fixed-bed tube bundle reactor and slurry bubble column reactor methanation revealed that a much higher GHSV is possible for fixed-bed reactors. However, for transient conditions, the slurry bubble column reactor is far more temperature stable, as the fixed-bed temperature fluctuations are immense [9, 51]. In summary, tube-bundle reactors are an attractive technology for large-scale PtG plants, as they allow for efficient heat removal at moderate costs. Nevertheless, promising research progress in the field of heat-transfer-optimized structured and microstructured systems might outweigh their complexity drawbacks in future.

In terms of operational pressure, higher pressure is beneficial for both thermodynamics (figure 15.3) and kinetics. The more the operational pressure is increased, the more the methane concentration and the CO/CO_2 conversion downstream from a catalytic methanation reactor converting co-SOEC syngas become higher [43]. However, higher pressures require more compression power, which influences the overall PtG efficiency. As a result, the pressure level should be held as low as possible to provide sufficient methanation performance. In addition, carbon deposition should be taken into account, which can be reduced by increased pressures. In addition, the downstream process, e.g. the gas grid pressure level, should be considered in this regard. The compression of process gas can be conducted at different stages within the process configuration. In section 15.5, the influence of compressor alignment on process efficiency is elaborated.

15.5 Energetic efficiency and the socioeconomic impact of co-SOEC syngas methanation

The successful implementation of a PtG technology depends on its economic feasibility in an existing energy market. The balance of the costs and revenues of a PtG system needs to compete with conventional energy provision technologies, such as fossil fuels, which have profited from decades of technical development. The current economic situation and the future outlook for a combined co-SOEC and methanation PtG plant was described by Böhm *et al* [52]. The authors considered learning and scaling effects on capital expenditure (CAPEX) to evaluate the systems competitiveness of renewable SNG in comparison to conventional natural gas. For a 10 MW plant in 2030, Böhm *et al* predict SNG costs of 8.5 c€/kWh, compared to an average natural gas price of ~3.03 c€/kWh for non-household consumers in the first half of 2021 for the EU-27 [53]. The most relevant ordering of the SNG production costs is (in chronological order of contribution): electrical energy price>CAPEX> OPEX>CO_2 supply costs. The revenue generated by the electrolysis-side product, oxygen, which can be directly used in the combined industrial processes, has a beneficial cost effect. The savings in terms of CO_2 certificates reduce the SNG production costs further, but to a lesser extent than the oxygen. Higher CO_2 certificate prices of at least 330 €/t_{CO2} would be necessary to achieve cost parity, even in the long run. Based on a sensitivity analysis, the authors investigated the influence of cost parameters on the SNG production costs. As electrical energy costs are dominant, enhancing the energetic system efficiency results in the highest positive impact on the SNG costs. A 10% improvement in efficiency results in about a 10% reduction in SNG production costs. The second most sensitive influence parameter is the system lifetime, which only leads to a single-digit percentage cost reduction for a 25% lifetime increase [52].

Therefore, the overall energetic efficiency of SNG production in a combined co-SOEC and methanation plant that uses electric energy is critical to its economic success. As the chemical equations show, the methanation process releases significant quantities of reaction heat (equation (15.1) and equation (15.2)). This thermal energy can be used to preheat and evaporate water before it is fed to the co-SOEC system. Since the temperature range of the co-SOEC (600 °C–850 °C [54]) exceeds the maximum methanation cooling temperature (300 °C [9]), only preheating and evaporation of the co-SOEC feed water is possible. However, the excess methanation heat can be used to cover part of the co-SOEC energy demand. Posdziech *et al* found efficiencies for a SOEC system based on a lower heating value (LHV) of 60.5% with a steam generator and compression, 72% if no steam generator was necessary due to waste steam integration, and 84.5% without a steam generator or a compressor. Based on the higher heating value (HHV) the authors found 84.4% system efficiency with heat integration [55]. In agreement with these results, Ancona *et al* reported efficiencies of between 80% and 85% for a combined co-SOEC and methanation system based on the HHV [54]. Wang *et al* targeted realistic HHV efficiencies for SOEC-based power-to-methane plants of between 70% and 75% [56].

The beneficial synergy effect of a combined co-SOEC and methanation system can be concretely demonstrated using efficiency calculations. Equation (15.10) gives the principal system efficiency of a combined co-SOEC and methanation PtG plant.

$$\eta_{\text{PtG}} = \eta_{\text{co-SOEC}} \cdot \eta_{\text{compression}} \cdot \eta_{\text{methanation}}$$

$$= \frac{P_{l,\text{syngas}}}{P_{\text{el,co-SOEC}}} \cdot \frac{P_{l,\text{syngas}}}{P_{l,\text{syngas}} + P_{\text{el,compr.}}} \cdot \frac{P_{l,\text{SNG}}}{P_{l,\text{syngas}}}$$

$$= \frac{H_{l,\text{syngas}}\,\dot{m}_{\text{syngas}}}{P_{\text{el,co-SOEC}}} \cdot \frac{H_{l,\text{syngas}}\,\dot{m}_{\text{syngas}}}{H_{l,\text{syngas}}\,\dot{m}_{\text{syngas}} + P_{\text{el,compr.}}} \cdot \frac{H_{l,\text{SNG}}\,\dot{m}_{\text{SNG}}}{H_{l,\text{syngas}}\,\dot{m}_{\text{syngas}}} \tag{15.10}$$

For an example case of a 10 MW PtG unit, we now calculate the resulting LHV system efficiency and SNG output with and without heat integration of a combined co-SOEC and methanation plant in detail. The unit capacity is based on the co-SOEC input power, which can be a combination of electric and thermal energy. Losses from electricity conversion or CO_2 removal from off-gases are not considered. Table 15.3 gives the nomenclature of all the calculation variables. Following the published efficiency values mentioned above for SOEC systems, a co-SOEC efficiency of 73% (which does not include waste heat integration or compression) was considered in the calculations. As a result, 975.8 kg/h syngas can be produced from 10 MW of power input, as demonstrated by equation (15.11), providing a syngas concentration of 76.8 vol% of H_2, 18.3 vol% of CO, and 4.9 vol% of CO_2 on a dry basis.

$$\dot{m}_{\text{syngas}} = \frac{\eta_{\text{co-SOEC}}\,P_{\text{in,co-SOEC}}}{H_{l,\text{syngas}}} = \frac{\eta_{\text{co-SOEC}}\,P_{\text{in,co-SOEC}}}{\frac{1}{M_{\text{mix,syngas}}}\left(H_{l,\text{H}_2}\,y_{\text{H}_2}\,M_{\text{H}_2} + H_{l,\text{CO}}\,y_{\text{CO}}\,M_{\text{CO}}\right)}$$

$$= \frac{0.73 \cdot 10 \cdot 10^6[\text{W}]}{\frac{1}{8.83\left[\frac{\text{g}}{\text{mol}}\right]}\left(119972\left[\frac{\text{J}}{\text{g}}\right] \cdot 0.768 \cdot 2.02\left[\frac{\text{g}}{\text{mol}}\right] + 10103\left[\frac{\text{J}}{\text{g}}\right] \cdot 0.183 \cdot 28.01\left[\frac{\text{g}}{\text{mol}}\right]\right)}$$

$$= 0.271\left[\frac{\text{kg}}{\text{s}}\right] = 975{,}8\left[\frac{\text{kg}}{\text{h}}\right] \tag{15.11}$$

Before it arrives at the methanation unit, the syngas needs to be compressed. For a given pressure increase and gas mass flow, the electrical power consumption of the compressor can be determined. The resulting temperature of isentropic compression due to the pressure increase from 1.1 bar to 10 bar is demonstrated in equation (15.12).

Table 15.3. Nomenclature.

$E_{a,j}$	J mol^{-1}	Activation energy of reaction j
$c_{p,mix}$	J K^{-1} mol^{-1}	Heat capacity of the syngas mixture
$GHSV_{STP}$	h^{-1}	Gas hourly space velocity
$H_{l,CO}$	kJ kg^{-1}	Lower heating value of CO at stp
H_{l,H_2}	kJ kg^{-1}	Lower heating value of H_2 at stp
$H_{l,SNG}$	kJ kg^{-1}	Lower heating value of SNG at stp
$H_{l,syngas}$	kJ kg^{-1}	Lower heating value of syngas at stp
K_i	bar$^{-a,b,c}$	Adsorption constant for gas species i
$K_{i,0}$	bar$^{-a,b,c}$	Pre-exponential factor of adsorption
k_j	varying unit	Reaction rate coefficient of reaction j
$k_{j,0}$	varying unit	Pre-exponential factor of reaction rate coefficient
K_{p,CO_2}-Meth	bar^{-2}	Equilibrium constant
M_{CO}	g mol^{-1}	Molar mass of CO
M_{H_2}	g mol^{-1}	Molar mass of H_2
$M_{mix,syngas}$	g mol^{-1}	Molar mass of syngas mixture
$\dot{m}_{SNG}$	kg s^{-1}	Mass flow of SNG
$\dot{m}_{syngas}$	kg s^{-1}	Mass flow of syngas
$\dot{n}_{CO}$	mol s^{-1}	Mole flow of CO
$\dot{n}_{CO_2}$	mol s^{-1}	Mole flow of CO_2
$P_{el,compr}$	W	Electric power consumed by compressor
$P_{el,Co-SOEC}$	W	Electric power consumed by Co-SOEC
p_i	bar	Partial pressure of gas species i
p_{in}	bar	Low pressure level at compressor input
$P_{isentr.}$	W	Power consumption for isentropic compress.
$P_{l,syngas}$	W	Calorific power of syngas based on lower heating value
$P_{l,SNG}$	W	Calorific power of SNG based on lower heating value
p_{out}	bar	High pressure level at compressor output
$P_{reaction\ heat}$	W	Thermal power by methanation reaction heat
p_{STP}	bar	Standard pressure
R	J K^{-1} mol^{-1}	Ideal gas constant
r_j	mol kg^{-1} s^{-1}	Reaction rate of reaction j
T	K	Temperature
T_{in}	K	Temperature at compressor input
$T_{out,isentr.}$	K	Temperature at compressor output for isentropic compression
T_{STP}	K	Standard temperature
U_{CO}		CO conversion
U_{CO_2}		CO_2 conversion
$V_{catalyst}$	m^3	Catalyst bed volume
$\dot{V}_{in}$	m^3 h^{-1}	Input operating volume flow
y_{CO}		Mole fraction of CO in syngas
y_{H_2}		Mole fraction of H_2 in syngas
$\Delta H_{ads,i}$	J mol^{-1}	Enthalpy of adsorption for gas species i
$\Delta H^0_{r,COM}$	J mol^{-1}	Reaction enthalpy of CO methanation (STP)

$\Delta H^0_{r,CO_2M}$	J mol^{-1}	Reaction enthalpy of CO_2 methanation (STP)
$\eta_{compression}$		Efficiency of compression
$\eta_{co-SOEC}$		Efficiency of co-SOEC
$\eta_{heat\ transf.}$		Efficiency of heat transfer from methanation to co-SOEC
$\eta_{isentr.compr.}$		Efficiency of isentropic compression of gas
$\eta_{methanation}$		Efficiency of methanation system
η_{PtG}		Efficiency of power to gas system

For a mass flow of 975.8 kg h^{-1} of process gas, the isentropic compression power is calculated using equation (15.13). The electrical power consumed by the compressor, assuming an isentropic compression efficiency of 65% (not to be confused with the compression efficiency of the PtG system), can be expressed as shown in equation (15.14). Equation (15.15) can be used to determine the contribution of the compression efficiency to the overall PtG efficiency (as used in equation (15.10)).

$$T_{out,isentr.} = T_{in} \cdot \exp\left(\frac{R}{c_{p,\,mix}} \cdot \ln\left(\frac{p_{out}}{p_{in}}\right)\right)$$
$$= 298.15[K] \cdot \exp\left(\frac{8.314\left[\frac{J}{K\ mol}\right]}{29.3\left[\frac{J}{K\ mol}\right]} \cdot \ln\left(\frac{10[bar]}{1.1[bar]}\right)\right) = 556.3[K] \tag{15.12}$$

$$P_{isentr.} = \frac{\dot{m}_{syngas}}{M_{mix,\ syngas}} c_{p,\,mix}\,(T_{out,\ isentr.} - T_{in})$$
$$= \frac{0.271\left[\frac{kg}{s}\right]}{8.83 \cdot 10^{-3}\left[\frac{kg}{mol}\right]} 29.4\left[\frac{J}{K\ mol}\right](556.3[K] - 298.15[K]) \tag{15.13}$$
$$= 233.1\ [kW]$$

$$P_{el.\ compr.} = \frac{P_{isentr.}}{\eta_{isentr.\ compr.}} = \frac{233.1\ [kW]}{0.65} = 358.6[kW] \tag{15.14}$$

$$\eta_{compression} = \frac{H_{l,\ syngas}\,\dot{m}_{syngas}}{H_{l,\ syngas}\,\dot{m}_{syngas} + P_{el,\ compr.}}$$
$$= \frac{26932\left[\frac{kJ}{kg}\right] \cdot 0.271\left[\frac{kg}{s}\right]}{26932\left[\frac{kJ}{kg}\right] \cdot 0.271\left[\frac{kg}{s}\right] + 358.6\ [kW]} = 0.953 \tag{15.15}$$

The loss of exergy in the methanation process is based on the exothermic heat released by the chemical reaction, which needs to be removed from the reactor system. At full conversion, the maximum possible amount of exothermic heat is produced. For full stoichiometric CO methanation, the thermodynamically determined efficiency based on the lower heating value is 80%; for stoichiometric CO_2 methanation, it is 83% [7]. The efficiency of co-SOEC syngas methanation is calculated as shown in equation (15.16).

$$\eta_{\text{methanation}} = \frac{H_{l,\,\text{SNG}}\,\dot{m}_{\text{SNG}}}{H_{l,\,\text{syngas}}\,\dot{m}_{\text{syngas}}} = \frac{H_{l,\,\text{syngas}}\,\dot{m}_{\text{syngas}} - P_{\text{reaction heat}}}{H_{l,\,\text{syngas}}\,\dot{m}_{\text{syngas}}}$$

$$= \frac{H_{l,\,\text{syngas}}\,\dot{m}_{\text{syngas}} - \dot{n}_{\text{CO}}\,\Delta H^0_{r,\,\text{COM}}\,U_{\text{CO}} - \dot{n}_{\text{CO}_2}\,\Delta H^0_{r,\,\text{CO}_2\text{M}}\,U_{\text{CO}_2}}{H_{l,\,\text{syngas}}\,\dot{m}_{\text{syngas}}}$$

$$= \frac{H_{l,\,\text{syngas}}\,\dot{m}_{\text{syngas}} - \frac{\dot{m}_{\text{syngas}}}{M_{\text{mix,syngas}}}\left(y_{\text{CO}}\,\Delta H^0_{r,\,\text{COM}}\,U_{\text{CO}} + y_{\text{CO}_2}\,\Delta H^0_{r,\,\text{CO}_2\text{M}}\,U_{\text{CO}_2}\right)}{H_{l,\,\text{syngas}}\,\dot{m}_{\text{syngas}}}$$

$$= \frac{26932\left[\frac{\text{kJ}}{\text{kg}}\right]\cdot 0.271\left[\frac{\text{kg}}{\text{s}}\right] - \frac{0.271\left[\frac{\text{kg}}{\text{s}}\right]}{8.83\cdot 10^{-3}\left[\frac{\text{kg}}{\text{mol}}\right]}\left(0.183\cdot 206\left[\frac{\text{kJ}}{\text{mol}}\right]\cdot 1 + 0.049\cdot 164\left[\frac{\text{kJ}}{\text{mol}}\right]\cdot 1\right)}{26932\left[\frac{\text{kJ}}{\text{kg}}\right]\cdot 0.271\left[\frac{\text{kg}}{\text{s}}\right]}$$

$$= 0.8078 \tag{15.16}$$

The efficiency values of co-SOEC, compression, and methanation without heat integration add up to an overall PtG efficiency of 56.2% according to equation (15.17).

$$\eta_{\text{PtG}} = \eta_{\text{co - SOEC}}\cdot\eta_{\text{compression}}\cdot\eta_{\text{methanation}}$$

$$= 0.73\cdot 0.953\cdot 0.808 = 0.562 \tag{15.17}$$

For heat-integrated PtG systems, the excess heat generated by the methanation system can be used to reduce the electrical energy input and produce the same amount of SNG. With heat integration, the system efficiency changes to equation (15.18).

$$\eta_{\text{PtG,integr}} = \eta_{\text{co-SOEC,integr}}\cdot\eta_{\text{compression}}\cdot\eta_{\text{methanation}}$$

$$= \frac{H_{l,\text{syngas}}\,\dot{m}_{\text{syngas}}}{P_{\text{in, co-SOEC}} - P_{\text{reaction heat}}}\cdot\eta_{\text{compression}}\cdot\eta_{\text{methanation}}$$

$$= \frac{7.3\cdot 10^6[\text{W}]}{10\cdot 10^6[\text{W}] - 1.4033\cdot 10^6[\text{W}]}\cdot 0.952\cdot 0.808 = 0.849\cdot 0.952\cdot 0.808$$

$$= 0.654 \tag{15.18}$$

An efficiency of 65.4% would therefore be possible for a co-SOEC and methanation PtG plant with heat integration, according to the assumptions used above; this efficiency is 9.2 percentage points higher than without heat integration. 1.4 MW of heat would be recycled back to the co-SOEC system, reducing the total electric energy demand for the electrolysis from 10 MW to 8.6 MW. The system efficiency could be increased even further if more waste heat from secondary industrial processes were integrated into the co-SOEC. According to Sapountzi *et al* [57], HHV efficiencies close to 100% are possible for SOECs with thermal integration.

As Böhm *et al* [52] found, system efficiency plays an essential role in reducing SNG costs. Accordingly, a sensitivity analysis was carried out to investigate the most important parameters that enhance the efficiency of PtG systems.

As shown in figure 15.6, heat integration plays a fundamental role in enhancing the efficiency of combined co-SOEC and methanation. The integration of 50% more heat, starting from a base value of the methanation reaction heat, leads to an increase of about six percentage points of system efficiency. Based on process flowcharts, the recycling of methanation reaction heat to the co-SOEC input is further explained in section 15.6. A reduction in CO/CO_2 methanation conversion or an increase in H_2 excess also improves the system efficiency, since more H_2 is simply passed through the methanation. However, the gas quality regulations for SNG grid injection require high methane and low hydrogen concentrations and therefore high conversion rates and moderate H_2 excesses. Furthermore, a reduction in compressor power leads to higher efficiencies, which is a promising approach for PtG enhancement. The compression energy can be reduced by the 'dual-pressure-level

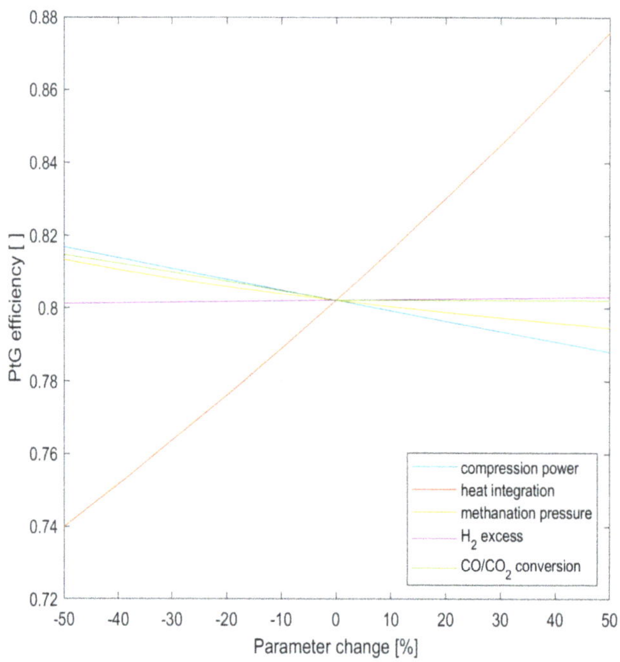

Figure 15.6. Sensitivity analysis of the overall co-SOEC and methanation PtG system.

methanation' concept further discussed in section 15.6. A reduction in methanation pressure by 40% from 10 bar to 6 bar led to an increase in overall efficiency of only one percentage point. Various other impacts such as methanation conversion and carbon deposition have to be considered with pressure changes.

15.6 Promising plant designs for efficient SNG production

The main technical benefit of combining endothermic co-SOEC with exothermic methanation lies in the possibility of transferring excess heat between these two process steps and therefore increasing the overall energy efficiency of the combined system. Using thermal integration, a combined co-SOEC and methanation system can convert 'green' electrical energy into methane at high LHV efficiencies of around 74%, based on the assumptions presented in section 15.5.

Figure 15.7 shows a possible plant design for a heat-integrated PtG system that includes co-SOEC and methanation. In addition to water, CO_2 from industrial sources is fed to the co-SOEC, producing a syngas mixture suitable for subsequent methanation. The water of the wet syngas needs to be condensed before it reaches the compressor. In order to reach a reactor pressure suitable for effective methanation at a low carbon deposition potential, the syngas is compressed to at least 4 bar, but usually up to 20 bar or higher. The methanation pressure level can be coordinated with the subsequent drying step and in particular with the necessary natural gas grid injection pressure. This PtG process design includes thermal integration, which is achieved by transferring methanation reaction heat to the co-SOEC water input. Thermal oil, pressurized boiling water, or molten salt can be used as cooling media in combination with a tube-bundle methanation reactor.

The compression of hydrogen and carbon monoxide requires gastight equipment that complies with increased safety standards. Furthermore, the position of the compressor significantly influences its compression capacity and, as a result, the overall plant efficiency. The compression of a higher volume flow of syngas at

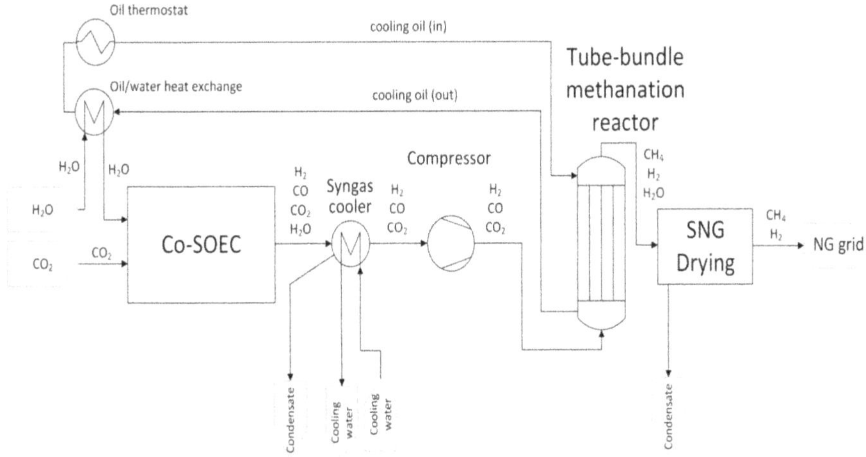

Figure 15.7. Possible plant design for a combined co-SOEC methanation PtG plant.

the co-SOEC outlet consumes a higher amount of electrical energy compared to other positions within the plant. Two alternative plant designs that result in reduced compression energy consumption are introduced below.

The 'dual-pressure-level methanation' concept is based on two sequential methanation reactors with intermediate water removal, as shown in figure 15.8. The compressor is placed between the first and the second reactor. As a result of a partial reaction in Reactor 1, less process gas needs to be compressed due to the volume-consuming reaction in the first reactor and intermediate water removal prior to compression. The use of intermediate compression can reduce the required compression capacity by up to 42% [43].

However, the carbon deposition risk within the second reactor is highly increased due to the removal of water. The tendency of the intermediate process gas to form solid carbon is shown in the ternary diagram of figure 15.4. Even at 10 bar of pressure, carbon deposition must be expected at certain temperatures. Higher pressures, higher hydrogen excesses, or excellent temperature control strategies would be necessary to reduce the carbon formation risk [43].

As an alternative to intermediate compression, CO_2 and H_2O can be fed at already increased pressures to the methanation system, as shown in figure 15.9. The benefit of this design for the overall process is reduced energy consumption and the technically simple and safe compression of liquid water and CO_2 compared to H_2- and CO-containing syngas. On the other hand, the co-SOEC needs to be operated under pressure. Brabandt and Poszdiech [58] overcame the mechanical problems of pressurized SOEC operation and successfully operated at 15 bar of pressure. The authors emphasized the efficiency benefits of the direct injection of pressurized steam from industrial processes and the combination with high-pressure downstream processes. Five percent to seven percent of the electrical energy demand

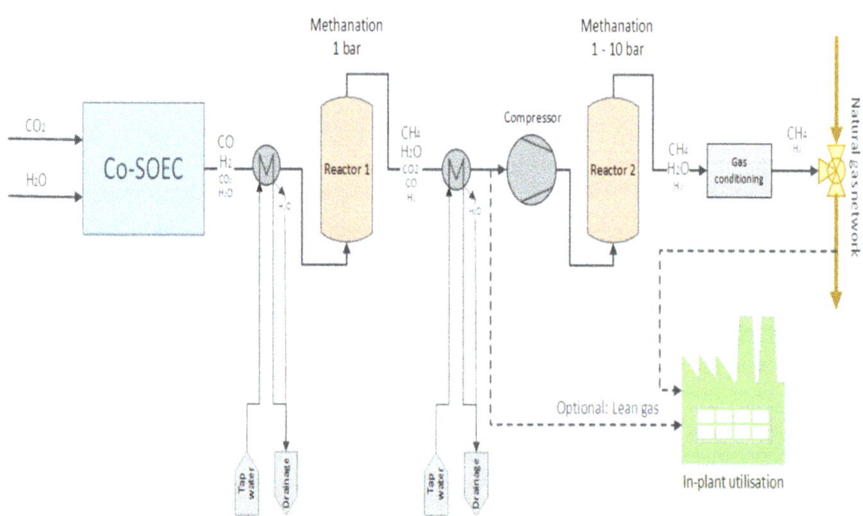

Figure 15.8. PtG plant design with dual-pressure-level methanation, including intermediate compression [43].

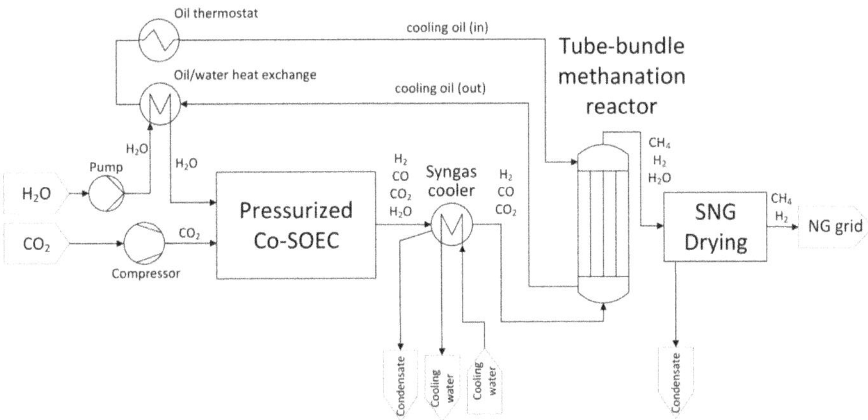

Figure 15.9. PtG process design including a pressurized co-SOEC.

can be saved by the direct provision of high-pressure steam instead of downstream hydrogen compression [58]. On the other hand, high-pressure (co-)SOEC is more complex due to the higher sealing effort required and suffers from increased cross permeation, higher temperatures, and a higher necessary operational voltage [58, 59].

In a fully integrated PtG plant, the direct integration of cooling water with the co-SOEC feed leads to the minimization of heat losses. Gruber *et al* [60] tried to directly couple a SOEC with a methanation system using steam from cooling water cycle as an SOEC feed. Unfortunately, it was not possible to control the steam mass flow to the extent necessary for full-load operation. However, the authors emphasized the positive effect on the overall efficiency. The use of water as a cooling medium would require pressures of more than 75 bar to guarantee a liquid cooling medium. Otherwise, the cooling water would evaporate, leading to a significant loss of heat transfer.

Although the main application scenarios of catalytic methanation will certainly be found in the area of industrial-scale energy provision, with the aim of preserving the living conditions on our globe, some potential purposes of this technology may go beyond the boundaries of planet Earth. An additional interesting aspect of the methanation process is its potential for long-term manned space missions. Hydrogen, a by-product of O_2 production by the electrolysis of water in order to provide air for breathing, can be used with exhaled CO_2 or CO_2 from Mars' atmosphere to produce methane and water by methanation. Water can then again be used for O_2 production and liquid methane can be used as propellant. As a result, closed-loop oxygen and water recycling is possible in space using methanation [61]. In 2011, a methanation system was installed at the International Space Station [62]. In 2012, a methanation system was demonstrated which was designed to produce methane on Mars from eighteen times lighter (and therefore cheaper) transportable hydrogen together with CO_2 harvested from Mars' atmosphere, leading to a significant reduction in transport expenses [63].

15.7 Concluding remarks

The combination of co-SOEC and methanation enables carbon dioxide to be recycled into a valuable natural gas substitute powered by renewable electrical energy. Co-SOEC syngas methanation benefits from reactor designs with intensified heat removal, such as tube-bundle reactors or microstructured reactors. Nickel catalysts offer high activity and high selectivity at a low price, whereas promoting agents, carrier materials, and form play substantial roles in the overall process. Both kinetic and thermodynamic considerations must be kept in mind not only to achieve the highest conversion rates, but also to prevent carbon deposition or catalyst sintering. Therefore, reactor temperature control is important for achieving low outlet temperatures which reduce the thermodynamic limitation and keep the maximum temperature just below the catalyst's limits, thus allowing high reaction rates. Furthermore, for the feed gas composition used for methanation, a moderate hydrogen excess should be considered to prevent carbon deposition and ensure high conversions. Additionally, SNG is produced at a sufficient output quality for direct grid injection in line with legal requirements and requires only a modest gas conditioning effort. Since the electric energy price has a dominant effect on the resulting SNG price, the overall system energy efficiency needs to be maximized. Heat integration from exothermal methanation or external industrial sources into the co-SOEC feed needs to be substantial to achieve high efficiencies. The reduction of compression power by dual-pressure-level methanation can contribute to economic competition against conventional energy carriers. In conclusion, for the successful realization of a combined co-SOEC and methanation facility, all system levels (catalyst, reactor, and plant design) should be optimized.

References

[1] Hashimoto K *et al* 1999 Global CO_2 recycling—novel materials and prospect for prevention of global warming and abundant energy supply *Mater. Sci. Eng.* A **267** 200–6

[2] Rönsch S, Schneider J, Matthischke S, Schlüter M, Götz M, Lefebvre J, Prabhakaran P and Bajohr S 2016 Review on methanation – from fundamentals to current projects *Fuel* **166** 276–96

[3] Lehner M, Tichler R, Steinmüller H and Koppe M (ed) 2014 *Power-to-Gas: Technology and Business Models SpringerBriefs in Energy* (Cham: Springer International Publishing)

[4] Sabatier P and Senderens J B 1902 New methane synthesis *J. Chem. Soc.* **82** 333

[5] Bartholomew C H 1982 Carbon deposition in steam reforming and methanation *Catalysis Reviews* **24** 67–112

[6] Kiewidt L and Thöming J 2015 Predicting optimal temperature profiles in single-stage fixed-bed reactors for CO_2-methanation *Chem. Eng. Sci.* **132** 59–71

[7] Neubert M 2019 *Catalytic Methanation for Small- and Mid-scale SNG Production* Friedrich-Alexander-Universität Erlangen-Nürnberg (FAU) https://opus4.kobv.de/opus4-fau/front-door/index/index/year/2020/docId/13118

[8] Guilera J, Boeltken T, Timm F, Mallol I, Alarcón A and Andreu T 2020 Pushing the limits of SNG process intensification: high GHSV operation at pilot scale *ACS Sustain. Chem. Eng.* **8** 8409–18

[9] Lefebvre J, Bajohr S and Kolb T 2020 Modeling of the transient behavior of a slurry bubble column reactor for CO_2 methanation, and comparison with a tube bundle reactor *Renew. Energy* **151** 118–36
[10] Schlautmann R, Böhm H, Zauner A, Mörs F, Tichler R, Graf F and Kolb T 2021 Renewable power-to-gas: a technical and economic evaluation of three demo sites within the STORE&GO project *Chem. Ing. Tech.* **93** 568–79
[11] Wulf C, Linßen J and Zapp P 2018 Review of power-to-gas projects in europe *Energy Proc.* **155** 367–78
[12] Younas M, Loong Kong L, Bashir M J K, Nadeem H, Shehzad A and Sethupathi S 2016 Recent advancements, fundamental challenges, and opportunities in catalytic methanation of CO_2 *Energy Fuels* **30** 8815–31
[13] Mills G A and Steffgen F W 1974 Catalytic methanation *Catal. Rev.* **8** 159–210
[14] Hu D, Gao J, Ping Y, Jia L, Gunawan P, Zhong Z, Xu G, Gu F and Su F 2012 Enhanced investigation of CO methanation over Ni/Al_2O_3 catalysts for synthetic natural gas production *Ind. Eng. Chem. Res.* **51** 4875–86
[15] Fan M-T, Miao K-P, Lin J-D, Zhang H-B and Liao D-W 2014 Mg–Al oxide supported Ni catalysts with enhanced stability for efficient synthetic natural gas from syngas *Appl. Surf. Sci.* **307** 682–8
[16] Qin H, Guo C, Wu Y and Zhang J 2014 Effect of La_2O_3 promoter on NiO/Al_2O_3 catalyst in CO methanation *Korean J. Chem. Eng.* **31** 1168–73
[17] Liu H, Zou X, Wang X, Lu X and Ding W 2012 Effect of CeO_2 addition on Ni/Al_2O_3 catalysts for methanation of carbon dioxide with hydrogen *J. Nat. Gas Chem.* **21** 703–7
[18] Liu Q, Gu F, Lu X, Liu Y, Li H, Zhong Z, Xu G and Su F 2014 Enhanced catalytic performances of Ni/Al_2O_3 catalyst via addition of V_2O_3 for CO methanation *Appl. Catal.,* A **488** 37–47
[19] Rönsch S, Köchermann J, Schneider J and Matthischke S 2016 Global reaction kinetics of CO and CO_2 methanation for dynamic process modeling *Chem. Eng. Technol.* **39** 208–18
[20] Koschany F, Schlereth D and Hinrichsen O 2016 On the kinetics of the methanation of carbon dioxide on coprecipitated $NiAl(O)_x$ *Appl. Catalysis* B **181** 504–16
[21] Kopyscinski J, Schildhauer T J, Vogel F, Biollaz S M and Wokaun A 2010 Applying spatially resolved concentration and temperature measurements in a catalytic plate reactor for the kinetic study of CO methanation *J. Catal.* **271** 262–79
[22] Coenen J, van Nisselrooy P, Croon M, de, van Dooren P and van Meerten R 1986 The dynamics of methanation of carbon monoxide on nickel catalysts *Applied Catalysis* **25** 1–8
[23] Ayers K E, Anderson E B, Capuano C, Carter B, Dalton L, Hanlon G, Manco J and Niedzwiecki M 2010 Research advances towards low cost, high efficiency PEM electrolysis *ECS Trans.* **33** 3–15
[24] Schlereth D and Hinrichsen O 2014 A fixed-bed reactor modeling study on the methanation of CO_2 *Chem. Eng. Res. Des.* **92** 702–12
[25] Try R, Bengaouer A, Baurens P and Jallut C 2018 Dynamic modeling and simulations of the behavior of a fixed-bed reactor-exchanger used for CO_2 methanation *AIChE J* **64** 468–80
[26] Gruber M 2020 Detaillierte Untersuchung des Wärme- und Stofftransports in einem Festbett-Methanisierungsreaktor für Power-to-Gas Anwendungen Dissertation Karlsruher Institut für Technologie (KIT) https://doi.org/10.5445/IR/1000105268
[27] Krier C, Hackel M, Hägele C, Urtel H, Querner C and Haas A 2013 Improving the methanation process *Chemie-Ingenieur-Technik* **85** 523–8
[28] Kai T, Takahashi T and Furusaki S 1988 Kinetics of the methanation of carbon dioxide over a supported $Ni–La_2O_3$ catalyst *Can. J. Chem. Eng.* **66** 343–7

[29] Martinez Molina M, Kern C and Jess A 2016 Catalytic hydrogenation of carbon dioxide to methane in wall-cooled fixed-bed reactors *Chem. Eng. Technol.* **39** 2404–15
[30] Schollenberger D, Bajohr S, Gruber M, Reimert R and Kolb T 2018 Scale-up of innovative honeycomb reactors for power-to-gas applications – The Project Store&Go *Chem. Ing. Tech.* **90** 696–702
[31] Weatherbee G and Barthomolew C 1982 Hydrogenation of CO_2 on group VIII metals II. kinetics and mechanism of CO_2 hydrogenation on nickel *J. Catal.* **77** 460–72
[32] Yang Lim J, McGregor J, Sederman A J and Dennis J S 2016 Kinetic studies of CO_2 methanation over a Ni/γ-Al_2O_3 catalyst using a batch reactor *Chem. Eng. Sci.* **141** 28–45
[33] Giglio E, Deorsola F A, Gruber M, Harth S R, Morosanu E A, Trimis D, Bensaid S and Pirone R 2018 Power-to-gas through high temperature electrolysis and carbon dioxide methanation: reactor design and process modeling *Ind. Eng. Chem. Res.* **57** 4007–18
[34] Xu J and Froment G F 1989 Methane steam reforming, methanation and water-gas shift: I. Intrinsic kinetics *AIChE J* **35** 88–96
[35] Ducamp J, Bengaouer A and Baurens P 2017 Modelling and experimental validation of a CO_2 methanation annular cooled fixed-bed reactor exchanger *Can. J. Chem. Eng.* **95** 241–52
[36] Parlikkad N R, Chambrey S, Fongarland P, Fatah N, Khodakov A, Capela S and Guerrini O 2013 Modeling of fixed bed methanation reactor for syngas production: operating window and performance characteristics *Fuel* **107** 254–60
[37] Kang W R and Lee K B 2013 Effect of operating parameters on methanation reaction for the production of synthetic natural gas *Korean J. Chem. Eng.* **30** 1386–94
[38] Klose J 1984 Kinetics of the methanation of carbon monoxide on an alumina-supported nickel catalyst *J. Catal.* **85** 105–16
[39] Zhang J, Fatah N, Capela S, Kara Y, Guerrini O and Khodakov A Y 2013 Kinetic investigation of carbon monoxide hydrogenation under realistic conditions of methanation of biomass derived syngas *Fuel* **111** 845–54
[40] Gruber M, Wiedmann D, Haas M, Harth S, Loukou A and Trimis D 2021 Insights into the catalytic CO_2 methanation of a boiling water cooled fixed-bed reactor: simulation-based analysis *Chem. Eng. J.* **406** 126788
[41] Sun D and Simakov D S 2017 Thermal management of a sabatier reactor for CO_2 conversion into CH_4: simulation-based analysis *J. CO2 Util.* **21** 368–82
[42] Kreitz B, Wehinger G and Turek T 2019 Dynamic simulation of the CO_2 methanation in a micro-structured fixed-bed reactor *Chem. Eng. Sci.* **195** 541–52
[43] Krammer A, Medved A, Peham M, Wolf-Zöllner P, Salbrechter K and Lehner M 2020 Dual pressure level methanation of Co-SOEC syngas *Energy Technol.* **9** 2000746
[44] Kusnezoff M, Megel S, Rix C, Adam P, Reichelt E, Herz G, Jahn M, Trofimenko N and Michaelis A 2019 Co-electrolysis CFY-stack operation and integration for carbon capture and utilization *ECS Trans.* **91** 2579–87
[45] Gantenbein A, Witte J, Biollaz S M, Kröcher O and Schildhauer T J 2021 Flexible application of biogas upgrading membranes for hydrogen recycle in power-to-methane processes *Chem. Eng. Sci.* **229** 116012
[46] ENTSOG, Gas Infrastructure Europe and Hydrogen Europe How to Transport and Store Hydrogen – Facts and Figures (ENTSOG, Gas Infrastructure Europe and Hydrogen Europe)
[47] GRTgaz, GRDF, elenegy, Geomethane, Teregea, REGAZ Bordaux, Stroengy and R-GDS 2019 Technical and economic conditions for injecting hydrogen into natural gas networks:

Final report June 2019 (GRTgaz, GRDF, elenegy, Geomethane, Teregea, REGAZ Bordaux, Stroengy and R-GDS)
[48] ÖVGW 2021 Richtlinie G B210: Gasbeschaffenheit 75.060
[49] DVGW 2021 G 260 Arbeitsblatt 09/2021: Gasbeschaffenheit (310700)
[50] El Sibai A, Rihko Struckmann L K and Sundmacher K 2017 Model-based optimal sabatier reactor design for power-to-gas applications *Energy Technol* **5** 911–21
[51] Lefebvre J, Bajohr S and Kolb T 2019 A comparison of two-phase and three-phase CO_2 methanation reaction kinetics *Fuel* **239** 896–904
[52] Böhm H, Lehner M and Kienberger T 2021 Techno-economic assessment of thermally integrated co-electrolysis and methanation for industrial closed carbon cycles *Front. Sustain* **2** 726332
[53] Eurostat 2021 Natural gas price statistics: Natural gas prices for non-household consumers https://ec.europa.eu/eurostat/statistics-explained/index.php?title=Natural_gas_price_statistics#Natural_gas_prices_for_non-household_consumers (accessed 25 Oct 2021)
[54] Ancona M A, Antonucci V, Branchini L, Catena F, Pascale A, de, Di Blasi A, Ferraro M, Italiano C, Melino F and Vita A 2019 Thermal integration of a high-temperature co-electrolyzer and experimental methanator for power-to-gas energy storage system *Energy Convers. Manage.* **186** 140–55
[55] Posdziech O, Schwarze K and Brabandt J 2019 Efficient hydrogen production for industry and electricity storage via high-temperature electrolysis *Int. J. Hydrogen Energy* **44** 19089–101
[56] Wang L, Pérez-Fortes M, Madi H, Diethelm S, van herle J and Maréchal F 2018 Optimal design of solid-oxide electrolyzer based power-to-methane systems: a comprehensive comparison between steam electrolysis and co-electrolysis *Appl. Energy* **211** 1060–79
[57] Sapountzi F M, Gracia J M, Weststrate C J, Fredriksson H O and Niemantsverdriet J W 2017 Electrocatalysts for the generation of hydrogen, oxygen and synthesis gas *Prog. Energy Combust. Sci.* **58** 1–35
[58] Brabandt J and Posdziech O 2017 System approach of a pressurized high-temperature electrolysis *ECS Trans.* **78** 2987–95
[59] Buttler A and Spliethoff H 2018 Current status of water electrolysis for energy storage, grid balancing and sector coupling via power-to-gas and power-to-liquids: a review *Renew. Sustain. Energy Rev.* **82** 2440–54
[60] Gruber M, Weinbrecht P, Biffar L, Harth S, Trimis D, Brabandt J, Posdziech O and Blumentritt R 2018 Power-to-gas through thermal integration of high-temperature steam electrolysis and carbon dioxide methanation – experimental results *Fuel Proc. Tech.* **181** 61–74
[61] Vogt C, Monai M, Kramer G J and Weckhuysen B M 2019 The renaissance of the Sabatier reaction and its applications on Earth and in space *Nat. Catal.* **2** 188–97
[62] NASA Content Administrator 2012 The Sabatier System: Producing Water on the Space Station (NASA) https://www.nasa.gov/mission_pages/station/research/news/sabatier.html
[63] Muscatello A and Santiago-Maldonado E 2012 Mars In Situ Resource Utilization Technology Evaluation *Mars In Situ Resource Utilization Technology Evaluation 50th AIAA Aerospace Sciences Meeting including the New Horizons Forum and Aerospace Exposition (Nashville, Tennessee)* American Institute of Aeronautics and Astronautics (Reston, VA: American Institute of Aeronautics and Astronautics)

IOP Publishing

High-Temperature Electrolysis
From fundamentals to applications
Werner Sitte and Rotraut Merkle

Chapter 16

CO_2 electrolysis

Christopher Graves, Theis L Skafte and Søren Højgaard Jensen

Solid oxide electrochemical cells (SOECs) are uniquely capable of direct electrolysis of CO_2 into CO and oxygen at high energy efficiencies (>90%) and high current densities (>1 A cm^{-2}). Powered by renewable or nuclear electricity, solid oxide CO_2 electrolyzers can sustainably produce CO and other valuable chemical products. While a solid oxide CO_2 electrolyzer is now reliably operating on Mars as of 2021, insufficient device lifetime remains an obstacle to widespread commercialization. This chapter provides an overview of the thermodynamics and reaction kinetics of CO_2 electrolysis, the history of solid oxide CO_2 electrolyzer development, degradation phenomena, and the major applications of this technology.

16.1 Introduction and fundamentals

A solid oxide, oxygen-ion-conducting electrochemical cell can flexibly electrolyze carbon dioxide, steam, or both simultaneously. Chapters 13, 14, 15, and 18 of this book focus on steam electrolysis for hydrogen production and the co-electrolysis of CO_2 and H_2O for syngas or methane production. This chapter focuses on the production of CO via direct high-temperature CO_2 electrolysis using SOECs.

It is worth briefly mentioning that CO_2 electrolysis can also be carried out via less direct reactions in alternative types of electrochemical cell—using a molten carbonate cell at a similarly high temperature to those used in SOECs or using aqueous or polymer electrolyte cells at ambient temperatures. The latter utilize H^+ or OH^- ions as charge carriers and therefore typically involve the simultaneous production of some hydrogen with the CO. Although high-temperature electrolyzers present some challenges in terms of lifetime and the increased system complexity required for heat management, their major advantage is that their measured cell and stack energy efficiency values are double those of the ambient-temperature systems. For example, SOECs typically operate at ~1.5 V, whereas ambient-temperature cells typically operate at ~3 V and lower current densities [1].

doi:10.1088/978-0-7503-3951-3ch16

The following sections provide an introduction to the thermodynamics, electrode kinetics, and cell performance of solid oxide CO_2 electrolyzers, followed by a summary of the development history of this technology.

16.1.1 Thermodynamics

The energy required for CO_2 electrolysis, i.e. the conversion of CO_2 to CO (and O_2) is presented in figure 16.1. The figure presents the total energy demand (ΔH, enthalpy) of the reaction as a function of temperature.

$$CO_2(g) \rightleftarrows CO(g) + \tfrac{1}{2}O_2(g) \tag{16.1}$$

The energy required for reaction (16.1) can be divided into the electrical energy demand (ΔG, Gibbs free energy) and the heat demand ($T\Delta S$, temperature times entropy). Since the entropy term ΔS is positive for reaction (16.1), the required heat increases with temperature. Hence, at elevated temperatures, an increasing fraction of the total energy demand is provided in the form of heat.

Figure 16.1 also plots the reaction energy demand in terms of voltage, calculated by reaction (16.2):

$$\Delta G^{\circ} = -nFE^{\circ}, \tag{16.2}$$

where n is the number of electrons involved in the reaction ($n = 2$ for direct reduction of CO_2 to CO), F is Faraday's constant, and E° is the reversible cell voltage under standard conditions, i.e. the minimum cell potential required for CO_2 splitting. At 25 °C, ΔG° is 257 kJ mol^{-1}, corresponding to a reversible voltage of 1.33 V. At 700 °C, ΔG° is only 198 kJ mol^{-1} or 1.03 V.

The E^0 values listed above have been calculated by assuming a CO_2/CO ratio of unity on the fuel side of the cell, and an oxygen partial pressure of one atmosphere

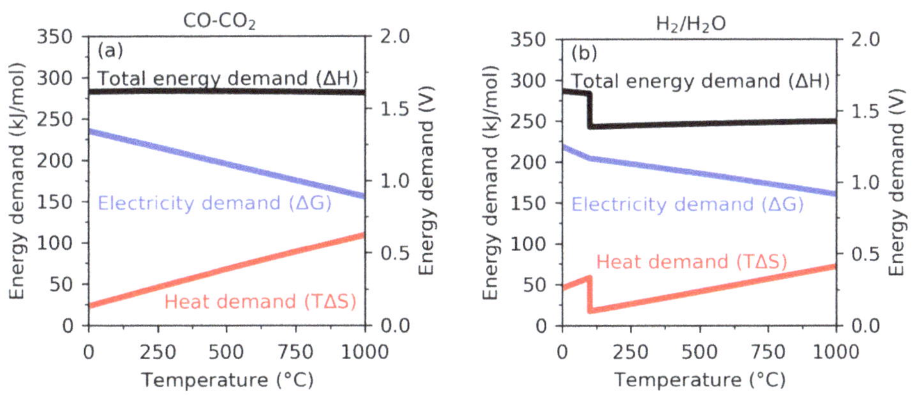

Figure 16.1. Thermodynamics of (a) CO_2 and (b) H_2O electrolysis. The energy requirement ΔH for CO_2 electrolysis is relatively independent of temperature. However, the electricity requirement ΔG drastically decreases with temperature. The thermodynamic data for H_2O electrolysis is shown for reference, and has very similar ΔG (and equivalent E) values across the whole temperature range.

on the oxygen side. The reversible cell voltage E depends on the reactants' partial pressures, as expressed by the Nernst equation, reaction (16.3).

$$E^0 = \Delta G/nF - RT/nF \ln(pCO_2/(pCO\, pO_2^{0.5})) \quad (16.3)$$

where R is the universal gas constant, T is the absolute temperature, pCO_2 is the partial pressure of CO_2 on the cathode, pCO is the partial pressure of CO on the cathode, and pO_2 is the partial pressure of oxygen on the anode side of the electrolysis cell. According to equation (16.3), when the CO_2 reduction reaction is carried out at a pressure of 10 bar (i.e. $p_{CO2} = p_{CO} = 5$ bar, $p_{O2} = 10$ bar), the reversible cell voltage increases slightly to 1.35 V at 25 °C and to 1.07 V at 700 °C. The higher voltage at the increased total pressure corresponds to the fact that reaction (16.1), which leads to a net production of gas molecules, is unfavorable at higher pressures. In practice, the local value of the thermodynamic parameters (ΔH, ΔG, $T\Delta S$) varies across the cell; it depends on local temperatures and partial pressures.

The minimum energy required to carry out CO_2 electrolysis, ΔH, can be provided to the cell by a combination of electricity and heat. If provided to the cell entirely by electricity, the cell must be operated at the thermoneutral voltage of 1.47 V ($\Delta H/nF$) at the typical operating temperature of about 700 °C. If the cell is operated exothermically, i.e. at more than 1.47 V, heat must be removed to prevent the temperature from rising. If the cell is operated endothermically, i.e. at less than 1.47 V, heat must be supplied externally to prevent the cell from cooling down. This means that operation at or below 1.47 V results in an electrical-to-chemical energy conversion efficiency of 100% for the cell. Today's cells and stacks are operated at less than 1.47 V per cell–at least in part to avoid carbon deposition (see section 16.2.1)—at practical current densities (see the next section and figure 16.2), and the necessary additional heat is

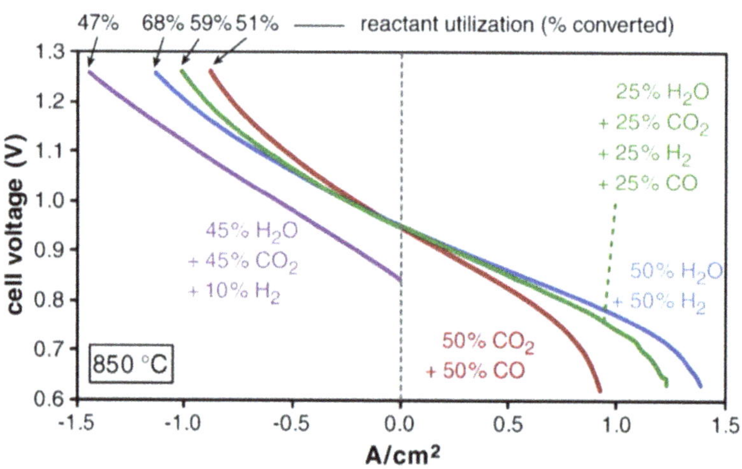

Figure 16.2. DC polarization (i–V) curves for the oxidation and reduction of chosen mixtures of H_2, H_2O, CO, and CO_2. Due to improvements in electrode materials and cell design, today's cell performance at lower temperatures (e.g. 700 °C–750 °C) is comparable to or better than these 850 °C measurements from more than a decade ago. Reprinted from [4] with permission, copyright Elsevier.

supplied externally. Additional system losses (due to e.g. heat transfer and the flow of the the reactants) result in a lower total system efficiency of up to 90%. Note that this high efficiency is rather unique to high-temperature electrolysis (of either CO_2 or H_2O), as low-temperature electrolysis typically results in impractically small current densities (leading to high capital costs) when operating below the thermoneutral voltage, and the fuel-cell mode of operation of solid oxide cells is inherently less efficient than electrolysis mode operation due to the thermodynamics of the reactions in that direction of operation.

16.1.2 Electrode kinetics and cell performance

The electrode reactions for CO_2 electrolysis in an SOEC are:

$$\text{Cathode: } CO_2 + 2e^- \rightarrow CO + O^{2-} \tag{16.4}$$

$$\text{Anode: } O^{2-} \rightarrow \tfrac{1}{2} O_2 + 2e^- \tag{16.5}$$

Compared with H_2O electrolysis, the CO_2 electrolysis reaction has slower reaction kinetics on typical SOEC cathodes made of composites of nickel and stabilized zirconia (Ni–SZ) or nickel and doped ceria (Ni–DC) [2–5]. The reaction mechanisms have been studied in model systems and in modeling [6, 7]. Model electrode measurements have provided some evidence that impurities that partially block electrochemical reaction sites may play a role in the differing electrode kinetics. Doped ceria and other alternative materials to the commonly used nickel-based electrodes present further variability in kinetics and reaction mechanisms [5]. In addition, the diffusion of CO/CO_2 gas mixtures present in the porous cathode during CO_2 electrolysis is slower than that of the H_2/H_2O gas mixtures present during H_2O electrolysis, which is independent of electrode material.

However, the impact of the higher electrode resistance and slower gas diffusion on the total cell performance during CO_2 electrolysis compared to that of H_2O electrolysis is, for many cells and operating conditions, not substantial—only a 20% decrease in current density, as shown in figure 16.2 [5]. The data used for figure 16.2 was obtained from measurements of a thin-electrolyte cell (~10 microns thick). If, instead, a thick-electrolyte, electrolyte-supported cell (commonly used in the SOEC field and with the same electrode performance) were to be used, the impact would be even less, because in this case the electrolyte's ohmic resistance would dominate the total cell resistance.

When nickel-based cathodes are used, part of the CO product gas must be recycled to the inlet and added to the CO_2 supply to prevent the nickel near the inlet from oxidizing, which occurs with a 100% CO_2 supply and results in damage due to the volume expansion of nickel to nickel oxide. Some alternative cathode materials, on the other hand, are able to operate with a 100% CO_2 supply, which can simplify SOEC system design [8].

16.1.3 History

The earliest reported work on high-temperature CO_2 electrolysis was funded by NASA in the early 1960s. The goal was to reduce CO_2 to CO and O_2 and utilize the O_2 for life support on extended space missions. The first projects aimed to develop, fabricate, and test small systems sized to provide respiratory support for the crew.

H. W. Chandler pointed out that CO_2 electrolysis is an ideal path for O_2 generation for space exploration, as no H_2O is involved, thereby eliminating liquid–gas separation in a zero-G environment [9]. Following early solid oxide CO_2 electrolyzer development by Weissbart and Smart [10], Elikan and Morris developed a reliable system that operated for more than 100 days in 1969 [11]. At this point, steam was added to the system as an inlet gas in addition to CO_2, which somewhat complicated the system by introducing the water–gas shift reaction and the need for a H_2 removal step. A 10YSZ electrolyte and Pt-based electrodes were used for five-cell stacks operating at 920 °C with a respectable current density of 0.2 A cm^{-2}. Research continued into the early 1970s, but then appeared to be abandoned at that point.

In the 1990s, research into high-temperature CO_2 electrolysis picked up again, as the 2001 Mars Surveyor Lander mission was being prepared [12, 13]. K. R. Sridhar from the University of Arizona led the research and after this Mars mission was canceled, Sridhar founded the solid oxide fuel cell (SOFC) company Bloom Energy in 2001. Bloom is currently the largest SOFC company in the world and has recently launched its first SOEC product, returning to its roots (although this SOEC product is used for steam electrolysis instead of CO_2 electrolysis).

In the 2000s and 2010s, the Risø National Laboratory in Denmark (which eventually became part of the Technical University of Denmark (DTU)) increased its research effort in the field of H_2O and CO_2 electrolysis, spurred on by the growing wind turbine research and market in Denmark [14]. Simultaneously, Ceramatec developed their SOFCs into SOECs and tested them for CO_2 electrolysis in collaboration with Idaho National Laboratory [3].

Much of the DTU know-how and technology was transferred to Haldor Topsoe A/S, which started the sister company Topsoe Fuel Cell A/S in 2004; this company focused on SOFC operation based on methane. When Topsoe Fuel Cell closed down in 2014, a small group of Haldor Topsoe engineers and technicians continued the work, but pivoted to CO_2 electrolysis. The same fuel-electrode-supported stack design was used to develop the eCOs™ system, which was the first commercial CO_2 electrolyzer; this electrolyzer produces high-quality CO with minor impurities, on site where the customer requires it [15]. This eliminates the hazardous, and thus expensive, transportation of CO gas from a centralized production plant. The first commercial demo plant was announced in 2016 and built in La Porte, Texas, USA. The second plant was announced in 2017 and was scaled to produce 96 $Nm^3\ h^{-1}$ of CO at a very high purity.

Also in 2014, a new NASA project was announced that led to the landing of a rover with a SOEC CO_2 electrolyzer system on Mars, thereby continuing the old

NASA missions from the 1960s and 1990s. See section 16.3.3 for more information about this project.

Research and development continues today; additional SOEC cell tests and systems are performing the co-electrolysis of CO_2 and H_2O for syngas production, such as those made by the German company Sunfire (see chapter 15 of this book).

16.2 Degradation

As mentioned earlier, the high operating temperatures of solid oxide CO_2 electrolyzers gives them major advantages in terms of energy efficiency and capital cost. Insufficient lifetime—which is mainly another result of the high temperature—is a challenge that inhibits the widespread commercialization of SOECs. However, device lifetime has been consistently improving over time in this field [16].

Two key degradation phenomena that are unique to CO_2 electrolysis (vs H_2O electrolysis) are carbon deposition and an extreme sensitivity to gas-phase impurities—both of which are most impactful with the commonly used Ni–SZ cathodes. Carbon deposition can be completely destructive and pull apart cell interfaces [17]. Gas-phase impurities (e.g. sulfur) that are present in trace quantities can play a critical role in gradual performance loss, especially in CO_2 electrolysis [18]. The following sections describe degradation by carbon deposition and impurity accumulation in more detail.

In addition to these unique degradation modes, other degradation phenomena, such as nickel migration, which are present during H_2O electrolysis are also commonly active during CO_2 electrolysis [19, 20]; see also chapter 4 of the book.

16.2.1 Carbon deposition

One of the most destructive failure mechanisms of high-temperature CO_2 electrolysis is carbon deposition. Operating with a mixture of CO and CO_2 at high temperature introduces the risk of crossing the equilibrium threshold of the Boudouard reaction:

$$2CO(g) \rightleftarrows CO_2(g) + C(s) \tag{16.6}$$

Given enough CO at high temperature, carbon can be deposited on any material, but since the most commonly utilized electrocatalyst for SOEC fuel electrodes, nickel, is also an excellent carbon deposition catalyst, this poses a unique challenge for high-temperature CO_2 electrolysis. This effectively limits the CO_2 conversion factor of this SOEC system. At higher temperatures, higher CO_2 utilization factors become possible, but the oxidation of steel components (e.g. interconnects) often limits the temperature at which the stack can run.

The thermodynamics of the Boudouard reaction (see figure 16.3(a)) dictates when carbon is likely to be deposited. When the CO concentration is increased beyond the threshold or the temperature is dropped sufficiently, carbon nanotubes (CNTs) start to grow from the Ni particles into the porosity of the electrode. CNTs are strong enough to push and pull the particles apart, and eventually the fuel electrode fractures (figure 16.3(b)), causing complete failure of the cell.

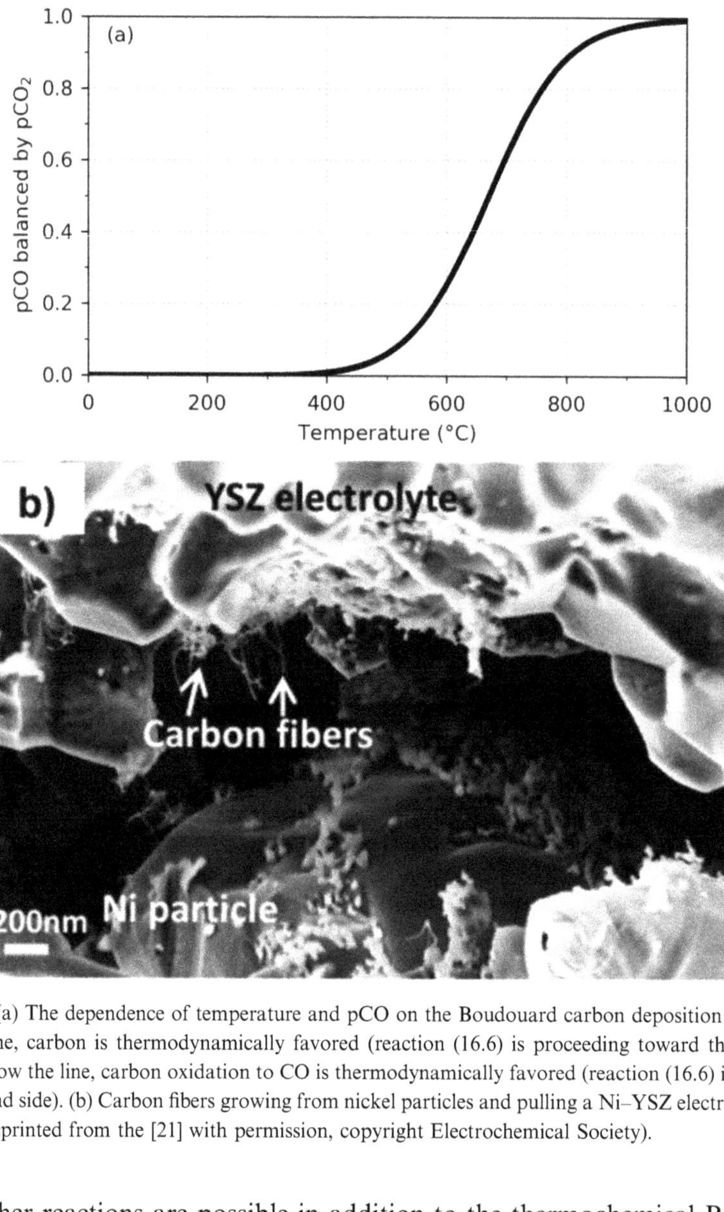

Figure 16.3. (a) The dependence of temperature and pCO on the Boudouard carbon deposition equilibrium. Above the line, carbon is thermodynamically favored (reaction (16.6) is proceeding toward the right-hand side), and below the line, carbon oxidation to CO is thermodynamically favored (reaction (16.6) is proceeding to the left-hand side). (b) Carbon fibers growing from nickel particles and pulling a Ni–YSZ electrode from the electrolyte (reprinted from the [21] with permission, copyright Electrochemical Society).

Two other reactions are possible in addition to the thermochemical Boudouard reaction, namely, the electrochemically driven reductions of CO_2 and CO [22, 23], as shown in equations (16.7) and (16.8), respectively. In these cases, the overvoltage on the electrode directly controls the deposition of carbon.

$$CO_2(g) + 4e^- \rightleftarrows 2O^{2-} + C(s) \tag{16.7}$$

$$CO(g) + 2e^- \rightleftarrows O^{2-} + C(s) \tag{16.8}$$

In addition to limiting the CO_2 utilization, dramatically increasing the risk of damaging the cell in the case of power failure or other operational errors, it has been shown that if only the average stack operating temperature and current density are considered, carbon deposition can occur earlier than one would expect. Due to the large gradients of temperature, current density, overpotential, and gas atmosphere inside the stack, the carbon threshold may be locally crossed in certain locations before one would expect [22, 24].

Numerous attempts have been made to improve the carbon tolerance of SOEC fuel electrodes [25], e.g. by replacing nickel with copper [26]. One of the more promising paths is to develop ceria-based electrodes, which have been shown to delay the carbon deposition onset and generally increase the carbon deposition tolerance, with and without the use of nickel [23].

16.2.2 Impurities

Another important degradation mechanism is poisoning of either of the electrodes. Impurities in the raw materials used to manufacture the cell or impurities in the gas streams can be deposited in the electrodes and block gas adsorption/desorption at the reaction sites. Some of the more notable impurities are sulfur and silica on the negative electrode and chromium on the positive electrode. Sulfur is a well-known poison for SOFCs and SOECs used to electrolyze water, but during CO_2 electrolysis it causes even more degradation, at least for conventional Ni-YSZ electrodes. As little as 10 ppb of H_2S or less in the CO/CO_2 gas stream can lead to additional degradation of Ni-YSZ electrodes [24], which means that CO_2 electrolysis imposes significantly more stringent requirements for inlet gas cleanup.

Due to the detrimental effect of impurities, the overpotential on the fuel electrode can increase so dramatically that it crosses the overpotential threshold for carbon deposition. Thus, operating on impure feed gases can directly lead to the complete failure of the cell and stack by triggering carbon deposition [19]. To improve the durability of an electrolysis system that fails due to carbon buildup, it is important to realize that the root cause of the failure may not be carbon deposition itself, but rather an impure gas stream.

16.3 Applications

16.3.1 Renewable CO_2-to-hydrocarbon fuels and other chemicals

Ultimately, the largest-scale potential application of CO_2 electrolysis is in the sustainable production of liquid and gaseous hydrocarbon fuels and other chemicals that can be used as drop-in replacements in the existing infrastructure and markets. This can be achieved by (i) CO_2 electrolysis and reacting the produced CO with H_2O; (ii) steam electrolysis and reacting the produced hydrogen with CO_2; (iii) the independent electrolysis of CO_2 and H_2O, which are mixed to obtain syngas; or (iv) single-step co-electrolysis of both CO_2 and H_2O in the same cell or stack [27]. Many system configurations have been examined for the production of different

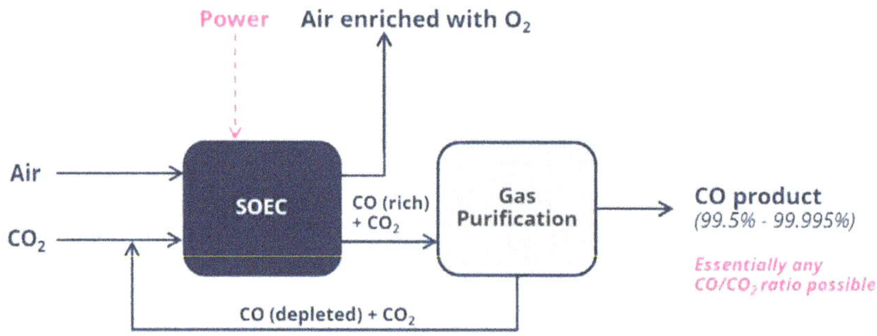

Figure 16.4. The conceptual design of the eCOs™ system (reprinted from [15] with permission, copyright Electrochemical Society).

hydrocarbon products. This CO_2-to-fuels concept, including the low-temperature electrolysis versions mentioned earlier, has been attracting more and more attention. For details, see chapters 14, 15, and 18.

16.3.2 Carbon monoxide production

Carbon monoxide is used in the specialty chemical industry and in the pharmaceutical industry [1]. The production of carbon monoxide is currently carried out at large centralized facilities; it is then transported to where it is needed. However, because carbon monoxide is toxic, the cost of handling and transporting it is high. SOECs can produce carbon monoxide from carbon dioxide with high efficiency and purity. Furthermore, since SOECs are modular, the production plant can be sized to the needs of each site. The heterogenous catalysis company, Topsoe A/S, seized the opportunity and was the first company to commercialize CO_2 electrolysis using SOECs. Their product, eCOs™, is a container-sized carbon monoxide production plant that primarily consists of a number of SOEC stacks and has the ability to produce carbon monoxide with a high purity of up to grade five (99.999%) (figure 16.4) [15].

16.3.3 Oxygen production on Mars (MOXIE)

Research into CO_2 electrolysis came full circle in the early 2020s by returning to the original objective, namely to produce O_2 for space exploration missions. Starting in 2014, NASA commenced a project to include an SOEC system on the Mars Rover scheduled for launch in 2020. The Martian atmosphere consists of 95% CO_2, which is an excellent opportunity to demonstrate the NASA concept of *in situ* Resource Utilization (ISRU), namely, the production of oxygen on the planet, as opposed to bringing it from Earth [28].

This would significantly reduce the required cargo. The oxygen used for propulsion outweighs the fuel by a factor of nearly four. The oxygen would initially be used for rocket propulsion for the return trip and in the longer term could also be utilized for humans to breathe.

In the Mars Oxygen *in situ* Resource Utilization Experiment (MOXIE), the objective was to demonstrate that the technology worked on Mars and to investigate any potential challenges for a future larger system. Ceramatec (now OxEon Energy, founded by a group of former Ceramatec people) produced the SOEC stack. MOXIE produced the first oxygen from CO_2 on another planet in April 2021 and has since gone on to operate nine more times as of July 2022 (figure 16.5).

Figure 16.5. (a) The Mars 2020 rover and (b) the MOXIE system (reprinted from [28] with permission, copyright Springer Nature).

16.4 Concluding remarks

High-temperature CO_2 electrolysis has seen tremendous development since the early 1960s and recently fulfilled the original purpose of producing oxygen in space and on another planet, intended for life support and for return trip propulsion. Other sustainable energy-driven applications on Earth are also in the process of being commercialized, e.g. the production of carbon monoxide and synthetic hydrocarbon fuels.

CO_2 electrolysis using SOEC still faces challenges related to lifetime and degradation, but the major degradation and failure mechanisms, such as carbon deposition and poisoning of the fuel electrode, have largely been understood. Universities and companies are working on how best to address these issues, for instance by replacing the cathode electrocatalyst, nickel. Overcoming these issues will leave SOEC technology with potentially the best prospects for cost-effective scale-up due to its advantageous high energy efficiency.

References

[1] Küngas R 2020 Review—electrochemical CO_2 reduction for CO production: comparison of low- and high-temperature electrolysis technologies *J. Electrochem. Soc.* **167** 044508

[2] Isenberg A O 1981 Energy conversion via solid oxide electrolyte electrochemical-cells at high-temperatures *Solid State Ionics* **3/4** 431–7

[3] Stoots C, O'Brien J and Hartvigsen J 2009 Results of recent high temperature coelectrolysis studies at the Idaho National Laboratory *Int. J. Hydrogen Energy* **34** 4208–15

[4] Graves C, Ebbesen S D and Mogensen M B 2011 Co-electrolysis of CO_2 and H_2O in solid oxide cells: performance and durability *Solid State Ionics* **192** 398–403

[5] Graves C, Chatzichristodoulou C and Mogensen M B 2015 Kinetics of CO/CO_2 and H_2/H_2O reactions at Ni-based and ceria-based solid-oxide-cell electrodes *Faraday Discuss.* **182** 75–95

[6] Shi Y, Luo Y, Cai N, Qian J, Wang S, Li W and Wang H 2013 Experimental characterization and modeling of the electrochemical reduction of CO_2 in solid oxide electrolysis cells *Electrochim. Acta* **88** 644–53

[7] Song Y, Zhang X, Xie K, Wang G and Bao X 2019 High-temperature CO_2 electrolysis in solid oxide electrolysis cells: developments, challenges, and prospects *Adv. Mater.* **1902033** 1–18

[8] Bidrawn F, Kim G, Corre G, Irvine J T S, Vohs J M and Gorte R J 2008 Efficient reduction of CO_2 in a solid oxide electrolyzer *Electrochem. Solid-State Lett.* **11** B167–70

[9] Chandler H W 1964 *Carbon dioxide reduction system* AMRL-TDR-64-42 Biomedical laboratory, Aerospace Medical Research Laboratories, Aerospace Medical Division, Air Force Systems Command, Wright-Patterson Air Force Base, Ohio see https://catalog.hathitrust.org/Record/009881701

[10] Weissbart J and Smart W H 1967 Study of electrolytic dissociation of CO_2–H_2O using a solid oxide electrolyte *NASA Contractor Report* **680** 1–92 https://ntrs.nasa.gov/api/citations/19670007847/downloads/19670007847.pdf

[11] Elikan L and Morris J P 1969 Solid electrolyte system for oxygen regeneration *NASA Contractor Report* **1359** 1-183 https://ntrs.nasa.gov/api/citations/19690018721/downloads/19690018721.pdf

[12] Tao G, Sridhar K R and Chan C L 2004 Study of carbon dioxide electrolysis at electrode/electrolyte interface: part I. Pt/YSZ interface *Solid State Ionics* **175** 615–9
[13] Tao G, Sridhar K R and Chan C L 2004 Study of carbon dioxide electrolysis at electrode/electrolyte interface: Part II. Pt-YSZ cermet/YSZ interface *Solid State Ionics* **175** 621–4
[14] Jensen S H, Larsen P H and Mogensen M B 2007 Hydrogen and synthetic fuel production from renewable energy sources *Int. J. Hydrogen Energy* **32** 3253–7
[15] Küngas R, Blennow P, Heiredal-Clausen T, Holt T, Rass-Hansen J, Primdahl S and Bøgild J H 2017 eCOs – a commercial CO_2 electrolysis system developed by haldor topsoe *ECS Trans.* **78** 2879–84
[16] Hauch A, Küngas R, Blennow P, Hansen A B, Hansen J B, Mathiesen B V and Mogensen M B 2020 Recent advances in solid oxide cell technology for electrolysis *Science* **370** eaba6118
[17] Tao Y, Ebbesen S D and Mogensen M B 2016 Degradation of solid oxide cells during co-electrolysis of steam and carbon dioxide at high current densities *J. Power Sources* **328** 452–62
[18] Ebbesen S D, Graves C, Hauch A, Jensen S H and Mogensen M B 2010 Poisoning of solid oxide electrolysis cells by impurities *J. Electrochem. Soc.* **157** B1419–29
[19] Hauch A, Traulsen M L, Küngas R and Skafte T L 2021 CO_2 electrolysis – gas impurities and electrode overpotential causing detrimental carbon deposition *J. Power Sources* **506** 230108
[20] Skafte T L *et al* 2022 Electrothermally balanced operation of solid oxide electrolysis cells *J. Power Sources* **523** 231040
[21] Tao Y, Ebbesen S D and Mogensen M B 2014 Carbon deposition in solid oxide cells during co-electrolysis of H_2O and CO_2 *J. Electrochem. Soc.* **161** F337–43
[22] Navasa M, Frandsen H L, Skafte T L, Sundén B and Graves C 2018 Localized carbon deposition in solid oxide electrolysis cells studied by multiphysics modeling *J. Power Sources* **394** 102–13
[23] Skafte T L *et al* 2019 Selective high-temperature CO_2 electrolysis enabled by oxidized carbon intermediates *Nat. Energy* **4** 846–55
[24] Skafte T L, Blennow P, Hjelm J and Graves C 2018 Carbon deposition and sulfur poisoning during CO_2 electrolysis in nickel-based solid oxide cell electrodes *J. Power Sources* **373** 54–60
[25] Boldrin P, Ruiz-Trejo E, Mermelstein J, Bermedez Menendez J M, Ramirez Reina T and Brandon N P 2016 Strategies for carbon and sulfur tolerant solid oxide fuel cell materials, incorporating lessons from heterogeneous catalysis *Chem. Rev.* **116** 13633–84
[26] Park S, Vohs J M and Gorte R J 2000 Direct oxidation of hydrocarbons in a solid-oxide fuel cell *Nature* **404** 265–7
[27] Graves C, Ebbesen S D, Mogensen M B and Lackner K S 2011 Sustainable hydrocarbon fuels by recycling CO_2 and H_2O with renewable or nuclear energy *Renew. Sustain. Energy Rev.* **15** 1–23
[28] Hecht M *et al* 2021 Mars oxygen ISRU experiment (MOXIE) *Space Sci. Rev.* **217** 9

IOP Publishing

High-Temperature Electrolysis
From fundamentals to applications
Werner Sitte and Rotraut Merkle

Chapter 17

Power-to-ammonia for fertilizers, chemicals, and as an energy vector

John Bøgild Hansen

The use of ammonia as an energy vector in sustainable energy scenarios has gained considerable attention lately, because ammonia is carbon free and can be produced from air, water, and renewable electricity. This chapter briefly reviews the literature on ammonia manufacture based on electrolysis, with a special emphasis on high-temperature electrolysis using solid oxide electrolysis cells (SOECs). Considerable energy savings can be achieved using SOECs compared to classical low-temperature electrolysis, due to the inherent high energy efficiency of SOEC technology and the synergy achieved by using the reaction heat from the ammonia synthesis to generate steam as a feedstock for the SOEC. Replacing fossil fuels with renewable ammonia will require an unprecedented expansion of renewable power production, electrolysis deployment, and ammonia synthesis technology.

17.1 Introduction

Ammonia is the second most common chemical produced in the world today. At a capacity of around 180 million tons per year, this amounts to 1% of global energy consumption as well as greenhouse gas emissions. Ammonia is mainly used as a fertilizer but also as a feedstock for a large number of basic chemicals. It is less well known that ammonia is also an excellent fuel. It is transported and stored in large quantities in liquid form at −33 °C (40 000 ton storage tanks are standard, corresponding to 206 GWh). In fossil-free future energy scenarios in which large proportions of power production are obtained using intermittent renewable sources such as wind and solar energy, storage and sector coupling have been identified as crucial enabling technologies [1]. One prominent sector in which ammonia could play a key role as an energy vector is global shipping. The International Maritime Organization (IMO) has pledged to reduce greenhouse gas emissions by 50% in 2050 compared to a 2008 baseline [2]. Shipping consumes approximately 3% of the world's energy consumption, mainly in the form of heavy oil products. Replacement of this fuel by carbon-based renewable fuels would require a large share of the biomass resources, or the implementation of direct air capture of CO_2. Both the International Energy Agency (IEA) and the Japanese government have also identified ammonia as a promising vector for the transport of carbon-free energy from areas with abundant, cheap, renewable power to resource-scarce, industrialized parts of the world.

17.2 Ammonia's properties

The volumetric energy density is of ammonia is higher than that of the carbon-free energy vector hydrogen, and the logistics of its handling is much simpler. Table 17.1 shows pertinent data for different fuels.

Ammonia is toxic and should be treated with respect, but safety analyses both by Risø and in the US have shown that it is no more dangerous than conventional fuels when handled professionally. The ammonia industry has an excellent safety record [3, 4].

17.3 Conventional ammonia production today

An authoritative description of ammonia production processes up until 2012 can be found in the very detailed work by Dybkjær and Appl [5–7]. Today, ammonia synthesis gas is predominantly generated by the steam reforming of natural gas, leading to an emission of 1.6–1.8 tons of CO_2 per ton of ammonia for state-of-the-art plants. A typical layout of such a plant is shown in figure 17.1.

The natural gas is preheated and thoroughly desulfurized before entering a fired tubular steam reformer. The nitrogen needed for ammonia synthesis is added in the form of air in a secondary reformer consisting of a burner followed by an adiabatic catalytic bed in a ceramic-lined steel vessel. High-pressure steam is generated in a waste heat boiler after the secondary reformer before the carbon monoxide content of the synthesis gas is converted to hydrogen by a water–gas shift in two reactors at different temperature levels. The majority of the carbon dioxide is removed in a chemical or physical wash, producing high-purity CO_2, which is commonly converted with the produced ammonia into urea, an important fertilizer. After the CO_2 wash, the last traces of carbon oxides are converted into methane because the

Table 17.1. Comparison of chemical electricity storage options.

Fuel	LHV (kWh kg^{-1})	Density (MWh m^{-3})	Temperature (°C)	Pressure (bar)
Ammonia	5.16	3.53	−33 or 20	1 or 10
Hydrogen, compressed	33.32	2.1	20	700
Hydrogen, liquid	33.32	2.4	−253	1
Methane, liquid	13.89	6.50	−162	1
Methanol	5.53	4.39	20	1
Marine gas oil	11.9	10.2	Ambient	Ambient
Very low sulfur fuel oil	11.2	10.6	Ambient	Ambient

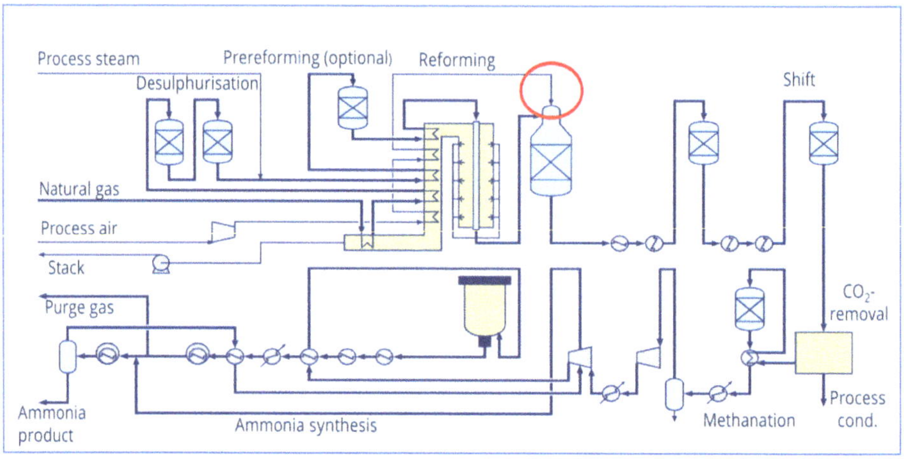

Figure 17.1. Typical ammonia plant layout based on natural gas feedstock. The red circle indicates where air is added to the secondary reformer.

ammonia synthesis catalyst is very sensitive to any oxygen compounds. The clean, stoichiometric synthesis gas is then compressed and sent to the ammonia synthesis recycle loop at production capacities of up to several thousand tons per day. The ammonia product is separated out as a liquid before the unconverted gas is recycled back to the ammonia converter.

The energy consumption of ammonia production from natural gas has dramatically decreased in the last 50 years to 28–33 GJ per metric ton and is now close to the thermodynamic limit of 18.6 GJ ton^{-1} (5.16 MWh ton^{-1}). Care should be exercised when evaluating efficiency; the actual thermodynamic limit is not the lower heating value, as ammonia is normally delivered in liquid form at −33 °C and the transformation from gas to liquid requires 0.9 GJ ton^{-1} (0.25 MWh ton^{-1}) of extra energy. This is equivalent to extra exergy in the product, which can be recuperated in e.g. an expander. It is often mentioned that ammonia production is very energy intensive, which is true, but most of the energy (>94%) used is spent in generating the synthesis gas, a stoichiometric 3:1 mixture of H_2/N_2. This fact also implies that the scope for saving energy by developing more active ammonia catalysts or processes is limited.

Ammonia produced in this way is termed 'gray' ammonia as opposed to 'black' ammonia produced from coal, which is the dominant technology in China. Approximately one quarter of the world's ammonia is produced in China, with a concomitant emission of up to three tons of CO_2 per ton of ammonia. However, it can also be sequestered by carbon capture and storage (CCS). The flue gas from the fired reformer is, however, diluted with uncombusted air and is at low pressure.

Recently, developments in autothermal reformers for gas–to–liquid (GtL) and methanol production have been applied to ammonia production. A flow scheme used by Haldor Topsøe is shown in figure 17.2.

In this layout, the primary reformer has been eliminated and replaced by an autothermal reformer that uses oxygen from an air separation unit, which also

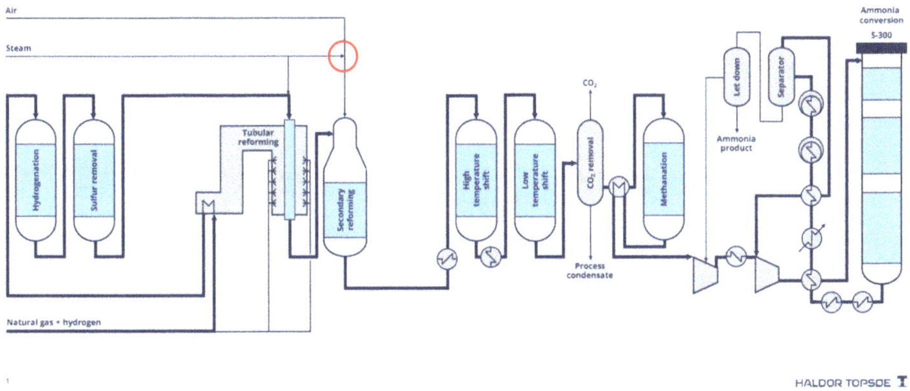

Figure 17.2. Ammonia production using Topsoe two-step reforming.

supplies nitrogen for the Haber–Bosch synthesis. The overall energy consumption is reduced by 3% compared to the classical layout of figure 17.1. Provided that the electric power used for air separation etc. is obtained from renewable sources, the overall CO_2 emissions can be reduced by up to 30%, and most of the emissions are in concentrated form ready for CCS.

This type of ammonia production is termed 'blue' and could be a serious competitor to an all electrical ammonia production, especially during a transition period. This concept can be taken even further in a hybrid configuration in which part of the hydrogen is produced by electrolyzers. This concept also offers several synergies for revamping or streamlining existing ammonia assets. The energy intensity and OPEX using classical, alkaline electrolysis will, however, be higher compared to natural gas based processes.

17.4 Electrified ammonia plant based on low-temperature electrolysis

In the past, the electrolysis of water by alkaline electrolysis coupled with an air separation plant was also used to produce ammonia. Interest in this technology is re-emerging, as concerns about CO_2 emissions are increasing at the same time as the price of electricity from intermittent, renewable sources decreases. Such a plant, suitable for a production capacity of 1000 tons per day, is illustrated in figure 17.3 [8].

The total power consumption of alkaline or polymer-electrolyte membrane (PEM)-electrolysis-based plants is about 10–10.5 MWh ton^{-1} ammonia due to the relatively low efficiency of the electrolysis step. The nitrogen used in the ammonia synthesis is provided by air separation. The preferred air separation units for large capacities are based on cryogenic separation. These plants are rather energy efficient, but very expensive. The energy consumed by the production of the nitrogen needed for one ton of ammonia is around 200 kWh [6, 7]. Such plants do not scale

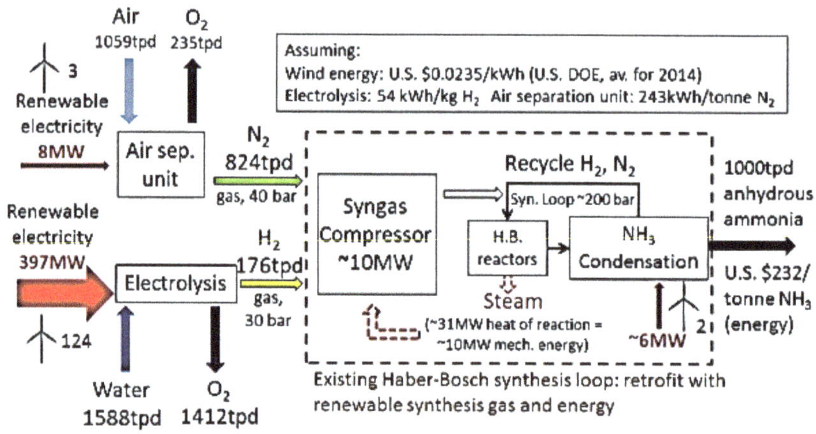

Figure 17.3. Hydrogen generation governs the energy demand for the synthesis of 1000 tons per day (tpd) of renewable ammonia in a 'bolt-on' concept. Renewable cryogenic N_2 and renewable H_2 from water electrolysis are fed to an existing Haber–Bosch synthesis loop. Electrolysis is assumed to produce 54 kWh kg^{-1} of H_2. Cryogenic N_2 is assumed to produce 243 kWh per ton of N_2. Wind turbines are assumed to operate at 3 MW each for the purposes of a simple illustration, disregarding the capacity factor. Reproduced from [8] with permission from AIP Publishing.

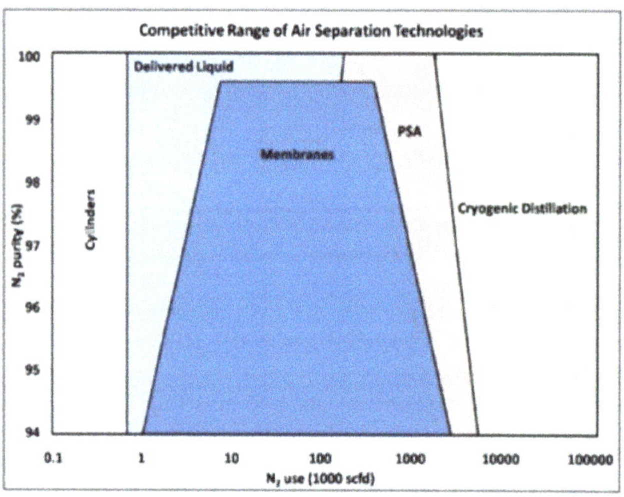

Figure 17.4. Competitive ranges of air separation technologies.

well downwards (they have a scaling exponent close to 0.5). Thus, for a smaller-scale, decentralized plant, the preferred air separation technology is pressure swing adsorption (PSA) or membranes, which are much less energy efficient, requiring 300–400 kWh per ton of ammonia for the nitrogen production [9]. See also figure 17.4.

Most of the literature dealing with ammonia production based on alkaline or PEM electrolysis is mainly concerned with the overall techno-economics, but some studies include design and bottom-up sizing of the equipment. One paper deals with actual operating experience from the Norsk Hydro plants using hydropower [10]. Norsk Hydro operated two plants at 160/165 MW capacity based on their own alkaline electrolyzer design. The active area of the cells was 2.1 m^2 and they operated at 0.17 A cm^{-2}. The unit had 235 cells in total, weighed 59 tons and had a length of 11 m and a diameter of 2.5 m. The faradaic efficiency was given as 98% at the start and degraded by an absolute value of 1% every four years. The cell voltage also increased from 1.67 V per cell at the start to 1.8 V after four years. The overall energy consumption given was 10 MWh MT^{-1} of ammonia, including the powers required for air separation (1%) and Haber–Bosch synthesis (7.5%). The power required for the electrolyzer unit was 4.4 kWh Nm^{-3} H_2 in the form of DC power, i.e. ignoring transformer losses.

17.5 Solid-oxide-electrolyzer-based ammonia production

The literature on power-to-X using SOECs [11] has grown exponentially since 2015, but papers that specifically address ammonia are relatively scarce. It is important to understand the basic thermodynamics of high-temperature water electrolysis because this delivers one of the main explanations for the high conversion efficiency achievable with SOECs. The water electrolysis reaction, i.e.

$$H_2O = H_2 + 0.5\,O_2\left(-\,\Delta H_{1023\,\mathrm{K}} = 248.1\frac{\mathrm{kJ}}{\mathrm{mol}}\right) \tag{17.1}$$

is highly endothermal, requiring 3.07 kWh Nm^{-3} of hydrogen produced at 750 °C. The enthalpy of the reaction is only weakly dependent on temperature, as illustrated in figure 17.5.

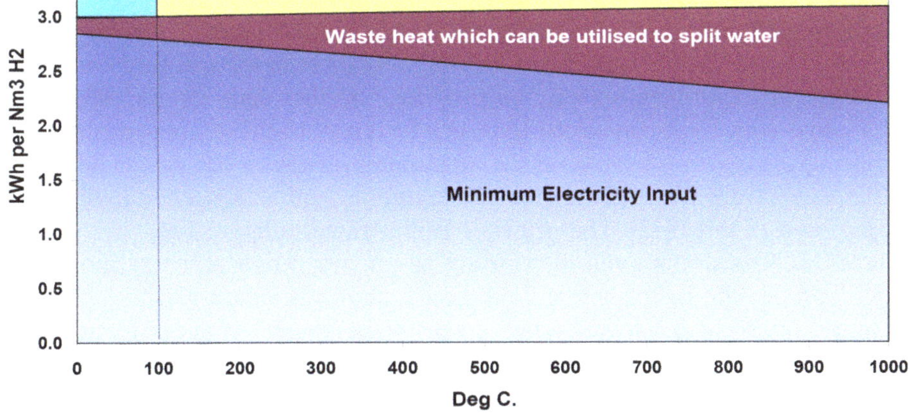

Figure 17.5. ΔH, ΔG, and $T\Delta S$ as a function of temperature for water splitting.

The reaction's enthalpy consists of two terms:

$$\Delta H = \Delta G + T\Delta S, \tag{17.2}$$

where ΔG is the Gibbs free energy change, which, as a minimum, has to be provided in the form of electrical energy, while the entropic part, $T\Delta S$, can be supplied as heat. ΔG decreases with temperature. At 80 °C, ΔG is 93% of ΔH but at 750 °C, it is only 77%. This heat can be provided by an outside source, such as burning part of the generated hydrogen between or in the stacks in air, thus supplying the nitrogen needed for ammonia. Loss mechanisms within the stack also generate Joule heating. These loss mechanisms are normally lumped together into the so-called area specific resistance (ASR), which is composed of ohmic resistance and activation overpotentials at both the anode and the cathode. The minimum operating voltage required for the reaction to take place is defined by:

$$V_O = \frac{\Delta G}{nF}, \tag{17.3}$$

where n is the number of electrons involved in the reaction (which is two for reaction (17.1)) and F is Faraday's constant 96 485 C mol^{-1}. V_O is the open-circuit voltage under standard conditions. The Joule heat generated is given by

$$Q_{\text{rest}} = i^2 \cdot \text{ASR} = i \cdot \left(V_{\text{op}} - V_{\text{nernst}}\right) \tag{17.4}$$

$$V_{\text{tn}} = \frac{\Delta H}{nF}. \tag{17.5}$$

At this voltage, the inlet and outlet temperatures of a stack are equal. Although the local current densities across the cells are not identical, operation at this voltage minimizes local temperature differences and thus mechanical stresses. For steam electrolysis, this voltage is 1.285 V at 750 °C.

SOECs indeed offer a much better energy efficiency than alkaline or PEM-based electrolyzers for hydrogen production, as demonstrated experimentally at the 50 kW scale in a Danish Energy Technology Development and Demonstration Program (EUDP)-funded project in which biogas was upgraded to pipeline-quality gas. The energy used by the SOEC stacks was measured by three independent methods to be 3.07 kWh Nm^s hydrogen, exactly as predicted by thermodynamics. SOECs can generate hydrogen by steam electrolysis, in which part of the steam can be supplied by utilizing the reaction heat from the ammonia synthesis loop. An air separation unit supplies nitrogen in order to generate ammonia synthesis gas at a hydrogen-to-nitrogen ratio of three [12]. The principle is illustrated in figure 17.6.

The SOEC-based ammonia plant using an air separation unit was modeled by Haldor Topsøe A/S; its schematic flow sheet is shown in figure 17.7:

Steam is mixed with a recycled hydrogen stream (to protect against oxidation of the fuel electrodes) and is preheated to the operating temperature of around 750 °C in feed/effluent exchangers that exchange heat with the effluent from the train of the stacks. The stacks operate at the thermoneutral voltage of 1.29 volts, so the stacks

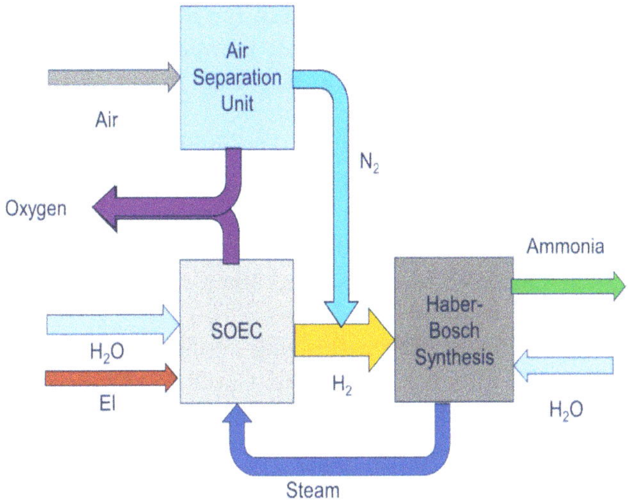

Figure 17.6. Ammonia production using an SOEC and an air separation unit.

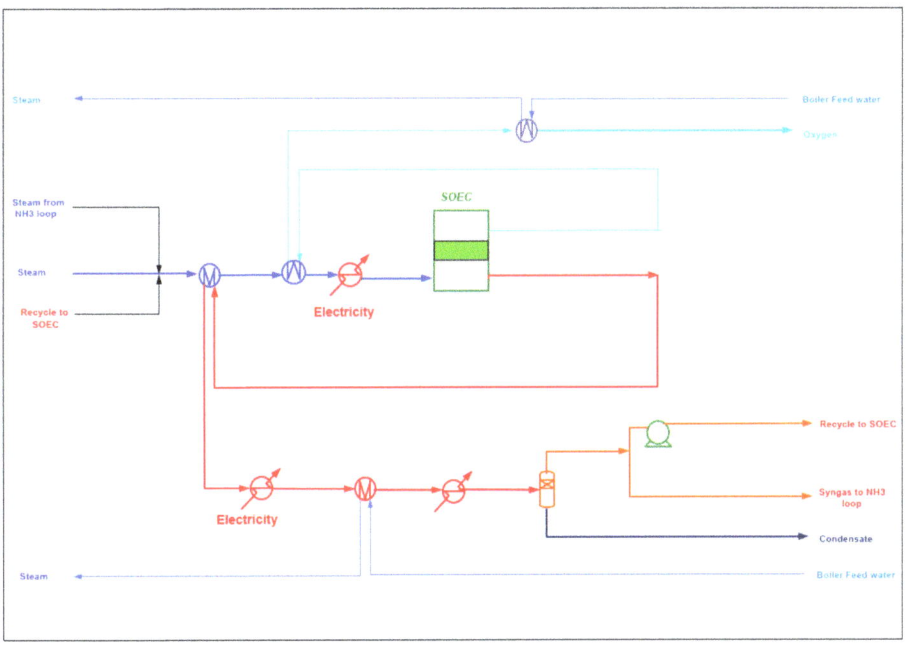

Figure 17.7. SOEC-based hydrogen production for ammonia synthesis.

are isothermal (they have same inlet and outlet temperatures). This operating strategy places the minimum mechanical strain on the SOEC cells. Steam from the ammonia synthesis loop (1 ton per ton of NH_3 produced) can be used directly in the SOEC plant, while the rest (1.25 ton per ton of NH_3 produced by steam conversion of 80% in the SOEC) has to be generated by electricity. This electricity input is provided at two positions: upstream of the SOEC stacks in the first electric heater, which provides a temperature increase that provide a reasonable minimum-temperature approach in the feed/effluent heat exchangers on the fuel and oxygen side; downstream of the feed/effluent exchangers, a second electric heater provides enough power to satisfy the steam balance. The total (DC) energy consumption of the SOEC plant is 6.7 MWh per metric ton (MT) of ammonia. To this energy consumption should be added the energy consumed by synthesis gas compression and the chiller compressor (a total of 0.31 MWh MT^{-1}), plus that used for the air separation unit (0.2–0.4 MWh MT^{-1}), so that the total energy consumption is 7.2–7.4 (DC) MWh MT^{-1} of ammonia produced.

The coupling of SOEC with Haber–Bosch ammonia synthesis was also discussed by Cinti *et al* [13]. Their work compared natural-gas-based, low-temperature electrolysis with SOEC-based ammonia synthesis based on some calculations performed by the commercial software Aspen. Unfortunately, their paper contained errors. The energy consumption of natural-gas-based plants was overestimated by more than 80% because a very high purge ratio from the ammonia loop was used. The low-temperature electrolysis plant's energy consumption was also too high by approximately 40%. For the SOEC case, it was assumed that the ammonia synthesis catalyst operates at 650 °C, which is more than 100 °C too high. Furthermore, the reaction heat from the ammonia synthesis was transferred in an unspecified manner to the SOEC, which also operated at 650 °C, without any temperature difference to drive the heat transfer. The energy efficiency definitions were wrong. Despite these errors, the possible synergy between the exothermal ammonia synthesis and the SOEC system was pointed out.

Zhang *et al* [14] carried out a careful bottom-up analysis of ammonia production from natural gas, biomass, and electrolysis by SOECs using Aspen modeling and pinch analyses [14]. They varied the operating parameters in order to optimize for either maximum efficiency or maximum economic return. The flow sheet for the SOEC unit is shown in figure 17.8.

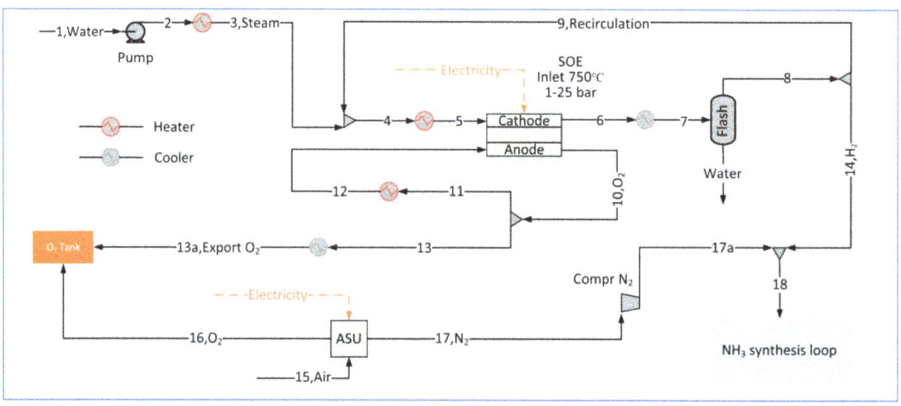

Figure 17.8. SOEC-based flow sheet, based on [14]. Reproduced with permission from Elsevier.

They operated the SOEC at 750 °C and controlled the temperature rise across the stack to a certain degree by recycling pure oxygen. They varied the SOEC operating pressure from 1.1 to 25 bar and the steam utilization from 0.3 to 0.8. A maximum temperature of 120 °C was allowed across the SOEC and high current densities of up to 0.96 A cm^{-2} were used. In effect, they thus used the SOEC to generate enough waste heat to close the steam balances. The resulting exit temperature of 870 °C appears rather high and leaves no room to compensate for ageing by increasing the inlet temperature as suggested by Hansen [15] for isothermal operation at the thermoneutral voltage. The maximum efficiency was calculated to be 75.3% and the most economic value was 74.1% (6.85 and 6.96 MWh MT^{-1} of NH_3, respectively), which appears somewhat high and is supposedly calculated without transformer losses. The calculated efficiencies for the natural-gas-based and biomass-based cases were 61–62% and 44%–45% respectively, which seem reasonable.

Although slightly off-topic for this chapter, the same group [16] also studied ammonia as an intermediate in a power-to-ammonia-to-power scheme, as shown in figure 17.9. The same SOC stacks were used, alternating between electrolysis and fuel cell mode in a closed loop, with only electricity entering and leaving. Nitrogen and oxygen were stored in high-pressure tanks. The layout used for the ammonia synthesis was similar to the scheme shown in figure 17.8 and had a calculated efficiency of 61%–69%, while the ammonia-to-power system had an efficiency of 42%–63%. It should be noted that the ammonia was cracked outside the stacks, so the advantage of the internal cracking of ammonia, as described by Hansen [17, 18], was not taken into account. The calculated round-trip efficiency was thus 27%–43%.

A research group at the Rheinisch-Westfälische Technische Hochschule (RWTH) in Aachen [19] addressed the same problem using a different layout, as indicated in figure 17.10. Instead of pressurized tanks for oxygen and nitrogen, they used an integrated refrigeration cycle. The system was simulated and costs estimated for a classical chemical engineering approach using gProms software. They arrived at a round-trip efficiency of 72%. Their scheme evidently presupposed that no leakages would occur between the nitrogen and oxygen sides. Using an electricity purchase price of $30 per MWh, they arrived at a price for the produced electricity of

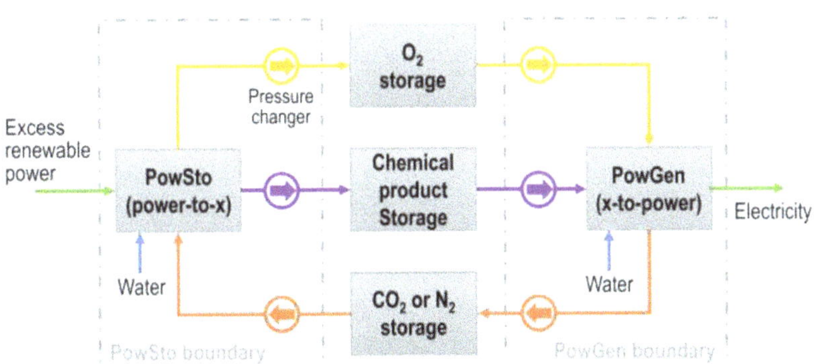

Figure 17.9. Power-to-ammonia-to-power according to [16]. Reproduced with permission from Elsevier.

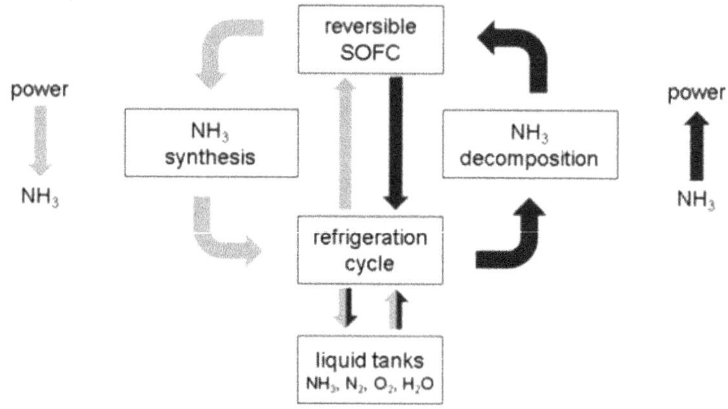

Figure 17.10. rSOC-scheme for power-to-power with an ammonia intermediate according to [19]. Reproduced with permission from AIChE.

$240 per MWh, which of course is not a good business proposition, and reflects the high investment cost. More work is obviously needed on the reversible solid oxide cell (rSOC) concepts.

The Aachen group has extended its work to study the synergies that could be obtained by combining a SOEC-based ammonia plant with a nitric acid plant [20]. For the stand-alone SOEC and Haber–Bosch plant, they calculated a total power consumption of 7.33 MWh MT^{-1} ammonia. They operated the SOEC stacks exothermally to close the steam balance. If coupled with a nitric acid plant, in which a lot of heat is generated by combusting ammonia with oxygen, there is enough steam available to allow for the better strategy of thermoneutral operation. The ammonia production cost also comes down from $610 to $570 per MT.

17.6 Novel electrified ammonia plant without an air separation unit

Haldor Topsøe A/S has developed a new process (patent pending) for ammonia synthesis gas generation. It utilizes the unique capabilities of an SOEC acting as an oxygen separation membrane and its ability to utilize heat in lieu of power, so that the expensive air separation unit can be eliminated. The concept is being demonstrated by a 50 kW demo unit located at Aarhus University in Foulum, Denmark. In addition to a number of advantages described in the following, this concept ensures far better scalability of the electrified ammonia plant. The principles are illustrated in figures 17.11 and 17.12.

Nitrogen is introduced to the synthesis gas by simply burning a part of the hydrogen generated by steam electrolysis. The heat evolved equals the extra power needed to generate the hydrogen burned. Another way of looking at the overall scheme is to note that the SOEC acts as an air separation unit because the electrolyte separates oxygen from the fuel side, which is done with very high efficiency. Instead of using electric heaters for the final preheating and to close the steam balance, as in the hydrogen-producing plant in figure 17.7, some air is added and burned in catalytic burners, as shown in figure 17.12. However, the main point is that this

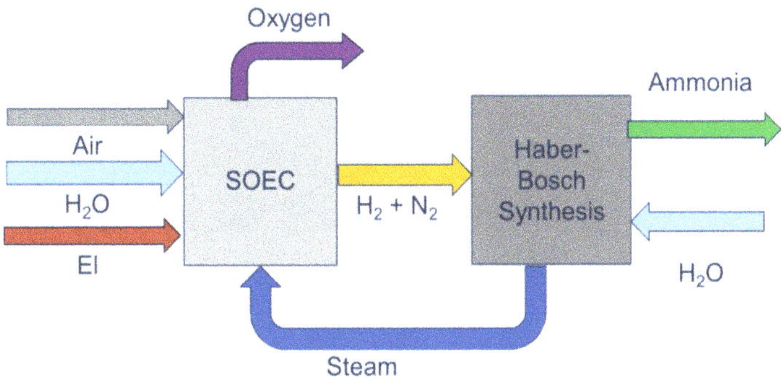

Figure 17.11. Principle of SOEC coupled to ammonia synthesis.

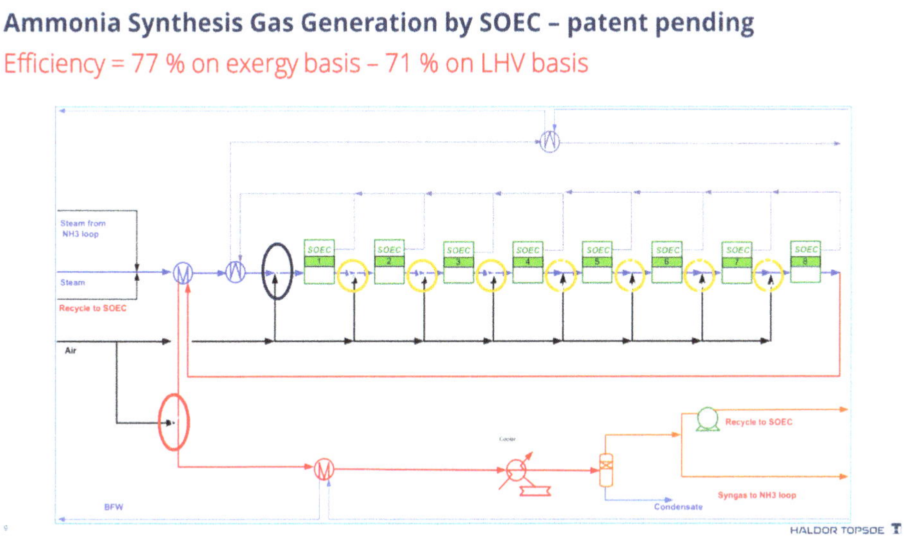

Figure 17.12. New SOEC-enabled ammonia synthesis process without a separate air separation unit. Air is added to catalytic burners at the positions indicated by the circles.

novel concept can be realized using a smaller investment due to the elimination of the air separation unit, which can be a big part of the CAPEX, especially at low capacities. Now, if the stacks are operated below the thermoneutral voltage, power is saved, but a temperature drop occurs across the stack. Air can then be burned between the stacks, bringing the inlet temperature of the next stack up again. This process can be repeated until a stoichiometric gas with a H_2/N_2 ratio of three is obtained for the ammonia synthesis. The burning of the air generates extra water, which needs to be electrolyzed, but due to the high efficiency of the SOEC—close to 100%—the net result is only a marginal increase in overall power consumption.

The simultaneous generation of hydrogen and nitrogen does, however, come at the cost of larger stack areas, because the average current density needs to be lower when heat is added externally in order to keep the stacks thermoneutral overall, as explained in figure 17.13. This fact, i.e. that heat added from the exterior directly to

$$T * \Delta S = Q_{ohm} + Q_{add}$$

$$Q_{ohm} = I^2 * ASR$$

$$I = \sqrt{\frac{T * \Delta S - Q_{add}}{ASR}}$$

$$T * \Delta S = 704\,Wh\,at\,923\,K$$

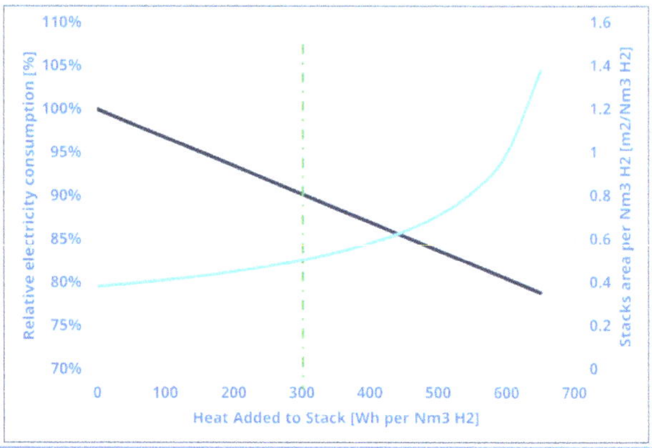

Figure 17.13. Tradeoff between heat added to the stacks and the stack area. The black line refers to the left vertical axis and the light blue line to the right axis.

the stacks results in larger stacks, is often overlooked in the SOEC literature. Providing feedstock steam from the outside does not incur this penalty.

There is thus a tradeoff between low energy consumption and investment in stack area. The optimum will depend on the electricity price and stack costs. Analyses carried out by Haldor Topsøe have, however, shown that the overall energy consumption for the new process is equal—within a few percent—to the energy consumption for the SOEC hydrogen production + air separation layout discussed above.

17.7 Techno-economic studies

The costs of producing green ammonia via electrolysis have been discussed in a number of papers, but the results diverge significantly. The costs depend on a number of factors, depending on whether the plant is grid connected or stand-alone (behind-the-meter):

- The CAPEX for the ammonia plant, which again can be broken down into the major blocks of electrolyzer, air separation unit, Haber–Bosch synthesis unit, storage units for hydrogen and nitrogen to alleviate the intermittency of the renewable power used, and possible power-generation units for the periods with low renewable power availability. The lifetime and degradation rate of the electrolyzers play a role and impact the OPEX.
- Electricity price: this is normally the dominant contributor to the cost. It can vary significantly if purchased from the grid. If the plant owner has also invested in the renewable source, it is only the CAPEX for these assets and their capacity factors that matter.
- The efficiency of the electrolyzer as function of load is obviously of great importance, due to its impact on the expenditure for electric power.
- The flexibility of the different production units, e.g. minimum and maximum design load, maximum ramping rates, etc.

- Scale of operation: The CAPEX of the electrolyzers scales almost linearly above 1 MW, but the air separation and Haber–Bosch units have a significantly lower scaling exponent of 0.5–0.67; this is mainly due to the characteristics of CAPEX for rotating machinery.
- Cost of financing and the required rate of return of capital.

It is evident that the production cost depends on many interrelated factors. The coupling of electrolysis-based ammonia plants was already studied by Lockheed as early as the 1970s, but no techno-economic studies were performed [21]. Many papers on the economics of electrified ammonia production have relied on the PhD thesis and subsequent papers of Morgan, who was the first to publish a bottom-up estimation of the cost of the equipment for islanded operation [22–24]. Space limitations does not allow a detailed description of all the other studies, but the reader is referred to appendix A for summaries. Some especially pertinent studies with respect to the coupling of electrified ammonia production and intermittent renewable power will, however, be briefly discussed.

Nayak-Luke and the research group at Oxford University [25–28] have studied the impact of the different factors on the cost of ammonia. In one study, they simulated a plant in Lerwick, Shetland Islands, Scotland, operating with renewable power from a mix of wind and solar resources. An overview of the concept is provided in figure 17.14.

The plant is equipped with hydrogen storage and a proton-exchange membrane fuel cell (PEMFC) that produces electricity with an efficiency of 50%. The air separation unit (ASU) and the Haber–Bosch (HB) loop are always operated above 20% of their rated capacity, as this minimum point is found to be the most optimistic in the literature. If the power obtained from the wind and sun is not sufficient for the electrolyzer, hydrogen is taken from storage. If there is not enough power to run the ASU and HB, the PEMFC is engaged to produce electricity.

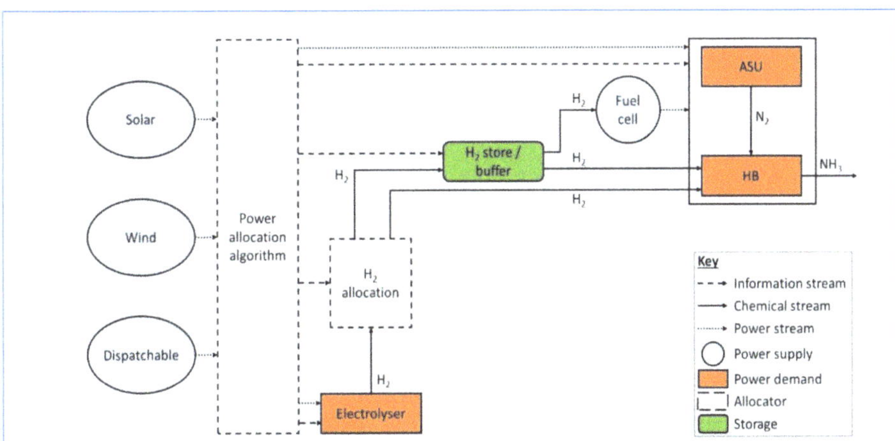

Figure 17.14. Flexible green ammonia plant. 'ASU' denotes the air separation unit. 'HB' denotes the Haber–Bosch ammonia loop. Reproduced from [28] with permission from the Royal Society of Chemistry.

In summary, there are four operating modes: (1) curtailment of excess power and storage of hydrogen; (2) storage of excess hydrogen and ammonia production; (3) withdrawal of hydrogen from storage; (4) withdrawal of hydrogen from storage and production of energy by the fuel cell to run the HB and ASU processes at the minimum load. All units are sized for 100 MW renewable power on average, and the relative sizes of the units are optimized based on the power price. It is assumed that the electrolyzer can respond instantaneously to load changes from 0% to 100%. The assumed consumption values are 53.4 kWh kg^{-1} for H_2, 0.119 kWh kg^{-1} for nitrogen, and 0.600 kWh kg^{-1} for the ammonia loop. The costs of electricity produced using wind and solar energy are $87 per MWh and $103 per MWh, respectively, in the base case, which are, of course, rather high. The CAPEX for the electrolyzer was £880 kW^{-1} = $1140 kW^{-1}. The optimum base-case configuration has a production of 83 220 tons per year and an ammonia cost of £1000 ton^{-1} = 1300 $ ton^{-1}.

The sensitivity analyses showed that for a 1% drop in production price (a) the electricity price should be 1.3% lower (b) the electrolyzer CAPEX should be 7.4% lower. Compared to an inflexible plant operating at 100%, being able to go to a 20% load lowered the ammonia cost by 9%, provided that the ramping rate can be higher than 4% per hour. The last result in particular is interesting and unique. The group also extended this methodology to a study of 534 locations in 70 countries for 2019 and 2030, and used meteorological data with a 30 min resolution to define the power profile and supply capacity factors. They switched to alkaline electrolyzers from NEL with a projected price of $341 kW^{-1} in 2030 and an energy efficiency of 70% (47.6 kWh kg^{-1} H_2). The levelized costs of electricity were estimated by fitting the historical cost data for wind and PV and extrapolating the trend of development to 2030. The two-parameter exponential fitted function gave excellent fits, with R^2 above 0.93, and the predicted levelized costs of electricity (LCOEs) in 2030 were $24.5 and $14.1 per kWh for wind and PV, respectively.

The sizes of the plants were determined by using 100 MW of renewable power on average, and the output of their calculations included the size and cost of each block as well as the levelized cost of ammonia. In 2030, numerous locations were expected to have production costs of less than $400 ton^{-1}. Using country-specific discount rates, somewhat surprisingly, such locations were found in Australia, Austria, Japan, the USA, Denmark, France, and the UK. If the lowest discount rate is used globally, for instance, by a multinational organization, the list is expanded to include countries in Africa, South America, the Middle East, and Asia. This shows that financial barriers can also be important in finding the optimum location. However, the paper does not address the point that even with the same discount rate, installation factors vary considerably by location. Another weakness, as also acknowledged in the paper, is that the model assumes perfect foresight of the weather conditions and that the efficiencies are constant, regardless of load. Nevertheless, the results show that even with a modest CO_2 tax, green ammonia can become competitive a decade ahead. Another interesting finding of the study is that a simplistic calculation using just an ideal capacity factor without allowing for storage, on-site power production, and individual sizing of the components can underestimate the levelized cost of ammonia (LCOA) by a factor of 1.56. It would also be very interesting to run the model with the better efficiency provided by the SOEC solution. SOEC electrolyzers are perfectly capable of transient operation, because if they operate at the thermoneutral point, as recommended by Hansen [15], the operational mode can be switched from idle or a few percent load to full power without any impact on the temperature profiles.

Fasihi *et al* [29] undertook a similar study that divided the globe into areas with a spatial resolution of 0.45° × 0.45° with respect to wind and PV resources. They considered both locating the power plant on site and locating it on the coast along with electric power transmission. They chose a base capacity of 400 000 MT of ammonia per year, and their design differed from that of Nayak-Luke *et al* in that they used batteries or hydrogen-fueled gas turbines to compensate for low renewable power production. Hydrogen storage in salt or rock caverns at feasible sites was also included. They were more conservative with respect to the flexibility of the ASU+HB synthesis, as they used a minimum load of 50%, a ramp-up limit of 2% and a ramp-down limit of 20% per hour with no major additional cost or efficiency loss. Considering the predicted evolution of the cost of equipment, they arrived at production costs of €320–€450 ton^{-1} with up 10 GT per year of ammonia in 2030 at the best available sites, which were found to be the Atacama Desert in Chile, the Horn of Africa, Yemen, and Southwest Niger, which were the least-cost regions. Costs below 500 €/tNH_3 become achievable in most regions of South America's coastlines, Mexico, the center and southwest of the US, Australia, Africa, the Middle East and North Africa (MENA) region, and Central Asia. In Europe, such cost levels are achievable in Spain, Portugal, the UK, Iceland, and Denmark. In 2050, the projected cost falls to to €260–€290 ton^{-1}, i.e. very competitive even with a low CO_2 tax. The base calculation assumes a weighted average capital cost of 7%. If this were 5% instead, which is feasible in Europe and the US, the cost of ammonia would drop by 14%–15%. A 10% change in the CAPEX of ammonia synthesis would only change the cost by up to 1.5%. On the other hand, an inflexible ammonia plant only operating at a 100% load could increase the cost by 50%. A 10% change in the CAPEX required for the electrolyzer would change the cost by 1% and 2% in the wind and PV cases, respectively.

The use of SOEC electrolyzers has only been sparingly investigated in techno-economic studies. The Institute for Sustainable Process Technology in the Netherlands has coordinated a study on power-to-ammonia using different technologies including PEMFC, solid-state ammonia synthesis, a special battery–electrolyzer hybrid called a Battolyzer, and SOECs [30]. They calculated power consumptions of 9.5 and 7.1 MWh ton^{-1} of ammonia for PEM- and SOEC-based production, respectively. Their assumptions about the flexibility of the system were rather conservative: 'With SOEC as the choice of electrolysis, power to NH_3 systems have a load range of 50% to 100% and can ramp up in 25 min and turn down in 13 min. The ramp-up time is limited by the NH_3 synthesis system, while the ramp down and the load range are limited by the electrolysers.' The economic evaluation showed that the ammonia price could not become competitive unless there was a high penetration of renewable power and the CAPEX of the SOEC came down from 1000 € kW^{-1} to 300 € kW^{-1}.

Zhang *et al* [14] also calculated the cost of ammonia using a SOEC-based front end. They calculated an investment cost of $139 million for a 50 000 ton per year plant with 7200 operating hours, i.e. a $2780 ton per year ammonia capacity. For comparison, the IEA suggested a CAPEX of around $900 per ton of ammonia per year for a natural gas ammonia plant. Zhang used a stack price of €2000 per stack (the stack power was not actually specified). The production cost was calculated to be $544 ton^{-1} with an electricity price of $73 per MWh. It should be noted that their calculations also included an income from selling oxygen at a price of $177 ton^{-1}, which can only be possible in very special cases.

The industrial group behind the Ammonfuel report [31], consisting of Alfa Laval, Haldor Topsøe, Hafnia, Siemens–Gamesa, and Vestas has made projections for the costs of different types of ammonia in 2025–30 and 2040–50 and compared them to the price of very low sulfur fuel oil (VLSFO). The results are presented as ranges in table 17.2 due the fact that both alkaline, PEM, and SOEC technologies were considered as well as two sizes of ammonia plant; the cost of fossil fuel was kept constant at the highest 2020 level. The sum of the CAPEX and fixed operating costs for 100 MW and 1 GW plants were estimated to be 375–475 US$ MT^{-1} and $190 MT^{-1} of ammonia capacity, respectively, in 2025–30. From 2040 onward, the CAPEX was predicted to come down to $150–$190 MT^{-1} of ammonia capacity. The power consumptions of alkaline and PEMFC electrolyzer-based units were estimated to be 10–10.5 MWh MT^{-1} of ammonia or as low as 7.6–7.8 MWh MT^{-1} ammonia for the SOEC-based plants.

It should be noted that the above prices for conventional and blue ammonia assume a natural gas price of around $7 GJ^{-1}. Predicting future prices is at best uncertain, and to underline this, it is interesting to consider the price of ammonia in Europe at the time of writing this review (January 2022). According to S&P Global Platts, the ammonia price was $1120 MT^{-1}. On the US Gulf Coast and in the Far East, the prices were $1090 and $1000 MT^{-1}, respectively. The high prices in Europe could be attributed to the all-time high natural gas prices at $41 GJ^{-1}.

In summary, it is clear that green ammonia will be two to three times more expensive than black ammonia to start with, but will achieve parity, even without a CO_2 tax, in the 2040–50 timeframe, provided that renewable power will be available in quantity at around €20 per MWh and that learning effects and scale will drive

Table 17.2. Predicted development of different ammonia prices. Fossil fuel prices are assumed to remain constant at 2020 levels.

	2025–30		2040–50	
	€30 per MWh		€20 per MWh	
Assumed renewable electricity price	Price per ton US ($ MT^{-1})	Price per GJ LHV US ($ GJ^{-1})	Price per ton US ($ MT^{-1})	Price per GJ LHV US ($ GJ^{-1})
VLSFO (<0.5%S)	500–600	12.5–15	500–600	12.5–15
Conventional ammonia	250	13.5	250	13.5
Blue ammonia	350–400	18.8–21.5	350–400	18.8–21.5
Green ammonia	400–850	21.5–45.7	275–450	14.8–24.1
Hybrid green ammonia	300–400	16.1–21.5	250	13.5

capital expenditure for the production plants. Another important factor will be the implementation of more efficient SOEC electrolyzers.

17.8 The future use of ammonia as an energy vector

Several international studies have identified ammonia as an interesting energy vector for enabling the transition to a carbon-free energy system. The IEA has included ammonia as an important fuel for combustion engines, gas turbines, and fuel cells in several publications. In its net zero scenario for 2050, the IEA foresees that 74 million tons of ammonia per year will be used for electricity production, and 250 million tons per year will be used for transport. In the IEA analysis of the future of hydrogen, it is concluded that electrolysis becomes competitive with natural gas with carbon capture, utilisation and storage (CCUS) for chemical production at electricity prices in the range of \$15–50 (US) per MWh for ammonia and in the range of \$10–65 per MWh for methanol, on the assumption of gas prices of \$3–10 per MBtu, provided that electrolysis CAPEX decreases by 50% and efficiency increases by 15%. The Japanese government has, in one of its Strategic Implementation Plans (SIPs), looked into the use of ammonia for different end uses, as shown in figure 17.15.

The program has demonstrated the different end uses in relevant demonstration tests. The sector for which the use of ammonia has attracted the most attention is shipping, which is responsible for 3% of global energy use and could be responsible for much more of the global greenhouse emissions in 2050 if business continues as usual.

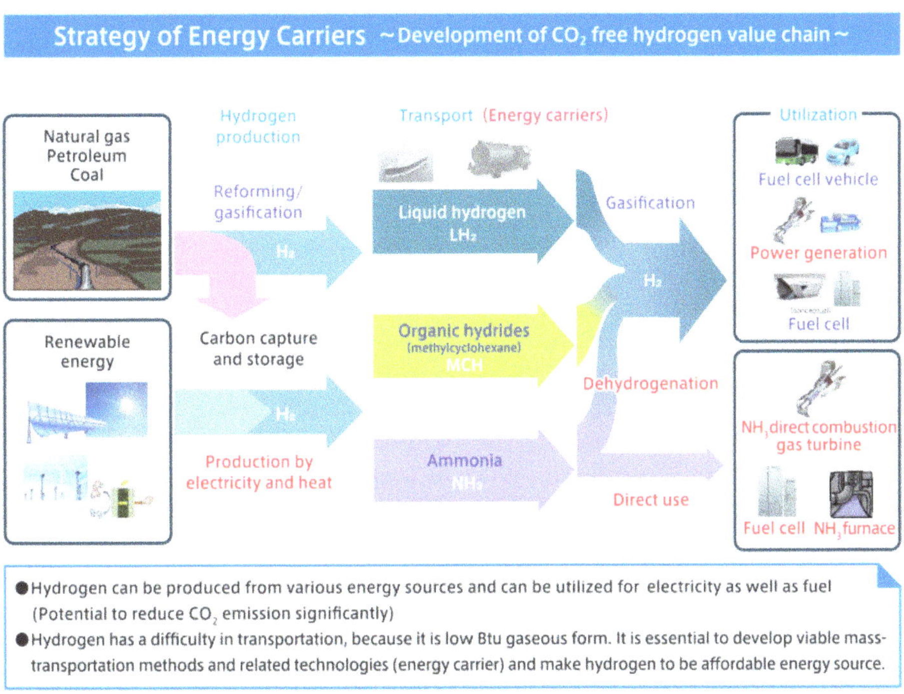

Figure 17.15. Strategy for energy carriers according to the Japanese Strategic Implementation Plan.

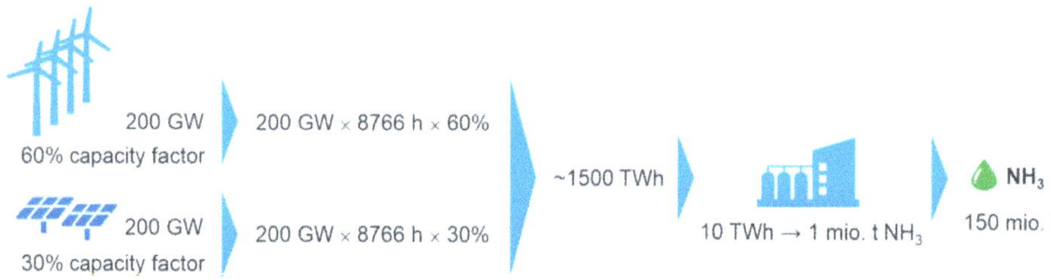

Figure 17.16. Overview of the scenario to cover 30% of world shipping fuel in 2050.

Table 17.3. Worldwide capacities of wind and solar power.

	Wind power (GW)	Solar PV (GW)
Cumulative 2020	733	714
Installed in 2020	111	124

The Ammonfuel report concludes that ammonia is an attractive and low-risk choice. The logistics chain is familiar with it, as more than 120 ports already have ammonia trading facilities in place. If 30% of the forecasted demand for shipping fuels in 2050 were to be met by ammonia, an extra production capacity of 150 million tons per year would have to be installed. Approximately 400 GW of renewable power would be needed to sustain this production, if it were based on alkaline electrolysis with an energy consumption of 10 MWh MT^{-1} of ammonia, as illustrated in figure 17.16.

To put this amount of wind and solar production into perspective, the worldwide current cumulative installed wind and solar power in 2020 [32, 33], as well as new additions in that year, are shown in table 17.3.

17.9 Concluding remarks

This review of ammonia production based on electrolysis, with a special emphasis on SOECs, has hopefully demonstrated that green ammonia produced from renewable sources can become an important, enabling factor in the endeavor to reach a global, zero-emission energy system. Its cost will initially be two to three times higher than that of fossil-fuel-based ammonia, but competitive parity is within reach, especially if a relevant CO_2 tax is implemented. Due to its 25%–30% energy advantage, SOEC technology will offer the opportunity to both bring down the cost and make green ammonia production possible at sufficient scale without using scarce renewable biomass resources. It will enable the accelerated deployment of wind and solar power resources with reduced land-area use. Finally, it will contribute to feeding the growing world population in a sustainable manner.

Appendix A

Reference	Renewable source ($ per MWh)	Grid connect	Electrolyzer	Electrolyzer consumption (kWh Nm^{-3} H_2)	ASU+ HB (kWh MT^{-1} NH_3)	Total (MWh MT^{-1} NH_3)	Plant Size (MT per day)	Electrolyzer (MW)	Electrolyzer cost ($ kW^{-1})	WACC (%)	LCOA ($ MT^{-1})
[34] Tunå	40–120	Yes	N.S.	4.25	N.S.	11.8	20.3	1	577	5–12	1015–2328
	–	–	–	–	–	–	6.1	3	–	–	1078–2392
[35] Ikäheimo	N.S.	(YES)	PEM	3.7–4.23	640	8.0–9.0	N.S.	N.S.	460	7	490–600
[36] Beerbühl	N.A.	YES	–	3.54–4.40	–	–	72–240	34–35	–	–	591
[26] Nayak-Luke	87 (W), 104(PV)	No	PEM	4.80	700	10.2	228	196	1142	0	750
[24] Morgan	Wind included in CAPEX	Yes	Alkaline	4.80	800	11.6	300	129	–	7	583–1224
[37] Espinola	10 Hydro	Yes	–	–	–	–	500	248	–	–	563
[30] ISTP 2017	Varying	(Yes)	SOEC	–	–	–	–	–	–	–	–
[38] Armijo	–	(Yes)	Alkaline	4.32	900	9.4	96	60	600	7	462–506
										10	462–571
[39] Sanchez	–	–	Alkaline	4.78	–	–	300	–	–	–	1535–1569
[40] Wu	55 (PV)	No	–	5.5	1137	9.13	567	–	–	–	863
[41] Osman	25 (PV)	No	Alkaline	4.4	420	10.43	1840	768	652	4	450–718
[14] Zhang	73	Yes	SOEC	–	–	6.9–7.0	152	45	–	10	–
[20] Wang	74	Yes	SOEC	–	–	7.33	500–1000	–	100	–	610–600

References

[1] Valera-Medina A (ed) 2021 *Techno-Economic Challenges of Green Ammonia as an Energy Vector* (London: Academic)

[2] IMO 2018 https://unfccc.int/sites/default/files/resource/250_IMO%20submission_Talanoa%20Dialogue_April%202018.pdf

[3] Duijm N J, Markert F and Paulsen J L 2005 http://orbit.dtu.dk/en/publications/safety-assessment-of-ammonia-as-a-transport-fuel(e233cd06-83eb-4fa0-981a-12a2074a927a).html

[4] Consultants Q 2009 https://nh3fuelassociation.org/wp-content/uploads/2013/01/nh3_riskanalysis_final.pdf

[5] Dybkjær I 2013 100 years of ammonia synthesis technology *AICHe Ammonia Safety Symp.* **54** 101–9

[6] Appl M 2011 Ammonia, 2. Production processes *Ullmann's Encyclopedia of Industrial Chemistry*

[7] Dybkjaer I 1995 *Ammonia: Catalysis and Manufacture* ed A Nielsen (Berlin: Springer) 199–327

[8] Pfromm P H 2017 Towards sustainable agriculture: Fossil-free ammonia *J. Renew. Sustain. Energy* **9** 034702

[9] Flowe M https://airbestpractices.com/system-assessments/air-treatmentn2/energy-costs-associated-nitrogen-specifications

[10] Grundt T and Christiansen K 1982 Hydrogen by water electrolysis as basis for small scale ammonia production. a comparison with hydrocarbon based technologies *Int. J. Hydrogen Energy* **7** 247–57

[11] Ince A C, Colpan C O, Hagen A and Serincan M F 2021 Modeling and simulation of Power-to-X systems: a review *Fuel* **304** 121354

[12] Hansen J B and Hendriksen P V 2019 The SOC4NH3 project. Production and use of ammonia by solid oxide cells *ECS Trans.* **91** 2455–65

[13] Cinti G, Frattini D, Jannelli E, Desideri U and Bidini G 2017 Coupling solid oxide electrolyser (SOE) and ammonia production plant *Appl. Energy* **192** 466–76

[14] Zhang H, Wang L, Van herle J, Maréchal F and Desideri U 2020 Techno-economic comparison of green ammonia production processes *Appl. Energy* **259** 114135

[15] Hansen J B 2015 Solid oxide electrolysis - a key enabling technology for sustainable energy scenarios *Faraday Discuss.* **182** 9–48

[16] Wang L, Zhang Y, Pérez-Fortes M, Aubin P, Lin T E, Yang Y, Maréchal F and Van herle J 2020 Reversible solid-oxide cell stack based Power-to-X-to-Power systems: comparison of thermodynamic performance *Appl. Energy* **275** 115330

[17] Hansen J B 2011 Direct Reforming Fuel Cells *Fuel Cells: Technologies for Fuel Processing* (Amsterdam: Elsevier) 13 409–50

[18] Hansen J B 2015 Fuel processing for fuel cells and power to fuels as seen from an industrial perspective *J. Catal.* **328** 280–96

[19] Wang G, Mitsos A and Marquardt W 2017 Conceptual design of ammonia-based energy storage system: system design and time-invariant performance *AlChE J.* **63** 1620–37

[20] Wang G, Mitsos A and Marquardt W 2020 Renewable production of ammonia and nitric acid *AlChE J.* **66** e16947

[21] Dubey M 1978 Technical and economic feasibility of making fertilizer from wind energy, water, and air *Sun, Mankind's Future Source of Energy, Proc. of the Int. Sol Energy Soc. Congr.* V (Oxford: Pergamon) pp 1812–21

[22] Morgan E, Manwell J and McGowan J 2014 Wind-powered ammonia fuel production for remote islands: a case study *Renewable Energy* **72** 51–61
[23] Morgan E R 2013 *Techno-economic Feasibility Study of Ammonia Plants Powered by Offshore Wind* (Amherst, MA: University of Massachusetts)
[24] Morgan E R, Manwell J F and McGowan J G 2017 Sustainable ammonia production from U.S. offshore wind Farms: a techno-economic review *ACS Sustainable Chemistry and Engineering* **5** 9554–67
[25] Nayak-Luke R, Bañares-Alcántara R and Collier S 2021 Quantifying network flexibility requirements in terms of energy storage *Renew. Energy* **167** 869–82
[26] Nayak-Luke R, Bañares-Alcántara R and Wilkinson I 2018 'green' ammonia: impact of renewable energy intermittency on plant sizing and levelized cost of ammonia *Ind. Eng. Chem. Res.* **57** 14607–16
[27] Nayak-Luke R M and Bañares-Alcántara R 2018 Long-term energy storage: what is the need and is ammonia a solution? *Comput. Aid. Chem. Eng.* **47** 1843–8
[28] Nayak-Luke R M and Bañares-Alcántara R 2020 Techno-economic viability of islanded green ammonia as a carbon-free energy vector and as a substitute for conventional production *Energy Environ. Sci.* **13** 2957–66
[29] Fasihi M, Weiss R, Savolainen J and Breyer C 2021 Global potential of green ammonia based on hybrid PV-wind power plants *Appl. Energy* **294** 116170
[30] ISTP 2017 https://ispt.eu/media/DR-20-09-Power-to-Ammonia-2017-publication.pdf
[31] Alfa Laval H, Haldor T, Vestas and Siemens G 2020 https://topsoe.com/hubfs/DOWNLOADS/DOWNLOADS%20-%20White%20papers/Ammonfuel%20Report%20Version%2009.9%20August%203_update.pdf
[32] IRENA 2021 Renewable Capacity Statistics 2021 https://irena.org/publications/2021/March/Renewable-Capacity-Statistics-2021
[33] IRENA 2020 Renewable Energy Statistics 2020 https://irena.org/publications/2020/Mar/Renewable-Capacity-Statistics-2020
[34] Tuna P, Hulteberg C and Ahlgren S 2014 Techno-economic assessment of nonfossil ammonia production *Environ. Prog. Sustain. Energy* **33** 1290–7
[35] Ikäheimo J, Kiviluoma J, Weiss R and Holttinen H 2018 Power-to-Ammonia in future North European 100% renewable power and heat system *Int. J. Hydrogen Energy* **43** 17295–308
[36] Beerbühl S, Kolbe B, Roosen C and Schultmann F 2014 Synthesis of ammonia from intermittently generated hydrogen *Chem. Ing. Tech.* **86** 649–57
[37] Espínola M O G 2011 Techno-economical study of the use of Itaipu dam spilled turbinable energy to produce ammonia *Int. J. Energy Envron. Eng.* **2** 57–64 https://ijeee.stb.iau.ir/article_510903.html
[38] Armijo J and Philibert C 2020 Flexible production of green hydrogen and ammonia from variable solar and wind energy: case study of Chile and Argentina *Int. J. Hydrogen Energy* **45** 1541–58
[39] Sánchez A and Martín M 2018 Optimal renewable production of ammonia from water and air *J. Clean. Prod.* **178** 325–42
[40] Wu C, Zheng S, Wang Z, Chen R, Hu X and Chen J 2021 Discussion on ammonia as one of the energy storage media of solar energy in China *Energy Strat. Rev.* **38** 100697
[41] Osman O, Sgouridis S and Sleptchenko A 2020 Scaling the production of renewable ammonia: a techno-economic optimization applied in regions with high insolation *J. Clean. Prod.* **271** 121627

Chapter 18

SOEC-based production of e-fuels via the Fischer–Tropsch route

Dorela Dhamo, Dominik Hess, Michael Rubin and Roland Dittmeyer

This chapter focuses on intensified technologies for the production of synthetic fuels from renewable electrical energy and CO_2. The relevance of this 'power-to-liquid (PtL)' approach for future energy systems is highlighted. Different ways to use a high-temperature solid oxide electrolyzer for synthesis gas production via a PtL process based on the Fischer–Tropsch (FT) route and CO_2 direct air capture (DAC) as the carbon source are described and compared (mainly from a process engineering perspective). The potential for heat integration between the FT unit, the electrolyzer, and the upstream DAC plant is illustrated using results from the German Kopernikus P2X project. The development status of modular PtL plants for decentralized applications is summarized, highlighting ongoing work on simplified refining of the FT syncrude to sustainable aviation fuel as well as transient operation. Finally, an outlook on the further scale-up of this technology is given.

18.1 Power-to-X from a systems perspective

It becomes clearer and clearer every day that mankind must quickly and greatly reduce its greenhouse gas emissions to limit global climate change. Net zero emission targets have been set by many countries worldwide, and strategies for achieving this have been worked out and are increasingly being implemented. CO_2 taxation is on the rise and can be expected to accelerate further. Therefore, great efforts are being made to increase the efficiency of use of primary energy in order to reduce the consumption of fossil fuels and the associated release of greenhouse gases into the atmosphere. In addition, many countries are expanding electricity generation from renewable sources, among which, onshore and offshore wind and photovoltaics are the most important technologies due to their strong cost reductions in recent years. However, due to their variable yields, which depend on the weather and time of day, the generating capacity of any system largely relying on renewables must be a

doi:10.1088/978-0-7503-3951-3ch18

multiple of the maximum electrical power demand, and new ways must be found to stabilize electrical grids that have a high proportion of renewable electricity generation. In order to counter the time-dependent imbalance between energy supply and energy demand, additional load-adaptable power generation systems are needed. Mechanical energy storage and battery storage systems are not suitable for storing large amounts of energy, for example to bridge dark lulls or even compensate for seasonal fluctuations; therefore, security of supply over longer periods can only be guaranteed using backup power generation from chemical energy carriers.

That being said, in most countries worldwide, electrical energy still represents a small portion of the total final energy consumption. Worldwide, it was 20.4% in 2021 (figure 18.1). Even if a clear trend towards electrification can be seen with today's energy transition and defossilisation, the change cannot take place as fast as needed to reduce greenhouse gas emissions, and this is already creating a strong demand for sustainably produced CO_2-free chemical energy carriers in the short and medium term. In addition, there is a long-term need for sustainable, high-energy-density chemical energy carriers, especially to provide high-temperature heat for industry and for certain transport applications, such as aviation, shipping, and long-haul heavy road transport, for which electrification is less viable.

Finally, it is not to be expected that populous industrialized countries in the northern hemisphere, such as Germany, will be able to sustainably cover all their energy needs from within their own national borders. It will therefore be necessary to import sustainably produced energy carriers from countries with high potential for renewable energy and lower demand. In principle, these could take the form of electrical energy, hydrogen, or chemical energy carriers derived from hydrogen, i.e. liquefied methane, methanol, ethanol, synthetic crude oil, or ammonia. However,

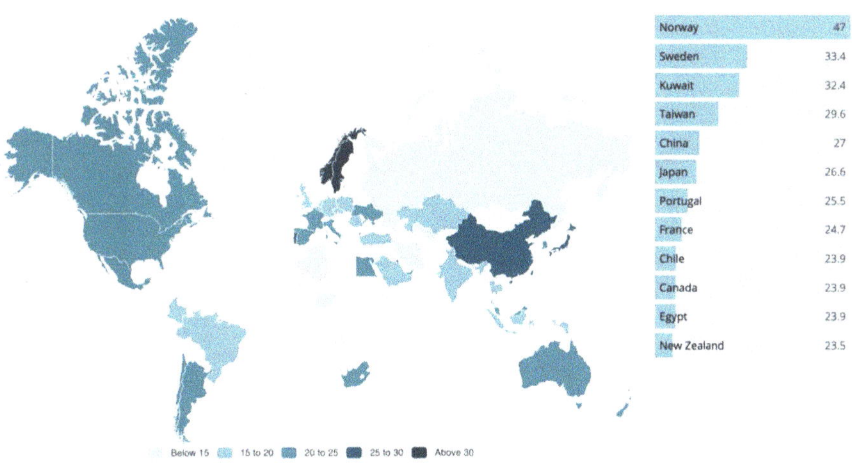

Figure 18.1. Electricty as a proportion of total final energy consumption in different countries in percentages. Data for 2021. The worldwide share in 2021 was 20.4%. Source: https://yearbook.enerdata.net/electricity/share-electricity-final-consumption.html (accessed on July 29, 2022). Reproduced with permission © Enerdata.

the infrastructure costs and energy losses created by transporting electrical energy over long distances are generally higher than for chemical energy carriers, relative to the energy content. The latter are therefore particularly suitable for energy import scenarios.

As a result, the hitherto unfamiliar concept of converting the combustion products CO_2 and H_2O back into chemical energy carriers using CO_2-free electrical energy, called 'power-to-X,' is attracting more and more attention. From a thermodynamic perspective, this is an energy downgrade, as the conversion of work into chemical energy or heat involves a loss of exergy. However, from a system perspective, power-to-X is necessary for all the reasons mentioned above and summarized in table 18.1.

Two different basic types of power-to-X plant can be imagined to fulfill the purposes mentioned in table 18.1. At locations with high potential for low-cost

Table 18.1. Reasons for the use of power-to-X in an energy system largely based on variable renewables.

Sector	Purpose	Challenges
Electrical power	• Prevent grid overload by conversion of excess electrical energy into chemical energy • Maximize yield from wind energy and solar photovoltaics • Reduce the need for power grid expansion by providing local conversion and a link to the gas grid • Provide sustainable fuels for electrical power generation when needed	• Network fees increase the costs for energy storage and power-to-X • Low number of full-load hours increases specific capital costs • Round-trip efficiency is lower than for mechanical or electrochemical storage • Dynamic operation creates technical challenges
Heat, fuels, and chemicals	• Provide sustainable fuels for generation of high-temperature heat • Provide sustainable fuels for aviation, shipping, and heavy-duty road transport • Provide sustainable carbon feedstocks for the chemical industry	• Conversion losses make fuels for heating less attractive than direct electrical heating or heat pumps • Conversion losses in power-to-X and power train make fuels less attractive than direct electrification
Overall	• Enable the import of large amounts of sustainable energy • Provide flexibility with regard to final energy carrier • Keep a national energy reserve	• Political and currency risks resulting from international dependencies • Additional costs for resilience options

renewable power and low population density, continuously operated large-scale power-to-X plants can be built, for which the corresponding wind energy or photovoltaic plants are additionally constructed without a connection to the local power grid. These plants are operated at the highest possible capacity, meaning that they must be integrated with storage facilities for electricity or chemical intermediates to compensate for the variability in power generation. Since there are usually no large CO_2 point sources at such locations and also a lack of a sufficient water supply and wastewater disposal, a supply of CO_2 and integrated water management must also be considered and implemented in addition to renewable power generation. Seawater desalination and direct CO_2 extraction from the atmosphere are often the preferred options here. The energy carriers produced can be shipped to other countries, and some can also be used locally. Because of the need for global defossilization, this also holds true for the additional renewable power generated by such a plant.

Apart from large-scale power-to-X plants at these so-called sweet spots, smaller distributed power-to-X plants that can be operated in a load-adaptable manner are needed. In countries with well-developed industrial infrastructure, they can be operated to utilize renewable power surpluses, relieve the power grid, and enable sector coupling. Depending on the location, large or small CO_2 point sources can be used in addition to the option of extracting CO_2 from the atmosphere. The supply of water for electrolysis and the treatment of wastewater do not pose any particular challenge for such plants, provided that existing industrial and municipal infrastructures can be used. Load-adaptable power-to-X plants are also attractive for stand-alone solutions in remote areas, where they can provide chemical fuels for local use. At such sites, the supply of water for electrolysis and the treatment of wastewater must also be realized, whereby particularly challenging environmental standards may have to be met. In addition, this concept requires that fuels can be produced in accordance with the specifications, and compliance with these specifications must be guaranteed by appropriate means.

Notwithstanding the need for sustainable chemical fuels with a view to climate change and the opportunities for power-to-X as outlined above, there are are also challenges, which mainly relate to economics and politics. Digging deeper into the costs and sustainability as well as the societal and international aspects of power-to-X would go beyond the objectives of this chapter. Instead, the authors refer the reader to a study by Frontier Economics from 2018 [1] and a scoping paper by experts from the international PtX-Hub Berlin from 2022 [2] as well as publications by Schnuell *et al* 2019 [3] and Koj *et al* 2019 [4].

In the above context, Fischer–Tropsch synthesis is an industrially established route for the production of clean-burning hydrocarbon fuels from synthesis gas; such fuels can be made from coal, natural gas, biomass, or even CO_2 and hydrogen [5, 6]. These fuels can be handled by existing infrastructures and do not require changes in the power train, which makes them very attractive as replacements for current fossil fuels in aviation, shipping, and long-haul heavy road transport. This chapter focuses on the integration of solid oxide electrolytic cell (SOEC) technology with Fischer–Tropsch synthesis to produce so-called electro-fuels (also known as e-fuels) from

CO_2, water, and renewable electricity. The choice of SOEC technology for integration with Fischer–Tropsch synthesis is motivated by a simplified process layout and higher overall efficiency. As liquid fuels are targeted, this approach is also known as PtL.

18.2 SOEC-based options for syngas generation for Fischer–Tropsch-based PtL plants

When synthesis gas is made from CO_2, the use of SOEC technology allows one to take advantage of the higher thermodynamic efficiency of splitting H_2O into H_2 and O_2 in the vapor state compared to that of the liquid state. Fischer–Tropsch synthesis is highly exothermic and runs at a temperature higher than 180 °C. This provides the opportunity to produce the steam needed for the electrolysis using the waste heat of the reaction, which is the basis for the high overall energy efficiency of the PtL process. Moreover, because conventional SOEC technology is based on a solid oxide-ion-conductive electrolyte rather than a proton-conductive one, CO_2 can be reduced to CO at the same time as the reduction of H_2O to H_2. This mode of operation is called 'co-electrolysis'; it avoids the requirement for a separate high-temperature reactor to convert CO_2 and H_2 into synthesis gas via the reverse water–gas shift reaction (rWGS).

18.2.1 Fischer–Tropsch synthesis process alternatives

Fischer–Tropsch (FT) synthesis is a surface polymerization reaction, in which synthesis gas (CO and H_2) is converted into hydrocarbons of different chain lengths and water as a side product (equation (18.1)). FT synthesis (FTS) has been commercially established for close to 100 years; the Sasol and Shell companies have the most practical experience today. The reaction is highly exothermic, with a heat release of approximately 140–160 kJ per mole of CO converted. The reaction mechanism is complex and still not resolved in all its details. Generally, the initiation of the reaction can be described as the adsorption of CO and H_2; chain propagation takes place in the polymerization reaction, and chain termination is performed in the desorption step [7, 8]. The water produced during the reaction is known to cause kinetic inhibition [9].

$$\mathrm{nCO} + 2\mathrm{nH_2} \rightarrow (-\mathrm{CH_2}-)_\mathrm{n} + \mathrm{nH_2O} \tag{18.1}$$

The FT products contain hydrocarbons, which are mostly alkanes and alkenes as well as oxygenates. The exact composition depends on the catalyst and the operating conditions. The hydrocarbon chain length distribution is determined by the probability of chain growth. It is often described by the Anderson–Schulz–Flory (ASF) equation (equation 18.2), which is derived from a simple chain growth mechanism in which the probability of chain growth is independent of the chain length [10].

$$w_n = n(1 - \alpha)\alpha^{(n-1)} \tag{18.2}$$

Here, w_n is the weight fraction of hydrocarbons with n carbon atoms, and α is the probability of chain growth. Taking the natural logarithm and rearranging leads to:

$$\ln\left(\frac{w_n}{n}\right) = \ln\left(\frac{1 - \alpha}{\alpha}\right) + n \cdot \ln(\alpha) \tag{18.3}$$

Equation (18.3) shows that plotting the natural logarithm of the weight fraction divided by the number of carbon atoms leads to a straight line that has a slope of ln (α). From this so-called ASF plot, the probability of chain growth can easily be determined as indicated in figure 18.2 for the results of a lab-scale reactor. While the long-chain hydrocarbons are typically reproduced more or less well, as in the case of the lab-scale data shown in figure 18.2, the short-chain hydrocarbons often deviate from the ASF line. More specifically, C_1 is higher than predicted whereas C_2 to C_4 are frequently lower than predicted by the ASF model [7]. It should be mentioned, however, that an accurate analysis of the product composition is generally far from trivial, particularly if the higher hydrocarbons are condensed and analyzed offline.

For FT process design, three important aspects should be considered: the catalyst, the operating conditions, and the reactor. Different metals, such as Fe, Co, Ru, Rh, and Ni are active for CO hydrogenation to hydrocarbons. They are typically used as nanoparticles on a porous inorganic support with a high surface area, e.g. γ-Al_2O_3, TiO_2, or SiO_2. It is generally agreed that Ni is not a good FT catalyst due to its high activity for hydrogenation and hydrogenolysis, which leads to high CH_4 and low alkene formation as well as the hydrocracking of long-chain hydrocarbons [9]. The metals most often used commercially are Fe and Co, due to their costs and availability. Fe is generally less active and less stable but cheaper. Different from

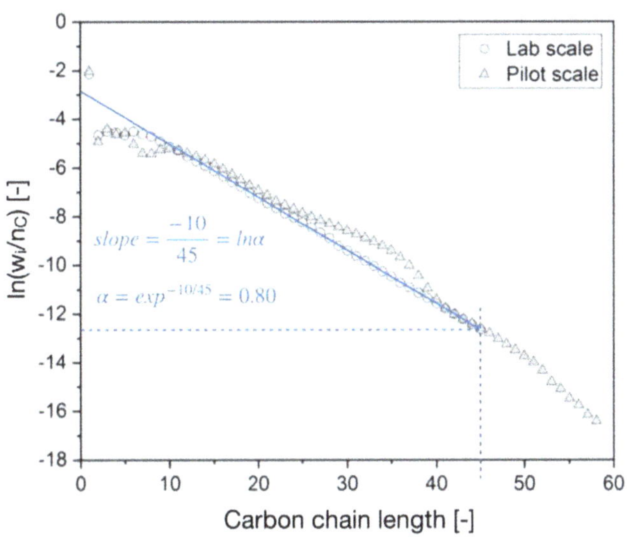

Figure 18.2. Exemplary ASF plot for a commercial Co-based FT synthesis (FTS) catalyst applied in two different reactors under similar conditions. Adapted from reference [11] (figure 10). Reproduced with permission. © Wiley.

Co, it also catalyzes the water–gas shift reaction, which lowers the selectivity by converting CO into CO_2 as a side product. To enhance its catalytic activity and selectivity as well as its stability, so-called promoters are added, which may have an impact on the catalyst's chemical or physical properties [7]. For example, Fe is stabilized with an alkali such as K_2O to achieve high selectivity, in particular when using an acidic support such as γ-Al_2O_3 [9].

The catalytic activity and product distribution depend on the operating conditions. The main parameters are temperature, pressure, synthesis gas composition, and space velocity. Increasing the temperature has a big impact on the desorption rate of the hydrocarbons from the catalyst surface. Therefore, shorter and less saturated hydrocarbons are produced at elevated temperatures. The pressures used are in the range of 15 to 60 bar; the pressure, in combination with the synthesis gas composition, influences the adsorption of CO on the catalyst. The synthesis gas composition, specifically the H_2/CO ratio, affects the product selectivity. Increasing the H_2/CO ratio leads to enhanced hydrogenation and therefore to a decreasing chain growth probability and a lower alkene fraction. Other than the above-mentioned operating conditions, space velocity does not have a big impact on product selectivity but rather on conversion.

Based on the operating temperature, there are two common operational modes, high-temperature (270 °C–350 °C) and low-temperature (200 °C–250 °C) FTS. Fe catalysts can be used for both. However, due to the advantages of Co in low-temperature FTS, Fe is mostly used at higher temperatures. High-temperature FTS mainly produces short-chain hydrocarbons in the C_2–C_4 range with a high fraction of alkenes; it also shows high CH_4 selectivity. The water–gas shift reaction increasingly sets in at higher temperatures, increasing the CO_2 selectivity. Longer hydrocarbon chains can be produced through low-temperature FT over supported Co catalysts [5, 7, 12]. This is generally the preferred option for the PtL process.

In principle, FTS starting from CO_2 instead of CO is desirable, as this avoids the need for the energy-consuming high-temperature synthesis gas production step upstream. In this case, the endothermic rWGS reaction (equation (18.4)) precedes the actual FTS (equation (18.1)).

$$CO_2 + H_2 \rightleftharpoons CO + H_2O \quad \Delta H_{298} = +41 \text{ kJ/mol} \tag{18.4}$$

It has been shown by several authors that CO_2-based FTS is possible with Fe catalysts and that a similar hydrocarbon distribution to that of CO-based FTS can be achieved using Fe catalysts [13, 14]. However, until now, this approach has not yet been implemented industrially. Compared to CO as a carbon source, the activity with CO_2 is rather low. This can be explained by the observation that FTS is rate limiting in this system, which is why the partial pressure of CO is mainly controlled by the equilibrium of the rWGS reaction [14]. At typical FTS temperatures, this equilibrium is on the left-hand side, leading to a low partial pressure of CO and therefore low activity. Moreover, starting with CO_2 means forming two moles of water per mole of carbon converted. This increases the partial pressure of water in the course of the reaction to a greater extent than when CO is used, which inhibits the kinetics and endangers catalyst stability.

For FTS, three main reactor types are to be considered: fixed bed, slurry bed, and fluidized bed. Important design requirements include avoiding the limitations of slow heat and mass transport, enabling catalyst–product separation, removal of the reaction heat, and the scalability of the system. Due to the high reaction heat, a cooling system and a significant heat exchange surface area are required between the reaction and the cooling side [7, 12]. This is precisely what inspired the development of microstructured reactors for FTS, which offer outstanding cooling power [11, 15, 16].

18.2.2 Syngas generation by SOEC and rWGS

According to equation (18.1), FTS needs a feed gas with an H_2:CO molar ratio of around two at a typical pressure of 20 bar. This can be produced from H_2 delivered by an SOEC unit and CO_2 via the rWGS reaction (equation (18.4)). A considerable number of scientific papers in the literature deal with the rWGS reaction in the context of PtL processes [17, 18]. Whereas it has not yet been practised industrially at a larger scale, several companies show it in their portfolio. The challenges of rWGS include catalyst stability, reactor design for high-temperature operation and efficient heat integration, as well as carbon formation and high-temperature corrosion. The Karlsruhe Institute of Technology (KIT) spin-off INERATEC (Karlsruhe, Germany) has developed very compact pressurized rWGS reactors for modular PtL plants based on a micro packed-bed concept, i.e. microstructured plates filled with fine catalyst particles. INERATEC has built several reactors with a design capacity of 500 kW H_2 input for two 1 MW PtL plants installed in northern Germany in 2021 (figure 18.3).

In addition to equation (18.4), a number of side reactions have to be considered for a rWGS reactor system. These include the methanation of CO_2 (equation (18.5)) as well as CO (equation (18.6)), the Boudouard reaction (equation (18.7)), and methane pyrolysis (equation (18.8)).

$$CO_2 + 4H_2 \rightleftharpoons CH_4 + 2H_2O \quad \Delta H_{298} = -165 \text{ kJ mol}^{-1} \tag{18.5}$$

$$CO + 3H_2 \rightleftharpoons CH_4 + H_2O \quad \Delta H_{298} = -206 \text{ kJ mol}^{-1} \tag{18.6}$$

Figure 18.3. Atmosfair's 1 MW Fischer–Tropsch synthesis plant built by INERATEC at the EWE biogas plant site in Werlte, Germany. Photographs (© R Dittmeyer) taken during the official inauguration on October 4, 2021. Left: one of the two rWGS reactors (open door). Right: two out of eight FTS reactor modules (left door).

$$2CO \rightleftharpoons CO_2 + C \quad \Delta H_{298} = -172 \text{ kJ mol}^{-1} \tag{18.7}$$

$$CH_4 \rightleftharpoons C + 2H_2 \quad \Delta H_{298} = +75 \text{ kJ mol}^{-1} \tag{18.8}$$

According to thermodynamics, high temperatures are needed to achieve high selectivity to CO and avoid solid carbon formation via the Boudouard reaction (figure 18.4). Several catalysts are under discussion, such as Ni/Al_2O_3, Ni/SiO_2, Rh/Al_2O_3, Pt/TiO_2, and K-Mo_2C. The challenges in catalyst development for the rWGS reaction are comparable to those encountered in steam methane reforming, in which sintering, hydrothermal ageing, and carbon deposition that blocks the active sites are important deactivation mechanisms which have to be countered by catalyst design and a careful choice of operating conditions [19–22].

To limit the size of the rWGS reactor, operation at the same pressure as that used for FTS is favorable. In addition, compressing the CO_2 instead of the synthesis gas requires less energy. However, elevated pressure is associated with an increased amount of CH_4 at equilibrium (figure 18.4). The kinetic inhibition of methane formation is therefore desirable. Pressurized H_2 can be provided either by an ambient-pressure SOEC unit followed by a H_2 compressor or by a pressurized SOEC unit. While the energy required to compress water prior to electrolysis is substantially lower than that required for the compression of the produced H_2, operating an SOEC unit at a pressure of 20 bar requires a pressure vessel around the stacks, which significantly complicates the system design and increases the equipment cost. Figure 18.4 gives the equilibrium compositions of all species according to equations (18.4)–(18.8) at 1 bar and 20 bar of total pressure for an H_2:CO_2 molar ratio of 2.1 in the feed in order to show the effects of pressure and temperature on the equilibrium. For this feed, an H_2:CO molar ratio of around two is predicted for the produced synthesis gas at 20 bar and 870 °C. The CO_2 conversion rate is 67% and

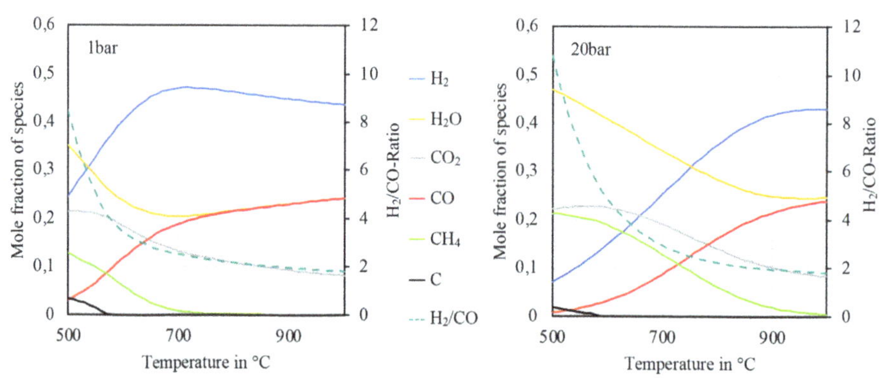

Figure 18.4. Equilibrium compositions for the rWGS reaction at pressure levels of 1 bar (left) and 20 bar (right), respectively, as a function of temperature. The H_2:CO_2 molar ratio in the feed is fixed at 2.1:1. The H_2:CO molar ratio in the product varies from 2.16 at 800 °C to 1.95 at 900 °C and 1 bar and from 2.21 at 800 °C to 1.94 at 900 °C and 20 bar. Calculated using Aspen Plus V10 (minimization of the Gibbs free energy; solid carbon modeled as graphite; all other products in vapor phase; property method used: Soave–Redlich–Kwong (SRK) [23]).

the methane selectivity is 12%, which give a CO yield of 60%. The upper temperature limit for solid carbon formation is 600 °C. Hence, from a thermodynamic point of view, operation at 800 °C or more seems safe. However, as both the species concentrations and the temperature vary along with the conversion rate, the local thermodynamic conditions everywhere inside the reactor have to be analyzed from the perspective of carbon formation. This analysis must also consider the effect of recycling unconverted synthesis gas and gaseous by-products of the FTS back into the rWGS reactor, which is necessary to reach a high carbon utilization.

Industrial catalysts for FTS do not tolerate O_2 impurities in the synthesis gas at levels of more than ppm. Here, SOECs have an advantage over PEM electrolyzers, as the oxide-ion-conducting ceramic membranes do not allow the crossover of molecular O_2. Therefore, as long as all cells retain their mechanical integrity, it can be assumed that the produced H_2 meets the O_2-free requirement.

FTS cannot be run at 100% CO conversion for reasons of catalyst activity and stability. Also, due to the mechanism used, short-chain hydrocarbons are formed as side products. Therefore, after FTS, a gaseous stream appears which can be recycled into the rWGS reactor as shown in figure 18.5 in the case of a process scheme that relies on an ambient-pressure SOEC unit to produce H_2. The reactor design must take this recycled stream into account, both with a view to internal heat exchange and also in terms of its effects on the equilibrium and the reaction kinetics, in addition to the heating required for the H_2 and CO_2 feed gas streams. Providing the reaction heat for the endothermic rWGS reaction at the required high temperature level is not a trivial design task. Hence, alternative reactor concepts have been developed, e.g. a concept for autothermal operation that uses O_2 injection [22]. O_2 injection increases the amount of water formed in the rWGS stage and obviously needs pure O_2. However, the latter is available from the electrolyzer, and avoiding the installation of a large heat-transfer area in the reactor simplifies the reactor design. It must be ensured that the O_2 is completely consumed in the rWGS reactor down to the ppm level in order to protect the FTS catalyst.

The SOEC shown in the schematic of figure 18.5 makes use of indirect heat exchange to superheat the evaporated water with the sensible heat of the H_2 product stream. Moreover, the reaction heat of the FTS can be used to evaporate the water

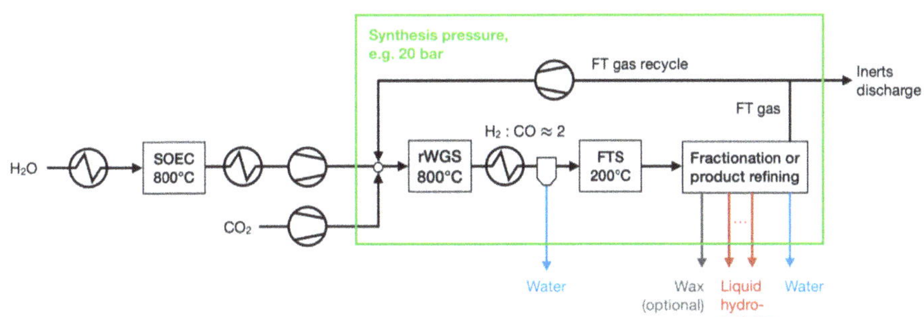

Figure 18.5. Basic scheme of a PtL plant that uses an ambient-pressure SOEC together with an rWGS reactor stage for the production of synthesis gas for FTS.

for the SOEC, as discussed in more detail in section 18.3. However, it still seems unsatisfactory from the point of view of the overall energy balance that the H_2 has to be cooled down before reaching the compressor and afterwards heated up again to a similar temperature for conversion in the rWGS reactor. This could be avoided using a pressurized SOEC thermally integrated with a downstream rWGS reactor.

18.2.3 Syngas generation by co-SOEC

Taking the idea of thermally integrating an SOEC with a downstream rWGS reactor one step further by including material integration leads to the concept of generating the synthesis gas directly in the SOEC by the simultaneous conversion of steam and CO_2 on the same cathode. Figure 18.6 shows the basic scheme, again for the case of ambient-pressure operation of the SOEC. Pressurized operation is possible as well, but comes with the same challenges as those encountered for H_2 generation via SOEC.

Given the high operating temperature and the fact that a Ni-containing cermet is used for the cathode, the SOEC is capable of converting CO_2 into CO in the presence of steam and H_2 derived from it in two ways. CO_2 can be reduced via the electrochemical pathway (equation (18.9)) [23], and it can also react via the rWGS pathway (equation (18.4)) with H_2 stemming from the steam electrolysis. While the commonly used terms 'co-electrolysis' or 'co-SOEC' suggest that the electrochemical pathway is dominant, this is actually only the case for a steam content lower than 20%, due to the fact that the electrochemical reduction of steam proceeds considerably faster than that of CO_2 at the voltages typically applied [24].

$$CO_2 + 2e^- \rightarrow CO + O^{2-} \tag{18.9}$$

18.2.4 Comparative assessment

The two different options for syngas generation were studied using process simulations performed by Aspen Plus V10. A pinch analysis using Aspen Energy Analyzer V10 was included to assess the potential for indirect heat transfer between the individual process units and streams to maximize the energy efficiency. More information on the procedure and the results can be found in [25]. The simulation

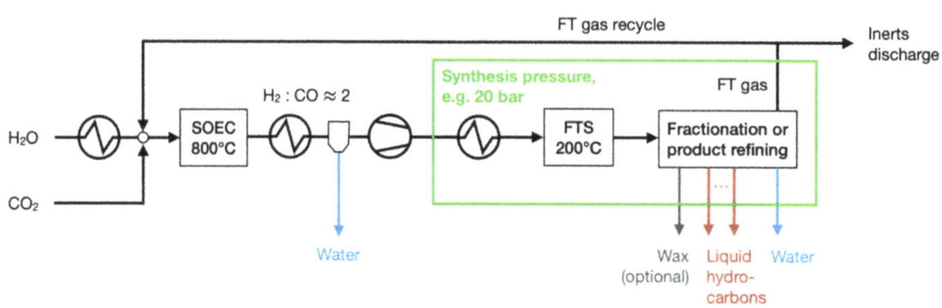

Figure 18.6. Basic scheme of an ambient-pressure SOEC in co-electrolysis mode to provide synthesis gas for FTS.

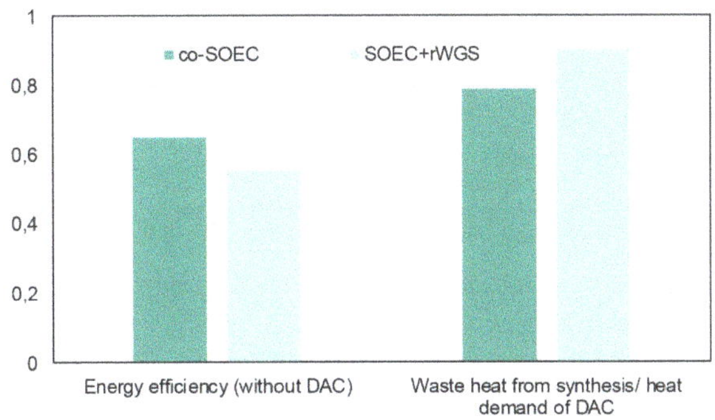

Figure 18.7. Comparison of two PtL process variants using co-SOEC or SOEC-rWGS for syngas generation for FTS in terms of (a) energy efficiency (chemical energy of the fuel / total energy input to the process) and (b) the waste heat produced by the synthesis divided by the heat demand of a low-temperature DAC.

was based on thermodynamic and stoichiometric models for the individual process units using conversion and product distribution data from the literature [26]. The reaction kinetics were not considered at this stage, as the main purpose of the simulation was to establish and compare the maximum efficiencies of the PtL process for the various syngas generation methods and examine the potential for providing heat to an optional upstream CO_2 capture process. The two key observations are reported in figure 18.7. The co-SOEC variant turns out to offer a substantially higher energy efficiency of up to 64% compared to the SOEC-rWGS variant, which only reaches 55%. However, the rWGS option can deliver more surplus heat for an optional CO_2 capture plant connected to the synthesis process, which eats up part of the efficiency advantage in the case of a combined process, particularly if low-temperature DAC is to be used (for more details, see the following section).

18.3 Process integration in SOEC-based PtL plants

Process integration in an SOEC-based PtL plant primarily aims at increased overall energy efficiency. This is possible by using the reaction heat generated by the synthesis step to produce steam for the electrolysis. The available options are determined by the reactor technology used and the specific layout of the synthesis process. Depending on the CO_2 source and the strategy used to process the FT crude product (a wide spectrum of hydrocarbons of different chain lengths, comprising saturated and unsaturated compounds as well as some oxygenates) into finished fuels, the PtL plant may also include other process stages in addition to electrolysis and FTS which need or produce heat at various temperature levels and therefore have to be considered for thermal integration at the plant level. The second aim of process integration is to simplify the process design and minimize the number of process units. This reduces the interstage temperature and pressure changes and may also reduce the investment requirements.

18.3.1 Potential CO_2 sources

The CO_2 needed for the production of synthetic hydrocarbons may be supplied to a PtL plant by pipeline. Alternatively, it may be obtained from a local point source or extracted from the ambient air by DAC. The specific energy demand for CO_2 capture generally increases with decreasing CO_2 content; table 18.2 provides ranges for industrial off-gases and ambient air. Note that for DAC, a significant portion of that energy is electrical, as a much larger amount of gas must be passed through a contactor to capture one ton of CO_2, due to the low CO_2 content. Some of the options for CO_2 supply do not allow the waste heat of the synthesis and the electrolysis to be used to meet the heat demand of CO_2 capture. This is obvious for a pipeline separating CO_2 capture and use in space. Moreover, for many industrial point sources, plenty of low-temperature waste heat is available on site. In such cases, no particular benefit arises if additional waste heat can be provided by the fuel synthesis process. DAC is differentiated according to the process used for CO_2 capture. In high-temperature DAC, CO_2 is absorbed in an aqueous KOH solution, which is regenerated by adding $Ca(OH)_2$, resulting in the formation of solid $CaCO_3$. The $CaCO_3$ is then split into CaO and CO_2 at temperatures in the range of 800 °C–900 °C, which are generated by burning natural gas in pure oxygen [25, 27]. The CO_2 resulting from the combustion mixes with the CO_2 released from $CaCO_3$ and is collected and purified if required. Given that this process needs high-temperature heat for regeneration, it would not profit from a nearby FT-based fuel synthesis process, which could deliver waste heat at a temperature level of up to 200 °C or even 300 °C. The situation is different when low-temperature DAC is used, as this requires waste heat at a temperature level between 90 °C and 120 °C, depending on the choice of adsorbent [25, 28, 29].

Regarding the specific costs, CO_2 from a large industrial point source such as a cement plant, a municipal waste incineration plant, or a bioethanol plant is cheaper than CO_2 collected from smaller distributed point sources, such as biogas, sewage gas plants, or DAC. Table 18.2 shows cost ranges reported in the literature [25, 30–32]. However, large industrial point sources are typically not found at locations with high

Table 18.2. Costs for CO_2 capture and purification in 2020 and predicted costs for 2050. The numbers for DAC are based on information provided by Carbon Engineering, Climeworks, and Global Thermostat. Data for point sources were taken from the literature. Given the maturity of technologies for CO_2 capture from point sources, no further cost reduction is expected there. Reproduced from [25] with permission.

		Point source capture					
		Cement		Combustion		DAC	
		2020	2050	2020	2050	2020	2050
CO_2 content	%	14–33		7–45		0,04	
Cost	€/t	15–140		30–130		125–800	41–82
Energy demand	kWh/t	>600		>330		N.I.	1400–2500

Table 18.3. Cost structure of PtL fuels for exemplary locations with different costs for the provision of CO_2 and renewable electrical energy. Calculations are based on König *et al* 2015 and their simulated Fischer–Tropsch synthesis process. PtL fuel costs were recalculated from published data by converting the costs for electrical energy and CO_2 accordingly. All costs given are pure production costs without taxes and transport fees. Reproduced from [25] with permission.

CO_2	Share of CO_2		Electricity	Share of electricity		PtL-Cost	Reference
€/t	€/L	%	€/MWh	€/L	%	€/L	
37.75	0.085	3.1	105	1.85	67.5	2.74	Albrecht *et al* 2017 [30]
11.29	0.060	2.3	105	1.87	73.1	2.56	Becker *et al* 2012 [31]
37.59	0.1	2.9	140	2.31	68.1	3.39	König *et al* 2015 [32]
37.59	0.1	3.6	105	1.73	62.5	2.77	Own calculations (based on Albrecht *et al* 2017 [30] and König *et al* 2015 [32])
37.59	0.1	7	20	0.33	24	1.37	
100	0.27	17	20	0.33	22	1.53	
250	0.67	34	20	0.33	17	1.93	

potential for low-cost electrical power generation from renewables. Because electrical power is the factor that has the greatest influence on the manufacturing costs of PtX products, DAC is often the most viable option for CO_2 supply at such 'sweet spot' locations. Table 18.3 lists the expected proportions of CO_2 and electricity in the total PtL costs for different CO_2 and electricity prices.

It turns out that for every euro cent per kWh spent on electricity, one has to reckon on a cost of 16–18 euro cents per liter of fuel. On the other hand, for every 10 euros spent per ton of CO_2, fuel costs amount to two to three euro cents per liter.

18.3.2 Thermal integration between FT and co-SOEC

Low-temperature FTS is operated at a reaction temperature of 200 °C–250 °C, while high-temperature FTS runs at 270 °C–350 °C. The reaction delivers about 140–160 kJ of heat per mole of CO converted. FTS reactors have to be cooled in order to prevent an undesirable increase in temperature, which would otherwise accelerate methane formation and may cause deactivation of the catalyst. According to the reaction stoichiometry of FTS (equation (18.1)), two moles of H_2 are needed per mole of CO. In addition, up to one mole of H_2 is needed per mole of CO_2 in the SOEC to produce CO, depending on how much of the CO is actually produced via the electrochemical (equation (18.9)) or thermocatalytic (equation (18.4)) pathway. Hence, up to three moles of water have to be evaporated to be converted into H_2 in the SOEC per mole of CO_2 converted first into CO and then into hydrocarbons. This needs up to 122.1 kJ at standard temperature and pressure. Comparing these numbers shows that the reaction heat of the FTS is more than sufficient to evaporate all the water needed for the SOEC.

For low-temperature FTS, the use of boiling water for cooling is an effective way to control the reactor temperature, as this is associated with a high heat-transfer coefficient on the cooling side and a uniform coolant temperature, while the pressure

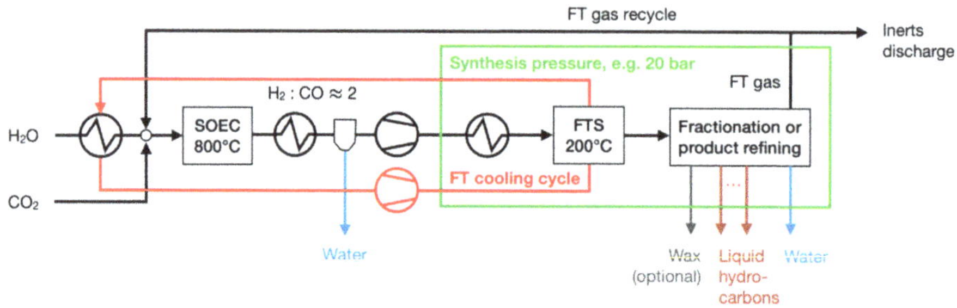

Figure 18.8. Scheme based on figure 18.6 in which a boiling-water FT cooling cycle produces the steam for the SOEC.

in the cooling cycle is still moderate, namely 14.5 bar at 200 °C and 38.7 bar at 250 °C. Typically, one would use a closed water cooling cycle to produce steam at the reactor temperature, which is then condensed via indirect heat exchange in an evaporator to produce superheated steam (at ambient pressure) for the SOEC. For high-temperature FTS reactors, molten salt cooling is more appropriate, as the boiling-water pressure at 350 °C would be as high as 164.3 bar. With molten salt cooling, one has to live with an increase in the coolant temperature in the FTS reactor. Moreover, the design of the evaporator that produces superheated steam (at ambient pressure) for the SOEC using hot molten salt differs from the design of an evaporator heated by condensing steam.

Stand-alone FTS and SOEC units typically already make use of internal heat exchange to maximize the energy efficiency and to provide the necessary functionalities to execute start-up, shutdown, or load-change procedures. Thermal integration in an SOEC-FTS plant may go beyond just using the FT cooling cycle to produce the steam for the SOEC. The available opportunities can be identified by a pinch analysis based on detailed piping and instrumentation diagrams (P&IDs) of the specific units. This will also show the potential for heat export to other installations on site. However, the requirements of start-up, shutdown, and load-change procedures for the single units as well as the integrated plant with FT gas recycling must nevertheless be taken into account (figure 18.8).

18.3.3 Thermal integration in DAC-based PtL plants using co-SOEC for syngas generation

As outlined above, low-temperature DAC is the preferred CO_2 source for PtL production when no industrial point source is available. For this technology, the electrical energy demand is in the range of 150–300 kWh per ton of CO_2 captured. In addition, 1.170–2.000 kWh of thermal energy are needed at a temperature level of 100 °C (see reference [25] and the references therein). The overall energy efficiency and the economics of the PtL plant can be improved if at least part of this heat can be provided using the waste heat remaining from the fuel synthesis process. The potential for heat integration depends on the specific process layout. Figure 18.9 exemplarily shows the major heat and electrical energy flows of an integrated PtL process combining ambient-pressure co-SOEC, FTS at 20 bar, and two-stage FT

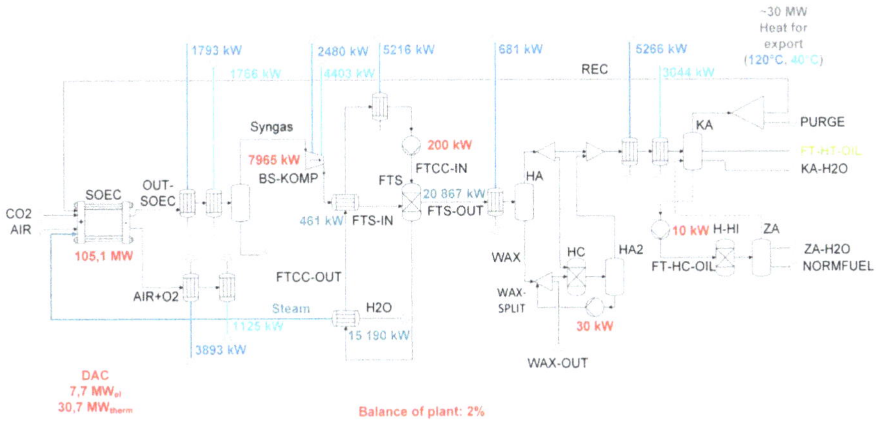

Figure 18.9. Process flow diagram of the synthesis part of a PtL process that uses low-temperature DAC as a CO_2 source, syngas generation by co-SOEC, FTS, and FT crude refining. The electrical energy input is marked in red. The waste heat to be used in the DAC plant is marked in blue. Dark blue indicates a waste heat temperature level of 120 °C, light blue denotes 40 °C. Table 18.4 reports the overall energy efficiency.

Table 18.4. Overall energy efficiency of the PtL process shown in figure 18.9.

	Energy efficiency including DAC [%]	
Output	Only waste heat at 120 °C used	All waste heat used (heat at 40 °C raised to 120 °C via high-temperature heat pump)
65 MW (LHV)	48.3	50.2
70 MW (HHV)	52.0	54.1

crude refining performed by the hydrocracking and isomerization of the FT wax and the joint hydrogenation of the primary and hydrocracked FT oils.

Not shown is the DAC plant, which needs 7.7 MW of electrical energy and 30.7 MW of heat at a temperature slightly above 100 °C to provide the 400 t/d of CO_2 required for the PtL plant. Calculations were performed using Aspen Plus V10. The plant needs 105 MW of electrical energy for the electrolysis, 8 MW for syngas compression, 0.2 MW for the pump in the cooling cycle of the FTS reactor, and 40 kW for two pumps in the FT crude refining section, which adds up to an electricity consumption of 113.3 MW. Two percent of that sum was added for the electricity consumption of the remaining plant components. The plant produces 138 t/d of hydrotreated FT oil, from which a kerosene fraction meeting the ASTM D7566 specifications for synthetic paraffinic kerosene can be obtained by distillation. About 20 MW of waste heat at 120 °C and 10 MW at 40 °C is available from the plant. Table 18.4 gives the chemical energy output of the PtL plant in terms of the lower and higher heating values of the FT oil as well as the overall energy efficiency, defined as the ratio of the chemical energy output divided by the total energy input including the energy demand of the DAC plant. In order to utilize the low-temperature heat at

40 °C for the DAC process, a high-temperature heat pump with a performance coefficient of two was assumed, which raises this waste heat to a temperature level of 120 °C, i.e. half of this heat is added to the electrical power consumption. With these assumptions, the PtL process (including DAC) reaches an overall efficiency of 50.2% if the lower heating value is considered and 54.1% if the higher heating value is considered.

18.3.4 Options for FT crude refining

To produce fuels compliant with specifications from FT crude, subsequent refining is necessary. The strategy used for the refining also depends on the regulation of the target product, which may be jet fuel, diesel, or motor gasoline. One option is to co-process the FT crude with fossil fuel materials in existing refineries and in this way reduce the CO_2 footprint of all the fuels produced by that pool. Alternatively, the FT crude can be upgraded to fuels compliant with specifications on site in a simplified mini-refinery. While the first option is generally perceived to be more cost-effective, the second also has advantages: it minimizes fuel transport. It enables sustainable fuel production in remote areas with limited infrastructure. This could even be used to produce fuels tailored for the specific location, e.g. in the arctic. Moreover, it is expected that future refineries will have a distinctly different product distribution than today's, with a higher proportion of jet fuel and much less fuel for road transport. Therefore, the process portfolio of the existing refineries would in any case have to be adapted to this changing demand. Last but not least, regulations may differ, either enforcing the admixture of CO_2-free fuels with their fossil analogs or a complete switch to 100% sustainable fuels, depending on the strategy for reducing emissions in certain applications.

For aviation jet fuel, the possible alternatives for introducing sustainable fuels are illustrated in figure 18.10. The first is the co-processing of the FT crude with fossil fuel materials. Here, the resulting fuel must fulfill the ASTM D1655 norm. The second is to produce a sustainable drop-in fuel in a dedicated plant for blending with jet fossil fuel, e.g. to meet admixing quotas. In this case, the underlying production process must be a certified process according to ASTM D7566, and the product must fulfill the ASTM D7566 norm. Synthetic paraffinic kerosene (SPK) produced by FTS has generally qualified on that basis since 2009. Blending SPK with fossil jet fuel is possible up to 50 vol%. The reason for this limitation is that FT-SPK does not contain aromatics which are needed to ensure seal swell and tightness of valves in the existing aircraft and aviation fuel handling systems. The specification therefore also states that the blend must have a minimum aromatics content of 8%. The third option is to develop a new simplified refining scheme for the processing of FT crude to 100% sustainable aviation fuel (SAF). Such a process would have to go through approval according to ASTM D4054. One possible strategy for synthesis could be to produce certain preferred aromatic compounds with a low potential for soot formation from the side products of FT-SPK production and add them to the FT-SPK at a level of 8–10 vol%, thereby making up a 100% SAF. This approach is being investigated in ongoing research.

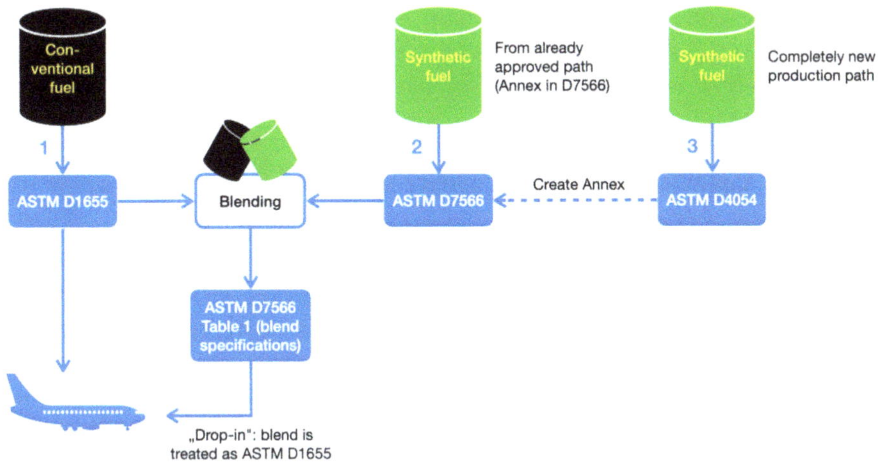

Figure 18.10. Different paths for the approval of FT-based sustainable aviation fuels (SAFs). Path 1 consists of the co-processing of SAF with conventional kerosene, in which the product must comply with the specifications of ASTM D1655 (as is the case for conventional kerosene). Path 2 consists of the blending of SPK with fuel from an existing certified synthesis route. Here, the synthesis route and the produced SAF must comply with the specifications of ASTM D7566. Blending is limited to 50 vol%. Path 3 involves the synthesis of SAF via a new route which must go through an approval process. Standard and fast-track approval processes exist and are described in ASTM D4054.

18.4 Modular technologies that enable decentralized PtL production

The decentralized production of PtL fuels needs process technologies that can be operated in a load-adaptable manner. As outlined in section 18.1, such plants could be operated in a grid-integrated manner in order to utilize local electricity surpluses and to fully exploit the potential yield of renewable energies. Moreover, they could also be directly connected to renewable electricity generation plants in isolated networks. Since electricity generated using wind energy and photovoltaics is naturally subject to time-of-day and seasonal fluctuations, PtL plants must either follow the dynamics of electricity generation exactly or intermediate storage facilities for the energy harvested must be provided to reduce the requirements for temporal changes in throughput. As storage systems are associated with considerable additional costs, an optimization problem arises, whereby the optimal system design is influenced by the dynamic capability of the PtL plant. Modular multistrand systems generally offer advantages over single-strand systems in terms of load flexibility, as individual strands can be successively brought into standby mode without having to shut down the entire system. However, this requires special concepts in order to maintain the operational state of strands in standby mode, which must take into account the thermal inertias in the system as well as the safety requirements. Thermal integration between individual process units, which is desirable for reasons of efficiency, must also be taken into account here and increases the complexity of the system. A compromise must be found between the dynamic capacity of the entire system, the overall energy efficiency in full-load and partial-load operation, and the equipment required. Information about the ability of individual stages of the PtL process chain to perform rapid load changes is only available to a very limited extent, particularly in the case of experimental data. In the following, some aspects related to the load-adaptable operation of the different units of a PtL plant are discussed.

18.4.1 Load-adaptable operation of low-temperature DAC in a PtL plant

Low-temperature DAC is a modular technology by nature. The process consists of successive adsorption and desorption cycles, whereby several to many identical modules are combined into a complete plant, depending on the size of the plant (see, for example reference [25]). In general, the systems can be operated very flexibly. In the adsorption cycle, electricity is needed to suck the air through the modules. In the desorption cycle, heat at approximately 100 °C is required to recover the CO_2. For an integrated PtL plant, the DAC unit would be interfaced with the downstream electrolysis unit via CO_2 storage. Moreover, heat storage at a temperature level of 100 °C–120 °C would be advisable to improve the thermal integration with the synthesis part of the PtL plant in transient operation.

18.4.2 Load-adaptable operation of SOECs in a PtL plant

All electrolysis technologies are modular by nature, as they are composed of a large number of individual cells of limited size, which are combined into stacks. Several stacks are combined into a unit, several of which make up a system. The SOEC does not support quick temperature changes for reasons related to the mechanical integrity of ceramic cells and stacks. The heat balance of a module is influenced by the ohmic losses that generate heat and the endothermicity of the electrochemical splitting of water and eventually CO_2 as well as the thermocatalytic rWGS reaction. This means that the strategy for load-adaptable operation must include heating or cooling, depending on the throughput, in order to keep the temperature of the module within a reasonable window; this is also required in complete standby mode. Thermal management at the module and unit levels has to consider indirect heat exchange between ingoing steam and eventually CO_2, as well as FT recycled gas, and outgoing H_2 or syngas and oxygen. As steam produced using the waste heat of the FT unit is used to evaporate water, load-adaptable operation of the PtL process needs an additional option to generate steam. One possibility would be joint heat storage that supplies both the SOEC and DAC units with heat when needed.

18.4.3 Load-adaptable operation of Fischer–Tropsch synthesis reactors

Advanced microstructured reactors for FTS have been developed from lab-scale devices to sizeable prototypes at the Institute for Micro Process Engineering (IMVT) over the past ten years [33–36]. The basis for these devices is an intensified packed-bed reactor approach that uses microstructured plates filled with fine catalyst particles. Two structured catalyst plates (see figure 18.11, left) stacked face to face are sandwiched between cooling plates with a patented cellular design of micro-channels for fluid distribution, providing a uniform 2D distribution of hot water for cooling by evaporation [16]. By stacking several of these arrangements and then

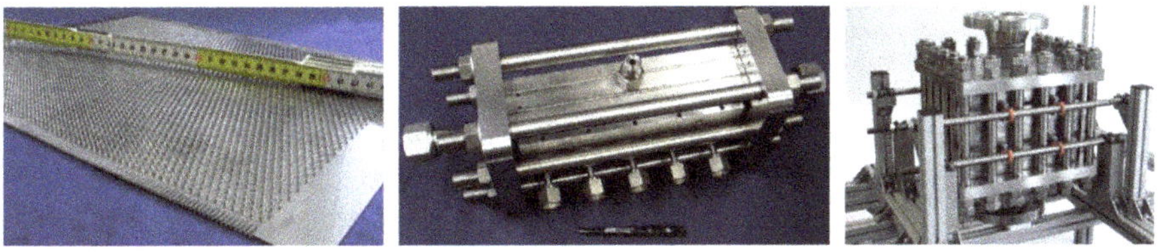

Figure 18.11. Advanced FTS reactor technology developed at the Institute for Micro Process Engineering at KIT. Left: Full-size microstructured catalyst plate. Middle: Short stack reactor module used to validate boiling-water cooling in transient operation in the lab. Right: first-generation full-sized reactor module for a FT crude production capacity of one barrel per day.

applying diffusion bonding to the stacked plates, a monolithic reactor block is created which afterwards receives flanges and fittings to connect it to the periphery (see figure 18.11, middle and right). This technology has been scaled up to modules with a design capacity of one barrel per day of FT crude (see figure 18.11, right) at IMVT and is currently being commercialized and further scaled up by the Institute's spin-off, INERATEC (see also figure 18.3).

The outstanding cooling power of this system avoids hot-spot formation, even at high reaction rates. Moreover, the design of the cooling system enables quick changes to be made to the catalyst bed temperature by adjusting the pressure on the cooling side, which defines the temperature of evaporation. This system was validated in a multitude of experiments with modules similar to the one shown in figure 18.11 (middle). Load-adaptable operation was also studied. For this purpose, the reactor was confronted with variable feed flow rates or, alternatively, a variable H_2/CO ratio in the feed. This approach was based on the assumed H_2 generation pattern of a polymer-electrolyte membrane (PEM) electrolyzer fed with renewable power from a real photovoltaic (PV) system at KIT. A varying feed flow rate at a constant H_2/CO ratio would be obtained in a system in which the appropriate amount of CO_2 is added to the H_2 flow from the electrolyzer before the syngas generation step. The other case represents syngas production from biomass gasification, in which the H_2 flow from the electrolyzer is added to the syngas to increase the yield of synthetic fuels. The FTS reactor was operated in two ways: first at a constant temperature, leading to variations in the conversion rate and selectivity, and second by adaption of the pressure on the cooling side to set the temperature so as to achieve constant conversion at the given feed flow rate and H_2/CO ratio. For the second approach, the required temperature was derived from a simplified kinetic model. It turned out that both the reactor operation as well as the catalyst itself seemed stable under such transient conditions. The total runtime was in the range of 1000 h. Accelerated deactivation was not observed. Moreover, it could be shown that the strategy of keeping the conversion constant via adjustment of the reactor temperature paid off, as it led to reduced methane selectivity. Details of these investigations can be found in references [37, 38]. A validation of load-adaptable operation at the plant level is planned, which will use the containerized pilot plants available in the Energy Lab 2.0 at KIT (see section 18.4.4 below).

18.4.4 Simplified FT crude refining to synthetic paraffinic kerosene

The combination of FTS with a subsequent upgrade of the FT products has also been studied. A simplified scheme could be worked out which is compatible with both the upgrading of FT crude in a refinery and with its on-site upgrading to SPK (see figure 18.12). The effluent from the FT reactor first undergoes hydrocracking and isomerisation on a commercial catalyst, followed by a two-stage separation system. The remaining FT wax is condensed in a hot trap for recycling back into the hydrocracking reactor. After removal of the wax, the FT oil and the produced water are condensed in a cold trap and the gases are recycled back into the synthesis gas preparation unit, which may use the rWGS or co-SOECs. The water and FT oil are phase separated, and the FT oil is subsequently hydrogenated on a second commercial catalyst to remove remaining unsaturated and oxygenated compounds. The resulting hydrocracked and hydrotreated FT oil is finally distilled to obtain kerosene, diesel, and naphtha fractions.

This scheme has been successfully tested in the lab, and a high potential for reaching all required specifications for FT-SPK according to the ASTM D7566 norm has been confirmed. Figure 18.13 shows the hydrocarbon product distribution of the FT oil at the different stages. For the kerosene fraction (figure 18.13, bottom right), the only two parameters not yet met are the acidity level, which is still slightly too high, and the temperature difference T90-T10, which is slightly too low, indicating an overnarrow boiling range. Fine tuning of the process parameters of the two catalytic stages and the distillation is ongoing to solve these issues. Moreover, an FT crude refining module which implements the simplified upgrading scheme of figure 18.12 at a scale of two barrels per day for the PtL pilot plant is currently under construction within the Energy Lab 2.0 at KIT (see also section 18.4.4). This module will also enable testing of the alternative option of separating

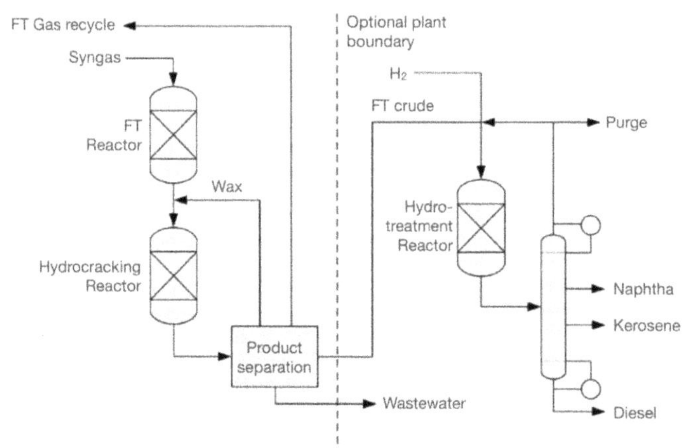

Figure 18.12. Simplified scheme for the refining of FT crude to SPK. The FT syncrude obtained via product separation after hydrocracking could also be shipped to a refinery for co-processing with fossil fuel materials or an upgrade to finished fuel.

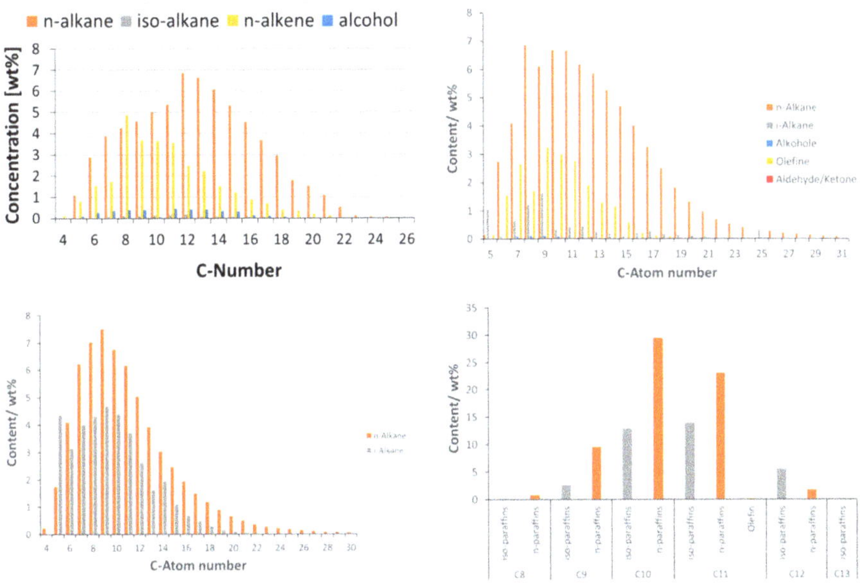

Figure 18.13. Results of studies of simplified FT crude refining. Top left: product distribution of the original FT oil. Top right: Product distribution of the hydrocracked FT product. Bottom left: product distribution after hydrotreatment of the previously hydrocracked FT oil. Bottom right: product distribution of the kerosene fraction after distillation.

the FT wax with a hot trap right after the FTS reactor and sending only this fraction to the hydrocracking reactor together with make-up H_2. In that scheme, the FT oil straight from the FTS reactor and the hydrocracked FT oil are combined and hydrotreated in the second step.

18.4.5 Integrated PtL plant based on a 250 kW co-SOEC at KIT's energy lab 2.0

In the framework of the ongoing second funding phase of the Kopernikus P2X project, Sunfire, INERATEC, and KIT are about to establish an integrated PtL plant for the production of one to two barrels per day of FT crude by May 2023. This will also contain a module that refines FT crude to SPK in order to validate the scheme shown in figure 18.12 at a larger scale. Sunfire will provide a 250 kW ambient-pressure co-SOEC system, which will be integrated into the Energy Lab 2.0 site at KIT. The syngas generated by this system will be compressed and sent via pipe to the FT section of an existing PtL plant (see figure 18.14). To enable this coupling, KIT will install a capacity upgrade of the FT section for one to two barrels per day and an additional module to refine the FT crude to SPK. Moreover, a steam pipe connecting the FT cooling cycle to the evaporator for the co-SOEC unit will be added as well as a pipe taking the gas from the FT crude refining section back into the co-SOEC to enable high carbon utilization. The DAC plant intended to deliver the CO_2 will not be included, but operated at Climeworks's site in Zürich. Hence, the connection of the DAC section and the synthesis section has to be made virtually

Figure 18.14. Two-stage PtL plant at KIT's Energy Lab 2.0, including rWGS for syngas production from CO_2 and H_2 and FTS. The plant was developed and built by INERATEC for a capacity of one barrel per day. An upgrade to a capacity of two barrels per day is planned as well as an extension via a second container on top to accommodate an FT crude refining module (currently under construction). Photograph: Cynthia Ruf, KIT. © KIT.

using CO_2 provided from a liquid CO_2 storage tank available in the Energy Lab 2.0. This plant will be the first worldwide to integrate an SOEC in co-electrolysis mode with an FTS unit at a semi-technical scale. A further scale-up of this configuration to 1–2 MW for the co-SOEC is planned for the third funding phase of the P2X project and will take place, if the project is approved, at the Industry Park Hoechst in Frankfurt, where INERATEC plans to build a 10 MW PtL plant.

18.5 Concluding remarks

Power-to-X technologies will be needed worldwide to reach net zero emissions as quickly as possible, which is a must in order to limit the damages expected as a consequence of global climate change. The most important target products are e-kerosene for aviation and e-diesel for shipping, heavy-duty traffic, and heavy-duty machinery.

In the context of power-to-X, SOEC is a preferred technology for electrolysis for several reasons. In thermodynamic terms, it can reach a higher energy efficiency compared to PEM and alkaline systems. Moreover, it is able to produce synthesis gas directly in the electrolysis step, so that no additional rWGS reactor is required. This, together with heat integration with the FTS, allows a further increase in the overall energy efficiency. Finally, yet importantly, SOEC does not need noble metals.

The load adaptation of power-to-X processes has mostly been studied in simulations so far and rarely in experiments. The ability to adapt to varying loads is an important feature for decentralized power-to-X plants. Advanced reactor technologies such as microstructured reactors seem to be key for transient operation. The results obtained using smaller reactor modules are promising, but load-adaptive operation has to be demonstrated at the plant level, and this requires much more work to be done.

It seems possible to upgrade FT crude to SPK using a rather simple scheme based on two catalytic steps followed by distillation. However, the yield of SPK still has to be quantified, and the robustness of the scheme must be established. In addition, utilization possibilities for the diesel and naphtha fractions have to be explored.

The world's first integrated PtL plant using co-SOEC, FTS, and FT crude refining is being built in a 250 kW format at KIT's Energy Lab 2.0. If all goes well, this plant should be operational by mid-2023 and deliver two barrels per day of FT crude as well as SPK for admixing with conventional kerosene. A further scale-up to 1–2 MW is already being planned.

Acknowledgments

The authors wish to acknowledge the funding of their work by the 'Kopernikus P2X' project (Grant No. 03SFK2K0–2) of the German Federal Ministry of Education and Research (BMBF) and the 'PowerFuel' project (Grant No. 03EIV071B) of the German Federal Ministry of Economic Affairs and Climate Action as well as the 'reFuels' project (Grant No. 4–8825/1/35 and 4–8825/1/45) of the State Ministry of Transport, Baden-Württemberg.

Moreover, they wish to thank all partners of these projects as well as all involved colleagues at IMVT for fruitful collaboration and in particular INERATEC GmbH, DLR-TT, and Aviation Project Consult GmbH for the joint work on FT crude refining.

References and additional reading

[1] Perner J and Bothe D 2018 International aspects of a Power-to-X roadmap *World Energy Council Germany* https://weltenergierat.de/wp-content/uploads/2018/10/20181018_WEC_Germany_PTXroadmap_Full-study-englisch.pdf.

[2] PtX Hub Berlin 2022 PtX.Sustainability – Dimensions and Concerns, https://ptx-hub.org/wp-content/uploads/2022/05/PtX-Hub-PtX.Sustainability-Dimensions-and-Concerns-Scoping-Paper.pdf

[3] Schnuell C, Thoeming J, Wassermann T, Their P, von Gleich A and Goessling-Reisemann S 2019 Socio-technical-economic assessment of power-to-X: potentials and limitations for an integration into the German energy system *Energy Res. Soc. Sci.* **51** 187–97

[4] Koj J C, Wulf C and Zapp P 2019 Environmental impacts of power-to-X systems – a review of technological and methodological choices in life cycle assessments *Renew. Sustain. Energy Rev.* **112** 865–79

[5] de Klerk A 2011 *Fischer–Tropsch-Refining* (Weinheim: Wiley)

[6] Maitlis P M and de Klerk A (ed) 2013 *Greener Fischer–Tropsch Processes for Fuels and Feedstocks* (Weinheim: Wiley)

[7] de Klerk A, Li Y and Zennaro R 2013 Fischer–Tropsch technology *Greener Fischer–Tropsch Processes for Fuels and Feedstocks* ed P M Maitlis and A de Klerk (Weinheim: Wiley) pp 53–80

[8] van de Loosdrecht J, Botes F, Ciobica I, Ferreira A, Gibson P, Moodley D, Saib A, Visagie J, Weststrate C and Niemantsverdriet J 2013 Fischer–Tropsch synthesis: catalysts and chemistry

Comprehensive Inorganic Chemistry II – From Elements to Applications ed J Reedijk and K Poeppelmeier 2nd edn (Amsterdam: Elsevier) pp 525–57
[9] Dry M E 1990 Fischer–Tropsch synthesis over iron catalysts *Catal. Lett.* **7** 242–51
[10] Förtsch D, Pabst K and Groß-Hardt E 2015 The product distribution in Fischer–Tropsch synthesis: an extension of the ASF model to describe common deviations *Chem. Eng. Sci.* **138** 333–46
[11] Loewert M, Hoffmann J, Piermartini P, Selinsek M, Dittmeyer R and Pfeifer P 2019 Microstructured Fischer–Tropsch reactor scale-up and opportunities for decentralized application *Chem. Eng. Technol.* **42** 2202–14
[12] Guettel R, Kunz U and Turek T 2008 Reactors for Fischer–Tropsch synthesis *Chem. Eng. Technol.* **31** 746–54
[13] Choi Y H, Jang Y J, Park H, Kim W Y, Lee Y H, Choi S H and Lee J S 2017 Carbon dioxide Fischer–Tropsch synthesis: a new path to carbon-neutral fuels *Appl. Catal* **202** 605–10
[14] Brübach L, Hodonj D and Pfeifer P 2022 Kinetic analysis of CO_2 hydrogenation to long-chain hydrocarbons on a supported iron catalyst *Ind. Eng. Chem. Res.* **61** 1644–54
[15] Deshmukh S R, Tonkovich A L Y, Jarosch K T, Schrader L, Fitzgerald S P, Kilanowski D R, Lerou J J and Mazanec T J 2010 Scale-up of microchannel reactors for Fischer–Tropsch synthesis *Ind. Eng. Chem. Res.* **49** 10883–8
[16] Pfeifer P, Schmidt S, Betzner F, Kollmann M, Loewert M, Böltken T and Piermartini P 2022 Scale-up of microstructured Fischer–Tropsch reactors – status and perspectives *Curr. Opin. Chem. Eng.* **36** 100776
[17] Daza Y A and Kuhn J N 2016 CO_2 conversion by reverse water gas shift catalysis: comparison of catalysts, mechanisms and their consequences for CO_2 conversion to liquid fuels *RSC Adv.* **6** 49675–91
[18] González-Castaño M, Dorneanu B and Arellano-García H 2021 The reverse water gas shift reaction: a process systems engineering perspective *React. Chem. Eng* **6** 954
[19] Juneau M *et al* 2020 Assessing the viability of K-Mo_2C for reverse water–gas shift scale-up: molecular to laboratory to pilot scale *Energy Environ. Sci.* **13** 2524–39
[20] Galhardo T S, Braga A H, Arpini B H, Szanyi J, Gonçalves R V, Zornio B F, Miranda C R and Rossi L M 2021 Optimizing active sites for high CO selectivity during CO_2 hydrogenation over supported nickel catalysts *J. Am. Chem. Soc.* **143** 4268–80
[21] Gioria E, Ingale P, Pohl F, Naumann d'Alnoncourt R, Thomas A and Rosowski F 2022 Boosting the performance of Ni/Al_2O_3 for the reverse water gas shift reaction through formation of CuNi nanoalloys *Catal. Sci. Technol.* **12** 474
[22] Hannula I, Kaisalo N and Simell P 2020 Preparation of synthesis gas from CO_2 for Fischer–Tropsch synthesis—comparison of alternative process configurations *C* **6** 55
[23] Khoshtinat Nikoo M and Amin N A S 2011 Thermodynamic analysis of carbon dioxide reforming of methane in view of solid carbon formation *Fuel Proc. Technol* **92** 678
[24] Küngas R 2020 Review-electrochemical CO_2 reduction for CO production: comparison of low- and high-temperature electrolysis technologies *J. Electrochem. Soc.* **167** 044508
[25] Dittrich L, Nohl M, Jaekel E E, Foit S, de Haart L G J and Eichel R A 2019 High-temperature co-electrolysis: a versatile method to sustainably produce tailored syngas compositions *J. Electrochem. Soc.* **166** F971–5
[26] König D H, Freiberg M, Dietrich R and Wörner A 2015 Techno-economic study of the storage of fluctuating renewable energy in liquid hydrocarbons *Fuel* **159** 289

[27] Kaneko T, Derbyshire F, Makino E, Gray D, Tamura M and Li K (ed) 2000 *Ullmann's Encyclopedia of Industrial Chemistry. Coal Liquefaction* (Weinheim: Wiley Verlag GmbH & Co. KGaA)

[28] Keith D W, Holmes G, St. Angelo D and Heidel K 2018 A process for capturing CO_2 from the atmosphere *Joule* **2** 1573

[29] Deutz S and Bardow A 2021 Life-cycle assessment of an industrial direct air capture process based on temperature–vacuum swing adsorption *Nat. Energy* **6** 203

[30] Heß D, Klumpp M and Dittmeyer R 2020 Nutzung von CO_2 aus Luft als Rohstoff für synthetische Kraftstoffe und Chemikalien *Studie im Auftrag des Ministeriums für Verkehr Baden-Württemberg, Institut für Mikroverfahrenstechnik, Karlsruher Institut für Technologie* https://vm.baden-wuerttemberg.de/fileadmin/redaktion/m-mvi/intern/Dateien/PDF/29-01-2021-DAC-Studie.pdf

[31] Albrecht F G, König D H, Baucks N and Dietrich R 2017 A standardized methodology for the techno-economic evaluation of alternative fuels – a case study *Fuel* **194** 511

[32] Becker W L, Braun R J, Penev M and Melaina M 2012 Production of Fischer–Tropsch liquid fuels from high temperature solid oxide co-electrolysis units *Energy* **47** 99

[33] Nezman I, Xie J, Golub K W, Carneiro J, Olsen K, Ping E W, Jones C W and Sakwa-Novak M A 2021 Chemical kinetics of the autooxidation of poly(ethylenimine) in CO_2 sorbents *ACS Sustain. Chem. Eng.* **9** 8477–86

[34] Myrstad R, Eri S, Pfeifer P, Rytter E and Holmen A 2009 Fischer–Tropsch synthesis in a microstructured reactor *Catal. Today* **147** S301

[35] Piermartini P, Boeltken T, Selinsek M and Pfeifer P 2017 Influence of channel geometry on Fischer–Tropsch synthesis in microstructured reactors *Chem. Eng. J.* **313** 328

[36] Loewert M, Hoffmann J, Piermartini P, Selinsek M, Dittmeyer R and Pfeifer P 2019 Microstructured Fischer-Tropsch reactor scale-up and opportunities for decentralized application *Chem. Eng. Technol.* **42** 2022

[37] Pfeifer P, Piermartini P and Wenka A 2017 *Mikrostrukturreaktor zur Durchführung exothermer, heterogen katalysierter Reaktionen miteffizienter Verdampfungskühlung* DE 10 2015 111 614 A1 https://register.dpma.de/DPMAregister/pat/register?AKZ=1020151116146&CURSOR=2

[38] Loewert M and Pfeifer P 2020 Dynamically operated Fischer–Tropsch synthesis in PtL – part 1: system response on intermittent feed *ChemEngineering* **4** 21

[39] Loewert M, Riedinger M and Pfeifer P 2020 Dynamically operated Fischer–Tropsch synthesis in PtL – part 2: coping with real PV profiles *ChemEngineering* **4** 27

IOP Publishing

High-Temperature Electrolysis
From fundamentals to applications
Werner Sitte and Rotraut Merkle

Chapter 19

Reversible solid oxide cell systems as key elements of achieving flexibility in future energy systems

David Paczona, Christoph Sejkora and Thomas Kienberger

The application of reversible solid oxide cell (rSOC) systems in energy infrastructure allows cells to be reversibly operated as fuel cells (FCs) or electrolysis cells (ECs). Different levels of energy system integration and flowsheet options are possible. This chapter first derives the need for such energy conversion units in future energy systems and extracts ideas for the systems' flowsheets from the literature. Second, the main part of this chapter focuses on the operational parameters' influence on system performance, which is discussed for a chosen set of system flowsheets and application scenarios. In this process, general insights into the behavior of the system are generated. Finally, the chapter concludes with design suggestions for different rSOC operations in future energy systems.

19.1 Introduction

To meet the United Nations 1.5 °C climate target [1], society has defined ambitious objectives all around the globe, aiming at significantly reducing human greenhouse gas (GHG) emissions. Today, around 77% of them are energy-related [2]. As a result, a massive transformation of global energy systems needs to be brought about in the upcoming centuries in order to reach net zero GHG emissions on time. Three general fields of action must therefore be applied worldwide:

i) First and foremost, switching from fossil-fuel-dominated energy supply systems to systems based on renewable energies (REs) is vital. Since the fossil share in the global energy supply is around 81% [3], this can only be accomplished by massive and quick action.
ii) Technologies for energy efficiency must be introduced at all stages of the entire energy conversion chain to reduce energy losses. Waste heat potentials must be used to supply low-temperature demands.

doi:10.1088/978-0-7503-3951-3ch19

iii) The behavior of energy-consuming services, especially in the Organisation for Economic Co-operation and Development (OECD) member countries, must change to a more sparing and considered use.

Taking up these actions, the European Union's (EU's) 'Green Deal' [4], for instance, strives for net zero emissions by 2050. The intermediate target for 2030 aims to reduce EU greenhouse gas emissions by 55% compared to their 1990 levels. This should be achieved by reducing the European gross domestic consumption by 36%–39% through efficiency measures and by expanding REs to 40% to supply it [5].

Most of the REs to be exploited are hydro, wind, and photovoltaic (PV) potentials, which will lead to volatility in electricity generation; balancing demand and generation with appropriate flexibility and storage technology will become crucial. Multi-energy systems [6] (MESs), which connect various economic sectors (electricity, gas, heat transport, etc.) via grid-connected energy carriers, allow for both. Interlinking volatile electricity production with the gas sector or the heat sector, for instance, enables long-term storage [7] or mitigates the strain imposed on electricity grids by PV or wind power [8]. MESs may also be beneficial in terms of energy efficiency. They enable highly efficient technologies to be used for both the final energy applications (heat pumps, battery electric vehicles (BEVs), etc.) and the energy-sector conversion units (combined heat and power (CHP) units, power-to-X units, industrial waste heat use, etc). Figure 19.1 shows the general structure of MESs.

Both RE expansion and energy efficiency enhancement require proper energy strategies and coordinated, concerted energy policies to reach net zero emissions. On the level of implementation-oriented, time-phased actions, such plans are widely missing. Scenario-based MES studies may help us to deeply understand the interdependencies of energy systems and therefore foster their development [9].

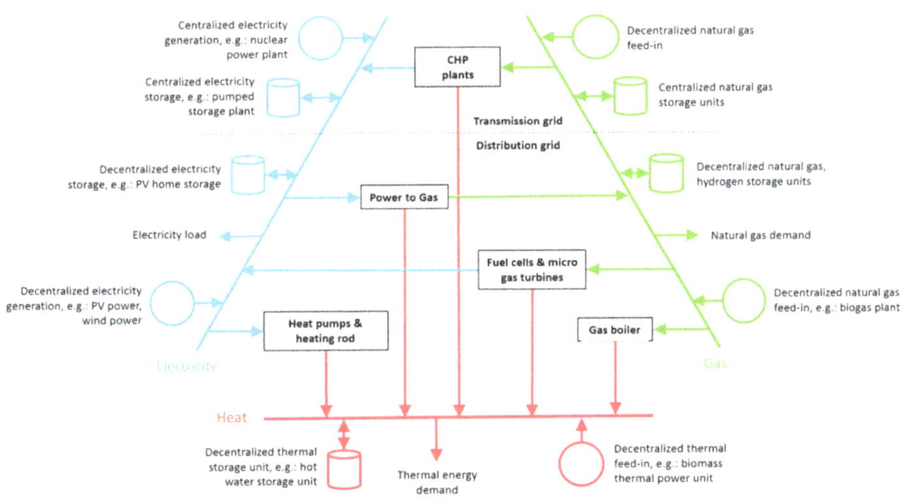

Figure 19.1. General structure of multi-energy systems (MES).

The authors of this work [10, 11] performed such investigations for Austria, an EU member state. Austria wants to reach net zero emissions by 2040 [12]—ten years ahead of the European Union. A quantum leap in this regard would be to achieve a decarbonized electricity system (on annual balance) by 2030 [13]. However, it is also true that no concrete action plans for Austria exist today. To facilitate policymakers in the process of identifying appropriate actions, the authors developed an exergy-based MES optimization methodology that minimizes the exegetic cost of supporting energy services for a given RE expansion. Exergy is the actual working capacity of any form of energy. Mechanical work and electricity are pure exergy. Chemical energy can be considered as 100% exergy. The exergy content of heat depends, according to Carnot's rule, on its temperature relative to the ambient temperature (see equation (19.7)). While energy is always conserved throughout all energy conversion processes (the first law of thermodynamics), exergy losses always occur whenever processes are irreversible (the second law of thermodynamics). In this sense, minimizing exegetic costs allows both the thermodynamic minimum energy demand of an energy system as well as the systemic location of system inefficiencies to be found. This approach uses data with high temporal resolution for both supply and demand. This enables us to show how future energy systems with high proportions of volatile REs can be fundamentally designed while considering an optimal technology mix that enables both energy efficiency and energy system flexibility.

For this calculation, the useful energy demands (space heating, process heat, light, mechanical work, etc) required to provide the energy services (space or process heat services, mobility services, lighting services, etc.) are first converted into useful exergy demands. Therefore, we apply exergy factors describing the thermodynamic energy quality required for the services [10]. Second, for REs, an analogous approach is taken: primary energies such as wind, PV, or biomass are considered to be exergy. For waste heat, we use corresponding temperature-dependent exergy factors. The subsequent minimization of exergy losses combines energy-efficient final energy application technologies with conversion and storage technologies in an optimal way to keep the exergy-related imports (in the case under consideration, electricity and gas imports) as low as possible (see equation (19.1)). In this case, we minimize the primary energy demand of the system under consideration.

$$\min\left(Ex_{\text{Loss,tot}}\right) = Ex_{\text{Sup,tot}} - Ex_{\text{UED,tot}} = \sum_i Ex_{\text{RES},i} + \left(\sum_j Ex_{\text{Imp},j} - \sum_k Ex_{\text{Exp},k}\right) - Ex_{\text{UED,tot}} \tag{19.1}$$

$Ex_{\text{Loss,tot}}$ The total exergy losses caused by both energy conversion and final energy applications

$Ex_{\text{Sup,tot}}$ The total exergy used to supply the considered energy system

$Ex_{\text{UED,tot}}$ The total useful exergy demand of all energy services to be covered

$Ex_{\text{RES},i}$ RE generation of resource i

$Ex_{\mathrm{Imp},j}$ The exergy import of energy carrier j
$Ex_{\mathrm{Exp},k}$ The exergy export of energy carrier k

Figure 19.2 shows the results obtained by applying this exergy-based approach to a possible RE expansion in Austria for the year 2040. Using these results, the RE

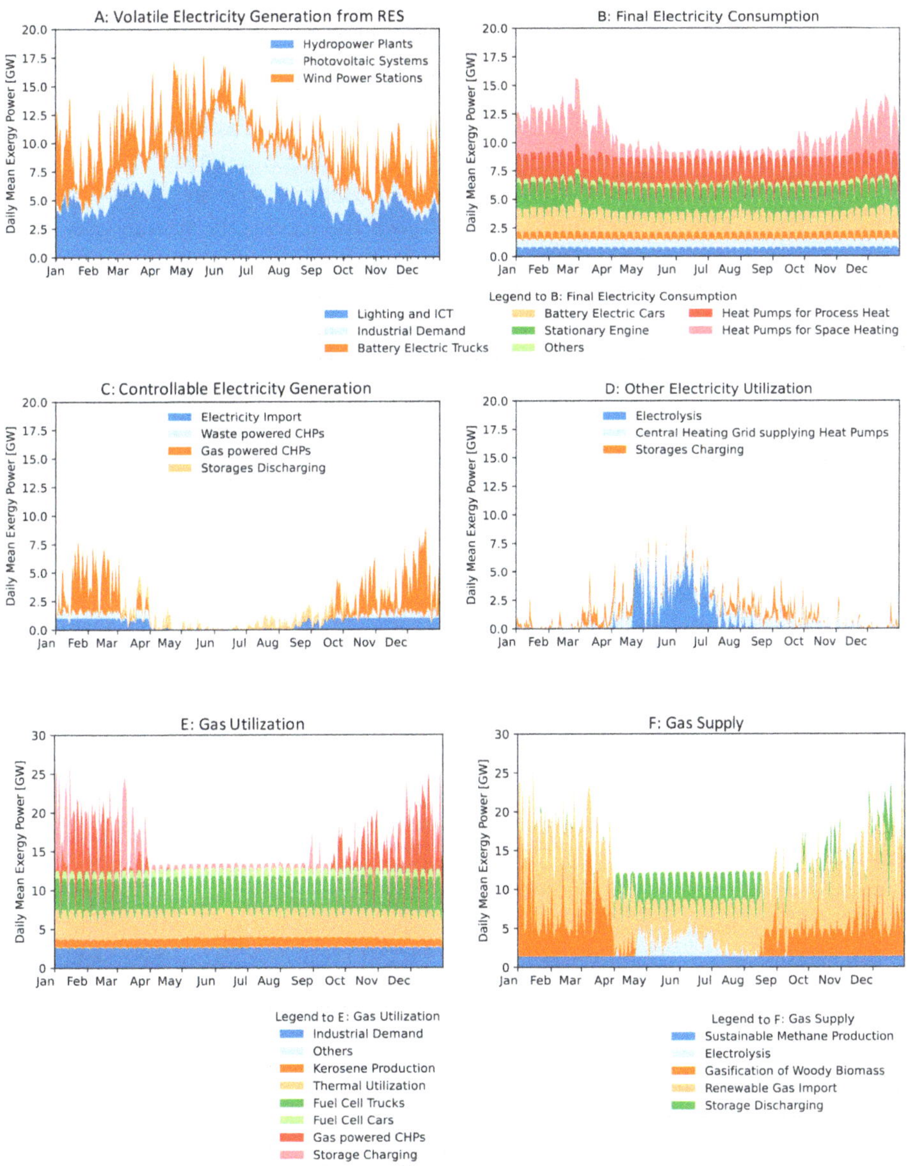

Figure 19.2. Exergy-optimized electricity system for 2040. (a) electricity generated by REs, (b) final electricity consumption, (c) deployment of undercoverages in the electrical energy system, (d) utilization of overcoverages in the electrical energy system, (e) final gas consumption, and (f) gas supply.

expansion target required to reach a decarbonized electricity system (on annual balance) in 2030 is linearly extrapolated. The useful exergy demand for 2040 is based upon the 2019 demand, adjusted by annual economic growth [14] and a decreased energy intensity within all sectors [15]. Figure 19.2(a) shows the expected seasonal effects, especially those due to PV and hydropower-based electricity generation. Panel (b) shows the final electricity consumption. Compared to today's figures, the exergy-optimized approach results in significant additional electricity demand, especially from BEVs and heat pumps used to supply both process heat and space heating. The latter naturally causes increased demand in the winter months, which results in an undercoverage of national REs (panel c). In the exergy-optimized case, national CHP units address this. They additionally provide waste heat that meets the space heat demand. In the summer months, RE overcoverage arises. This is used to operate electrolyzers and, to a minor extent, for pumped storage and process heat pumps (panel d). Due to the interaction between the electricity system and the gas system, there is a seasonal gas demand created by the CHP units (panels c and e). In addition, gas is used in an exergy-optimized system, in particular to supply high-temperature applications in industry and for various mobility needs in heavy transport (panel e). Those demands have a baseload character. Industrial waste heat and waste heat from CHP units reduce the primary energy demand for space heating in winter. Overall, optimizing exergy efficiency for final energy application and energy conversion and storage units can reduce primary energy use from about 400 TWh/a to about 240 TWh/a. Despite all this, imports, especially renewable gases, are to be expected in the future (panel f).

The results for Austria may be qualitatively valid for the future energy systems of other central European countries. Their RE-potential structures, as well as the structures of their energy service demands, are similar. In such modern energy systems with high proportions of REs, the application of both FCs and electrolysis is beneficial. Reversible FC systems and, in particular, rSOC systems combine their energy system advantages. For their future application, we can take away the following messages from the investigations shown above:

As a **conversion technology in the energy sector**, a possible rSOC application is first and foremost strongly influenced by the seasonal balance of RE generation and demand:

- FC operation is used to generate electricity mainly during the winter season. The high electric efficiency of the FC operational mode, compared to that of a classic combined cycle gas turbine (CCGT) power plant, is beneficial in this regard.
- EC operation is used to produce hydrogen during the summer season, using electricity from RE overcoverage. The hydrogen thus produced reduces renewable gas imports.

In an energy-sector application, an rSOC system must address rapid changes of load, but not fast switching between the electrolysis and FC operational modes. The combined-operation fuel–electrolysis cell allows for a large number of full-load hours (>5000 h/year), which improves the economics. Selling the rSOC's waste heat to cover space heat demands at temperature levels below 100 °C can lead to further economic gains. This would mainly, but not exclusively, occur during winter time.

As a **conversion technology in the industrial sector**, a possible rSOC application is strongly influenced by the industrial baseload demands for electricity and gas as well by price signals in the electricity markets:

- FC operation generates electricity mainly during times of high electricity prices, or during times when energy system support measures (e.g. in the form of positive secondary control capacity) are needed.
- EC operation produces hydrogen during times of low electricity prices or during times when energy system support measures (e.g. in the form of negative secondary control capacity) are needed.

In an industry-sector application, an rSOC system also faces fast load changes, switching between the electrolysis and FC operational modes. The combined-operation fuel–electrolysis cell, together with the baseload demands of the industry, allows for an even higher number of full-load hours (>7000 h/year), compared to the energy-sector application. The rSOC's waste heat can supply process heat demands at temperature levels above 200 °C (e.g. process steam), reducing the gas demand on-site.

19.2 The state of research into rSOC systems

Reversible systems based on solid oxide cells (i.e. rSOCs) that can operate in FC mode and electrolysis mode have already been the subject of several investigations. Various ideas have been proposed for the system configuration. This subsection attempts to categorize current developments.

Pure hydrogen and mixtures with methane, carbon monoxide, and carbon dioxide have been investigated for use as fuels in FC operation mode. The oxidant can be pure oxygen (from storage vessels) or ambient air. In the electrolysis mode of operation, pure steam and mixtures with carbon dioxide are possible supply gases on the fuel side. Depending on the fuel composition, the system can either be open [16] (e.g. methane from the grid and ambient air), half open [17–24] (e.g. hydrogen from storage and ambient air), or closed [25–27] (e.g. hydrogen and oxygen storage). The layouts in figure 19.3, which will be further discussed, are of the half-open type, applying stored hydrogen and ambient air.

In many publications about half-open systems, the thermal management of the stack is ensured by the regulation of air temperature and air mass flow. Thermoneutral operation in the electrolysis mode and air cooling in the FC operational mode has been considered by others [21–24]. Different approaches have been followed by considering cooling by diathermic oils [25, 26], thermal integration of methanization reactors [27] and metal hydride storage systems [18].

Most flowsheets in the literature include a recirculation path for the fuel. This can be done in different ways, as shown in figure 19.3. The hot exhaust fuel of the stack is mixed with either the cold [16] (A2) or hot fuel (A1), which is called 'hot gas recirculation' (hgr) throughout this chapter [19]. A different option is 'cold gas recirculation' (cgr), in which cooled exhaust fuel is mixed with the cold fuel (after

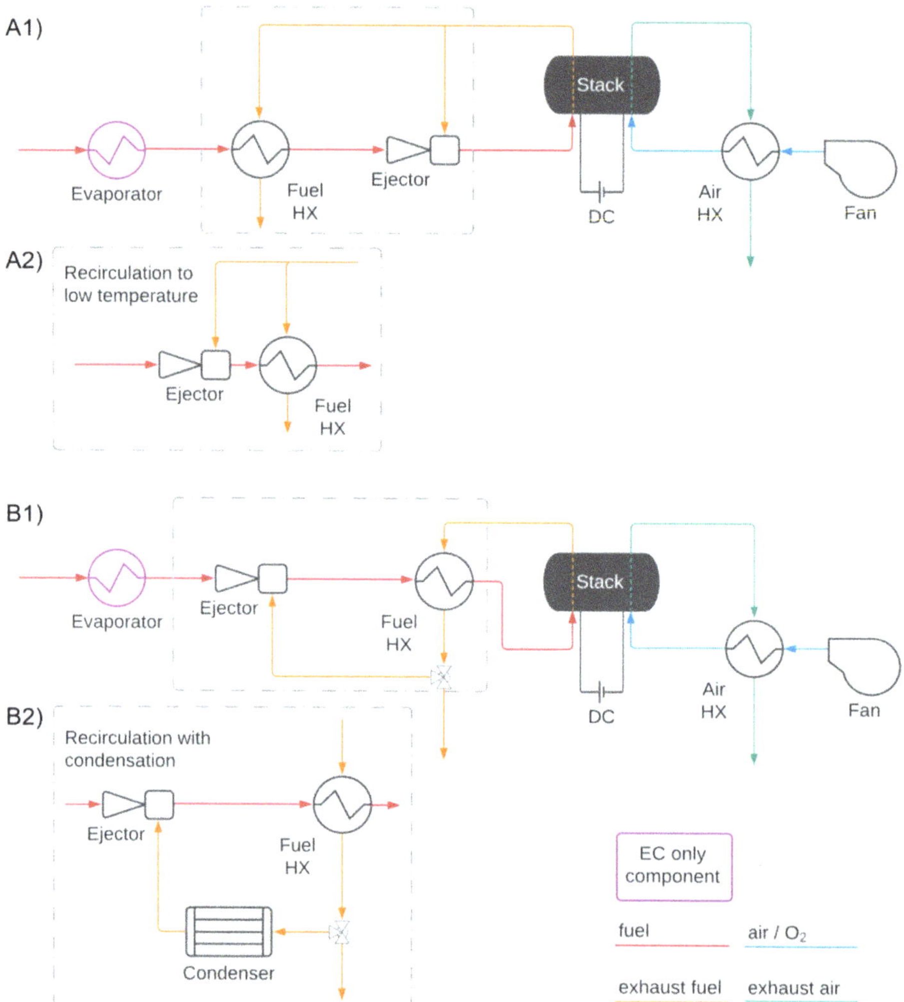

Figure 19.3. System layout ideas described in the literature: (A1) hot gas recirculation to cold (hgr), (A2) hot gas recirculation to the hot side of the fuel heat exchanger (HX), (B1) cold gas recirculation (cgr), (B2) cgr with condensation (cgr-cond).

passing through an evaporator) [19] (B1), which can be modified by adding a condenser (cgr-cond) of the recirculated stream [17, 19, 25, 26] (figure 19.3 B2). In the case of systems that operate with a synthesis gas mixture, which flows through the stack from one storage vessel to another, no circulation is used on the fuel side [27]. In this system, recirculation on the air side has been proposed by some studies [21–24, 27].

Commercial rSOC systems are already offered by SunFire [28], but little knowledge about their system layouts is publicly available. These systems are designed for applications in the energy, industry, and building sectors. In the latter two fields, thermal coupling to heat sources and consumers is proposed to achieve high system efficiencies.

The abovementioned flowsheets have been investigated by research groups in different ways. Frank *et al* [17] studied how the system performance was influenced by changes of the recirculation rate at fixed values of the stack fuel utilization. In the research performed by Giap *et al* [19], the influence of the fuel's hydrogen concentration on efficiency was studied. These investigations are helpful in understanding the system behavior.

However, when talking about rSOC systems, it is more helpful to think in terms of system parameters. Low stack fuel utilization, a parameter which is often used in the literature, does not mean that a system with recirculation also has low fuel utilization. The difference in the definitions of fuel utilization at the stack and system levels can be seen from equations (19.2) and (19.3). Similarly, it is not easy to draw conclusions about the system from the effect of hydrogen concentration in the stack.

$$fu_{\text{stack}} = 1 - \frac{\dot{m}_{\text{fuel,stack out}} \cdot y_{\text{H}_2\text{,fuel,stack out}}}{\dot{m}_{\text{fuel,stack in}} \cdot y_{\text{H}_2\text{,fuel,stack in}}} \tag{19.2}$$

$$fu = 1 - \frac{\dot{m}_{\text{exhaustf uel}} \cdot y_{\text{H}_2\text{,exhaust fuel}}}{\dot{m}_{\text{fuel}} \cdot y_{\text{H}_2\text{,fuel}}} \tag{19.3}$$

So far, the literature has yielded little insight into the combined effects of parameter variations and their consequences for system design. The approach described in this chapter starts from the assumption that it is not enough to vary one parameter to investigate the effect on the system performance. Parameters should be varied across the whole multidimensional space in order to study combined effects and get deeper insights into system behavior. Instead of going into the details of possible changes in the configuration of one specific system layout, the influence of operational parameters should be investigated for a selection of different flowsheets, as displayed in figure 19.3. These configurations cover the core concepts of rSOC systems that include recirculation. The non-recirculation case can be seen as the limit with a zero recirculation rate. The performance dependence on operational parameters is studied for different operational modes and flowsheets. In this way one can learn about the preferred setup for round-trip operation and system limitations. The main outcomes of the investigations presented in this chapter are:

I) Understanding the combined influence of operational parameters on the system performance
II) Determination of the best operating points for the best system performance in different operational modes
III) Quantification of efficiency-increasing measures for different system flowsheets
IV) Determination of the best system flowsheets, for the use cases of the energy sector and the industry sector, to meet the demand for flexibility with the best system performance

19.3 Methodology

This section describes the steps and methods used to produce the results of section 19.4. First, the method used to choose the exemplary layouts for further investigations is described. This is followed by an explanation of the modeling approach. A description of the ideas used to progress from the analysis of individual operational modes to round-trip operation concludes this section.

19.3.1 Choice of system layouts

The system flowsheet used for further investigation was chosen in accordance with the basic concepts described in the literature (figure 19.3). These flowsheets are of the half-open type (i.e. hydrogen is stored in a vessel and ambient air acts as oxidant). The basic concepts are extended to include all the necessary components, as can be seen in figure 19.4. As before, these flowsheets can be divided into:

- Hot gas recirculation (hgr)
- Cold gas recirculation (cgr)
- Cold gas recirculation with condensation (cgr-cond)

Reversible solid oxide cell (SOC) systems can operate in two different modes:

- **The FC mode of operation:** the stack produces electricity from the reaction of hydrogen and oxygen to produce water. The fuel entering the system is pure

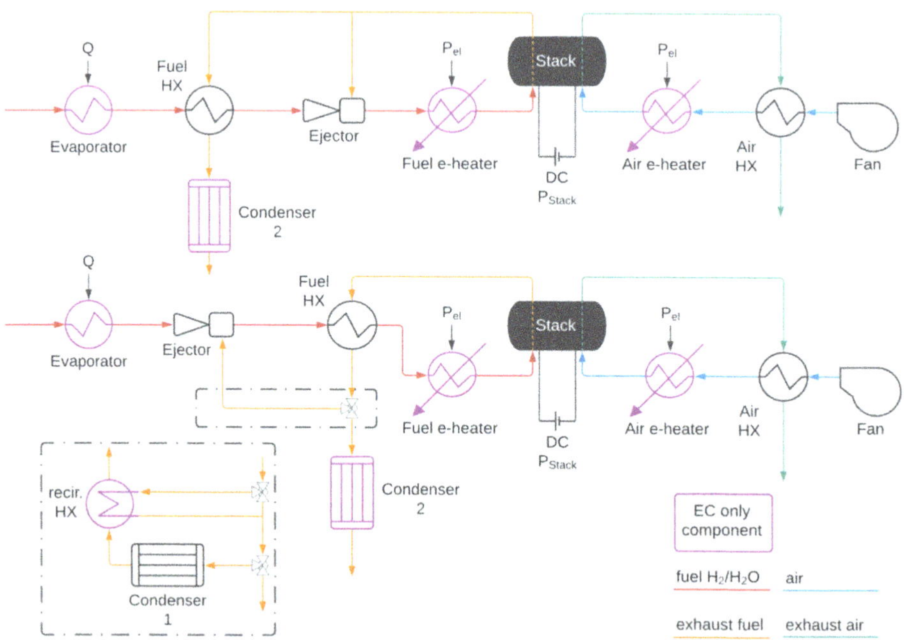

Figure 19.4. Extended system layout ideas: (A) hot gas recirculation (hgr) to the cold side of the fuel HX, (B1) cold gas recirculation (cgr), (B2) cgr with condensation in the cold gas recirculation (cgr-cond) path (Condenser 1) and a recirculation reheater after the condenser (recir. HX).

hydrogen, which is mixed with the recirculated exhaust fuel (which contains steam) before it enters the stack. Cooling of the stack is ensured by control of the air mass flow.

- **The EC mode of operation:** the stack consumes electricity for the water reaction, which produces hydrogen and oxygen. The fuel is pure water, which is mixed with the recirculated exhaust fuel (which contains hydrogen) before it enters the stack. The stack is operated below the thermoneutral voltage. For thermally stable operation, heating is required, which is ensured by electric heaters in the fuel (the fuel e-heater) and air streams (the air e-heater).

For the application of the system, two scenarios were chosen:

- **Energy-sector scenario (E):** no thermal interaction with other processes is considered. The rSOC waste heat, which may be used in other applications, does not influence the efficiency of the rSOC system and no industrial waste heat is available for integration. This reflects the possible real application of the system in the energy sector, in which no other facilities might be located nearby.
- **Industry-sector scenario (I):** in the FC operational mode, the system can provide generated waste heat to industrial consumers. Industrial waste heat sources are used for the evaporation of water in the electrolysis operational mode.

In the following paragraphs, the flowsheets of the rSOC systems capable of the two operational modes and ready for both application scenarios are separately described for hot and cold gas recirculation.

19.3.1.1 Hot gas recirculation

In systems that use hot gas recirculation (see figure 19.3(A1)), the exhaust fuel stream is recirculated before being cooled by the fuel heat exchanger (HX). The fuel HX only needs to handle the mass flow entering the stack, and an increased recirculation rate does not require a bigger heat exchanger area. This configuration means that the ejector is at almost the same temperature as the stack, which excludes the use of other components, such as mechanical fans, for recirculation. The best point at which to include the recirculated exhaust fuel is after the fuel heat exchanger, so that gases of similar temperatures are mixed. In figure 19.3(A2), we can see that systems have been proposed which recirculate the hot exhaust fuel to the cold incoming fuel. This is disadvantageous for two reasons. First, the ejector is operated with a big temperature difference, which causes thermal stress. Second, the recirculation increases the temperature of the fuel entering the fuel HX. This means that less of the heat can be recovered here and that the exhaust fuel entering Condenser 2 has higher temperatures. For these reasons we do not treat this recirculation option in the studies shown here.

19.3.1.2 Cold gas recirculation

Systems that use cold gas recirculation (see figure 19.3(B1)) employ an ejector to recirculate the cooled exhaust fuel after the fuel HX and mix it with the cold fuel after the evaporator. In this configuration, fewer components are operated at high temperatures, but the mass flow through the fuel HX increases with higher recirculation rates, which inflicts higher requirements on this component. The cold gas recirculation can be modified by adding a condenser downstream from the recirculation (see figure 19.3(B2)), which lowers the temperature to less than the condensation point of water and separates the condensed water from the gas stream. Since even lower temperatures occur in this case, a fan can be used instead of an ejector. In the FC operational mode, the cooling in this condenser (Condenser 1) helps to cool the stack and consequently less airflow is needed. However, in the electrolysis operational mode, a loss of heat is disadvantageous. A simple additional heat exchanger in the recirculation that recovers heat from the exhaust fuel can compensate this heat loss, as also shown in the detail in figure 19.4. In the FC mode, the condenser has another very important advantage. Because it lowers the steam content, the hydrogen concentration in the fuel is less affected by the recirculation, and very high fuel utilization rates are possible at the system level .

Basically, to increase the fuel utilization in the FC mode, it would be possible to include a condenser in the high-temperature recirculation. To ensure permissible temperature differences between the gas entering the stack and the stack itself, it would be necessary to monitor the fuel temperature at the stack entry. Consequently, the recirculation flow could not be higher than just a fraction of the feed fuel flow. Overall, this flowsheet creates additional hurdles for system control without promising benefits. Therefore, it will not be discussed further.

19.3.2 Modeling

Thermodynamic zero-dimensional, time-resolved, steady-state models of the different rSOC system flowsheets according to figure 19.4 were set up in the software from Dassault Systèmes, Dymola. The model of the stack was provided by AVL List GmbH and is a virtual representation of a solid oxide cell stack from Fraunhofer IKTS. The important input and output variables of this model are summarized in table 19.1.

Table 19.1. Stack model: inputs and outputs.

Inputs	Outputs
Electric current	Electric voltage
	Electric power
	Stack temperature
Fuel composition	Exhaust fuel composition
Fuel temperature	Exhaust fuel temperature
Fuel mass flow	Exhaust fuel mass flow
Air composition	Exhaust air composition
Air temperature	Exhaust air temperature
Air mass flow	Exhaust air mass flow

In the simulations of all the other components, the Dymola-compatible Modelica library 'MixtureGasNasa' for media is used to determine the fluid properties. The approach used to model these components is described in the following paragraphs. A summary of the important parameters that must be set up in the calculation is given in table 19.2.

The heat exchangers (fuel HX, air HX, recirculation HX) are chosen to operate in counterflow, and they are modeled as 0D objects that exchange heat between two non-phase-changing fluids. The heat exchange is limited by the minimum set temperature difference at the pinch point ($\Delta T_{\mathrm{Pinch}}$) and the maximum transferable heat is given by the medium with the lower heat capacity flow. Given two of the four temperatures (those of the inputs and outputs) and the two gas compositions, the missing temperatures can be calculated. Additionally, the heat exchanger constant, which combines the area and the transfer coefficient, is calculated as a design dimension. In the off-design calculation, this constant can be set and $\Delta T_{\mathrm{Pinch}}$ calculated for this specific design.

In the evaporator model (see figure 19.3), the incoming fluid is heated to more than the boiling point of water at a pressure of 1 bar. The superheating temperature must be specified. The energy required for this change is calculated in terms of sensible and latent heat.

The condenser model (used for Condenser 1 and Condenser 2) cools down the incoming fluid to below the boiling point of water. The subcooling value (ΔT_{SC}) must be set. The heat released in the process is then calculated by considering the sensible and latent contributions. The latent heat released is calculated by the condensation heat of the condensed share of the steam. The steam content leaving the condenser is determined using the saturation pressure of water at the subcooled temperature. A condensation efficiency of 95% is used and this share of the maximum condensable amount is liquified and drained in the component.

Table 19.2. Overview of the component and system parameters in the rSOC system model.

Component	Parameters
Heat exchanger	Pinch-point temperature difference or heat exchanger constant
Evaporator	Superheating temperature difference
Condenser	Subcooling temperature difference and efficiency
Electric heater	In electrolysis operational mode only: outlet temperature
Fan	Pressure rise, efficiency
Ejector	–
System parameter	**Description**
Electric current	Operational current of the stack
Stack temperature	Temperature that the stack shall operate at
Fuel composition	Mass fraction of species in fuel
Air composition	Mass fraction of species in air
Recirculation rate	Ratio of fuel volume flow and recirculated exhaust fuel flow
Fuel utilization	System fuel utilization as defined in equation (19.3)
Air mass flow	Set in electrolysis mode, controlled in FC mode

In EC mode, the electric heaters provide the stack with thermal energy. They heat the incoming fluid to a temperature above the desired operational temperature of the stack. This temperature difference is calculated such that the desired stack temperature is reached. The energy required to heat the fluid is calculated from the sensible heat difference.

The energy consumption of the fan on the air side is calculated from the volume flow, the pressure rise, and the efficiency, as can be seen in equation (19.4).

$$P_{\text{fan}} = \dot{V} \cdot \Delta p \cdot \eta \tag{19.4}$$

The ejector is assumed to provide the desired pressure drop in the recirculation for the set recirculation flow rate. This assumption is made so that component limitations, which are not physical limitations and that could be overcome by clever design, do not impair the theoretically best possible system efficiencies. The recirculation rate is calculated as defined in equation (19.5).

$$rr = \frac{\dot{V}_{\text{recirculation}}}{\dot{V}_{\text{fuel}}} \tag{19.5}$$

During the operation of the system, some parameters are controlled in order to operate in steady-state conditions. The control strategies and parameters depend on the operation, which can be FC mode or EC mode, as described below.

In the FC operational mode, the heat produced by the stack must be discharged. This can be achieved by varying the air mass flow. The lower limit for the air mass flow is given by the minimum oxygen stream that is required for the reaction in the stack. This minimum mass flow can be calculated from the set air composition and the set electric current. The air excess (λ) is defined in equation (19.6) as the ratio of the oxygen mass flow of the air stream and the reaction mass flow.

$$\lambda = \frac{\dot{m}_{\text{air}} \cdot y_{\text{air},\, O_2}}{\dot{m}_{O_2,\, \text{reaction}}} \tag{19.6}$$

The EC mode consumes heat, and λ can be set almost freely in this case. As the flow is lowered the oxygen concentration on the air side of the stack increases, which reduces the efficiency. This is not a hard limit, but another limitation is that the heating of the stack is ensured by the sensible heat of the air stream. The temperature difference at the stack is limited to around 100 °C. This means that below certain air flow rates, the desired stack temperature cannot be maintained. In EC operation, an air mass flow is set that obeys these limitations.

In the system simulation, in addition to all the state variables of the fluids leaving and entering components, the important outputs summarized in table 19.3 are calculated.

$$\text{Exergy} = \text{Heat} \cdot \left(1 - \frac{T_{\text{ambient}}}{T_{\text{stream, hot}}}\right) \tag{19.7}$$

The system efficiency can be calculated in different ways, which reflect the different application scenarios (i.e. the energy and industry sectors) of the system. In

Table 19.3. System outputs in the rSOC system model.

Output	Calculation
Electric power	Stack model output
Electric heater power consumption	Electric heater model
Fan power consumption	Fan model
Fuel power content	Hydrogen flow of the fuel stream (FC) or exhaust fuel stream (EC) multiplied by the lower heating value (LHV) of hydrogen ($LHV_{H2} = 33.3$ kWh kg^{-1})
Condenser heat release	Condenser model
Evaporator required heat	Evaporator model
Exhaust fuel heat	Calculated using the enthalpic difference from a gas at ambient temperature using the 'MixtureGasNasa' library of Modelica
Exhaust air heat	Calculated using the enthalpic difference from a gas at ambient temperature using the 'MixtureGasNasa' library of Modelica
Condenser exergy release	Calculated from equation (19.7) using the condenser heat and the boiling temperature of water
Evaporator required exergy	Calculated from equation (19.7) using the evaporation heat and the boiling temperature of water
Exhaust fuel exergy	Calculated from equation (19.7) using the exhaust fuel heat and the exhaust gas temperature
Exhaust air exergy	Calculated from equation (19.7) using the exhaust air heat and the exhaust gas temperature

both scenarios, it is possible to operate the rSOC system in EC mode and FC mode. The different ways of defining the efficiency for both operational modes and application scenarios can be seen in equations (19.8)–(19.11).

Energy-sector:

$$\eta_{E,\,\mathrm{EC}} = \frac{P_{\mathrm{fuel}}}{P_{\mathrm{Stack}} + P_{\mathrm{fan}} + P_{\mathrm{e-heater}} + Q_{\mathrm{evaporator}}} \tag{19.8}$$

$$\eta_{E,\,\mathrm{FC}} = \frac{P_{\mathrm{Stack}}}{P_{\mathrm{Fuel}} + P_{\mathrm{fan}}} \tag{19.9}$$

Industry-sector:

$$\eta_{I,\,\mathrm{EC}} = \frac{P_{fuel}}{P_{\mathrm{Stack}} + P_{\mathrm{fan}} + P_{\mathrm{e-heater}} + Ex_{\mathrm{evaporator}}} \tag{19.10}$$

$$\eta_{I,\,\mathrm{FC}} = \frac{P_{\mathrm{Stack}} + Ex_{\mathrm{air}} + Ex_{\mathrm{fuel}} + Ex_{\mathrm{Condenserl}}}{P_{\mathrm{Fuel}} + P_{\mathrm{fan}}} \tag{19.11}$$

19.3.3 Round-trip operation of an rSOC system

The idea behind an rSOC system is to run the same system in both operational modes. This round-trip operation requires the air FX and the fuel HX (see figure 19.4) to be designed for this purpose. The changed system design modifies the maximum efficiencies for both operations, compared to a single mode of operation (either EC or FC). The basic idea behind examining the round-trip operation is that the amount of hydrogen produced during electrolysis is the same as that consumed in the FC operational mode, as indicated in figure 19.5.

For this purpose, the heat exchanger constant of both HXs is determined in the respective critical operation mode by setting the pinch-point temperature difference to 5 °C in all cases except for the air HX in FC mode, for which 90 °C was used. In the FC mode, the reaction in the stack reduces the mass flow on the side of the air HX. This means that the air cannot be fully preheated to the stack exhaust temperature, because the heat capacity of the flow of air entering the system is higher than that of the exhaust air leaving the stack. The stack limitations for temperature differences must be fulfilled in any case. With this limitation it turns out that at all calculated points, the air HX in the FC operational mode has a higher exchanger constant than in the EC mode, even though a pinch-point temperature difference of 90 °C is used. Therefore, its geometry is determined for the best operational point in FC mode. Similarly, due to the increase in fuel flow in EC mode, the fuel HX is defined for the best operational point in EC mode. In the FC mode, the requirements for this heat exchange are much lower.

To obtain the round-trip efficiency, the energy content of the fuel (P_{fuel}) in equations (19.9) and (19.11) must be divided by the efficiency in the corresponding EC mode (equations (19.8) and (19.10)). This results in equations (19.12) and (19.13). It can also be interpreted in such a way that the fuel's energy content is replaced by the energy demand of the electrolysis scaled by the ratio of produced versus consumed fuel.

$$\text{Energy sector round-trip}$$
$$\eta_{E,\text{RT}} = \frac{P_{\text{Stack,FC}}}{\left(P_{\text{Stack,EC}} + P_{\text{fan,EC}} + P_{\text{e-heater,EC}} + Q_{\text{evaporator,EC}}\right)\cdot P_{\text{fuel,FC}} / P_{\text{fuel,EC}} + P_{\text{fan,FC}}} = \eta_{E,\text{EC}} \cdot \frac{P_{\text{Stack,FC}}}{P_{\text{fuel,FC}} + \eta_{E,\text{EC}} \cdot P_{\text{fan,FC}}} \tag{19.12}$$

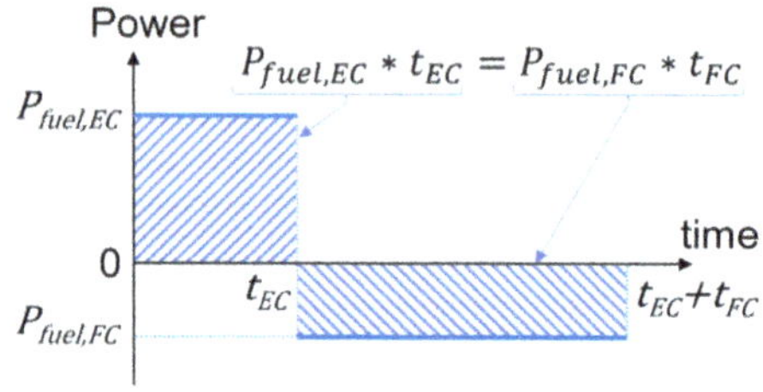

Figure 19.5. Full operating cycle of a reversible system that produces as much hydrogen in electrolysis as it uses in the FC operational mode.

Industry sector round–trip

$$\eta_{I,\mathrm{RT}} = \frac{P_{\mathrm{Stack,FC}} + Ex_{\mathrm{air,FC}} + Ex_{\mathrm{fuel,FC}} + Ex_{\mathrm{Condenser1,FC}}}{\left(P_{\mathrm{Stack,EC}} + P_{\mathrm{fan,EC}} + P_{\mathrm{e-heater,EC}} + Ex_{\mathrm{evaporator,EC}}\right)\cdot P_{\mathrm{fuel,FC}} / P_{\mathrm{fuel,EC}} + P_{\mathrm{fan,FC}}}$$
$$= \eta_{I,\mathrm{EC}} \cdot \frac{P_{\mathrm{Stack,FC}} + Ex_{\mathrm{air,FC}} + Ex_{\mathrm{fuel,FC}} + Ex_{\mathrm{Condenser1,FC}}}{P_{\mathrm{fuel,FC}} + \eta_{I,\mathrm{EC}}\cdot P_{\mathrm{fan,FC}}} \quad (19.13)$$

In these equations, we can see that the common way of calculating the round-trip efficiency (equation (19.14)) is only valid if either η_{EC} is equal to one or if $P_{\mathrm{fan,FC}}$ equals zero, which is a good approximation in all cases in which the product of η_{EC} and $P_{\mathrm{fan,FC}}$ is small compared to $P_{\mathrm{fuel,FC}}$.

$$\eta_{\mathrm{RT,\,approx}} = \eta_{\mathrm{FC}} \cdot \eta_{\mathrm{EC}} \quad (19.14)$$

The round-trip efficiency calculated in this approximative way is smaller than in the precise calculation, since $P_{\mathrm{fan,FC}}$ in the calculation of η_{FC} is not scaled by the electrolysis efficiency η_{EC} (compare the denominator of equations (19.9) and (19.11) with (19.12) and (19.13)).

19.4 Results and discussion of rSOC system behavior

In this section, the simulation results of the model described in section 19.3 are shown and analyzed. The content of the presented studies reaches from the effect of the stack temperature and an investigation of system layouts with cold and hot gas recirculation to an evaluation of efficiency measures and a discussion of round-trip operation.

19.4.1 Operational parameters for high efficiency in EC and FC mode

In this subsection, we try to understand the behavior of the rSOC system with respect to changing operational parameters. This is done separately for EC and FC mode, for both scenarios, and for the flowsheet options defined in section 19.3.2.

Before getting deeper into the effects of parameter changes, we investigate the influence of the rSOC temperature levels of figure 19.4. In the simulations, the ambient temperature at which fuel and air enter the system is 20 °C and the stack temperature is chosen to be 750 °C. The pinch-point temperature difference of the heat exchangers (air and fuel HXs) is set to 5 °C and in the FC operational mode, 90 °C is specified for the air HX. In EC mode and for all the flowsheet configurations of figure 19.4, this results in temperatures of around 110 °C at the fuel exhaust and 260 °C on the air exhaust side. In FC mode, the fuel exhaust temperature is the lowest in the cgr-cond flowsheet (figure 19.4(A)) at 105 °C and the highest in the cgr flowsheet (figure 19.4(B)) at 235 °C. The air exhaust temperature is 110 °C in all cases, since the pinch point of the air heat exchanger (air HX) is at its cold end and determines this temperature.

19.4.1.1 Operational parameter sensitivity analysis

19.4.1.1.1 Influence of the stack temperature
The main mechanisms through which the stack temperature affects the system efficiency are given by processes in the stack. One can see the change of the stack electric voltage (ΔG) and total reaction energy (ΔH) with its (mean) temperature in figure 19.6. In the steam region, an increased temperature reduces the electric stack voltage, which is given by the change in the Gibbs free energy (ΔG), while the total energy of the reaction increases slightly (ΔH). More heat ($T\Delta S$) is released as the share of ΔG decreases while ΔH grows.

$$\Delta G_r = \Delta H_r - T\Delta S_r \tag{19.15}$$

In FC mode, this means that the electric efficiency decreases with increased temperature, as can be observed in figure 19.7 (FC cgr and FC cgr-cond). Nevertheless, a high system efficiency can be achieved if an increased proportion of the high-temperature waste heat is utilized. In EC mode, the decrease in stack voltage would be beneficial for the efficiency. However, in SOCs, the reaction heat must be provided at a high temperature. Available high-temperature heat sources that can meet this demand can only be found in specific settings, such as steel mills or cement plants. Therefore, it is assumed in this chapter that the high-temperature heat must be provided using electric heaters. As a result, the advantage of a decreased stack voltage in EC mode is lost and the disadvantage of the increased total reaction energy demand (see figure 19.6 ΔH) remains.

In figure 19.6 we can see a step in thermal energy ($T\Delta S$) at the transition from the water region to steam region, which reflects the heat of evaporation of water. Nevertheless, in the steam region this energy must be also provided (EC) or can be extracted (FC) at some point in the system. Due to the use of two separate processes (stack reaction and vaporization), the heat of evaporation can be provided non-electrically. Low-temperature (>100°C) waste heat sources or coupling to exothermic processes (e.g. methanation) can be utilized (see chapter 15). The replacement of

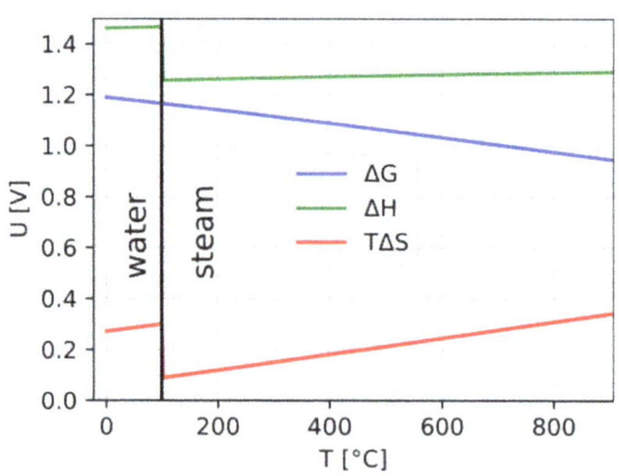

Figure 19.6. Change of the composition of the reaction energy with temperature for the reaction $2H_2 + O_2 \rightarrow 2H_2O$.

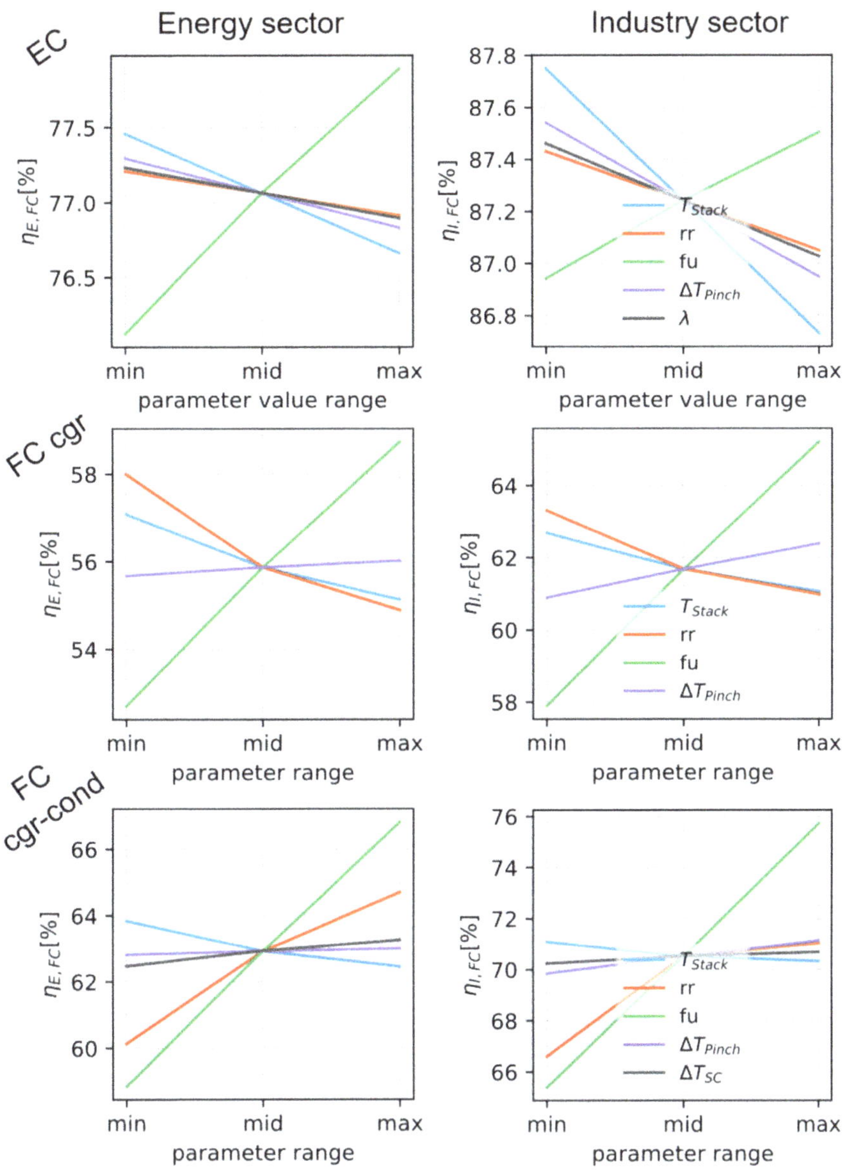

Figure 19.7. System sensitivity to parameters for the cold gas recirculation system; in the case of hot gas recirculation, the picture is similar. The range of efficiency variation in EC mode is factor of about five lower than in FC mode. The range of parameter variations can be seen in table 19.4.

high-quality energy (electricity) by lower-quality energy (heat) results in an increase in exothermic efficiency.

Another important process that affects the system performance through the stack temperature is the loss of heat to the environment. This loss increases with increasing process temperature, which reduces the efficiency in both operational modes but especially in EC mode.

19.4.1.1.2 The influences of recirculation rate (rr), fuel utilization (fu), HX pinch-point temperature difference (ΔT_{Pinch}), air excess ratio (λ), and subcooling temperature (ΔT_{SC})

In figure 19.7 a decrease in system efficiency with increased stack temperature can be seen in all cases. Furthermore, it can be observed that in EC mode, the efficiency decreases with an increase in any of the following parameters: rr, ΔT_{Pinch}, and λ. From this analysis, one might expect that small recirculation rates are preferred, and that systems without recirculation can be competitive with the ones that employ recirculation. However, the next two sections will disprove this conclusion and clarify the picture. The maximum value for the rr was chosen such (table 19.4) that it could be realized by ejectors and does not result in complete mismatch of system dimensions. While in FC mode, the efficiency increases with a higher ΔT_{Pinch} of the heat exchangers, it decreases in EC mode. The chosen ΔT_{Pinch} values in the air HX are much higher in the FC mode than in the EC mode (table 19.4). This is caused by the need for the air stream to provide cooling. In the condensed recirculation flowsheet, the effect of the recirculation rate is reversed, and higher subcooling temperatures (ΔT_{SC}) also increase the efficiency.

While, as mentioned, the stack temperature can only be varied within the limitations of the stack, the parameters λ, ΔT_{Pinch}, and ΔT_{SC} do not have strict limitations on their values. The fuel utilization (fu) is limited by the allowed range of stack fuel utilizations (fu_{Stack}, see equation (19.2)). Since fu_{Stack} depends on fu (see equation (19.3)) and rr, these two parameters cannot be investigated independently. Furthermore, the recirculation must provide a suitable gas composition for the stack. As can be seen in figure 19.7, in FC mode, these two parameters have the main impact on the efficiency and can be chosen from a wide range.

19.4.1.2 Cold gas recirculation: the influence of operational parameters

In this subsection, the parameter values are determined which lead to the best performance in a system with cold gas recirculation. Contour plots are used for the analysis. The region of the stack in which operation is infeasible, due to the limitations of fu_{Stack}, is separated by a black line and slightly grayed out. In the simulations used to generate figures 19.8 and 19.9, the stack temperature and ΔT_{Pinch} were set to the middle values of table 19.4. The minimum values of this table were chosen for the air excess ratio and the subcooling temperature.

Table 19.4. Parameter variation values

Parameter	Minimal value	Middle value	Maximal value
Stack temperature (T_{Stack}) [°C]	700	750	800
Recirculation rate (rr) [–]	0.5	2.75	5
Fuel utilization (fu) [–]	0.85	0.918	0.985
Fuel HX (ΔT_{Pinch}) [°C]	5	10	15
Air HX (ΔT_{Pinch}) [°C]	5(EC)/70(FC)	10(EC)/80(FC)	15(EC)/90(FC)
Air excess ratio (λ) [–]	0.5 (EC)	1.0 (EC)	1.5 (EC)
Subcooling temp. (ΔT_{SC}) [°C]	50 (FC)	60 (FC)	70 (FC)

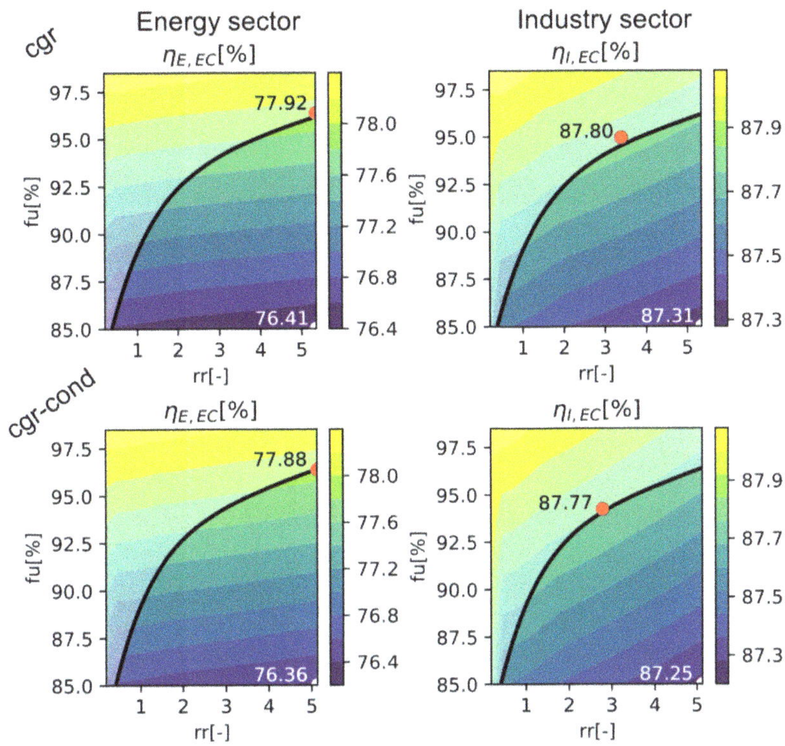

Figure 19.8. Efficiency of the cold gas recirculation flowsheet in EC mode, with (industry sector) and without (energy sector) heat integration/utilization and with (cgr-cond)/without (cgr) condensation in the recirculation. The black line separates the infeasible and feasible regions for the stack. The red points indicate the conditions of maximum efficiency.

Figure 19.7 gives a first overview of the system behavior, and we can see the difference in the magnitude of the performance influence for the EC and FC operational modes. Figures 19.8 and 19.9 provide closer insights. Here, we limit our investigation to the plane of the two parameters recirculation rate (rr) and fuel utilization (fu), because they strongly depend on each other and have a high impact on efficiency in all scenarios (see figure 19.7), as discussed in the previous section 19.4.1.1. By comparing both figures, we can see that in EC mode in figure 19.8 (white—minimum and black—maximum value), the influence of both parameters on the performance (0.5% ($\eta_{I,EC}$) and 1.5% ($\eta_{E,EC}$)) is much lower than in FC mode in figure 19.9 (4% ($\eta_{E,FC}$) to 14% ($\eta_{I,FC}$)). In the industry-sector scenario in particular, the effect in EC mode is very small. The reason is that as the stack power consumption (P_{Stack}) increases the heat produced in the stack increases as well; thus, the stack demands less electric energy from the electric heaters ($P_{e\text{-heater}}$). The increase in stack power and drop in electrical heating energy cancel almost exactly, which means that there is no influence on the efficiency, which can be concluded from equations (19.8) and (19.10).

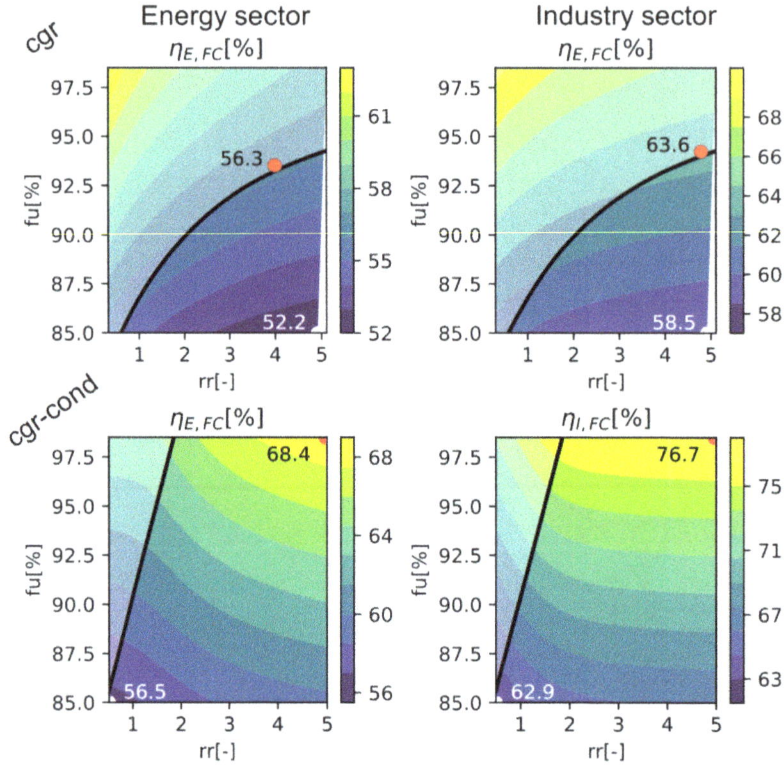

Figure 19.9. Efficiency of the cold gas recirculation flowsheet in FC mode, with (industry sector) and without (energy sector) heat integration/utilization and with (cgr-cond)/without (cgr) condensation in the recirculation. The black line separates the infeasible and feasible regions for the stack. The red points indicate the conditions of maximum efficiency.

19.4.1.2.1 Electrolysis cell operation mode

Although in EC mode, an increase in rr decreases the efficiency, the effect of better performance due to a higher possible fu moves the best operating points to considerable recirculation values. In the energy-sector scenario, the best fuel utilization and recirculation rates are higher than in the industry-sector scenario. One can see that this is caused by a smaller sensitivity to the rr (the slope of the contour lines). It is interesting that although the layout with condensation (figure 19.4(B2)) loses additional heat in the condensation step, the recovery of the fuel exhaust heat (recirculation HX, figure 19.4(B2)) can almost fully compensate this loss. The reason for this is that in the layout without condensation (figure 19.4(B1)), the heat of the exhaust fuel leaving the system is unused. In contrast, in the layout with condensation (figure 19.4(B2)), it is necessary to recover the heat to prevent mixing of a subcooled recirculation gas with the incoming steam from the evaporator. This is an important insight if a fan is used for recirculation instead of an ejector, since the temperature limit for fans is usually below 80 °C. From figure 19.8, we can see that the increase in efficiency from the grid scenario to the industry scenario is around 10% for both system flowsheets. That means that thermal integration with ambient aggregates in the industry-sector scenario is highly beneficial.

19.4.1.2.2 Fuel-cell operational mode
The efficiency pattern in FC mode without condensation is similar to that in EC mode, as can be seen in figure 19.9. An increase of rr alone decreases the efficiency but allows higher fuel utilization (fu). This decrease due to higher rr is stronger in the energy-sector scenario ($\eta_{E,FC}$) than in the industry-sector scenario ($\eta_{I,FC}$). Therefore, the optimum operating point in the energy-sector operation is found at lower values of rr. However, in the case of the flowsheet with condensation (cgr-cond), a totally different picture can be seen. Here, in the energy-sector scenario, the efficiency increases in the direction of higher rr, as can be observed in figure 19.9. This is due to processes in the stack. The advantage of a spatially more homogeneous fuel composition over the length of the gas channels in the stack outweighs the disadvantage of lowering the average hydrogen concentration. In the flowsheets without condensation (cgr) the recirculation causes a larger drop in the hydrogen concentration, which is not compensated by the described mechanism. Another important difference can be observed in the flowsheet with condensation. The reduction of the steam content in the recirculation due to condensation almost cancels the fuel utilization limitations of the stack. The grayed-out zone for this case in figure 19.9 is small and does not exclude the theoretically best operational region of the system, which is not the case for all the other investigated flowsheet configurations and operation modes. The fuel utilization is, in this case, limited only for low rr. Using figure 19.9, the effect of condensation in recirculation can be quantified. It can increase the efficiency by 13% ($\eta_{E,FC}$) in the energy-sector scenario and 12% ($\eta_{I,FC}$) in the industry-sector scenario. Furthermore, the efficiencies of all scenarios can be compared. Upon changing from the energy-sector scenario to the industry-sector scenario, the efficiency with the cgr flowsheet (figure 19.4(B1)) increases by 7% and that of cgr-cond (figure 19.4(B2)) increases by 8%.

19.4.1.3 Hot gas recirculation: the influence of operational parameters
In this subsection, the hot gas recirculation flowsheet is examined similarly to the previous subsection on cold gas recirculation. The contour plots in figure 19.10 were made in the same way as described in section 19.4.1.2 for cold gas recirculation. Likewise, they show the minimum and maximum values (white and black numbers, respectively). From this we can see that the variation in efficiency in EC mode (0.6% ($\eta_{I,EC}$) and 1.6% ($\eta_{E,EC}$)) is much smaller than in FC mode (around 6% ($\eta_{I,EC}$) and 5% ($\eta_{E,EC}$)). The reason for this lower sensitivity in EC mode was already discussed in section 19.4.1.2. In figure 19.10, we can see that the efficiency of the best operation point (red dot, black number) grows by 10% in EC mode and by 7% in FC mode if we change from the energy-sector scenario to the industry-sector scenario.

19.4.1.3.1 Electrolysis cell operational mode
In both scenarios, the recirculation rate has a very small impact on system performance. An increased recirculation rate (rr) permits higher fuel utilization (fu).

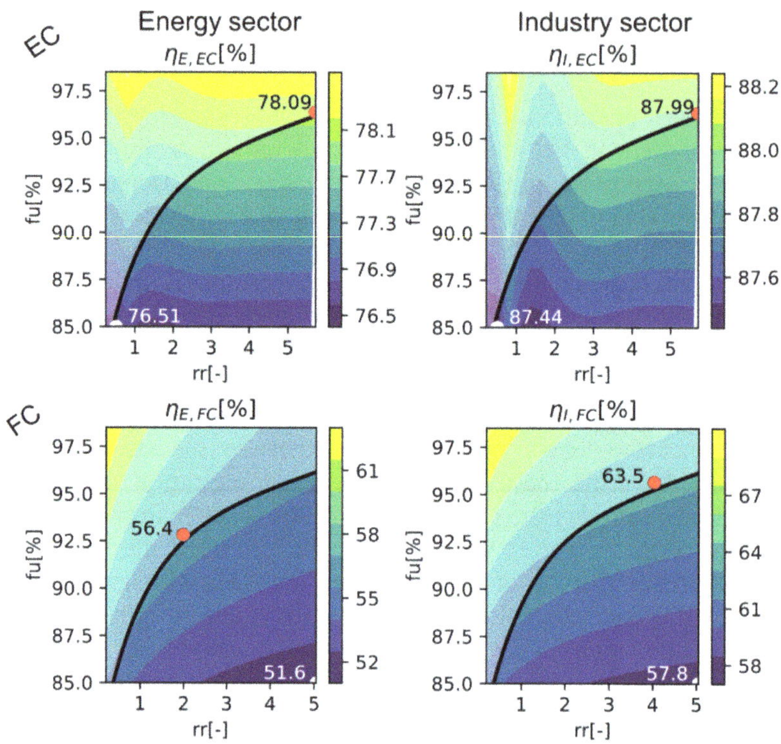

Figure 19.10. Efficiency of hot gas recirculation in EC (top) and FC (bottom) modes, with (industry sector) and without (energy sector) heat integration/utilization and with (cgr-cond)/without (cgr) condensation in the recirculation. The black line separates the non-feasible and feasible regions for stack operation. The red points indicate the conditions of maximum efficiency.

Consequently, the most efficient operation is found at high values of rr. The reasons for this behavior are two mechanisms that act on the efficiency in opposite ways. As is the case in any other flowsheet configuration, an increase in rr leads to a lower average steam concentration, which decreases efficiency. This lowered efficiency causes more heat generation in the stack, which reduces the need for external electric heaters by the same amount. Therefore, high rr are beneficial, as there is no downside, but they allow high values of fu. The electrolysis efficiency in this flowsheet configuration is slightly higher than with cold gas recirculation. In addition to the changed influence of the rr, another reason for this may be that less heat is lost from the system, as the recirculated gas stream does not have to pass the fuel HX.

19.4.1.3.2 Fuel-cell operational mode

The graphs of FC mode in figure 19.10 look almost identical to those of cold gas recirculation without condensation (cgr) in figure 19.9. Even the efficiencies of the most efficient points nearly match, although they lie at different values of the parameters. Thus, the discussion of cgr (section 19.4.1.2) applies here as well.

19.4.2 Evaluation of measures to increase efficiency

In this subsection, we explore the effect of two different options that extend the flowsheet in figure 19.4 to increase system efficiency in EC mode, and one possibility for FC mode:

- Heat recovery from exhaust air for preheating and evaporation in EC mode
- Heat recovery from exhaust fuel for preheating and exhaust air for superheating in EC mode
- Use of high-temperature heat potentials in FC mode

19.4.2.1 Heat recovery from exhaust air for preheating and evaporation in electrolysis mode

In the flowsheets of figure 19.4, the heat exchangers for internal heat recovery (the fluid HX and air HX) work with gas only on the respective system side. The fuel HX works with fuel and exhaust fuel and the air HX works with air and exhaust air. In EC mode, the mass flow of the air is increased by the transferred oxygen in the stack; thus, the exhaust air exceeds the incoming air mass flow. As a result, the pinch point in the air HX lies on the hot side and the exhaust air leaves the system at temperatures of up to 300 °C. For the fuel HX, the exact opposite applies: the pinch point lies on the cold side. To use this exhaust air heat, an additional heat exchanger can be added on the fuel side, as can be seen in figure 19.11.

In this configuration, the heat of the exhaust air stream can be recovered internally. This heat suffices to preheat the feed fuel (water) and produce a part of the steam needed (preheater/evaporator). In figure 19.12, one can see the increases in efficiency for the different scenarios (the energy and industry sectors) and flowsheets (cgr and hgr) in EC mode. By comparing this graphic with figures 19.8 and 19.10, we can see that an increase of around 2.8% ($\Delta\eta_{E,EC}$) is achieved in the energy-sector scenario and an increase of 0.8% ($\Delta\eta_{I,EC}$) is achieved in the industry-sector scenario in both flowsheet configurations (cgr and hgr).

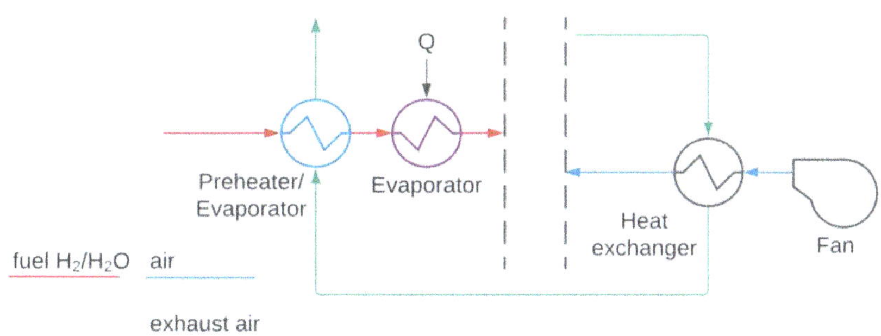

Figure 19.11. Flowsheet with heat recovery from exhaust air for preheating and evaporation in the electrolysis operational mode (an extension of figure 19.4).

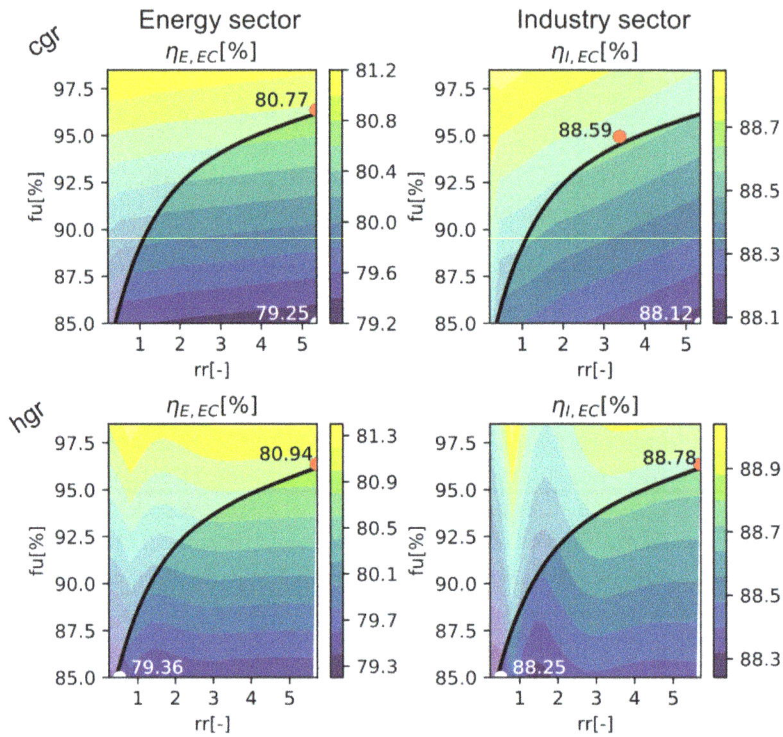

Figure 19.12. Changed efficiency pattern for the flowsheet of figure 19.11 in EC mode, with (industry sector) and without (energy sector) heat integration/utilization.

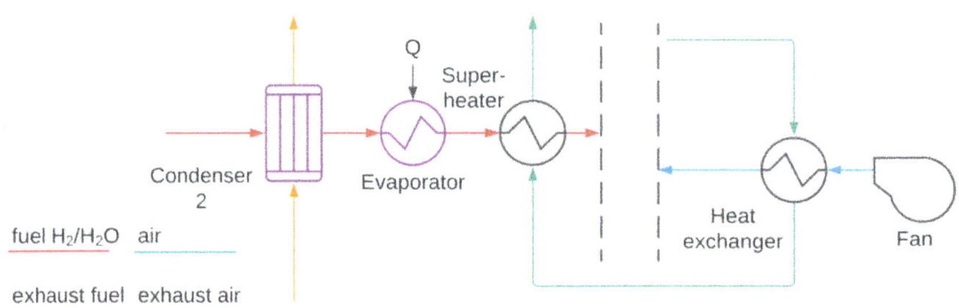

Figure 19.13. Flowsheet of heat recovery from the exhaust for preheating and exhaust air for superheating in EC mode (an extension of figure 19.4).

19.4.2.2 Heat recovery from exhaust fuel for preheating and exhaust air for superheating in electrolysis mode

A different version of heat recovery from the exhaust streams can also be realized. The condensing exhaust fuel stream contains more heat than that needed for preheating the feed fuel, so its heat can be used to preheat the fuel feed water to a temperature close to its boiling point (see figure 19.13 Condenser 2). The exhaust air has a temperature of more than 100 °C and it has a higher heat capacity stream than the fuel feed gas. Therefore, part of its heat content can be used to superheat the evaporated fuel close to the exhaust air temperature (superheater). Since the internal fuel heat exchanger has its pinch point on the cold side, increasing the cold-side temperature can slightly increase the temperature of heat recovery for the fuel entering the electric heater on the fuel side. Consequently, less electric energy is used in this heater.

For the efficiencies at the most efficient points, with patterns like those shown in figure 19.12, increases of about 1.0% ($\Delta\eta_{E,EC}$) and 0.5% ($\Delta\eta_{I,EC}$) can be achieved in the energy-sector scenario and the industry-sector scenario, respectively. The efficiency increase is lower than that obtained using the previously proposed system improvement. The reason for this is that the exhaust fuel can only preheat the feed fuel and not evaporate it as in the previous case. Thus, the electricity demand in the evaporator is decreased much less. The reduction in the electric energy used to meet the high-temperature demand due to the superheating with exhaust air is only minor and cannot outweigh this disadvantage.

19.4.2.3 Use of high-temperature heat potentials in fuel-cell mode

The changes to the base scenarios related to both hot and cold gas recirculation shown in figure 19.4 consist of additional heat exchangers (HT-HX) that are added to extract high-temperature heat from the exhaust gases of the stack. This new configuration is shown in figure 19.14.

The usable high-temperature heat is calculated so as to ensure that the stack entrance temperature is exactly 100 °C lower than the stack temperature. The exergy content of this heat is calculated according to equation (19.7), and it is an additional contribution to the denominator in equation (19.11). If the amount of heat generated during operation stays the same, an increase in efficiency can be achieved by increasing the temperature and therefore the exergy. The changed efficiency patterns are shown in figure 19.15. We can see that in a cgr-cond system (according to figure 19.4(B2)), when compared to the results shown in figures 19.9 and 19.10, the efficiency has increased ($\Delta\eta_{I,EC}$) by 3.4%. The increase is 2.0% if heat is only extracted at the air side. In the case of a system with hot gas recirculation, the efficiency can be increased by 7.0% compared to the base case (figures 19.9 and

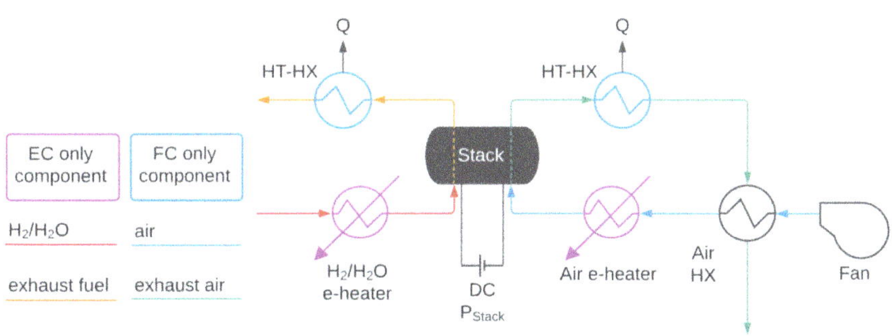

Figure 19.14. Flowsheet with high-temperature heat decoupling in FC mode (an extension of figure 19.4).

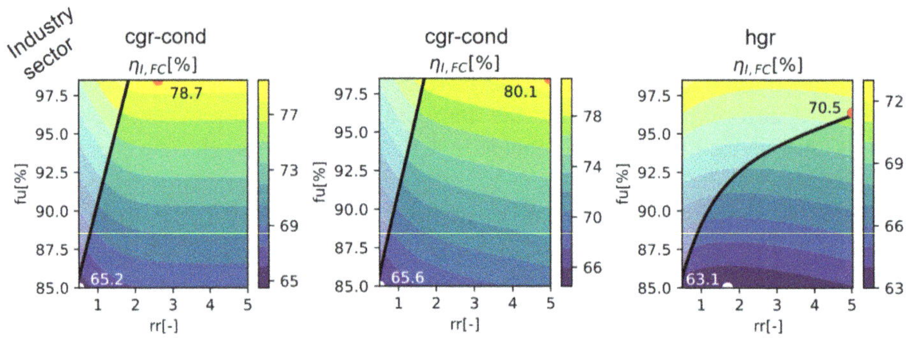

Figure 19.15. Efficiency patterns in industrial-sector application with high-temperature heat consumers. The left-hand graph for cgr-cond represents the case of high-temperature heat extraction on the air side only, while the other graphs refer to figure 19.14 and represent high-temperature heat extraction on both the air and fuel sides.

19.10). The amount of high-temperature heat was quantified. Its proportion of the stack's electric power is 15% for the cold gas and 30% for hot gas recirculation.

The configuration with high-temperature heat extraction can significantly increase the system efficiency in the industry-sector scenario ($\eta_{\mathrm{I,FC}}$), according to equation (19.11) if there is a consumer for the high-temperature heat. For large systems, this could also be a steam boiler in a steam turbine cycle. At times without high-temperature heat extraction, other stack cooling mechanisms must be implemented. The simplest option is a valve in the exhaust air stream that bypasses the heat exchanger (air HX), so that the air entering the stack is heated less. In this case, the air exhaust system must be able to deal with temperatures above 300 °C.

19.4.3 Round-trip operation and the design of heat exchangers

Up until now, the two operational modes of the system (EC and FC) have been investigated independently. This means that the geometry of heat exchangers was chosen separately for each mode. We now analyze a system with maximized efficiency as it is switched from EC mode to FC mode as shown in figure 19.5. The system flowsheets for the cases under discussion are cold gas recirculation (cgr) in EC mode and condensation with condensation (cgr-cond) in FC mode.

Figure 19.16 shows the changed efficiency patterns if the heat exchanger constants are determined for the respective scenario (energy sector and industry sector). These results can be compared to the independent analyses of the EC and FC modes in figures 19.8 and 19.9.

The efficiency increased slightly in EC mode (by around 0.1% for both scenarios) due to the optimization of the heat exchangers (fuel HX and air HX) for the best operational point. The patterns changed, since the pinch-point temperature difference now varies with the recirculation rate (mass flows through the fuel HX). As the recirculation rate increases, the mass flow through the fuel heat exchanger increases as well. Thus, the pinch-point temperature difference increases, causing a higher electricity demand in the electric heater and lowering the efficiency. At scant

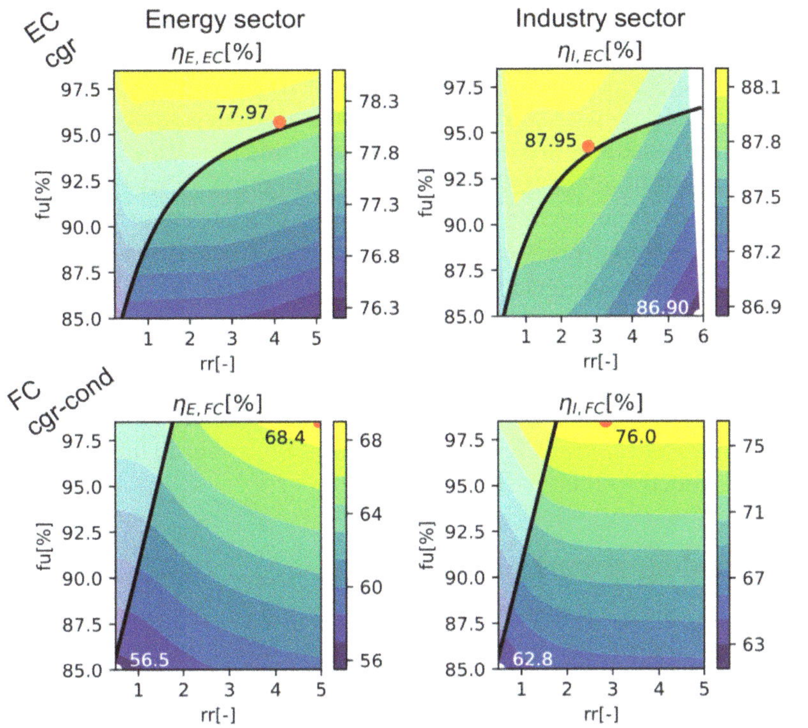

Figure 19.16. Efficiency pattern for system round-trip operation with cold gas recirculation flowsheets (FC with condensation in recirculation).

recirculation rates, the stack reactions produce too little energy and the stack temperature cannot be maintained. The latter effect causes the rise in efficiency at the left-hand ends of the EC graphs in figure 19.16.

In FC mode for the energy-sector scenario, the efficiency is unchanged compared to the pure FC mode in figure 19.9. However, the efficiency in the industry scenario is slightly decreased, by 0.7% ($\Delta\eta_{I,FC}$). The reasons for this are the changed heat exchanger pinch-point differences, which decrease the exhaust temperature and therefore the waste heat exergy content of both the fuel and air streams. In this scenario, the design should probably be executed in a different way by making a better design compromise in the EC mode.

The efficiencies of the round-trip operation can be calculated using equations (19.11) and (19.12); they are 53.5% ($\eta_{E,RT}$) in the energy-sector scenario and 67.0% ($\eta_{I,RT}$) in the industry-sector scenario. Using the approximation of equation (19.14) we get the approximative efficiency values 53.3% ($\eta_{E,RT}$) and 66.8% ($\eta_{I,RT}$). This shows that the approximation is good and delivers (as expected) slightly lower efficiencies. To be precise, by taking the values for single-mode operation from sections 19.4.1 and 19.4.2, an additional approximation is used as a result of neglecting the effect of the round-trip heat exchanger design. This is a legitimate method, since the optimum round-trip design is close to the previously investigated single-mode designs.

The measures used to increase the system efficiency in the previous section can also be applied to the system in round-trip operation. If we limit ourselves to analyses of the efficiency at the best operational point, the approximation for the round-trip efficiency can be used to calculate the efficiency values for all combinations of different system flowsheets, improvements, and scenarios. This is justified by the small deviations that can be observed in the paragraph above for the case of cold gas recirculation. As an example, the round-trip efficiencies of the basic hot gas recirculation flowsheet from section 19.4.1.3 can be calculated in this way, which gives the values 44.0% ($\eta_{E,RT}$) and 55.9% ($\eta_{I,RT}$).

One hurdle in setting up a reversible system with recirculation shall be mentioned here. The volume flow for the system flowsheets of figure 19.16 in EC mode is larger by a factor of 2.5 than in FC mode. This means that two different ejectors or a mechanical fan must be used to achieve the required recirculation rates in both operations (EC and FC). There may be other solutions, but in any case, additional complexity is added to the system.

19.5 Concluding remarks

In the following, the main results are summarized and conclusions are drawn for the design of rSOC systems for different applications. During this process, questions I–IV of section 19.2 are considered and used to structure this section.

Message I. Understanding the combined influence of operational parameters on the system performance:

From the parameter sensitivity analysis in section 19.4.1.1, we can conclude that the stack temperature and the air excess ratio should be at the lower boundary of the possible operational range. For the system flowsheets of figure 19.4, the lower boundary for the air excess ratio could theoretically be removed by using a different method of stack heating. The heat exchanger pinch-point differences have opposite effects in the EC and FC modes. In the round-trip operation of the rSOC system, a good way of determining the heat exchanger area is to use the EC mode for the fuel HX and the FC mode for the air HX. There is a need for a more detailed investigation of the best heat exchanger dimensions for round-trip operation.

Message II. Determining the best operating points for the best system performance in different operational modes:

Figures 19.8–19.16 show the optimum operating points for both operational modes (EC and FC), application scenarios (the energy and industry sectors), and flowsheets (cgr, cgr-cond, hgr). These graphs strongly support the conclusion that high recirculation rates (>1) are beneficial in all cases, and are always better than a lack of recirculation. Therefore, the system flowsheet with fuel recirculation is preferred.

Message III. Quantification of efficiency-increasing measures for different system flowsheets:

Measures for increasing system efficiency are discussed in section 19.4.2. In the energy-sector scenario in particular, the options for EC mode have a significant impact on the system performance. Furthermore, it was found that simple heat recovery from exhaust air has the best outcome. In this case, the efficiency can only be increased by providing the heat of evaporation for the steam and by using high-temperature waste heat sources. For the FC mode, high-temperature extraction can make remarkable improvements to the flowsheet with hot gas recirculation. The effect is also noteworthy in the cold gas recirculation flowsheet with condensation. However, a limitation for this measure is implied by the ambient demand for high-temperature heat.

Message IV. Determination of the best system flowsheets, for the use cases of the energy sector and the industry sector, to meet the flexibility demand with best system performance:

Depending on the main operating mode, the best system flowsheet varies. If the EC mode is dominant, high-temperature recirculation may be used. Since, in this case, the recirculation rate can be higher in EC mode than in FC mode (see section 19.4.1.3), it is promising that the same ejector could be applied for both operating modes. This should be investigated further in simulations using detailed models of the ejector recirculation system. The cold gas recirculation flowsheet with condensation in the recirculation path is superior in terms of efficiency in FC mode. This is the preferred layout for systems without high-temperature extraction that spend a considerable proportion of their operational time in FC mode.

References

[1] 2015 United Nations Treaty Collection: Chapter XXVII Environment: 7. d Paris Agreement vol. 3156 Accessed 29.02.2022

[2] European Environment Agency, European Topic Centre on Climate Change Mitigation 2021 Annual European Union greenhouse gas inventory 1990–2019 and inventory report 2021: Submission to the UNFCCC Secretariat EEA/PUBL/20 Accessed 15 Dec 2021

[3] IEA 2021 Total primary energy supply by fuel, 1971 and 2019 https://iea.org/data-and-statistics/charts/total-primary-energy-supply-by-fuel-1971-and-2019 Accessed 29.02.2022

[4] European Commission 2019 The European Green Deal COM/2019/640 final Accessed 29.02.2022

[5] European Commission 2021 Fit for 55: delivering the EU's 2030 Climate Target on the way to climate neutrality COM/2021/550 final Accessed 28 Feb 2022

[6] Mancarella P 2014 MES (multi-energy systems): an overview of concepts and evaluation models *Energy* **65** 1–17

[7] Greiml M, Traupmann A, Sejkora C, Kriechbaum L, Böckl B and Pichler P *et al* 2020 Modelling and model assessment of grid based multi-energy systems *Int. J. Sustain. Energy Plan. Manage.* **29** 7–24

[8] Greiml M, Fritz F and Kienberger T 2021 Increasing installable photovoltaic power by implementing power-to-gas as electricity grid relief – a techno-economic assessment *Energy* **235** 121307

[9] Pfenninger S, Hawkes A and Keirstead J 2014 Energy systems modeling for twenty-first century energy challenges *Renew. Sustain. Energy Rev.* **33** 74–86

[10] Sejkora C, Kühberger L, Radner F, Trattner A and Kienberger T 2020 Exergy as criteria for efficient energy systems—a spatially resolved comparison of the current exergy consumption, the current useful exergy demand and renewable exergy potential *Energies* **13** 843

[11] Sejkora C, Lindorfer J, Kühberger L and Kienberger T 2021 Interlinking the renewable electricity and gas sectors: a techno-economic case study for Austria *Energies* **14** 6289

[12] Bundeskanzleramt Österreich 2020 Aus Verantwortung für Österreich (Out of a sense of responsibility for Austria): Regierungsprogramm 2020-24 https://bundeskanzleramt.gv.at/bundeskanzleramt/die-bundesregierung/regierungsdokumente.html (https://bundeskanzler-amt.gv.at/en/federal-chancellery/the-austrian-federal-government/government-documents.html) Accessed 18 Feb 2022

[13] Bundesministerium für Klimaschutz, Umwelt, Energie, Mobilität, Innovation und Technologie 2022 Bundesgesetz über den Ausbau von Energie aus erneuerbaren Quellen (Erneuerbaren-Ausbau-Gesetz – EAG) (The Renewable Energy Expansion Act) StF: BGBl. I Nr. 150/2021 (NR: GP XXVII RV 733 AB 982 S. 115. BR: 10690 AB 10724 S. 929.) [CELEX-Nr.: 32018L2001, 32019L0944, 32019L0692]: EAG https://ris.bka.gv.at/GeltendeFassung.wxe?Abfrage=Bundesnormen&Gesetzesnummer=20011619 Accessed 18 Feb 2022

[14] Krutzler T, Kellner M, Heller C, Gallauner T, Stranner G and Wiesenberger H *et al* Energiewirtschaftliche Szenarien im Hinblick auf die Klimaziele 2030 und 2050 (Energy management scenarios with regard to climate targets 2030 and 2050): Synthesebericht 2015 Perspektiven für Umwelt und Gesellschaft REP-0535 Umweltbundesamt Wien

[15] Bundesministerium für Klimaschutz, Umwelt, Energie, Mobilität, Innovation und Technologie 2020 Energie in Österreich: Zahlen, Daten, Fakten (Energy in Austria 2020: Numbers, data and facts) https://bmk.gv.at/themen/energie/publikationen/zahlen.html Accessed 28 Feb 2022

[16] Singer D V 2017 Reversible solid oxide cells for bidirectional energy conversion in spot electricity and fuel markets *Doctoral Thesis* (Columbia University)

[17] Frank M, Deja R, Peters R, Blum L and Stolten D 2018 Bypassing renewable variability with a reversible solid oxide cell plant *Appl. Energy* **217** 101–12

[18] Giap V-T, Lee Y D, Kim Y S and Ahn K Y 2020 A novel electrical energy storage system based on a reversible solid oxide fuel cell coupled with metal hydrides and waste steam *Appl. Energy* **262** 114522

[19] Giap V-T, Kim Y S, Lee Y D and Ahn K Y 2020 Waste heat utilization in reversible solid oxide fuel cell systems for electrical energy storage: fuel recirculation design and feasibility analysis *J. Energy Stor.* **29** 101434

[20] Giap V-T, Kang S and Ahn K Y 2019 High-efficient reversible solid oxide fuel cell coupled with waste steam for distributed electrical energy storage system *Renew. Energy* **144** 129–38

[21] Reznicek E P and Braun R J 2018 Techno-economic and off-design analysis of stand-alone, distributed-scale reversible solid oxide cell energy storage systems *Energy Convers. Manage.* **175** 263–77

[22] Reznicek E P 2016 Design and simulation of reversible solid oxide cell systems for distributed scale energy storage *Master Thesis* (Golden, CO: Colorado School of Mines) https://hdl.handle.net/11124/170645
[23] Wendel C H and Braun R J 2016 Design and techno-economic analysis of high efficiency reversible solid oxide cell systems for distributed energy storage *Appl. Energy* **172** 118–31
[24] Wendel C H, Kazempoor P and Braun R J 2016 A thermodynamic approach for selecting operating conditions in the design of reversible solid oxide cell energy systems *J. Power Sources* **301** 93–104
[25] Perna A, Minutillo M and Jannelli E 2018 Designing and analyzing an electric energy storage system based on reversible solid oxide cells *Energy Convers. Manage.* **159** 381–95
[26] Spath S 2019 *Modellierung einer Power-to-Gas-(to-Power)-Anlage auf Basis einer RSOC mit Ebsilon® Professional* Montanuniversität Leoben https://pure.unileoben.ac.at/portal/en/publications/modellierung-einer-powertogastopoweranlage-auf-basis-einer-rsoc-mit-ebsilon-professional(f276dad8-81cd-4c3d-a2e3-0da0727a910f).html
[27] Mottaghizadeh P, Santhanam S, Heddrich M P, Friedrich K A and Rinaldi F 2017 Process modeling of a reversible solid oxide cell (r-SOC) energy storage system utilizing commercially available SOC reactor *Energy Convers. Manage.* **142** 477–93
[28] SunFire GmBH 2016 SunFire supplies Boeing with world's largest commercial reversible electrolysis (RSOC) system https://sunfire.de/de/news/detail/sunfire-liefert-weltgroesste-kommerzielle-reversible-elektrolyse-rsoc-an-boeing Accessed 12 Dec 2021

Chapter 20

Economic aspects of power-to-gas

Hans Böhm and Robert Tichler

This chapter covers the economic aspects of the electrolysis and power-to-gas (PtG) processes and has a special focus on high-temperature electrolysis. In this context, projected market potentials for PtG are discussed on a global scale alongside with an estimation of the resulting technology cost reductions induced by economies of scale. It is shown that this is a major driver for reducing the product generation costs of hydrogen and synthetic natural gas (SNG), as electricity supply costs become dominant. Therefore, we outline the importance of high electric efficiencies and the utilization of internal and external synergy effects.

20.1 Market perspectives

The production of hydrogen from renewable electricity is expected to play a major role in the decarbonization of industrial processes and the energy system as a whole. This is also illustrated by the number of governments that have already defined national hydrogen strategies [1]. Consequently, the European Union also defined a dedicated strategy for renewable hydrogen in Europe, in line with the European Green Deal. The strategic vision therein aims at installing electrolysis capacities of at least 6 GW in the EU by 2024 and 40 GW by 2030. Additionally, 40 GW of renewable hydrogen electrolyzer capacity is planned to be installed in the EU neighborhood before 2030 to provide appropriate import capacity to the EU [2]. It also becomes apparent that hydrogen and PtG are expected to play significant roles in future energy systems when one looks at the number of related projects (including power-to-X) that became operational in recent decades (see figure 20.1).

The following subsections outline the future demand for hydrogen and PtG[1] products in general (section 20.1.1) and the specific demand potentials for high-temperature electrolysis, including co-electrolysis (section 20.1.1.2).

[1] In the context of this chapter, the term *PtG* is used as a collective term for *power-to-hydrogen* and *power-to-methane*.

doi:10.1088/978-0-7503-3951-3ch20

Figure 20.1. Global overview of power-to-gas (and power-to-X) projects. Based on information from [3].

20.1.1 Hydrogen and power-to-gas

Future market potentials for hydrogen and PtG may be derived from different objectives. On the one hand, electro-catalytically produced gas may be discussed as an important part of future energy supply as a whole, either as a renewable energy carrier and storage medium or an alternative to existing fossil fuels. On the other hand, hydrogen is already an important feedstock for industrial processes and chemical synthesis that is still almost completely based on fossil resources, and thus has to be replaced by carbon-neutral alternatives. Beyond that, the defossilization of industrial processes may reveal additional demands for which transformation to a hydrogen-based route is the currently most promising alternative. While the direct replacement of today's fossil streams is largely dependent on the design of future energy systems, an indication of additional demands may already be derived from foreseeable technology pathways and today's industrial conditions. Nonetheless, both approaches are discussed separately in the following.

20.1.1.1 Demands for power-to-gas from an energy system perspective

National and international commitments to long-term climate and energy targets have revealed corresponding demands for renewable hydrogen in different sectors, which are depicted by the abovementioned hydrogen strategies and roadmaps. Corresponding demand potentials have been evaluated and published in the relevant literature in recent years. In the following, some of those are discussed, but the discussion is not exhaustive. Furthermore, due to restraints and differences in the individual studies—regional, temporal, and sectoral—these can not be used to derive a total global demand potential. This collection of data is mainly based on and taken from [4].

For the German power sector, a demand potential for power-to-hydrogen capacities of 26–36 GW_{el} was evaluated as a dependency of the amount of short-term storage capacity, such as batteries, that will be installed before 2050 [5, 6]. However, the demand for flexible options has even been assessed at a significantly higher level of 89–134 GW_{el} [7]. Other estimations based on economic aspects range between these extremes [8]. These values may rise significantly if non-power sectors, namely mobility, industry, and (residential) heat supply, are taken into account and thus hydrogen is not only considered as an energy carrier, but also as a feedstock for downstream synthesis processes, subsumed under the term power-to-X [8–10]. Comparable results have been published for other European countries. For instance, the transition of the Spanish energy system to renewable energy sources (RESs) is expected to result in a demand for PtG storage capacities of 7–19.5 GW_{el} by 2050 [11]. However, for complete defossilization, this demand may be more than four times as high as the upper limit of the aforementioned range [12]. Calculated demands for the Italian energy system show a comparable value [13].

However, to assess supra-regional or international demands, for instance those of a low-carbon European energy system, a more comprehensive perspective of the energy system has to be considered, instead of individual national demands. Corresponding EU-related studies have shown that the future demands for power-to-hydrogen and power-to-methane mainly depend on the provided boundary conditions, such as underground carbon storage options and positive drivers for SNG. Thus, the resulting capacity demands for PtG or electrolysis cover a wide range from 70–1000 GW_{el} [14, 15]. Based on the given cost-optimization approach, the upper range for SNG from PtG could reach a capacity of 546 GW, which would thus imply a coverage of 75% of the gas demand expected for the EU by 2050 [15]. Similar calculations for a global scale are predictably scarce, due to the complexity and variety of interacting regional and supra-regional energy systems. However, single estimations suggest a demand for a long-term power-to-methane storage capacities of about 2360 GW_{el} for a global, decentralized, and 100% renewable electricity supply scenario [16]. The variation in the individual estimations for future demands for PtG from national European to international levels is shown in figure 20.2.

20.1.1.2 Future industrial demands for hydrogen

The vast majority of today's global hydrogen production, which amounts to about 115 Mt per year, is utilized in industrial applications. Within this total, the greatest individual uses are in oil refining (33%), ammonia (27%) and methanol (11%) production, and direct reduction in steelmaking (3%). While the use in oil refining, despite increasing consumption, may even decrease in the context of defossilization, the demands in the other sectors mentioned are expected to rise. Even though the demand for ammonia in the production of fertilizers is expected to grow by 1.7% per year until 2030 [17], the increase in the other industrial uses of ammonia, which currently account for only 10%–20% of the global demand [18], will be more significant. Due to the increasing use of methanol as a fuel additive and additional demands for methanol as a base chemical for the renewable production of high-value chemicals (HVCs), the production of methanol is expected to increase by more than 50% by 2030 and almost double by 2050 in relation to the abovementioned amounts [18]. Thus, the respective annual demand for hydrogen as a feedstock for the chemical industry is about to increase by more than 22 Mt by the year 2050, just because of these two base chemicals, ammonia and methanol [17].

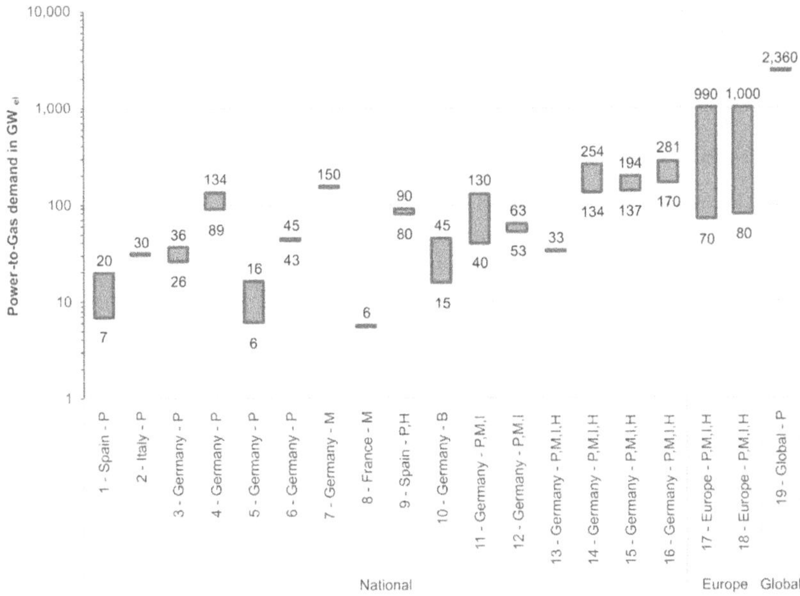

Figure 20.2. National and international PtG demand potentials for the year 2050 according to relevant literature (P—power sector, M—mobility sector, H—heat sector, I—industry sector, B—based on biogas). Reproduced from [4].

In the iron and steel industry, the transition to a hydrogen-based direct reduced iron (DRI) route to process iron ore is widely discussed as the most promising path for significantly reducing the emissions of primary steel production. Today's implementation of the process is limited, leading to a global hydrogen demand of 4 Mt per year. The development of future demands is highly dependent on the future weighting of different steelmaking routes. In addition to the transition rate of primary steel production to a DRI route, the general ratio of primary steelmaking versus secondary steel produced via electric arc furnace (EAF), using high proportions of scrap metal, will have a relevant impact on the resulting hydrogen demand of the sector. In the short- to mid-term, radical adaptions on a global scale cannot be expected, thus the annual hydrogen demand is predicted to approximately double by 2030. Long-term estimations forecast an annual global hydrogen demand of up to 62 Mt by 2050 for a moderate increase in secondary production to a 29% share. If the secondary production percentage rises to 47%, accompanied by a simultaneous complete switch from gas-based to pure hydrogen-based DRI, the annual hydrogen demand in 2050 is expected to fall within the range of 47–67 Mt [17].

Due to the fact that today's hydrogen production is almost completely based on fossil resources, the abovementioned industrial demands have to be met by renewables to meet long-term climate targets; thus, according to current perspectives, these demands must be met by electrolysis powered by renewable electricity. Under the presumption of an average efficiency of 70%, related to the lower heating value

(LHV), the demands placed on future electrolysis systems by the two depicted sectors alone, i.e. chemical and steel industries, would amount to global electrolysis capacity demands of 6640–7590 TWh_{el} and 1900–2170 GW_{el}, respectively [4].

20.1.2 High-temperature (co-)electrolysis

Compared to low-temperature electrolysis technologies, such as alkaline electrolysis cells (AEC) and proton exchange membrane electrolysis cells (PEMECs), the properties of high-temperature or solid oxide electrolysis cells (SOECs) offer potential advantages for certain use cases. While the elevated temperatures allow for significantly increased electrical efficiencies by partially replacing the electric energy input with thermal energy, the resulting heat demand has to be met. However, in industrial applications in particular, the potential for the integration of process waste heat is often present. Therefore, despite the fact that its technology status still lags behind those of AECs and PEMECs, industrial experts expect high-temperature electrolysis to be preferentially implemented in energy-intense application fields, such as the steel and chemical industries [19]. Thus, a significant share of the abovementioned long-term industrial demands may be met using SOECs.

Apart from industrial waste heat, various power-to-X processes include exothermal downstream synthesis processes. For instance, thermal integration between endothermal SOEC operation and catalytic methanation, which is usually operated at temperatures of 250 °C–700 °C, allows overall process efficiencies to be increased to >80%. [20]. In addition, high-temperature electrolysis provides the ability to perform the co-electrolysis of H_2O and CO_2, thus allowing for the generation of a suitable syngas composition for various downstream processes in the power-to-X process chain [21, 22]. The resulting gains in efficiency in the production of e-fuels by Fischer–Tropsch synthesis [23] or chemical feedstocks, such as methanol [24], make high-temperature electrolysis the preferable option for the chemical industry—as long as the lower operating costs outweigh the much higher investment costs of SOECs compared to AECs and PEMECs.

20.2 Technology cost-reduction potentials

Despite high market potentials for power-to-X applications in general, and hydrogen from electrolysis in particular, the broad application of this technology is partially hindered by its ongoing high technology costs compared to the fossil-based production of hydrogen and hydrogen-based materials. Therefore, cost reductions are a critical success factor. Appropriate cost reductions result from advancements in the technology itself—e.g. from materials savings, simplifications, or the production process and are usually gathered under the term *economies of scale*. Even though different scaling effects usually overlap and often cannot be separated from each other, two main drivers for reducing the costs of novel technologies need to be distinguished:

- cost effects driven by the increasing numbers of produced units and thus constantly growing production experience—also referred to as *economies of manufacturing scale*

- cost effects resulting from the actual scale of implementation of individual projects due to savings in peripheral equipment—also referred to as *economies of unit scale*

With the depicted role of PtG in future energy systems, and thus a required rapid development of capacity at a number of installations as well as on the individual scale, both effects can be expected to have a significant impact on future technology costs. However, an ex-ante estimation of the expected cost reductions is important in order to support the early implementation of this technology. On the one hand, potential operators have to decide on the investments required for implementations by ensuring long-term economic viability. On the other hand, from a political perspective, these projections are needed to identify regulatory measures and funding gaps in order to support early adoption. In the following, these long-term projections for electrolysis are discussed in terms of both of the described economy-of-scale effects.

20.2.1 Economies of manufacturing scale

As outlined above, economies of manufacturing scale are related to the cost reductions resulting from the overall experience gained from the cumulative number of produced units. Therefore, this effect is also known as technological learning and can be represented by a learning curve. The original theory of technological learning describes the empirical finding that costs decrease by a certain constant percentage for each doubling of cumulative production. This constant percentage is commonly known as the learning rate. Even if the related methods used to determine learning curves have seen different advancements and adaptions—e.g. component-wise evaluation or the integration of additional drivers apart from production volumes alone, such as patents or R&D spending—this basic concept is still scientific practice [25]. What remains to be considered is that even though the learning curve is often related to the time of implementation, it is actually only dependent on the cumulative number of installed or produced units (including R&D efforts). Thus, it is particularly the case that projections of future costs from an annual reference are always dependent on a certain presumed market potential and rely on the fulfillment of such. As a consequence, early investments in a technology at higher—and hence probably non-competitive—costs are necessary in order to initiate the learning curve. An observant postponement is thus not an option, since the learning investment in the technology has to be made (see figure 20.3).

Recent studies project significant cost reductions for the common electrolysis technologies under the presumption that the abovementioned industrial hydrogen demands (of the chemical and steel industries) will be completely served by power-to-hydrogen. For a better visualization and in line with current climate targets, the figures shown presume the fulfillment of these demands by 2050. Starting from a cumulative global electrolysis production volume of about 21 GW_{el} by 2018 [25], mainly based on chlor-alkali electrolysis, of which about 8 GW_{el} is actually operational, this would imply a compound average annual growth rate (CAAGR) of 17%–21% [4]. While these values seem reasonable in comparison to growth rates seen for other energy technologies such as wind power, which has seen a CAAGR of

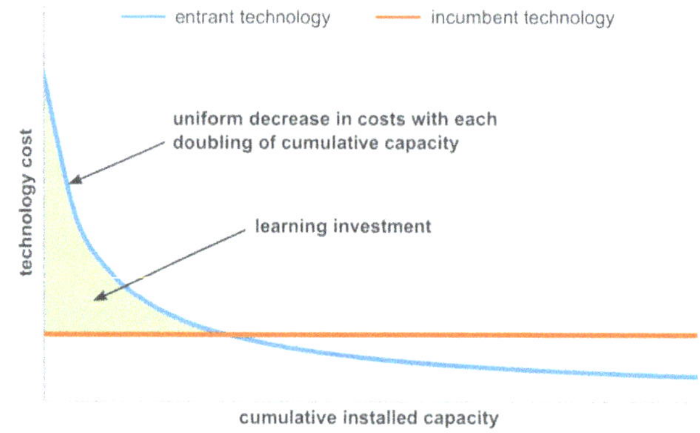

Figure 20.3. Closing the cost gap between entrant and incumbent technologies. Reproduced from [25].

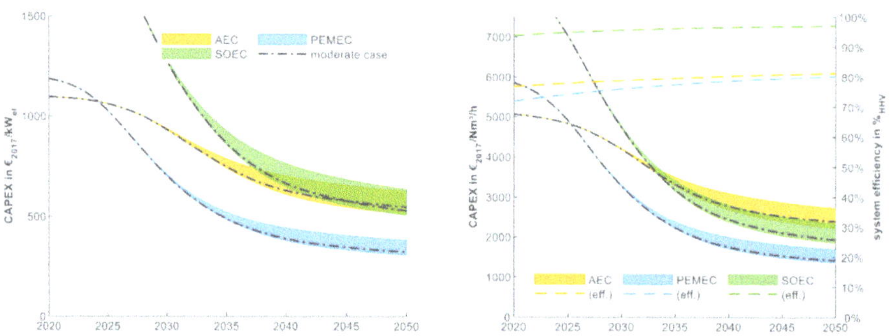

Figure 20.4. Projected system costs for the development for electrolysis based on economies of manufacturing scale; capacities related to electric input (left) and hydrogen output (right). Reproduced from [4].

19% from 2007 to 2017, or PV, which grew by as much as 48% per year between 2012 and 2017, maintaining such high rates over more than three decades is highly ambitious [26, 27]. However, based on these numbers, appropriate learning curves can be derived for alkaline, PEM, and solid oxide electrolysis, as shown in figure 20.4[2]. While from today's perspective, the capital expenditures (CAPEXes) for alkaline and PEM electrolysis are comparable, SOEC-based systems are currently uncompetitive [4], which is mostly due to the lower maturity of the technology. However, this cost relation is expected to change significantly as the implementation of the technology increases. PEMEC-based systems are expected to clearly undercut the costs of their alkaline competitors within the next few years because of their higher potential for technological learning and expected increasing overall market share due to their advantages of efficiency and flexibility [28, 29]. Even though PEM electrolyzers are expected to be most cost competitive in the long term, given an adequate market share, high-temperature electrolysis is expected to

[2] The cost ranges shown relate to different demand fulfillments: the demands of the chemical industry only (upper range); including global energy storage (moderate case line); and the demands of the steel industry (lower range).

reach a similar cost level to that of alkaline systems. Overall, the capital costs for all electrolysis technologies are expected to decrease by 12%–15% for each doubling of individual cumulative production [4]. Therefore, given an adequate market share, the cost-reduction potential of SOECs is the most significant, simply based on the lower starting value of the installed capacity.

As the cost ranges of the individual technologies are so close, the factor of electric efficiency becomes highly relevant. As the right-hand side of figure 20.4 shows, high-temperature electrolysis may play to its strength of high electric efficiency when it comes to the costs related to actual hydrogen output rather than electric capacity. In this case, SOEC systems could economically already outperform AECs in the mid-term, and the remaining gap from PEMECs is small. Thus, depending on the actual market share of high-temperature electrolysis, especially with regard to the discussed potential preference in industrial environments, it is expected to become a serious competitor to the current leading technologies.

20.2.2 Economies of unit scale

In addition to the technological learning effects that result from widespread usage, the implementation of large-scale systems is expected to come with additional cost reductions compared to the costs of today's usual scales. These unit scaling effects arise for different reasons, such as optimized surface-to-volume ratios (e.g. reduced housing costs), cost savings for peripheral components (e.g. reduced piping expenses), or synergies with connected plant components. Consideration of the scaling effects is a common cost estimation practice when chemical equipment and plants are scaled up [30]. As in the case of the calculation of learning curves, different approaches are available and different levels of detail can be applied in the estimation of the relevant effects. Current projections for PtG systems use a per-component approach to consider the potential differences between technologies and their common system structures [4]. Figure 20.5 shows the appropriate cost effects of

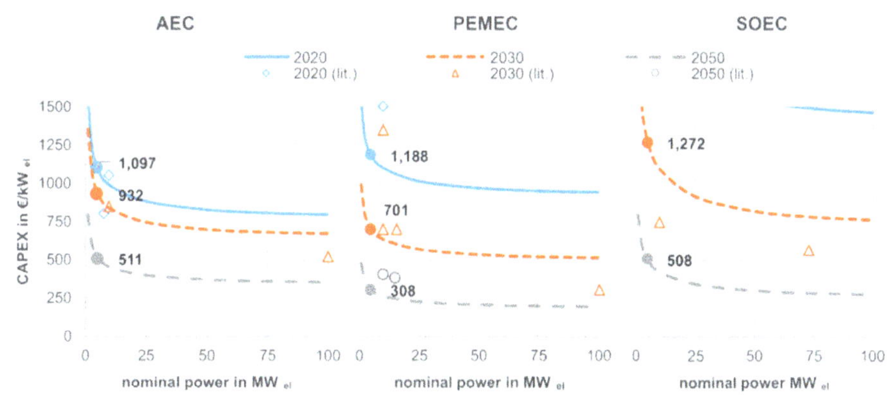

Figure 20.5. Capital costs for the development of electrolysis systems based on component-wise economies of scale in comparison to estimations from the literature (marked and labeled values refer to the reference size of 5 MW_{el}). Individual lines indicate the experience-curve-related values for the years 2020 (blue), 2030 (orange), and 2050 (gray). Reproduced from [4].

a potential upscaling up to 100 MW_{el} (starting from a reference value of 5 MW_{el}) in combination with the learning curves presented in the previous section. Comparing the cost curves for the three technologies, it can be seen that the unit scaling effects for AECs and PEMECs seem to be quite limited, especially when it comes to larger scales. In contrast, SOEC shows steeper curves and thus additional leverage towards economic competitiveness. This can be explained by the smaller impact of the stack module (which has a low potential for cost reduction) on the overall system costs, compared to its low-temperature competitors, which use expensive catalyst materials. However, the currently complex and thus cost-intensive peripheral equipment required for heat and fluid management benefits considerably from unit scaling effects. All in all, cost-reduction potentials of >75% from general upscaling are projected for large-scale applications with a broad distribution [4].

Even though the combination of both scaling effects already reveals significant cost-reduction potentials for all the common electrolysis technologies, figure 20.5 also shows that the individual studies and estimations found in the literature partially suggest even higher potentials, especially for mid-term projections.

20.2.3 Cost-reduction potentials for high-temperature electrolysis

As discussed in the previous section, economies of scale are expected to reveal the highest cost-reduction effects for high-temperature electrolysis across all common electrolysis technologies based on their current technology states. From the stack perspective, the requirements imposed on the individual materials are generally high due to the elevated operating temperatures and the temperature cycles involved in changing from standby or shutdown to an operational state, and vice versa. Therefore, the respective costs are uniformly distributed over the relevant stack components [28, 31], as shown in figure 20.6. Since common solid oxide cell designs do not rely on platinum-group metals (PGMs) as catalysts, cost-reduction effects are mainly driven by general material savings, especially those due to rare-earth materials, the improvement of production processes such as the sintering of the electrode–electrolyte assembly (EEA), and the automation of the cell stacking and assembly processes [28].

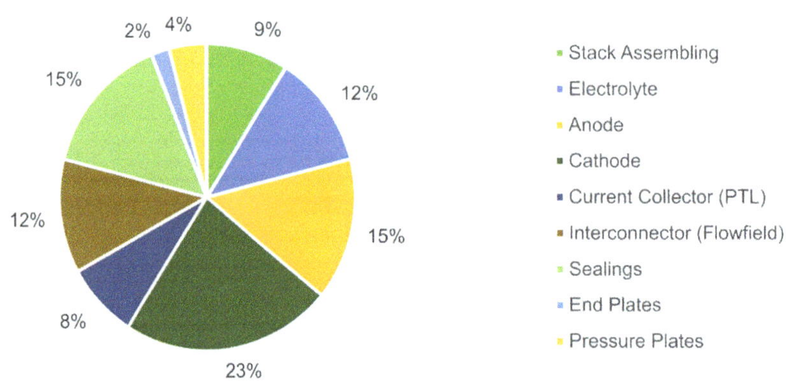

Figure 20.6. SOEC stack cost structure. Based on [25].

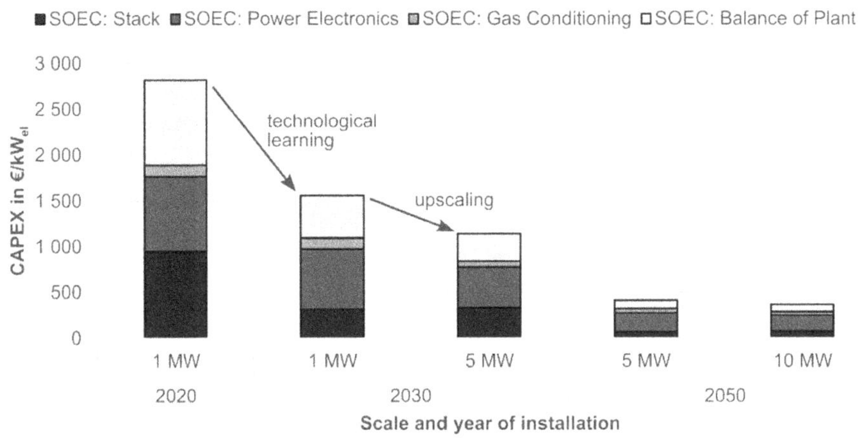

Figure 20.7. Expected economies of scale for SOECs. Based on [20].

However, as already mentioned in section 20.2.2, due to the complexity of the overall system in relation to high-temperature electrolysis, peripheral system components beside the stack are just as relevant for cost reduction. Assuming that these peripheral modules (primarily power electronics, gas and fluid conditioning, and the remaining balance of plant components) mainly include common and well-established standard parts, cost reductions driven by increasing SOEC production volumes are expected to be minor [25]. On the contrary, the complexity of these modules due to high-temperature management, the supply of water in a vapor state, and potential heat exchange with external processes increase the potential for achieving cost reductions by upscaling individual electrolysis units. In particular, heat exchanger equipment offers advantageous scaling factors [30, 32] and the generation and supply of steam also profits significantly from implementation at larger scales. In addition, with respect to heat losses, lower surface-to-volume ratios are preferable. Figure 20.7 illustrates the expected economies of scale in a combined view of technological learning and unit upscaling for SOECs at the low MW scale up to 2050. The impact of upscaling—beyond the significant learning curve—on the peripheral components can clearly be seen in the 2030 comparison, in contrast to the negligible scaling effects for the stack module [4, 20].

20.3 Product generation costs

Even though a variety of indicators is used in the literature to indicate and assess the economic performance and viability of conversion processes such as PtG, the explanations used herein focus on product generation costs. In short, that means the resulting cost per unit of product, namely hydrogen or SNG, generated in the PtG process. As already described, the products generated are considered relevant in a variety of transformation processes, be it as energy carriers or feedstock materials and in the transition of the energy system as a whole. Therefore, indicators that describe the profitability of individual implementations, such as net present value (NPV) or payback time, are of less relevance. Additionally, as a part of the long-term replacement of fossil fuels, the target is to assure economic competitiveness at the product level, either by support of the renewable alternative or by taxation of fossils. A comparison at the product cost level supports the identification of the appropriate regulation gaps.

Due to the still early industrialization stage of the PtG technology as a whole, the product generation costs for upcoming installations at relevant scales can only be evaluated by relying on the abovementioned projections of cost developments. Furthermore, funding efforts to support learning investments (see figure 20.3) and respective regulations are constantly adapted, apply competing measures for energy transition, and strongly depend on the implementation location. The same applies to the generation and supply of renewable electricity, which are affected by the same implications in terms of energy transition, and concurrently represent the major share of the operational costs in electrochemical hydrogen production [1, 20]. Therefore, given these significant impact factors, a universal statement about current and future hydrogen or SNG generation costs cannot seriously be made from a current perspective. The following sections are meant to provide an overview of the expected cost potentials for PtG products. Moreover, cost-reduction potentials are also discussed and exploitable synergies and by-products are considered.

20.3.1 Electricity costs and efficiency impact

As mentioned above, the cost of the electricity supply represents a significant fraction of the overall production costs of hydrogen from electrolysis. For today's implementations, it may account for 50%–90%, since the availability of renewable electricity also impacts the achievable full-load hours of operation [1]. Consequently, optimization of the electricity supply costs is often the focus of potential operational concepts for PtG applications [33]. With grid-service-oriented operation in mind, such as the utilization of electricity production peaks and the provision of storage capacity, operation of the PtG application only takes place at the lowest electricity costs [34]. Furthermore, recent studies show that hydrogen production in countries with higher potential for renewable electricity (e.g. PV in northern Africa or offshore wind in northern Europe) and import to central European countries is potentially more competitive than local production, if transport can be realized by pipelines. On the contrary, imports that take place without the respective infrastructure, which thus require liquefaction and transport by ship (e.g. from Saudi Arabia or Chile) are expected to attract higher costs than production from local RESs [35, 36].

When one considers the whole PtG or power-to-X process chain, less electricity-hungry downstream processes shift the cost balance towards capital-related costs, especially for novel technologies. Due to the still-high capital costs (CAPEX), the cost optimum for hydrogen production may fall somewhere between 2500 and 6000 annual full-load hours [17], depending on the presumed electricity and electrolyzer costs. However, given additional CAPEX for downstream synthesis, e.g. methanation, the relevant evaluations have shown that this optimum shifts towards maximizing the annual full-load hours (see figure 20.8) [4, 20, 37]. Furthermore, with the given economies of scale, despite potentially decreasing costs for renewable

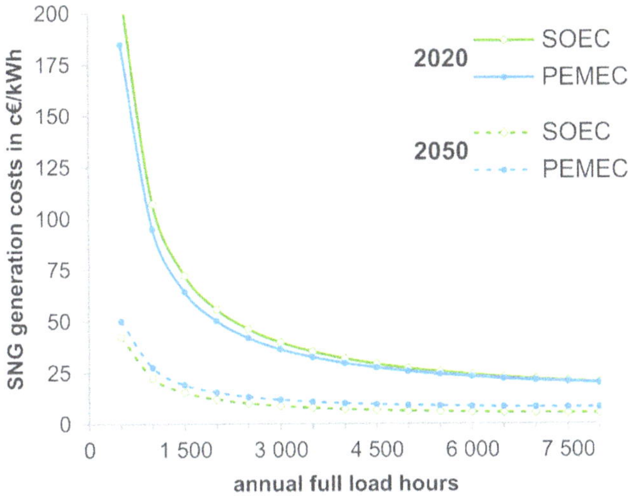

Figure 20.8. Dependence of SNG generation costs on annual full-load hours of operation and the impact of electric efficiency (presumed efficiencies: $\eta_{SOEC} = 79\%_{el,LHV}$, $\eta_{PEMEC} = 54\%_{el,LHV}$). Based on calculations performed in [20] and [4].

electricity generation [38], electric energy costs are expected to become the dominant part of PtG product generation costs in the long term [20].

As a consequence of the persistently high cost share of electricity supply in PtG and power-to-X applications, an corresponding reduction in the electricity demand and thus an increase in electric efficiency is apparently necessary for the reduction of overall product generation costs. As shown in figure 20.8, even with the highest efficiencies achievable from process optimizations today [39], the significantly higher technology costs of high-temperature electrolysis do not yet allow their operation to be competitive with less cost-intensive technologies. However, with the long-term projected economies of scale (see section 20.2.3) generation costs are minimized by the use of SOECs, especially for power-to-X processes that allow a high degree of thermal integration [20].

20.3.2 Exploitation of by-products and synergy effects

Another important way to increase the economic performance of PtG and power-to-X applications is to exploit the unavoidable by-products. The electrochemical conversion of water to hydrogen inevitably generates significant amounts of oxygen. Oxygen is an important resource for a variety of industrial processes, e.g. steel-making [40, 41], but may also gain importance in the path of energy transition, for instance in oxyfuel combustion processes [42]. Since the conventional production of oxygen at larger scales by cryogenic air separation (CAS) is accompanied by high electricity demands of about 250–300 kWh per ton of oxygen [43], failing to use this by-product is also counterproductive in terms of energy efficiency. Furthermore, the achievable oxygen sale revenues are estimated to range from €50 to €150 per of ton O_2 [20]. As shown in figure 20.9, exploiting this resource could have a significant

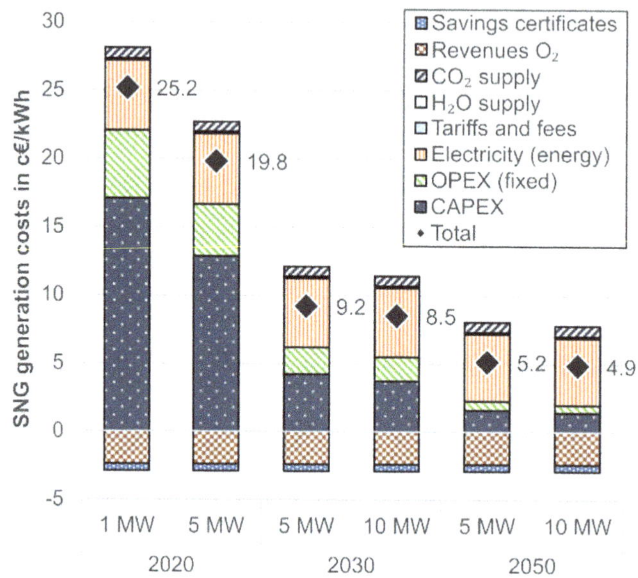

Figure 20.9. SNG generation costs for different plant capacities and years of implementation. Reproduced from [20].

impact on the effective net production costs of power-to-X products, especially in the case of the long-term costs.

In addition to the direct utilization of the oxygen by-product, exploitation of the waste heat produced by the electrolysis process is worth consideration. In particular, given the decreasing temperatures in modern and future district heating networks, even the integration of low-temperature waste heat from electrolysis becomes relevant, if no other direct use is possible. While such waste heat utilization apparently increases the overall process efficiency of the electrolysis, the high expected demand for electrolyzer capacities (see section 20.1) will create a significant potential for waste heat. Recent studies have shown that 6%–10% of the district heating demands within the EU—or 2.5%–4% of the total heat demand at <100 °C, including industrial use—could be met by waste heat from electrolysis by 2030 [19]. However, these estimations still exclude the additional heat potentials of downstream synthesis processes, e.g., methanation, which in many cases result in the additional generation of waste heat, potentially at higher amounts and costs. Furthermore, the given explanations show that for an effective integration of power-to-X processes, the identification of potential synergies, both within the process chain itself as well as in exchange with external facilities, is an essential process that requires additional consideration for future implementations.

20.3.3 Carbon costs and circular economies

Since hydrogen is considered to be an important feedstock for various downstream synthesis processes, this role of hydrogen is also widely discussed in the context of the future carbon capture and utilization (CCU) concept. Methanation, and

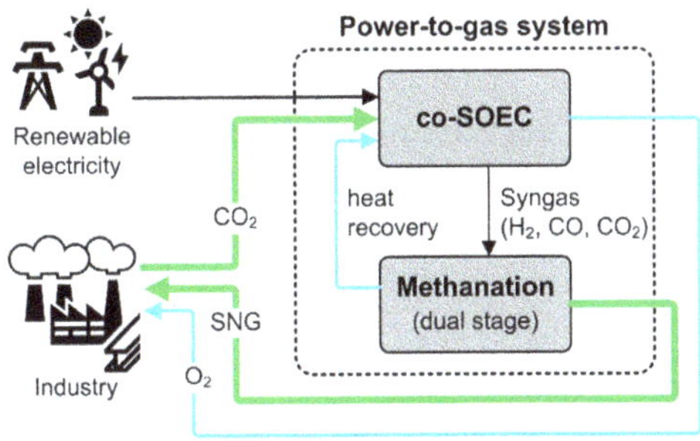

Figure 20.10. Possible configuration of a power-to-X process implemented with a closed carbon circle (green) and utilization of synergies (blue). Reproduced from [20].

therefore power-to-methane, is the simplest form of CCU in this regard. For the required CO_2, additional supply costs are to be expected. These costs are mainly dependent on the source and the costs of separation or capture from the source stream. Therefore, high-purity streams from biogenic processes, such as fermentation, are associated with lower costs than those of diluted off-gases from combustion processes [44]. However, the appropriate cost share of final generation costs is low compared to the capital and electricity expenditures (see also figure 20.9) [20].

Even though direct costs for CO_2 supply are mostly negligible in current power-to-X considerations, CCU processes may impact net carbon emissions, either for the supplier of the carbon source or the consumer of the product, and thus potentially affect costs for carbon taxes or EU Emissions Trading System (ETS) certificates. Even though the effect is low at current certificate costs, as seen in figure 20.9, it can become significant when cost levels reach values of €150 or more per ton of CO_2 [20]. While the ecological benefit over fossil fuels is obvious for the use of renewable carbon sources (e.g. CO_2 from biogas upgrading) and thus the reduction in fossil emissions is also clear, the situation for the reuse of fossil CO_2 emissions is less clear. However, for implementations in industrial environments, the establishment of closed carbon cycles is an option that is worth discussing. In this context, CO_2 emitted by industrial processes is utilized in a power-to-X application to produce a hydrocarbon-based fuel or feedstock. This product is then reintroduced into the originating industrial process (see figure 20.10). Additionally, the possibility of storing the hydrocarbons thus generated allows the supply of renewable electricity to be decoupled from the respective demand in the industry. The techno-economic performance of such circular economies, particularly those featuring thermal integration and the utilization of by-product oxygen, has been investigated in recent studies [20].

20.4 Concluding remarks

In this chapter, economic theory was used to discuss the direct economic impacts that can be expected in the future from the broad international introduction of PtG applications. The importance of PtG as part of the decarbonization of the energy system is undisputed, especially in the segment of renewable hydrogen production;

the number of worldwide installations of production PtG facilities is steadily increasing and will continue to grow strongly.

This analysis is based on the calculation of learning curves and economies of scale, which in turn is based on an analysis of the expected future demand for these systems that quantifies the expected cost intervals of the investment costs of the plants. As outlined, the expected demand for corresponding production capacity is projected to be significant, especially for industrial use, since the long-term global hydrogen demand is expected to increase by up to 90 million tons per year for the chemical and steel industries alone. The increase in electrolyzer capacities thus required will, as a consequence, bring about an enormous reduction in technological fixed costs of 75% and more. On the one hand, these cost reductions are driven by technological learning and thus the experience gained from increasing cumulative production. On the other hand, they benefit from increasing system sizes per installed unit and thus from corresponding economies of scale. In this context, SOEC technologies are also expected to reach investment cost levels competitive with those of established low-temperature electrolysis technologies in the medium to long term, albeit with significant advantages in terms of electrical efficiency. In particular, for thermal integration with downstream exothermic synthesis processes, SOEC and co-SOEC implementations are likely to outperform lower-cost solutions in the long term. However, to achieve the optimal operation of electrolysis and power-to-X processes, the exploitation of potential by-products and synergy effects is essential to be cost-effective.

This chapter did not explicitly address the realization of the ramp-up curves required to unlock the demand potential of electrolysis technologies in particular. However, it should be explicitly pointed out that technologies and systems such as PtG applications, which offer positive system services for the entire energy system that are difficult to map directly via the market—such as the seasonal storage of electrical energy—must receive positive regulatory support. Only then will the significant cost reductions, which can be generated through technological learning and economies of scale, be initiated and made possible in the first place.

Acknowledgment

The contents included in this chapter are part of the national Austrian flagship project HYDROMETHA (contract # 864578), which has received funding from the Austrian Climate and Energy Fund under the framework of the 'Energy Research Program.' Their support is gratefully acknowledged.

References

[1] International Energy Agency 2021 *Global Hydrogen Review 2021* (Paris: International Energy Agency) https://www.iea.org/reports/global-hydrogen-review-2021

[2] European Commission 2020 *A Hydrogen Strategy for a Climate-Neutral Europe (COM (2020) 301 Final)* (Brussels: European Commission) https://eur-lex.europa.eu/legal-content/EN/TXT/?uri=CELEX%3A52020DC0301

[3] Thema M, Bauer F and Sterner M 2019 Power-to-gas: electrolysis and methanation status review *Renew. Sustain. Energy Rev.* **112** 775–87

[4] Böhm H, Zauner A, Rosenfeld D C and Tichler R 2020 Projecting cost development for future large-scale power-to-gas implementations by scaling effects *Appl. Energy* **264** 114780

[5] Henning H-M and Palzer A 2013 *Energiesystem Deutschland 2050: Sektor- und Energieträgerübergreifende, modellbasierte, ganzheitliche Untersuchung zur langfristigen Reduktion energie-bedingter CO_2-Emissionen durch Energieeffizienz und den Einsatz Erneuerbarer Energien* (Freiburg: Fraunhofer-Institut für Solare Energiesysteme ISE) https://www.ise.fraunhofer.de/de/veroeffentlichungen/studien/studie-energiesystem-deutschland-2050.html

[6] Agora Energiewende and Öko-Institut e.V 2017 *Erneuerbare vs. fossile Stromsysteme: ein Kostenvergleich: Stromwelten 2050—Analyse von Erneuerbaren, kohle- und gasbasierten Elektrizitätssystemen* (Berlin: Agora Energiewende and Öko-Institut e.V.) https://www.agora-energiewende.de/veroeffentlichungen/erneuerbare-vs-fossile-stromsysteme-ein-kostenvergleich/

[7] Götz P, Huneke F, Lenck T and Linkenheil C P 2016 *Minimaler Bedarf an langfristiger Flexibilität im Stromsystem bis 2050 – Studienerweiterung* (Berlin: Energy Brainpool) https://www.energy-brainpool.com/fileadmin/download/Studien/Studie_2016-02-23_Minimaler_Windgasbedarf_GreenpeaceEnergy_EnergyBrainpool.pdf

[8] Breyer C, Müller B, Möller C, Gaudchau E, Schneider L, Gajkowski K, Resch M and Pleßmann G 2013 *Vergleich und Optimierung von zentral und dezentral orientierten Ausbaupfaden zu einer Stromversorgung aus Erneuerbaren Energien in Deutschland* (Berlin: Reiner Lemoine Institut gGmbH) https://www.bvmw.de/fileadmin/pdf-archiv/Studie_zur_dezentralen_Energiewende.pdf.pdf

[9] Hecking H, Hintermayer M, Lencz D and Wagner J 2017 *Energiemarkt 2030 und 2050—Der Beitrag von Gas- und Wärmeinfrastruktur zu einer effizienten CO_2-Minderung* (Köln: ewi Energy Research & Scenarios gGmbH) https://www.ewi.research-scenarios.de/cms/wp-content/uploads/2017/11/ewi_ERS_Energiemarkt_2030_2050.pdf

[10] Klein S, Klein S W, Steinert T, Fricke A and Peschel D 2017 *Erneuerbare Gase – ein Systemupdate der Energiewende* (Berlin: enervis energy advisors GmbH) https://www.wind-energie.de/fileadmin/redaktion/dokumente/publikationen-oeffentlich/themen/03-sektoren-kopplung/20171212_studie_erneuerbare_gase.pdf

[11] Bailera M and Lisbona P 2018 Energy storage in Spain: forecasting electricity excess and assessment of power-to-gas potential up to 2050 *Energy* **143** 900–10

[12] Lisbona P, Frate G F, Bailera M and Desideri U 2018 Power-to-gas: analysis of potential decarbonization of Spanish electrical system in long-term prospective *Energy* **159** 656–68

[13] Guandalini G, Robinius M, Grube T, Campanari S and Stolten D 2017 Long-term power-to-gas potential from wind and solar power: a country analysis for Italy *Int. J. Hydrogen Energy* **42** 13389–406

[14] Blanco H, Nijs W, Ruf J and Faaij A 2018 Potential for hydrogen and power-to-liquid in a low-carbon EU energy system using cost optimization *Appl. Energy* **232** 617–39

[15] Blanco H, Nijs W, Ruf J and Faaij A 2018 Potential of power-to-methane in the EU energy transition to a low carbon system using cost optimization *Appl. Energy* **232** 323–40

[16] Pleßmann G, Erdmann M, Hlusiak M and Breyer C 2014 Global energy storage demand for a 100% renewable electricity supply *Energy Proc.* **46** 22–31

[17] International Energy Agency 2019 *The Future of Hydrogen* (Paris: International Energy Agency) https://www.iea.org/reports/the-future-of-hydrogen

[18] International Energy Agency 2018 *The Future of Petrochemicals* (Paris: International Energy Agency) https://www.iea.org/reports/the-future-of-petrochemicals

[19] Böhm H, Moser S, Puschnigg S and Zauner A 2021 Power-to-hydrogen and district heating: technology-based and infrastructure-oriented analysis of (future) sector coupling potentials *Int. J. Hydrogen Energy* **46** 31938–51

[20] Böhm H, Lehner M and Kienberger T 2021 Techno-economic assessment of thermally integrated co-electrolysis and methanation for industrial closed carbon cycles *Front. Sustain.* **2** 726332

[21] Banerjee A, Wang Y, Diercks J and Deutschmann O 2018 Hierarchical modeling of solid oxide cells and stacks producing syngas via H_2O/CO_2 co-electrolysis for industrial applications *Appl. Energy* **230** 996–1013

[22] Biswas S, Kulkarni A P, Giddey S and Bhattacharya S 2020 A review on synthesis of methane as a pathway for renewable energy storage with a focus on solid oxide electrolytic cell-based processes *Front. Energy Res.* **8** 570112

[23] Herz G, Reichelt E and Jahn M 2018 Techno-economic analysis of a co-electrolysis-based synthesis process for the production of hydrocarbons *Appl. Energy* **215** 309–20

[24] Zhang H and Desideri U 2020 Techno-economic optimization of power-to-methanol with co-electrolysis of CO_2 and H_2O in solid-oxide electrolyzers *Energy* **199** 117498

[25] Böhm H, Goers S and Zauner A 2019 Estimating future costs of power-to-gas—a component-based approach for technological learning *Int. J. Hydrogen Energy* **44** 30789–805

[26] International Energy Agency 2019 *Global Energy and CO_2 Status Report 2018: The Latest Trends in Energy and Emissions 2018* (Paris: International Energy Agency) https://www.iea.org/reports/global-energy-co2-status-report-2019

[27] REN21 2018 *Renewables 2018 Global Status Report* (Paris: REN21) https://www.ren21.net/wp-content/uploads/2019/05/GSR2018_Full-Report_English.pdf

[28] Smolinka T, Wiebe N, Sterchele P, Palzer A, Lehner F, Jansen M, Kiemel S, Miehe R, Wahren S and Zimmermann F 2018 *Studie IndWEDe: Industrialisierung der Wasserelektrolyse in Deutschland: Chancen und Herausforderungen für nachhaltigen Wasserstoff für Verkehr, Strom und Wärme* (Berlin: Nationale Organisation Wasserstoff- und Brennstoffzellentechnologie—NOW GmbH) https://www.now-gmbh.de/wp-content/uploads/2020/09/indwede-studie_v04.1.pdf

[29] Buttler A and Spliethoff H 2018 Current status of water electrolysis for energy storage, grid balancing and sector coupling via power-to-gas and power-to-liquids: a review *Renew. Sustain. Energy Rev.* **82** 2440–54

[30] Peters M S and Timmerhaus K D 1991 *Plant Design and Economics for Chemical Engineers (McGraw-Hill Chemical Engineering Series)* 4th edn (New York: McGraw-Hill)

[31] Scataglini R, Wei M, Mayyas A, Chan S H, Lipman T and Santarelli M 2017 A direct manufacturing cost model for solid-oxide fuel cell stacks *Fuel Cells* **17** 825–42

[32] Perry R H, Green D W and Maloney J O 1997 *Perry's Chemical Engineers' Handbook* 7th edn (New York: McGraw-Hill)

[33] Gorre J, Ruoss F, Karjunen H, Schaffert J and Tynjälä T 2020 Cost benefits of optimizing hydrogen storage and methanation capacities for power-to-gas plants in dynamic operation *Appl. Energy* **257** 113967

[34] Tichler R, Bauer S and Böhm H 2022 Power-to-gas *Storing Energy* ed T M Letcher (Amsterdam: Elsevier) **25** 595–612

[35] Brändle G, Schönfisch M and Schulte S 2021 Estimating long-term global supply costs for low-carbon hydrogen *Appl. Energy* **302** 117481
[36] World Energy Council 2021 *Decarbonised Hydrogen Imports into the European Union: Challenges and Opportunities* (London: World Energy Council)
[37] Schlautmann R, Böhm H, Zauner A, Mörs F, Tichler R, Graf F and Kolb T 2021 Renewable power to gas: a technical and economic evaluation of three demo sites within the STORE&GO Project *Chem. Ing. Tech.* **93** 568–79
[38] International Energy Agency 2021 *World Energy Outlook 2021* (Paris: International Energy Agency) https://www.iea.org/reports/world-energy-outlook-2021
[39] 2018 FFG - Österreichische Forschungsförderungsgesellschaft mbH FFG Projektdatenbank -HydroMetha (accessed 13 Dec 2018) https://projekte.ffg.at/projekt/2903976
[40] Bieda B 2012 Life cycle inventory processes of the arcelorMittal poland (AMP) S.A. in Kraków, poland—basic oxygen furnace steel production *Int. J. Life Cycle Assess.* **17** 463–70
[41] Kirschen M, Badr K and Pfeifer H 2011 Influence of direct reduced iron on the energy balance of the electric arc furnace in steel industry *Energy* **36** 6146–55
[42] Carrasco-Maldonado F, Spörl R, Fleiger K, Hoenig V, Maier J and Scheffknecht G 2016 Oxy-fuel combustion technology for cement production—state of the art research and technology development *Int. J. Greenhouse Gas Control* **45** 189–99
[43] Wu F, Argyle M D, Dellenback P A and Fan M 2018 Progress in O_2 separation for oxy-fuel combustion–a promising way for cost-effective CO_2 capture: a review *Prog. Energy Combust. Sci.* **67** 188–205
[44] Rodin V, Lindorfer J, Böhm H and Vieira L 2020 Assessing the potential of carbon dioxide valorisation in europe with focus on biogenic CO_2 *J. CO2 Util.* **41** 101219

www.ingramcontent.com/pod-product-compliance
Ingram Content Group UK Ltd.
Pitfield, Milton Keynes, MK11 3LW, UK
UKHW051909100926
8556IPUK00005B/822

9 780750 339490